Lecture Notes in Mathematics

Edited by A. Dold and B.

1362

P. Diaconis E. Nelson
D. Elworthy G. Papanicolaou
H. Föllmer S. R. S. Varadhan

École d'Été de Probabilités de Saint-Flour XV–XVII, 1985–87

Editor: P.L. Hennequin

Springer-Verlag
Berlin Heidelberg New York London Paris Tokyo

Authors

Persi Diaconis
Harvard University, Department of Mathematics
One Oxford St., Cambridge, MA 02138, USA

David Elworthy
Mathematics Institute, University of Warwick
Coventry, CV4 7AL, Great Britain

Hans Föllmer
Universität Bonn, Institut für angewandte Mathematik
Wegelerstr. 6, D-5300 Bonn 1, Federal Republic of Germany

Edward Nelson
Department of Mathematics, Princeton University
Fine Hall, Washington Road, Princeton, NJ 08544, USA

George Papanicolaou
New York University, Courant Institute of Mathematical Sciences
251 Mercer Street, New York, NY 10012, USA

Srinivasa R.S. Varadhan
New York University, Courant Institute of Mathematical Sciences
251 Mercer Street, New York, NY 10012, USA

Editor

Paul-Louis Hennequin
Université Blaise Pascal, Complexe S
Mathématiques Appliquées
63177 Aubière Cedex, France

Mathematics Subject Classification (1985): 60-02, 20 C 30, 58 G 32, 60 B 15, 60 F 10, 60 G 10, 60 G 15, 60 G 60, 60 H 10, 60 H 15, 60 J 15, 60 J 25, 60 J 60, 60 K 35, 65 C 10, 73 D 70, 81 C 20

ISBN 3-540-50549-0 Springer-Verlag Berlin Heidelberg New York
ISBN 0-387-50549-0 Springer-Verlag New York Berlin Heidelberg

© Springer-Verlag Berlin Heidelberg 1988
Printed in Germany

Printing and binding: Druckhaus Beltz, Hemsbach/Bergstr.
2146/3140-543210

INTRODUCTION

Ce volume rassemble six des cours qui ont été donnés à l'Ecole de Calcul des Probabilités de Saint-Flour durant les trois années 1985 (2 au 19 Juillet), 1986 (17 Août au 3 Septembre) et 1987 (1er au 18 Juillet).

Le cours donné en 1986 par O.E. BARNDORFF-NIELSEN est publié par ailleurs dans la collection "Lecture Notes in Statistics".

Les cours de P. CARTIER (1985) et L. ELIE (1987) ne nous sont pas encore parvenus et il nous est apparu qu'il n'était plus possible de retarder la publication des textes dont les auteurs avaient respecté les délais convenus. Nous espérons pouvoir publier ces deux cours avec ceux de l'Ecole 1988 et reprendre dès 1989 le rythme annuel de publication. Nous remercions les éditions SPRINGER de leur compréhension.

Chaque année, l'Ecole est l'occasion, pour les participants, de présenter leurs travaux de recherche. Nous donnons ci-dessous la liste de ces exposés, un résumé peut être obtenu sur demande. Certains ont été publiés dans les Annales Scientifiques de l'Université Blaise Pascal (Clermont II).

P.L. HENNEQUIN

TABLE DES MATIERES

(Une table détaillée se trouve au début de chaque cours)

LARGE DEVIATIONS AND APPLICATIONS

Srinivasa R.S. VARADHAN

TABLE DES MATIERES

S.R.S. VARADHAN : "LARGE DEVIATIONS AND APPLICATIONS"

Large Deviations and Applications

S.R.S. Varadhan
New York University
Courant Institute of Mathematical Sciences
251 Mercer Street
New York, NY 10012

Section 1. Introduction, Definitions and Examples.

There are many situations arising in analysis, physics and other areas where solutions to certain problems are naturally expressed in terms of function space integrals or expectations of certain functionals with respect to specific stochastic processes. This representation of the solution can be used for many purposes. First of all it can be used to prove the existence of the solution, and it can also be used to establish some of the qualitative properties of the solution. One may also be able to perform some Monte Carlo simulations to evaluate the integral.

However, we will take a different point of view in these lectures. Often there are one or more parameters in the problem and we have a situation where the functional to be integrated as well as the probability measure to be used in the integration may depend on these parameters. Asymptotic evaluation of these integrals when the parameter becomes large or small is rather useful. Whereas in a single integral contributions come from the entire range of integration it is quite conceivable that as the parameters approach their extreme values the integration process becomes singular in the sense that the major contribution to the integral comes from a set whose measure is becoming extremely small. The principle of large deviation is the art of determining how small the probabilities of these rare events really are. It is then used to identify where the major contribution to the integral comes from and leads to a precise estimation of the integral itself.

Let us examine this by means of a simple example. Let $x_1, x_2, \cdots, x_n, \cdots$ be a sequence of independent positive random variable with a common distribution α. We will assume for simplicity that $1 \leq x_i \leq 2$ with probability 1. Let $\xi_n = x_1 x_2 \cdots x_n$. Then $\log \xi_n = \sum_1^n \log x_i$ and $\lim_{n\to\infty} \frac{1}{n} \log \xi_n = \int_1^2 \log x \, d\alpha(x)$, i.e. with probability nearly one we expect $\xi_n \sim a^n$ where $\log a = \int_1^2 \log x \, d\alpha(x)$. On the other hand $E\xi_n = (Ex_1)^n = (\int_1^2 x \, d\alpha(x))^n$. By Jensen's inequality $Ex_1 \geq \exp(E\log x_1)$ and if α is nondegenerate the inequality is strict. Therefore the contribution to $E\xi_n$ from typical sequences which grow like a^n does not account for the growth in $E\xi_n$. Where does the contribution come from? We might try to analyze what the probability is for ξ_n to grow like λ^n. Of course unless $1 \leq \lambda \leq 2$ this probability is zero. Unless $\lambda = \exp(E\log x_1)$ this probability goes to zero. It turns out that the probability goes to zero exponentially rapidly, i.e. like $[\rho(\lambda)]^n$ where $\rho(\lambda)$ can be explicitly determined as a function of λ. The quantity $\rho(\lambda)$ is one if $\lambda = \exp(E\log x_1)$, $0 \leq \rho(\lambda) \leq 1$ for all values of λ and $\rho(\lambda) = 0$ unless $0 \leq \lambda \leq 1$. The contribution to the integral $E\xi_n$ from those ξ_n which are like λ^n is then $\lambda^n[\rho(\lambda)]^n$. The maximum contribution comes from that value of λ_0 such that

$$\lambda_0\rho(\lambda_0) = \sup_{0\leq\lambda\leq 1} \lambda\rho(\lambda)$$

and $Ex_1 = \lambda_0\rho(\lambda_0)$. The principle of large deviation tells us which values of ξ_n contribute to $E\xi_n$ as $n \to \infty$.

A more complicated example is provided by considering a Markov chain on a finite state space X. For $x,y \in X$ let $\pi(x,y)$ be the probability of transition from state x to state y in a single step. We will assume for simplicity that $\pi(x,y) > 0$ for all $x,y \in X$. Under this condition we have a unique invariant set of probabilities $\alpha(x)$, $x \in X$ and we have the usual set of ergodic theorems on the long term behavior of irreducible finite state space Markov Chains. Let us take a function $V: X \to R$ and consider $E_x\exp[V(X_1) + \cdots V(X_n)] = J_n(x)$ where x is the starting point or X_0 and $X_1,\cdots X_n$ are the states of the Markov Chain at times 1 through n. If we take a frequency count $f_n(y)$ of the number of visits to the state y by the string $X_1,\cdots,X_n$ then

$$\begin{aligned} J_n(x) &= E_x\exp[\sum_y V(y)f_n(y)] \\ &= E_x\exp[n \sum_y V(y) \frac{f_n(y)}{n}]. \end{aligned}$$

By the ergodic theorem $\frac{f_n(y)}{n}$ is very close to $\alpha(y)$ for all states y with a very high probability. If $\beta(y)$; $y \in X$ is any other probability vector on the state space $P_x[\frac{f_n(y)}{n} \sim \beta(y)$ for all $y \in X]$ is likely to be very small if $\beta \neq \alpha$. In fact the probability is exponentially small with a rate $\exp[-nI(\beta)]$ which is independent of the starting point to within the exponential constant. The contribution to $I_n(x)$

from the strings $X_1,\cdots,X_n$ whose relative proportion of visits to the states is close to $\beta(\cdot)$ is $\exp[n\Sigma V(y)\beta(y) - nI(\beta)]$ so that one expects

$$(1.1)\qquad \lim_{n\to\infty} \frac{1}{n} \log J_n(x) = \sup_{\beta} [\Sigma V(y)\beta(y) - I(\beta)]$$

where the supremum is taken over all probability vectors β on X. $I(\alpha)$ of course is zero and $I(\beta) > 0$ for $\beta \neq \alpha$. We can of course evaluate $J_n(x) = \langle\delta_x,(\pi e^V)^n e\rangle$ where e^V is the diagonal matrix with $\exp\{V(x)\}$ on the diagonal place corresponding x. Here e is the vector of units and δ_x is the vector with 1 at x and 0 everywhere else. Frobenius theory of positive matrices evaluates

$$(1.2)\qquad \lim_{n\to\infty} \frac{1}{n} \log J_n(x) = \log \lambda(V)$$

where $\lambda(V)$ is the spectral radius of the positive matrix $\{\pi(x,y)\exp[V(y)]\}_{x,y\in X}$. So we obtain a relation

$$(1.3)\qquad \log \lambda(V) = \sup_{\beta} [\Sigma V(y)\beta(y) - I(\beta)] .$$

It turns out that $I(\beta)$ and $\log \lambda(V)$ are convex functions of β and V respectively and we have the dual relationship

$$(1.4)\qquad I(\beta) = \sup_{V} [\Sigma V(y)\beta(y) - \log \lambda(V)] .$$

Technically one proves the exponential rates of decay for sets of interest by variants of the above formula.

We will conclude this lecture by providing a formal definition of what we mean by the principle of large deviation. The definition is formulated in terms of exponential rates of decay.

Let X be a complete separable metric space and let P_n be a sequence of probability measures on the Borel σ-field of X. A function $I(x)$ from $X \to R$ is called a rate function if

(i) $0 \leq I(x) \leq \infty$

(ii) $I(x)$ is lower semi-continuous on X, and

(iii) The level sets $A_\ell = \{x: I(x) \leq \ell\}$ are compact sets in X.

The sequence $\{P_n\}$ is said to obey a principle of large deviation with rate function $I(\cdot)$ if

a) For every closed set $C \subset X$

$$\limsup_{n\to\infty} \frac{1}{n} \log P_n(C) \leq - \inf_{x\in C} I(x)$$

and

b) For every open set $G \subset X$

$$\liminf_{n\to\infty} \frac{1}{n} \log P_n(G) \geq -\inf_{x\in G} I(x) .$$

It follows then that if A is a Borel set with

$$\inf_{x\in A^0} I(x) = \inf_{x\in \bar{A}} I(x)$$

then

$$\lim_{n\to\infty} \frac{1}{n} \log P_n(A) = -\inf_{x\in A} I(x) .$$

Here A^0 and $\bar{A}$ are respectively the interior and closure of A.

Sometimes one runs into a situation where the rate function satisfies only (i) and (ii) and only a weak form of the large deviation principle with

a) replaced by

a') For every compact set $K \subset X$

$$\limsup_{n\to\infty} \frac{1}{n} \log P_n(K) \leq -\inf_{x\in K} I(x)$$

holds.

b) Stays the same. In such situations it is difficult to aggregate the local estimates provided by the rate function $I(\cdot)$ into a global estimate. Of course if the space X is compact then there is no difference between the two.

Section 2. Basic General Facts.

In this lecture we will establish some of the basic implications that follow from the principle of large deviations.

Let P_n satisfy the principle of large devitions on a complete separable metric space with a rate function $I(\cdot)$. Let $F(\cdot)$ be a bounded continuous function on X. Then

Theorem 2.1

$$\lim_{n\to\infty} \frac{1}{n} \log \int \exp[nF(x)]dP_n(x) = \sup_x [F(x) - I(x)] .$$

Proof: Given any $\delta > 0$ we can find a finite number $C_1, C_2, \cdots, C_N$ of closed sets such that the oscillation of $F(x)$ on C_j is less than δ for every j and such that $C_1, \cdots, C_N$ cover X. Then

$$\int_X \exp[nF(x)]dP_n \le \sum_{j=1}^{N} \int_{C_j} \exp[nF(x)]dP_n$$

$$\le \sum_{j=1}^{N} \exp[n \sup_{x\in C_j} F(x)]\cdot P_n(C_j)$$

$$\le \sum_{j=1}^{N} \exp[n \inf_{x\in C_j} F(x) + n\delta]\cdot P_n(C_j) .$$

Therefore

$$\limsup_{n\to\infty} \frac{1}{n} \log \int_X \exp[n\, F(x)]dP_n$$

$$\le \sup_{1\le j\le N} [\inf_{x\in C_j} F(x) + \delta - \inf_{x\in C_j} I(x)]$$

$$\le \sup_{1\le j\le N} [\sup_{x\in C_j} \{F(x) - I(x)\} + \delta\}$$

$$= \sup_{x\in X} [F(x) - I(x)] + \delta .$$

Since $\delta > 0$ is arbitrary we have

$$(2.1) \qquad \limsup_{n\to\infty} \frac{1}{n} \log \int_X \exp[n\, F(x)]dP_n \le \sup_{x\in X} [F(x) - I(x)]$$

On the other hand if $x \in X$ is arbitrary and U is a neighborhood of x with $F(y) \ge F(x) - \varepsilon$ for $y \in U$ then

$$\int_X \exp[n\, F(y)]dP_n \ge \int_U \exp[n\, F(y)]dP_n$$

$$\ge \exp[n\, (F(x) - \varepsilon)]P_n(U) .$$

Therefore

$$\liminf_{n\to\infty} \frac{1}{n} \int_X \exp[n\, F(y)]dP_n$$

$$\ge F(x) - \varepsilon - \inf_{y\in U} I(y)$$

$$\ge F(x) - I(x) - \varepsilon .$$

Since $x \in X$ and $\varepsilon > 0$ are arbitrary we obtain

$$(2.2) \qquad \liminf_{n\to\infty} \frac{1}{n} \log \int_X \exp[n\, F(y)]dP_n(y) \ge \sup_x [F(x) - I(x)]$$

(2.1) and (2.2) establish the theorem.

Sometimes one has to use a more complex version of theorem 2.1. We will state and prove the part of this version relevant for upper bounds.

<u>Theorem 2.2</u>. Let $F_n(x)$ be a sequence of nonnegative functions such that for some lower semicontinuous non negative function $F(x)$

$$\liminf_{n\to\infty} F_n(x_n) \geq F(x)$$

for every sequence $x_n \to x$, and every $x \in X$. Then

$$\limsup_{n\to\infty} \frac{1}{n} \log \int_X \exp[-n\, F_n(x)]dP_n$$
$$\leq - \inf_{x\in X} [F(x) + I(x)] .$$

<u>Proof</u>: Let $\ell = \inf_{x\in X}[F(x) + I(x)]$. For any $\delta > 0$ and $x \in X$ there is a neighborhood $U_{\delta,x}$ of x such that

$$\inf_{y\in \overline{U}_{\delta,x}} I(y) \geq I(x) - \delta$$

and

$$\liminf_{n\to\infty} \inf_{y\in U_{\delta,x}} F(y) \geq F(x) - \delta$$

Therefore as $n \to \infty$

$$\int_{U_{\delta,x}} \exp[-n\, F_n(y)]dP_n \leq P_n(\overline{U}_{\delta,x})\exp[-n(F(x) - \delta) + o(n)] .$$

Since $U_{\delta,x}$ is a closed set,

$$\limsup_{n\to\infty} \frac{1}{n} \log \int_{U_{\delta,x}} \exp[-nF_n(y)]dP_n$$
$$\leq -[F(x) - \delta] - [I(x) - \delta]$$
$$\leq -\inf_{y\in U_{\delta,x}} [F(x) + I(x)] + 2\delta .$$

If K is any compact set in X, then a finite union of $U_{\delta,x}$ as x varies will cover K. Let us call this finite union U_δ. We have

$$\limsup_{n\to\infty} \frac{1}{n} \log \int_{U_\delta} \exp[-nF_n(y)]dP_n$$
$$\leq -\inf_{y\in U_\delta} [F(x) + I(x)] + 2\delta$$
$$\leq -\ell + 2\delta .$$

Let us pick $K = \{x: I(x) \leq k\}$ where $k >> \ell$. Then

$$\limsup_{n\to\infty} \frac{1}{n} \log \int_{X-U_\delta} \exp[-n\, F_n(y)]dP_n$$
$$\leq \limsup_{n\to\infty} \frac{1}{n} \log P_n[X - U_\delta]$$
$$\leq -\inf_{x\in U_\delta} I(x) \leq -k .$$

Therefore

$$\limsup_{n\to\infty} \frac{1}{n} \log \int_X \exp[-n\, F_n(y)]dP_n$$
$$\leq \text{Max}[-\ell + 2\delta, -k] .$$

Since $k < \infty$ is arbitrary and $\delta > 0$ is arbitrary we let $k \to \infty$ and $\delta \to 0$ to obtain our theorem.

A situation that comes up often in applications is the following: P_n is a sequence of probability measures on X satisfying a large deivation principle with a rate function $I(\cdot)$. We have a continuous map $F: X \to Y$ into another complete separable metric space. We denote by $Q_n = P_n \circ F^{-1}$, the image of P_n on Y udner F. One can ask if Q_n satisfies a large deviation principle and if so what is the relation of its rate function to $I(\cdot)$.

Theorem 2.3. Q_n satisfies a large deviation principle with a rate function $I'(y)$ given by

$$I'(y) = \inf_{x:F(x)=y} I(x) .$$

Proof: It follows from the properties of the rate function $I(\cdot)$ that $I'(\cdot)$ is a rate function on Y. Moreover for any closed set $C \subset Y$

$$\limsup_{n\to\infty} \frac{1}{n} \log Q_n(C) = \limsup_{n\to\infty} \frac{1}{n} \log P_n(F^{-1}C)$$
$$\leq -\inf_{x\in F^{-1}(C)} I(x)$$
$$= -\inf_{y\in C} I'(y) .$$

The lower bound for open sets is similar. We will refer to this theorem as the contraction principle.

The theorems in the theory of large deviations are fairly stable under reasonable perturbations; for instance if we assume that P_n satisfies a large deviation principle with rate $I(\cdot)$ and F_n are continuous maps from $X \to Y$ converging uniformly on compact sets of X to F then for the image $Q_n = P_n \circ F_n^{-1}$ we have again a theorem.

Theorem 2.4. $Q_n = P_n \circ F_n^{-1}$ satisfies a large deivation principle with the rate function

$$I'(y) = \inf_{x:F(x)=y} I(x) .$$

Proof: Let $A \subset Y$ be a closed. Let $C_n = \{x : F_n(x) \in A\}$ then

$$P_n(C_n) = Q_n(A) .$$

If we let $C = \{x : F(x) \in A\}$, from the uniform convergence of F_n to F on compact sets it follows that given any open set U containing C and any compact $K \subset X$ there is a neighorhood K^δ of K such that

$$C_n \cap K^\delta \subset U \quad \text{for sufficiently large } n .$$

Therefore for n large enough

$$P_n(C_n) \leq P_n(U) + P_n(X - K^\delta)$$
$$\leq P_n(\bar{U}) + P_n(X - K^\delta) .$$

If we take $K = \{x : I(x) \leq \ell\}$ we obtain

$$\limsup_{n\to\infty} \frac{1}{n} \log P_n(C_n) \leq \operatorname{Max}[- \inf_{x\in\bar{U}} I(x) , -\ell] .$$

We let $\ell \to \infty$ and $\bar{U} \downarrow C$ so that

$$\lim_{\bar{U}\downarrow C} \inf_{x\in\bar{U}} I(x) = \inf_{x\in C} I(x)$$

and we obtain our desired result.

Let us now take $G \subset Y$ to be open. Let $y \in G$ be arbitrary and $x \in X$ be such that $F(x) = y$. Since $F_n(x)$ tends to $F(x)$ uniformly on compact sets we can find a neighborhood V of x in X such that $F_n(V) \subset G$ for n large enough. Therefore for sufficiently large n

$$Q_n(G) \geq P_n(V)$$

and

$$\liminf_{n\to\infty} \frac{1}{n} \log Q_n(G) \geq \liminf_{n\to\infty} \frac{1}{n} \log P_n(V)$$
$$\geq -I(x) .$$

Since this is true for every $x \in X$ such that $y = F(x) \in G$ we have our theorem.

Section 3. Large deviations for stationary stochastic processes.

Our lectures deal mainly with large deviations of various ergodic phenomena. Let us take a sequence $X_1, X_2, \cdots X_n, \cdots$ of real valued random variables which form an ergodic stationary sequence in the strict sense. We can extend the process to nonpositive integers and obtain a stationary process $\{X_j\}$ $-\infty < j < \infty$. The measure P corresponding to such a process is a translation invariant ergodic measure on the doubly infinite sequences of real numbers. For every n we have the random variable $\xi_n = \frac{X_1+\cdots X_n}{n}$ and this will have a distribution Q_n under P. The ergodic theorem asserts that if $E^P|X_1| < \infty$ then Q_n for large n is close to the degernate distribution at $a = E^P X_1 \in R$. Hopefully under suitable susmptions on the underlying process P, Q_n will satisfy a large deviation principle on R with a rate function $h(x)$, $x \in R$. Since the entire probability is getting concentrated at $x = a$ we expect $h(a) = 0$ and $h(x) > 0$ for $x \neq a$.

We can consider a more general situation in which the stationary process $\{X_j\}$ takes values in an arbitrary Polish space X. We can take for our ξ_n the random varible $\xi_n(f) = \frac{f(X_1)+\cdots f(X_n)}{n}$ where f is a bounded continuous function on X. Again we can expect to strengthen the ergodic theorem by establishing a large deviation principle for the distributions Q_n^f of $\xi_n(f)$ under P on R with a rate function $h_f(y)$, $y \in R$. We cn be more ambitious and consider a vector $f_1,\cdots,f_d$ of functions on X and expect for the random vector $\xi_n(f)$ or for its distribution Q_n^f on R^d under P a large deviation principle with a rate function $h_f(y)$, $y \in R^d$.

In fact one should be even more ambitious and consider the random variable

$$\xi_n = \frac{\delta_{X_1} + \cdots \delta_{X_n}}{n}$$

as a map from the space Ω of infinite sequences into the space of probability measure on X. ξ_n is then the empirical distribution of the process based on the first n random variables. The distribution of ξ_n is then a measure Q_n on the spce M of probability measures on X. The ergodic theorem is still valid and asserts that ξ_n converges almost surely (in the sense of weak convergence in the space M) with respect to P to the measure α which is the one dimensional distribution of the random variable X_1 under P. We may again expect for Q_n a principle of large deviation to hold in the space M with a rate function $I(\mu)$, $\mu \in M$. Needless to say, we expect $I(\alpha) = 0$ and $I(\mu) > 0$ for $\mu \neq \alpha$. If f is a bounded continuous vector valued function on X (with values in R^d) then

$$\frac{f(X_1) + \cdots f(X_n)}{n} = \int f \, d\xi_n .$$

Therefore the map $\mu \to \int f d\mu$ from $M \to R^d$ maps Q_n onto Q_n^f. By the contraction principle one may deduce the large deviation principle for Q_n^f from that of Q_n. The relation between the rate functions is provided by Theorem 2.2. Therefore

$$h_f(y) = \inf_{\mu:\int f d\mu = y} I(\mu) .$$

We have so far looked at the ergodic theorems for random variables of the form

$$\xi_n(f) = \frac{f(X_1) + \cdots, f(X_n)}{n} .$$

But the ergodic theorems apply equally well for random variables of the form

$$(3.1) \qquad \xi_n(g) = \frac{g(X_1,X_2) + \cdots g(X_n,X_{n+1})}{n} .$$

Continuing on in the same spirit we might want to look at the map

$$\xi_n^{(2)} = \frac{\delta_{X_1,X_2} + \cdots \delta_{X_n,X_{n+1}}}{n} .$$

from Ω into $M^{(2)}$ the space of measures on $X \times X$ or $X^{(2)}$. We may expect again a large deviation principle with a rate function $I^{(2)}(\beta)$ for $\beta \in M^{(2)}$. There is nothing special about 2 and we may take

$$\xi_n^{(k)} = \frac{\delta_{X_1,\cdots X_K} + \delta_{X_2,\cdots X_{k+1}} + \cdots + \delta_{X_n,\cdots X_{n+k-1}}}{n} .$$

In fact we should abandon all restraint and consider $\xi_n^{(k)}$ for the totality of all possible values of k. This has to be done with a little care. For every sequence

$$\omega = (\cdots x_{-1}, x_0, x_1, \cdots) \in \Omega$$

and every n let us consider the sequence

$$\omega^{(n)} = (\cdots x_1, \cdots, x_n, x_1, \cdots x_n, x_1, \cdots, x_n, \cdots) .$$

Formally the ith coordinate of $\omega^{(n)}$ is given by

$$\omega_i^{(n)} = x_i \quad \text{if } 1 \le i \le n$$

$$\omega_{i+n}^{(n)} = \omega_i^{(n)} \quad \text{for all i and n} .$$

In other words we keep the chunk of ω from x_1, through x_n and make it periodic outside of period n. If we look at all the periodic sequences of period n in Ω and denote this set by $\Omega^{(n)}$ then the map $\omega \to \omega^{(n)}$ defines a map π_n from $\Omega \to \Omega^{(n)}$. Given any point $\omega^{(n)}$ in $\Omega^{(n)}$, denoting by T the shift in Ω, $\omega^{(n)}, T\omega^{(n)}, \cdots, T^{n-1}\omega^{(n)}$ is a periodic orbit in $\Omega^{(n)}$ and $\dfrac{\delta_{\omega^{(n)}}+\delta_{T\omega^{(n)}}+\cdots\delta_{T^{n-1}\omega^{(n)}}}{n}$ defines a T invariant measure on $\Omega^{(n)}$. This is of course a stationary stochastic process on $\Omega^{(n)}$ and since

$\Omega^{(n)} \to \Omega$ this is a stationary process on Ω. In this manner for each n and ω we have defined a stationary process

$$R_{n,\omega} = \frac{1}{n}(\delta_{\omega^{(n)}} + \delta_{T\omega^{(n)}} + \cdots + \delta_{T^{n-1}\omega^{(n)}})$$

If $g(x_1,x_2)$ is viewed as a map from $\Omega \to R$ then

$$\int g(x_1,x_2)dR_{n,\omega} = \frac{1}{n}[g(x_1,x_2) + \cdots g(x_{n-1},x_n) + g(x_n,x_1)] .$$

This is not quite the same as what we have in (3.1) but the difference is just one term in n and becomes negligible as $n \to \infty$. The ergodic theorem again tells us that

$$P[\lim_{n\to\infty} R_{n,\omega} = P] = 1 .$$

We might as well expect a large deviation principle for the distribution $\hat{Q}_n$ of $R_{n,\omega}$ under P. Now the state space is the space of all stationary stochastic processes and we expect a rate function $H(Q)$ for $Q \in \mathcal{M}_S$ which is equal to zero only when $Q = P$. There is of course a natural map from $\mathcal{M}_S \to M$ which assigns to any stationary process its common one dimensional marginal distribution. If we call this map τ

$$\tau R_{n,\omega} = \xi_n = \frac{1}{n}(\delta_{x_1} + \cdots \delta_{x_n})$$

Since map τ is continuous from $\mathcal{M}_S \to M$ the contraction principle applies and we can have a large deviation principle in $\mathcal{M}$ if we have on in $\mathcal{M}_S$. The rate functions are of course related by

$$I(\mu) = \inf_{Q:\tau Q=\mu} H(Q) .$$

Of course to actually carry all of this out requires serious assumptions on the nature of the underlying stationary process P. We will, during these lectures, start with the special case of independent random variables or product measure for P. Then, we will look at the case of a Stationary Markov Chain. We will also look at Stationary Gaussian Processes. We will then extend these results to the case of continuous time Markov Process. Towards the end we will look at some applications of the theories developed here.

Section 4. Independent Random Variables.

Throughout these lectures, in all instances the rate functions will have a close connection with some sort of entropy. It is therefore important for us to spend some time establishing some of the properties of entropy.

Definition: Given any two probability measures β and α on a measure space (X,Σ) we define the entropy of β with respect to α as

$$(4.1) \qquad h(\beta;\alpha) = \sup_{V\in B_0} [\int V(x)d\beta(x) - \log\int e^{V(x)}d\alpha(x)]$$

where B_0 is the space of all bounded measurable functions on X.

This definition is the same as relative entropy or Kullback-Liebler information number: This is the content of the following theorem:

Theorem 4.1. The following two statements are equivalent:

a) $h(\beta,\alpha) = \ell < \infty$.

b) $\beta << \alpha$ and if $f(x) = \frac{d\beta}{d\alpha}$

then $f(x)\log f(x)$ is integrable with respect to α and

$$\int f(x)\log f(x)d\alpha(x) = \ell .$$

Proof: Let us first assume that b) holds for some finite ℓ. Then using $xy \le y\log y + e^{x-1}$ valid for x real and $y > 0$

$$\int V(x)d\beta(x) = \int V(x)f(x)d\alpha(x)$$

$$\le \int f(x)\log f(x)d\alpha(x) + \frac{1}{e}\int e^{V(x)}d\alpha(x)$$

we can write $V(x) = (V(x) - k) + k$. Then

$$\int V(x)d\beta(x) \le \int f(x)\log f(x)d\alpha(x) + e^{-(k+1)}\int e^{V(x)}d\alpha(x) + k .$$

We pick $k = \log\int e^{V(x)}d\alpha(x) - 1$. This yields

$$\int V(x)d\beta(x) \le \ell + \log \int e^{V(x)}d\alpha(x) .$$

Since V is arbitrary we establish a).

Let us now assume that a) is true. First we want to show that $\beta << \alpha$. let A be any set. Take $V(x) = k\chi_A(x)$. We get from a)

$$k\beta(A) \le \ell + \log(1 - \alpha(A) + \alpha(A)e^k)$$

or

$$\beta(A) \le \frac{1}{k}[\ell + \log(1 + \alpha(A)(e^k - 1)]$$

$$\le \inf_{k>0} \frac{1}{k}[\ell + \log(1 + \alpha(A)(e^k - 1))]$$

$$= \psi(\ell,\alpha(A))$$

where

(4.2) $$\psi(\ell,\delta) = \inf_{k>0} \frac{1}{k}[\ell + \log(1 + \delta(e^k - 1))] .$$

it is clear that $\psi(\ell,\delta) \to 0$ as $\delta \to 0$ for each fixed $\ell < \infty$ so that not only is $\beta << \alpha$ but any time ℓ is controlled such β are uniformly absolutely continuous with respect to α. We now have

$$\sup_{V\varepsilon B_0} \int V(x)f(x)d\alpha(x) \leq \ell + \log \int e^{V(\alpha)}d\alpha(x) .$$

We would like to take $V(x) = \log f(x)$ to otain b). But we do not know that $f(x)\log f(x)$ is integrable with respect to α. In any case we may only take bounded V. We pick

$$V_{\varepsilon,k} = \log[(f\wedge k)\vee\varepsilon] .$$

We let $\varepsilon \to 0$ and then $k \to \infty$. Since $f\log f$ is bounded near $f = 0$, letting $\varepsilon \to 0$ is no problem. Finally as $k \to \infty$ we use the monotone convergence theorem to establish our result. The details are left as an exercise.

Sometimes when we are dealing with a polish space X and its Borel sets for Σ, it is convenient to have the following lemma:

Lemma 4.2. If X is any Polish space the supremum in definition (4.1) can be taken over the class of bounded continuous functions and we will still have the same supremum as over all bounded measurable functions:

Proof: a trivial application of Lusin's theorem.

Suppose we have two probability measures β and α on (X,Σ) and a sub σ-field $\Sigma_0 \subset \Sigma$. Then we may just look at β and α on Σ_0 and restricting V to be bounded and measurable with respect to Σ_0 we obtain what we might call $h_{\Sigma_0}(\beta;\alpha)$. In other words the relative entropy is also with respect to a specified σ-field which may only be a sub σ-field. Obviously if $\Sigma_1 \subset \Sigma_2$ then $h_{\Sigma_1}(\beta;\alpha) \leq h_{\Sigma_2}(\beta;\alpha)$. We want to interpret the difference again as an entropy. Let us suppose that β and α possess regular conditional probabilities β_ω and α_ω given the σ-field Σ_1. Then

Lemma 4.3. We have the following identity

$$h_{\Sigma_2}(\beta;\alpha) = h_{\Sigma_1}(\beta,\alpha) + E^\beta h_{\Sigma_2}(\beta_\omega,\alpha_\omega) .$$

Proof: We can assume without loss of generality that $h_{\Sigma_1}(\beta,\alpha) < \infty$. Otherwise $h_{\Sigma_2}(\beta,\alpha) \geq h_{\Sigma_1}(\beta,\alpha) = \infty$ and the identity is valid because both sides are infinite. If $h_{\Sigma_1}(\beta,\alpha) < \infty$ then $\beta << \alpha$ on Σ_1 an therefore α_ω is defined not only almost everywhere with respect to α, but β as well. Therefore the second term on the right makes sense.

For any V bounded and Σ_2 measurable

$$\begin{aligned} E^\beta[V(x)] &= E^\beta E^{\beta_\omega} V(x) \\ &\leq E^\beta[\log E^{\alpha_\omega} e^{V(x)}] + E^\beta h_{\Sigma_2}(\beta_\omega,\alpha_\omega) \\ &\leq h_{\Sigma_1}(\beta,\alpha) + \log E^\alpha E^{\alpha_\omega} e^{V(x)} + E^\beta h_{\Sigma_2}(\beta_\omega,\alpha_\omega) \\ &= \log E^\alpha e^{V(x)} + h_{\Sigma_1}(\beta,\alpha) + E^\beta h_{\Sigma_2}(\beta_\omega,\alpha_\omega) . \end{aligned}$$

This proves one half of the lemma: As for the other half we note that if $\beta < \alpha$ on Σ_2 then

$$\frac{d\beta}{d\alpha}\Big|_{\Sigma_2} = \frac{d\beta}{d\alpha}\Big|_{\Sigma_1} \cdot \frac{d\beta_\omega}{d\alpha_\omega}\Big|_{\Sigma_2} .$$

Let us now suppose that we have a product measure P on $\Omega = \prod_{-\infty}^{\infty} X_i$ with marginal distribution α on each X_i which are copies of X. Let Q be any statinary process on Ω. We denote by F_m^n the σ-field generated by the coordinates x_i of $\omega \in \Omega$ for $n \leq i \leq m$. If $m = \infty$ the σ-field is denoted by F^n and if $n = -\infty$ by F_m. We denotely Q_ω the regular conditional probability distribution of x_1 given F_0 under Q. The rate function that will play a role here is

Definition 4.4. $H(Q;P) = E^Q h_{F_1^1}(Q_\omega,\alpha)$.

Although the functional $H(Q;P)$ is defined through entropy it has several variational formulae as well and we need them in order to establish some of its properties as well as in the proof of the large deviation principle.

Let A_n be the class of bounded measurable functions on $X^{(n)} = X\times\cdots\times X$ i.e. functions $F(x_1,\cdots,x_n)$, satisfying the condition

$$\int_X \exp[F(x_1,\cdots,x_n)]d\alpha(x_n) \leq 1 \quad \forall\, x_1,\cdots,x_{n-1} .$$

Theorem 4.5.

$$H(Q;P) = \sup_n \sup_{F\in A_n} E^Q(F)] .$$

Moreover we can replace A_n by the set of bounded continuous function C_n of n variables instead and we also have

$$H(Q;P) = \sup_n \sup_{F\in C_n} E^Q[F] .$$

Proof: Let $F \in A_n$. Then

$$\begin{aligned} E^Q[F] &= E^Q[F(x_1,\cdots,x_n)] \\ &= E^Q E^Q[F(x_1,\cdots,x_n)|F_{n-1}] \\ &= E^Q \log E^P[e^{F(x_1,\cdots x_{n-1})}|F_{n-1}] + H(Q;P) \\ &\leq H(Q;P) . \end{aligned}$$

To establish the reverse inequality we define $\bar{Q}$ on the σ-field F_1 by making $\bar{Q} = Q$ on F_0 and making the first coordinate independent of the past F_0 and having the distribution α. Then $\bar{Q}_\omega = \alpha$. Then we see that

$$\begin{aligned} H(Q;P) &= h_{F_1}(Q,\bar{Q}) \\ &= \sup_{F\in F_1} E^Q[F] - \log E^{\bar{Q}}[e^F] \\ &= \sup_n \sup_{F=F(x_{-n},\cdots,x_{-1},x_0,x_1)} E^Q[F] - \log E^{\bar{Q}}[e^F] \\ &\quad \text{(by a variant of Lusin's theorem)} \\ &\leq \sup_n \sup_{F\in F_1^{-n}} E^Q[F] - E^Q[\log \int e^{F(\cdots x_1)} d\alpha(x_1)] \\ &= \sup_n \sup_{F\in A_n} E^Q[F] . \end{aligned}$$

Another way of calculating $H(Q,P)$ is by the following Theorem.

Theorem 4.6.

$$H(Q;P) = \sup_n \frac{1}{n} \sup_{F\in D_n} E^Q[F]$$

where

$$D_n = \{F: F = F(x_1,\cdots,x_n) \; ; \int e^{F(x_1,\cdots,x_n)} dP \leq 1\} .$$

Proof: Let us call $F(x_1,\cdots,x_n)$ by $F_n(x_1,\cdots,x_n)$ and define successively

$$F_k(x_1,\cdots,x_k) = \log \int e^{F_{k+1}(x_1,\cdots,x_{k+1})} d\alpha(x_{k+1}) .$$

Then we verify that $F_0 \leq 0$ and

$$\int e^{F_{k+1}(x_1,\cdots,x_{k+1}) - F_k(x_1,\cdots,x_k)} d\alpha(x_{k+1}) \leq 1 .$$

Therefore by Theorem 4.5

$$E^Q F_{k+1} - E^Q F_k \le H(Q,P)$$

adding over $k = 0,1,\cdots,n-1$ and using $F_0 \le 0$ we have

$$E^Q F_n - E^Q F \le nH(Q;P) .$$

To prove the converse we note that by definition

$$\sup_{F\in D_n} E^Q[F] = h_{F_n^1}(Q,P) .$$

We compute

$$h_{F_n^1}(Q;P) - h_{F_{n-1}^1}(Q,p)$$

$$= h_{F_1^{2-n}}(Q;P) - h_{F_1^{3-n}}(Q;P) \text{ (by stationarity)}$$

$$= h_{F_1^1}(Q_\omega^{-(n-2)};\alpha) .$$

As $n \to \infty$ the above term has a limit inferior of at least $h_{F_1^1}(Q_\omega,\alpha) = H(Q;P)$. This establishes the theorem:

We are now ready to prove our upper and lower bounds for establishing the large deviation principle. We have the measure $\hat{Q}_n$ which is the distribution of $R_{n,\omega}$ under P. We divide the theorem into many lemmas.

<u>Theorem 4.7</u>. For any closed set $C \subset M$

(4.3) $$\limsup_{n\to\infty} \frac{1}{n} \log \hat{Q}_n(C) \le - \inf_{Q\in C} H(Q,P)$$

and for $G \subset M$, open

(4.4) $$\liminf_{n\to\infty} \frac{1}{n} \log \hat{Q}_n(G) \ge - \inf_{Q\in G} H(Q;P) .$$

<u>Proof</u>: Let us denote for any Borel set A

(4.5) $$J(A) = \limsup_{n\to\infty} \frac{1}{n} \log \hat{Q}_n(A) .$$

We will establish several properties of J(A) as lemmas leading up to the proof of (4.3). The lower bound (4.4) will be dealt with later.

<u>Lemma 4.8</u>. Let $F(x_1,\cdots,x_k) \in C_k$. Then for $n \ge 1$

$$E^P\{\exp[\tfrac{1}{k}(F(x_1,\cdots,x_k) + F(x_2,\cdots,x_{k+1}) + \cdots F(x_n,\cdots,x_{n+k-1}))\} \le 1 .$$

<u>Proof</u>: We write

$$F(x_1,\cdots,x_k) + \cdots F(x_n,\cdots x_{n+k-1})$$

$$= G_1(x_1,x_2,\cdots) + G_2(x_1,x_2,\cdots) + \cdots G_k(x_1,x_2,\cdots)$$

where

$$G_1 = F(x_1,\cdots,x_k) + F(x_{k+1},\cdots,x_{2k}) + \cdots$$

$$G_2 = F(x_2,\cdots x_{k+1}) + F(x_{k+2},\cdots,x_{2k+1}) + \cdots$$

$$G_k = F(x_k,\cdots,x_{2k-1}) + F(x_{2k},\cdots,x_{3k-1}) + \cdots .$$

Then

$$E^P\{\exp[\tfrac{1}{k}(F(x_1,\cdots,x_k) + \cdots \quad F(x_n,\cdots,x_{n+k-1}))]\}$$

$$= E^P\{\exp[\tfrac{1}{k}(G_1 + \cdots + G_k)]\}$$

$$\leq E^P[\tfrac{1}{k}(e^{G_1} + \cdots + e^{G_k})]$$

$$\leq \frac{1}{k}\sum_1^k E^P[e^{G_i}] \leq 1 .$$

Lemma 4.9. For any Borel set $A \subset M$

$$J(A) \leq -\sup_k \frac{1}{k} \sup_{F\epsilon C_k} \inf_{Q\epsilon A} E^Q\{F\} .$$

Proof:

$$\left|\frac{1}{n}[F(x_1,x_2,x_k) + \cdots + F(x_n,\cdots,x_{n+k-1})] - E^{R_{n,\omega}}\{F\}\right| \leq \frac{k-1}{n} ||F||$$

where $||F||$ is the sup norm on F.

Therefore from lemma 4.8

$$E^P[\exp[\tfrac{1}{k}(F(x_1,\cdots,x_k) + \cdots F(x_n,\cdots,x_{n+k-1}))]] \leq C$$

or

$$E^{\hat{Q}_n}[\exp[\tfrac{n}{k} \textstyle\int F dQ]] \leq C$$

or

$$\hat{Q}_n(A) \leq C e^{-\frac{n}{k} \inf_{Q\epsilon A} \int F dQ} .$$

Taking logs, dividing by n and taking limsup as $n \to \infty$

$$J(A) \leq -\frac{1}{k} \inf_{Q\epsilon A} \int F\, dQ .$$

Since $F \epsilon C_k$ and k are arbitrary we have our result.

Lemma 4.10. Let $K \subset M$ be compact and let $\epsilon > 0$ be given. Then there exists an open set G_ϵ in M such that $K \subset G_\epsilon$ and

$$J(G_\varepsilon) \leq - \inf_{Q \varepsilon K} H(Q;P) + \varepsilon .$$

Proof: Let $\varepsilon > 0$ and k be given. For each $Q \varepsilon K$ there exists an integer $k(Q)$ and $F_Q \varepsilon C_{K(Q)}$ such that

$$\frac{1}{k(Q)} \int F_Q dQ \geq H(Q,P) - \varepsilon/2 .$$

Since $\int F_Q dQ$ is a continuous linear functional of Q, there is a neighborhood N_Q of Q such that for $Q' \varepsilon N_Q$

$$\frac{1}{k(Q)} \int F_Q dQ' \geq H(Q,P) - \varepsilon .$$

Therefore from lemma 4.9

$$J(N_Q) \leq - H(Q,P) + \varepsilon .$$

N_Q as Q varies over K is an open covering of K and let $N_{Q_1}, \cdots, N_{Q_L}$ be a finite subcover. Denoting by G_ε such a finite subcover clearly G_ε K and

$$\begin{aligned} J(G_\varepsilon) &\leq - \min_{1 \leq j \leq L} J(N_{Q_j}) \\ &\leq - \min_{1 \leq j \leq L} [H(Q_j,P) - \varepsilon] \\ &\leq - \min_{1 \leq j \leq L} H(Q_j;P) + \varepsilon \\ &\leq - \inf_{Q \varepsilon K} H(Q;P) + \varepsilon . \end{aligned}$$

Lemma 4.11. Given any $\ell < \infty$ there exists a compact set $K_\ell \subset M$ such that

$$J(K_\ell^c) \leq - \ell .$$

Proof: For each $Q \varepsilon M$ let us denote by q the marginal distribution of x_0. Then if B_ℓ M is a compact set of probability measures on X. Then

$$K_\ell = \{Q : q \varepsilon B_\ell\} \text{ is compact in } M .$$

We need therefore only estimate

$$\hat{Q}_n \{Q : q \varepsilon B_\ell^c\}$$

or

$$P\{\omega : \frac{\delta_{x_1} + \cdots \delta_{x_n}}{n} \varepsilon B_\ell^c\} .$$

From Prohorov's theorem we can take

$$B_\ell = \{q : q(D_j^\ell) \geq 1 - \frac{1}{j} \quad \text{for } j = 2, \cdots .\}$$

where $D_j^\ell \subset X$ are compact subsets of X for each j and ℓ. Therefore

$$\begin{aligned} P\{\omega: & \frac{\delta_{x_1} + \cdots \delta_{x_n}}{n} \in B_\ell^c\} \\ & \leq \sum_{j=2}^{\infty} P\{\omega: \frac{\chi_{D_j^\ell,c}(x_1) + \cdots \chi_{D_j^\ell,c}(x_n)}{n} \geq \frac{1}{j}] \\ & \leq \sum_{j=2}^{\infty} 2^n e^{-n\ell j} \quad \text{(by Lemma 4.12 with } \theta = \ell j^2) \end{aligned}$$

if we pick D_j^ℓ such that $\alpha(D_j^\ell) \geq 1 - e^{-\ell j^2}$ which is possible by Prohorov's theorem. If we assume $\ell \geq 1$, then the lemma follows.

<u>Lemma 4.12</u>. Let $x_1, \cdots, x_n$ be n independent random variables with values 0 or 1 with probability ε and 1 − ε. Then for any θ > 0

$$\text{Prob}[\frac{x_1 + \cdots, x_n}{n} \geq \delta] \leq e^{-n\theta\delta}(\varepsilon e^\theta + (1 - \varepsilon))^n .$$

<u>Proof</u>: Apply Tchebechev's inequality to

$$E[e^{x_1 + \cdots, x_n}] = (\varepsilon e^\theta + 1 - \varepsilon)^n .$$

Proof of (4.3). Given any closed set C we pick an $\ell < \infty$ and the comapct set K_ℓ of lemma 4.11. Then K_ℓ C is compact and

$$\begin{aligned} J(C) & \leq - \text{Min}[J(K_\ell \cap C), J(K_\ell^c)] \\ & \leq - [\inf_{Q \in C} H(Q;P), \ell] . \end{aligned}$$

Letting $\ell \to \infty$ we obtain our result. We now work on the lower bound:

Let us denote by $\psi(\omega)$ the Radon-Nikodym derivative of Q_ω with respect to α on the σ-field F_1^1. Then $\psi(\omega)$ can be thought of as a measurable function on F_1 and $E^Q \log\psi(\omega) = H(Q;P)$. Moreover if we denote by T the shift on the space of sequences then $\frac{dQ_\omega}{dP}$ on F_n^1 is $\exp[\psi(\omega) + \psi(T\omega) + \cdots\psi(T_\omega^{n-1})]$.

<u>Lemma 4.13</u>. Let Q be any ergodic element in M. Then for any neighborhood N of Q in M

$$\liminf_{n\to\infty} \frac{1}{n} \log \hat{Q}_n[N] \geq - H(Q;P) .$$

<u>Proof</u>: Assume $H(Q;P) < \infty$.

$$\begin{aligned} \hat{Q}_n(N) & = P\{R_{n,\omega} \in N\} \\ & = \int_{R_{n,\omega} \in N} dP \end{aligned}$$

$$\geq \int_{R_{n,\omega}\in N} \frac{dP}{dQ}\, dQ$$

$$= \int_{R_{n,\omega}\in N} e^{-(\log\psi(\omega)+\cdots\log\psi(T^{n-1}\omega))}\, dQ$$

$$\geq \int_{R_{n,\omega}\in N \cap E_\delta} e^{-n(H(Q,P)+\delta)}\, dQ$$

where $E_\delta = \{\omega : \frac{1}{n}\sum_{j=1}^{n} \log \psi(T^j_\omega) \leq H(Q;P) + \delta\}$. By ergodic theorem $Q\{\omega : R_{n,\omega} \in N\}$ as well as $Q(E_\delta)$ tend to 1 as $n \to \infty$ and we are done:

<u>Lemma 4.14.</u> If $Q \in M$ is arbitrary then Q has an integral representation

$$Q = \int_{M_e} Q' \pi_Q(dQ')$$

over the ergodic measures and

$$H(Q,P) = \int_{M_e} H(Q',P)\Pi_Q(dQ') \; .$$

<u>Proof</u>: From standard results in ergodic theory we know that the integral representation is valid. Moreover the regular conditional probability has a version Θ_ω such that

$$\Theta_\omega = Q_\omega \text{ a e } \omega \; Q \quad \forall \in M \; .$$

Therefore

$$H(Q;P) = \int h_{F_1^1}(\Theta_\omega,\alpha)dQ$$

clearly satisfies the lemma:

Now we prove the lower bound (4.4). If $Q \in M$ is arbitrary we can approximate it by a finite linear çombination of ergodic ones such that $Q \sim \Sigma\pi_j Q_j$ and $H(Q;P) \sim \Sigma\pi_j HJ(Q_j,P)$. We can therefore assume without loss of generality that $Q = \Sigma\pi_j Q_j$ with Q_j ergodic. Let n be given and define $n_j = \pi_j n$. For a given $\omega = \omega_1$ let $\omega_j = T^{n_1+\cdots n_{j-1}}\omega_1$. Then $R_{n,\omega} \sim \Sigma\pi_j R_{n_j,\omega_j}$ since the only difference is the periodization at the end. Since the topoloogy on M is essentially convergence of finite dimensional distributions for a given finite dimensional range the effect of the periodization goes to zero as $n \to \infty$. Therefore given a neighborhood N of Q, there are neighborhoods N_j of Q_j such that

$$R_{n_j,\omega_j} \in N_j \text{ for } \forall j \;\Rightarrow\; R_{n,\omega} \in N \; .$$

Therefore

$$P[R_{n,\omega} \in N] \geq \Pi\, P[R_{n_j,\omega_j} \in N_j] .$$

Taking logs, dividing by n and taking limint as $n \to \infty$

$$\underline{\lim}_{n\to\infty} \frac{1}{n} \hat{Q}_n(N) \geq - \Sigma\pi_j H(Q_j,P) \geq - H(Q) - \varepsilon$$

and we are done: We want to conclude this section by stating some properties of various entropy functions:

Lemma 4.15. For fixed α, $h(\beta;\alpha)$ is a lower semicontinuous convex function of β in the weak topology.

Proof:

$$h(\beta;\alpha) = \sup_{F\in C(X)} [\int F d\beta - \log\int e^F d\alpha] .$$

The properties are now obvious:

Lemma 4.16. For fixed P, a product measure based on α, $H(Q;P)$ is lower semicontinuous in Q:

Proof:

$$H(Q,P) = \sup_k \sup_{F\in C_k} [\int F dQ]$$

and the lemma is obvious:

Lemma 4.17. For fixed β

$$\inf_{Q:q=\beta} H(Q;P) = h(\beta;\alpha)$$

and the inf is attained at the product measure:

Proof: It is clear that for the product measure Q_β

$$H(Q_\beta;P) = h(\beta;\alpha) .$$

C_k contains functions $e^{F(x_1)+\cdots F(x_k)}$ with $\int e^{F(x)} d\alpha(x) \leq 1$. Therefore for any Q, $H(Q;P) \geq h(q;\alpha)$.

Lemma 4.18. For any $\ell < \infty$ and any α

$$\{q: h(q;\alpha) \leq \ell\} \text{ is compact in } M .$$

Proof: From inequality (4.2) if $\alpha(A) < \delta$ then $q(A) < \eta$ where $\eta = \eta(\delta,\ell) \to 0$ as $\delta \to 0$ for each ℓ. Since α is tight by Prohorov's theorem $\{q: h(q,\alpha) \leq \ell\}$ is uniformly tight and hence compact.

Lemma 4.19. For any ℓ, $\{Q:H(Q,P) \le \ell\}$ is compact in M.

Proof: As Q varies over our set q varies over a set contained in $\{q:h(q;\alpha) \le \ell\}$ and is therefore conditionally compact. Therefore so is the set of measures Q.

Section 5. Markov Chains

In this section we will assume the base measure P to be based on a Markov chain rather than a product measure. One difference however is that the P measure will be defined only on F° which is all that is needed. Moreover instead of a single P we have a family P_{x_0} depending on the starting point x_0 at time zero. The transition probability is $\pi(x,dy)$. We make the following assumption on $\pi(x,dy)$.

Hypothesis 1.

$\pi(x,dy)$ has the Feller property or for any bounded continuous function $f(\cdot)$ on X,

$$(\pi f)(x) = \int f(y)\pi(x,dy)$$

is bounded and continuous on X.

The random processes $R_{n,\omega}$ are defined as before and instead of $\hat{Q}_n$ we now have $\hat{Q}_{n,x_0}$ depending on the starting point x_0 as well. We will describe the results in this case and indicate in the proof only modifications needed in the earlier proof for the independent case:

Definition 5.1. Given π and $Q \in M$ we define

$$H(Q;\pi) = E^Q\{h_{F_1^1}(Q_\omega,\pi(x_0,\cdot))\} .$$

Here x_0 is thought of as a function of ω and then the relative entropy of Q_ω and $\pi(x_0,\cdot)$ is calculated on the σ-field F_1^1 corresponding to x_1. The answer that depends on ω is averaged with respect to Q.

In theorem 4.5 we should modify A_n so that

$$A_n = \{F:F = F(x_1,\cdots,x_n) \text{ and } \int_X \exp[F(x_1,\cdots,x_n)]\pi(x_{n-1},dx_n) \le 1 \quad \forall\ x_1,\cdots x_{n-1}\}$$

C_n is defined accordingly:

Then we have

Theorem 5.2.

$$H(Q;\pi) = \sup_n \sup_{F\in A_n} E^Q[F]$$

$$= \sup_n \sup_{F \in C_n} E^Q[F] .$$

Proof:

Proceeds in a manner identical to theorem 4.5 with minor obvious modifications. We next define D_n as

$$D_n = \{F : F = F(x_1,\cdots,x_n) \text{ and } \int e^{F(x_1,\cdots,x_n)} dP_{x_0} \leq 1 \quad \forall x_0\} .$$

Then we have the analog of theorem 4.6.

Theorem 5.3.

$$H(Q,P) = \sup_k \frac{1}{k} \sup_{F \in D_k} E^Q[F] .$$

Proof: We define for given $F = F(x_1,\cdots,x_n)$, $F_k = F_k(x_1,\cdots,x_k)$ successively for $k < n$ by

$$F_k(x_1,\cdots,x_k) = \log \int e^{F_{k+1}(x_1,\cdots,x_{k+1})} \pi(x_k,dx_{k+1}) .$$

Then $E^Q[F_{k+1} - F_k] \leq H(Q,\pi)$ by theorem 5.2. The proof is completed as before. Now we start establishing the large deviation principle.

First we define for any $x_0 \in X$

$$(5.1) \qquad J_{x_0}(A) = \limsup_{n\to\infty} \frac{1}{n} \log \hat{Q}_{n,x_0}(A) .$$

We then have

Lemma 5.4. For any compact set $K \subset M$ and any $\varepsilon > 0$ there exists a neighborhood G_ε K such that

$$J_{x_0}(G_\varepsilon) \leq - \inf_{Q \in K} H(Q;\pi) + \varepsilon .$$

Proof:

Identical to lemmas 4.8, 4.9 and 4.10.

The main difference is only at this point. In order to go from compact K to closed C we need to make a strong positive recurrence assumption on the transition probability $\pi(x,dy)$.

Hypothesis 2

Let us suppose that there are functions U(x) and V(x) on X with the following properties:

a) $U(x) \geq 1$ for all x and $(\pi U)(x)$ is bounded on compact subsets of X.

b) $V(x) = \log U(x) - \log(\pi U)(x)$ is bounded below (away from $-\infty$) and for any ℓ $\{x: V(x) \leq \ell\}$ is a totally bounded subset of X.

Under hypothesis 2 we can establish the analog of lemma 4.11.

Lemma 5.5. Given any $\ell < \infty$, there is a compact set K_ℓ M such that

$$J_{x_0}(K_\ell^c) \leq -\ell \quad \text{for every } x_0 \in X .$$

Proof: The proof of lemma 5.5 will depend on lemma 5.6 and will follow the lines of the proof of lemma 4.11.

Lemma 5.6. Given any ℓ and j there is a compact set D_j^ℓ X such that

$$P_{x_0}\{\omega: \frac{1}{n}[\chi_{D_0^\ell,c}(x_1) + \cdots \chi_{D_j^\ell,c}(x_n)] > \frac{1}{j}\} \leq C^n e^{-\ell n j}$$

for all j and n. Here C is some fixed constant.

Proof: From hypothesis 5.5 we have

$$E^{P_{x_0}}\{e^{V(x_p)+\cdots V(x_n)} U(x_{n+1})\} = \pi U(x_0) .$$

Since $U \geq 1$ and $\pi U(n_0) < \infty$ we have

$$E^{P_{x_0}}\{e^{V(n_1) + \cdots V(x_n)}\} \leq C .$$

If we take $D_j^\ell = \{x: V(x) \leq \lambda\}$ for some λ then

$$V(x_1) + \cdots V(x_n) \geq \lambda \sum_{r=1}^{n} \chi_{D_j^\ell,c}(x_r)]] - nC_1 .$$

where $-C_1$ is a lowerbound for $V(\cdot)$. Therefore

$$E^{P_{x_0}}[\exp[\lambda \sum_{r=1}^{n} \chi_{D_j^\ell,c}(x_r)]] \leq C^n .$$

Therefore $P_{x_0}[\frac{1}{n} \sum_{r=1}^{n} \chi_{D_j^\ell,c}(x_r) \geq \frac{1}{j}] \leq C^n e^{-\frac{\lambda n}{j}}$.

If we choose $\lambda = \ell j^2$ in D_j^ℓ we have our estimate. If we therefore have hypothesis 1 and 2 we can get the upper bound part of the large deviation principle. Moreover one can check through the proof that all estimates are valid uniformly provided x_0 varies over a compact subset of X. Now we start working on the lower bound. We need another hypothesis.

Hypothesis 3. The transition probability $\pi(x,dy)$ has a density $\pi(x,y)$ with respect to a reference measure α such that

a) $\pi(x,y) > 0$ a e α for each $x \in X$ and

b) the map $x \to \pi(x,\cdot)$ is continuous as a map of X into $L_1(\alpha)$.

Theorem 5.7. Let $Q \in M$ be ergodic then for any open N containing Q

$$\lim_{n\to\infty} \frac{1}{n} \log P_{x_0}[R_{n,\omega} \in N \text{ and } x_n \in K_2] \geq - H(Q;\pi)$$

where K_2 is any compact set in X with $\alpha(K_2) > 0$; and the limit is uniform for x_0 varying over any compact set in X.

Proof: Let us pick K_3 such that $q(K_3) \geq \frac{1}{2}$, where q is the marginal of Q. Denoting by

$$\frac{dQ_\omega}{dP_{x_0(\omega)}\big|F_1^1} = \psi(\omega)$$

where $x_0(\omega)$ is the coordinate of ω corresponding to zero we have

$$\frac{dQ_\omega}{dP_{x_0(\omega)}\big|F_1^1} = \exp[\sum_{j=0}^{n-1} \log \psi(T^j\omega)] .$$

Therefore if we take the set

$$E_{N,n} = \{\omega: R_{n,\omega} \in N \text{ and } x_n(\omega) \in K_3\}$$

then

$$P_{x_0(\omega)}(E_{N,n}) \geq \int_{E_{N,n}} \exp[-\sum_{j=0}^{n-1} \log \psi(T^j\omega)]dQ_\omega$$

$$\geq e^{-[H(Q,P)+\delta]n} Q_\omega[E_{N,n} \cap \{\omega: \frac{1}{n}\sum_{1}^{n-1} \log \psi(T^j\omega) \leq H(Q;P) + \delta\}] .$$

Let us denote by $\phi(n,x)$ the quantity

$$\phi(n,x) = P_x(E_{N,n})e^{[H(Q;P)+\delta]n} \wedge 1 .$$

Then

$$\phi(n,x_0(\omega)) \geq Q_\omega[E_{N,n} \cap D_{n,\delta}]$$

where

$$D_{n,\delta} = \{\omega: \frac{1}{n}\sum_{1}^{n-1} \log \psi(T^j\omega) \leq H(Q;P) + \delta\} .$$

Taking expectations with respect to Q we have

$$\int\phi(n,x)dq(x) \geq Q[E_{N,n} \cap D_{n,\delta}] \tag{5.2}$$

$$\to q(K_3) \geq \frac{1}{2} \text{ as } n \to \infty .$$

Therefore

$$\liminf_{n\to\infty} \int\phi(n,x)dq(x) \geq \frac{1}{2} .$$

We can find a smaller neighborhood $N_1 \subset N$ such that if $R_{n-2,T\omega} \in N_1$ then $R_{n,\omega} \in N_1$ for n sufficiently large. Therefore for large n

$$\begin{aligned} &P_{x_0}(R_{n,\omega} \in N, x_n \in K_2) \\ &\quad \geq P_{x_0}(R_{n-2,\omega} \in N, x_{n-1} \in K_3, x_n \in K_2) \\ &\quad \geq (\int P_x(E_{N_1,n-2})\pi(x_0,dx)) \inf_{x \in K_3} \pi(x,K_2) . \end{aligned}$$

From our assumptions the last factor is strictly positive. We denote by θ some lower bound for it. Then

$$\begin{aligned} &P_{x_0}(R_{n,\omega} \in N, x_n \in K_2) \\ &\quad \geq \theta\, e^{-[H(Q;P)+\delta]n} \int \phi(n,x)\pi(x_0,dx) . \end{aligned}$$

It is now an elementary exercise that our assumptions and (5.2) imply that

$$\liminf_{n\to\infty} \int \phi(n,x)\pi(x_0,dx) > 0$$

and in fact uniformly over compact sets of starting points x_0. We finally have

Theorem 5.8. For C closed in M and G open in M

$$\limsup_{n\to\infty} \frac{1}{n} \log \hat{Q}_{n,x_0}(C) \leq - \inf_{Q\in C} H(Q;P)$$
$$\liminf_{n\to\infty} \frac{1}{n} \log \hat{Q}_{n,x_0}(G) \geq - \inf_{Q\in G} H(Q,P) .$$

Proof: All that remains is to pass from ergodic Q to non-ergodic Q. This is carried out exactly like the independent case. Instead of independence we make in each one of time periods the process to have its $R_{n_j,\omega}$ closet to Q_j and end up in a compact set K_2. Since we can afford to take the infimum over the starting point in $x_0 \in K_2$ at the next step it is almost the same as independence.

We also have the results, which are analogs of 4.15 through 4.19.

Lemma 5.9. For each fixed π, $H(Q;\pi)$ is lower semicontinuous in $Q \in M$.

Proof:

$$H(Q;\pi) = \sup_k \sup_{F\in C_k} [\int F dQ]$$

and the Feller property ensures that the normalization procedure that defined F_k inductively leave the $\{C_k\}$ invariant. Since the functionals on the right are continuous we have lower semi-continuity of $H(Q;\pi)$.

Lemma 5.10. For any $\beta \in M$

$$\inf_{Q:\,q=\beta} H(Q;P) = \inf_{\lambda\in M_\beta^{(2)}} h_{X^{(2)}}(\lambda;\lambda_0)$$

where λ,λ_0 are probability measures on $X\times X$ and $\lambda_0(dx,dy) = \beta(dx)\pi(x,dy)$. $M_\beta^{(2)}$ consists of all $\lambda \in M^{(2)}$ such that the marginals of both components are β.

Proof: Starting from λ we can construct a unique Markov chain (stationary) whose two dimensional distribution is λ at two consecutive time points. For such a Markov chain Q_λ one can compute

$$H(Q_\lambda;P) = h_{X^{(2)}}(\lambda;\lambda_0) .$$

If Q is not Markov then the $\bar{Q}$ associated to the two dimensional marginal of Q is always Markov and

$$H(\bar{Q};P) \leq H(Q;P) .$$

Lemma 5.11. For any $\beta \in M$

$$\inf_{\lambda\in M_\beta^{(2)}} H_{X^{(2)}}(\lambda,\lambda_0) = \sup_u \int \log \frac{u(x)}{(\pi u)(x)}\, d\beta(x)$$

where the supremum is taken over all bounded uniformly positive measurable functions:

Proof: For the proof we do not need the Feller condition on π so by a standard result on Polish spaces we might as well assume that X is compact. Then we may restrict u to bounded continuous functions.

Suppose for some u, $\int \log \frac{u(x)}{(\pi u)(x)}\, d\beta(x) = \ell$. Then,

$$\int \log \frac{u(y)}{(\pi u)(x)}\, \lambda(dx,dy) = \ell \quad \text{because } \lambda \in M_\beta^{(2)}$$

on the other hand

$$\log \int \frac{u(y)}{(\pi u)(x)}\, \lambda_0(dx,dy) = \log \int \frac{u(y)}{(\pi u)(x)}\, d\beta(x)\pi(x,dy)$$

$$= \log 1$$

$$= 0 .$$

By definition of $H_{X^{(2)}}(\lambda;\lambda_0)$ we have now

$$\lambda \in M_\beta^{(2)} \Rightarrow h_{X^{(2)}}(\lambda;\lambda_0) \geq \ell$$

and we have the easy half. For the other half what we have to show is that if

$$\inf_{\lambda\in M_\beta^{(2)}} h_{X^{(2)}}(\lambda;\lambda_0) \geq \ell \tag{5.3}$$

then we have to produce a continuous u for which $\int \log \frac{u(x)}{(\pi u)(x)} d\beta(x) \geq \lambda - \varepsilon$, where $\varepsilon > 0$ is given. this requires the use of the minimax theorem. From (5.3) we have

$$\inf_{\lambda \in M_\beta^{(2)}} \sup_V [\int V(x,y)\lambda(dx,dy) - \log \int e^{V(x,y)}\lambda_0(dx,dy)] \geq \ell .$$

By standard minimax theorem we can interchange sup and inf so that

$$\sup_V \inf_{\lambda \in M_\beta^{(2)}} [\int V(x,y)\lambda(dx,dy) - \log \int e^{V(x,y)}\lambda_0(dx,dy)] \geq \ell .$$

In other words given $\varepsilon > 0$ there is a V such that

$$\inf_{\lambda \in M_\beta^{(2)}} \int V(x,y)\lambda(dx,dy) \geq \ell + \log \int e^{V(x,y)}\lambda_0(dx,dy) - \varepsilon .$$

By normalization we may assume the existence of a V such that

$$(5.4) \qquad \ell + \log \int e^{V(x,y)}\lambda_0(dx,dy) \leq \varepsilon$$

and

$$(5.5) \qquad \int V(x,y)\lambda(dx,dy) \geq 0 \qquad \forall \lambda \in M_\beta^{(2)} .$$

We may rewrite (5.5) as

$$\inf_\lambda \sup_{\phi,\psi} [\int V(x,y)\lambda(dx,dy) + \int[\phi(y) + \psi(x)]\lambda(dx,dy) - \int[\phi(x) + \psi(x)]\beta(dx)] \geq 0$$

because the sup is 0 if $\lambda \in M_\beta^{(2)}$ and ∞ otherwise. Again by minimax theorem (5.5) implies

$$(5.6) \quad \sup_{\phi,\psi} \inf_\lambda [\int V(x,y)\lambda(dx,dy) + \int[\phi(y) + \psi(x)]\lambda(dx,dy) - \int[\phi(x) + \psi(x)]d\beta(x)] \geq 0$$

which means that given any $\varepsilon > 0$, there is pair ϕ,ψ such that (again by normalization)

$$(5.7) \qquad \int \phi d\beta = \int \psi d\beta = 0$$

and

$$(5.8) \qquad V(x,y) \geq \phi(x) + \psi(y) - \varepsilon \qquad \forall \ x \text{ and } y$$

(5.4) and (5.8) yield

$$(5.9) \qquad \ell + \log \int e^{\phi(x)+\psi(y)}\beta(dx)\pi(x,dy) \leq 2\varepsilon .$$

If we call $e^\psi = u$ then (5.9) is the same as

$$(5.10) \qquad \log \int e^{\phi(x)}(\pi u)(x)\beta(dx) \leq 2\varepsilon - \ell .$$

By Jensen's inquality we get

$$\int \phi(x)\beta(dx) + \int \log(\pi u)(x)\beta(dx) \leq 2\varepsilon - \ell$$

since $\log u = \psi$ from (5.7) we obtain

$$\int \log \frac{u(x)}{(\pi u)(x)} \beta(dx) \geq \ell - 2\varepsilon$$

and we are done.

If we define

$$I_\pi(\beta) = \sup_u \int \log \frac{u(x)}{(\pi u)(x)} \beta(dx) = \inf_{Q:q=\beta} H(Q;\pi)$$

then

Lemma 5.12. $I_\pi(\beta)$ is lower semi-continuous and convex. Under Hypothesis 2 the set $\{\beta : I_\pi(\beta) \leq \ell\}$ is compact in M. And under the same hypothesis $\{Q : H(Q;\pi) \leq \ell\}$ is compact in M.

Proof: By standard truncation we will have

$$\int V(x)d\beta \leq \ell$$

where $V(x)$ is the function of hypothesis 2). By Tchebyshev bounds we obtain the first part of our lemma. The second part follows trivially from the first part.

Section 6. Stationary Gaussian Process

For P we take a stationary Gaussian process with mean 0 and covariance

$$E\{x_n x_{n+k}\} = \rho_k = \frac{1}{2\pi}\int_0^{2\pi} e^{ik\theta} f(\theta)d\theta$$

where $f(\theta)$ is a continuous nonnegative function with $f(0) = f(2\pi)$. We assume that the process is nondeterministic so that $\int_0^{2\pi} \log f(\theta)d\theta$ is greater than $-\infty$.

We construct $R_{n,\omega}$ and $\hat{Q}_n$ and we aim to show that a large deviation principle is valid for $\hat{Q}_n$ with a rate function

$$H(Q;f) = E^Q\{\int_{-\infty}^{\infty} q(y/\omega)\log q(y/\omega)dy\} + \frac{1}{2}\log 2\pi + \frac{1}{4\pi}\int_0^{2\pi} \frac{dG(\theta)}{f(\theta)} + \frac{1}{4\pi}\int_0^{2\pi} \log f(\theta)d\theta \tag{6.1}$$

where $dG(\theta)$ is the spectral measure of Q i.e.

$$E^Q x_0 x_k = \frac{1}{2\pi}\int e^{ik\theta} dG(\theta) .$$

We will outline the basic steps involved in the proof of the large deviation principle for $\hat{Q}_n$ with the rate function provided by (6.1).

Step 1. We represent the random process $\{x_n\}$ as a moving average of the form

$$x_k = \sum_{n=-\infty}^{\infty} a_{n-k}\xi_n$$

where ξ_k are independent random variables and

$$\sqrt{f(\theta)} = \sum_{n=-\infty}^{\infty} a_n e^{in\theta} .$$

The sequence $\{a_n\}$ is in $\ell_2(z)$.

Step 2. We approximate a_n by $a_n^N = a_n(1 - \frac{|n|}{N})$ for $|n| \leq N$ and $a_n^N = 0$ otherwise. If we write

$$g_N(\theta) = \sum_{n=-\infty}^{\infty} a_n^N e^{in\theta}$$

then $g_N(\theta) \to \sqrt{f(\theta)}$ uniformly by Fejer's theorem:

Step 3. Let us define a map on $\Omega = \prod_{i\in z} R$ by

$$(\tau\omega)(k) = \Sigma\, a_{n-k}\omega(n)$$

then $P = P_0\tau^{-1}$ where P_0 is the product measure based on standard Guassians. If we define τ_N by

$$(\tau_N\omega)(k) = \Sigma\, a_{n-k}^N\omega(n)$$

then $P_N = P_0\tau_N^{-1}$ is Gaussian with mean 0 and spectral density

$$f_N(\theta) = |g_N(\theta)|^2 .$$

Step 4. For each N, $R_{n,\tau_N\omega}$ and $R_{n,\omega}\tau_n^{-1}$ are very close, as $n \to \infty$. In fact any difference between them is only due to periodization. They are both random stationary process and the large deviation principle for $R_{n,\omega}\tau_N^{-1}$ implies the large deviation principle for $R_{n,\tau_N\omega}$. Moreover since we have a large deviation principle for $R_{n,\omega}$ when the basic distribution is P_0, we have one for $R_{n,\omega}\tau_N^{-1}$ since $Q \to Q\tau_N^{-1}$ is a continuous map of M into M. The rate function for $R_{n,\tau_N\omega}$ whose distribution we call $\hat{Q}_{n,N}$ is given by

$$\inf_{Q':Q'\tau_N^{-1}=Q} H(Q',1) .$$

Step 5. We calculate

$$\inf_{Q':Q'\tau_N^{-1}=Q} H(Q',1) = H(Q,f_N)$$

$H(Q,f)$ for any f is given by formula (6.1). Step 5 is mainly a calculation.

Step 6.

$$\lim_{N\to\infty} \limsup_{n\to\infty} \frac{1}{n} \log P_0\{d(R_{n,\tau_N\omega}, R_{\tau\omega}) \geq \varepsilon\} = -\infty$$

for every $\varepsilon > 0$. This is again a calculation based on routine estimates for $\tau_N\omega - \tau\omega$.

Step 7.

$$\hat{Q}_n[C] = P[R_{n,\omega} \varepsilon C] = P_0[R_{n,\tau\omega} \varepsilon C]$$

$$\leq P_0[R_{n,\tau_N\omega} \varepsilon \bar{C}^{\varepsilon}] + P_0[d(R_{n,\tau\omega}, R_{n,\tau\omega}) \geq \varepsilon] .$$

Taking logs dividing by n, taking limsup and then letting $N \to \infty$ first and then $\varepsilon \to 0$, we obtain

$$\limsup_{n\to\infty} \frac{1}{n} \log \hat{Q}_n(C) \leq - \lim_{\varepsilon\to 0} \liminf_{N\to\infty} \inf_{Q\varepsilon \bar{C}_\varepsilon} H(Q;f_N)$$

and similarly for G open

$$\liminf_{n\to\infty} \frac{1}{n} \log \hat{Q}_n(G) \geq - \liminf_{N\to\infty} \inf_{Q\varepsilon G} H(Q,f_N) .$$

Step 8.

$$\lim_{\varepsilon\to 0} \liminf_{N\to\infty} \inf_{Q\varepsilon \bar{C}_\varepsilon} H(Q;f_N) \geq \inf_{Q\varepsilon C} H(Q;f)$$

$$\lim_{N\to\infty} \inf_{Q\varepsilon G} H(Q,f_N) \leq \inf_{Q\varepsilon G} H(Q;f) .$$

These two statements are proved by the explicit formulas for $H(Q;f_N)$ and $H(Q,f)$ and the explicit definition of f_N in terms of f. Finally

Step 9. $H(Q;f)$ is a rate function.

Section 7. Continuous Time Markov Processes

We will now assume that we have a Markov process with transition probabilities $p(t,x,dy)$ on a state space X with the following properties: The state space X is Polish. Moreover:

Hypothesis 1. The semigroup $(T_tf)(x) = \int f(y)p(t,x,dy)$ maps bounded continuous functions $C(X)$ into itself. For any starting point the measure P_x on the space of trajectories lives on $\Omega_0 = D[0,\infty)$ which is given the topology of Skorohod convergence on finite intervals. The map $x \to P_x$ is continuous.

Hypothesis 2. There exists a sequence u_n of functions in the domain D of the infinitesimal generator of the process with the following properties

a) $u_n(x) \geq 1$

b) $\sup_{x\in K} \sup_n u_n(x) < \infty$ for each compact K X

c) $\lim_{n\to\infty} u_n(x) = u(x)$ exists for each x. If $V_n(x) = -\frac{LU_n(x)}{U_n(x)}$ then

d) $V_n(x) \geq -C$ for some C for all n,x.

e) $V(x) = \lim_{n\to\infty} v_n(x)$ exists

f) for each $\ell < \infty$, $\{x:V(x) \leq \ell\}$ has compact closure in X.

Hypothesis 3. $p(1,x,dy)$ has density $p(1,x,y)$ for every x with respect to a reference measure α on X. Moreover $p(1,x,y) > 0$ a e α for each x. In addition the map $x \to p(1,x,\cdot)$ is continuous as a map of X into $L_1(\alpha)$.

We denote by Ω the Skorohod space $D(-\infty,\infty)$ and by M the space of stationary processes on Ω. M is a Polish space under weak convergence of processes on finite intervals and the projection map $\omega \to \omega(t)$ while not continuous in general is continuous at almost all points with respect to every $Q \in M$. [Q has no fixed points of discontinuity] for each $\omega \in \Omega_0$ we define $R_{t,\omega}$ by the continuous analog of $R_{n,\omega}$. We extend the trajectory $\omega(s)$, $0 \leq s \leq t$ periodically on either side to get a periodic orbit under the shift θ_s of period t and take $R_{t,\omega}$ as the orbital measure. For ech $x \in X$ we have the distribution $\hat{Q}_{t,x}$ of $R_{t,\omega}$ under P_x. We are interested in a large deviation principle for $\hat{Q}_{t,x}$ on M with some rate function H(Q). We will suppress the dependence of H(Q) on p(t,x,dy), which will be a fixed semigroup for our discussion.

The proof follows the discrete case very closely and we outline the proof giving details only where there are new aspects in the proof.

Definition 7.1. Given $Q \in M$ we define

$$H(Q,T) = E^Q h_{F_T^0}(Q_\omega, P_{\omega(0)}) .$$

Lemma 7.2.

$H(Q,T) = TH(Q)$ for some $0 \leq H(Q) \leq \infty$.

Proof: One checks by stationarity of Q and $p(t,x,dy)$ that $H(Q,T_1 + T_2) = H(Q,T_1) + H(Q,T_2)$. Since $H(Q,T) \geq 0$ it follows that $H(Q,T)$ is linear in T.

For each T we define A_T by

$$A_T = \{F : F \text{ is } F_T^Q \text{ measurable and } E^{P_x}\{\exp[F(\omega)]\} \leq 1 \quad \forall\, x\}$$

$$C_T = A_T \quad \{F : E^Q\{F\} \text{ is a continuous linear functional of } Q \text{ in } M\} .$$

We then have

Theorem 7.3.

$$H(Q) = \sup_{T>0} \frac{1}{T} \sup_{F \in A_T} E^Q\{F\} = \sup_{T>0} \frac{1}{T} \sup_{F \in C_T} E^Q\{F\} .$$

Proof: Same as theorem 5.3.

We can define

$$B_T = \{F : F \in F_1^{-T} \text{ and } E^{P_{\omega(0)}} e^{F(\omega)} \leq 1 \text{ everywhere } \} .$$

In the above definition integration with respect to $E^{P_{\omega(0)}}$ is carried out over F_1^Q only on each fiber of F_0^{-T}. We also have the analog of theorem 5.2 proved in exactly the same manner.

Theorem 7.4

$$H(Q) = \sup_{T>0} \sup_{F \in B_T} E^Q\{F\} .$$

We now start proving the large deviation principle.

We define for $x_0 \in X$ and $A \subset M$

$$J_{x_0}(A) = \limsup_{T \to \infty} \frac{1}{T} \log \hat{Q}_{T,x_0}(A) .$$

We then have

Theorem 7.5. For any compact set $K \subset M$ and any $\varepsilon > 0$ there exists a neighborhood $G_\varepsilon \supset K$ such that

$$J_{x_0}(G_\varepsilon) \leq -\inf_{Q \in K} H(Q) + \varepsilon .$$

Proof: Identical to the discrete case.

One main difference in the continuous case is that processes whose marginals vary over a compact set is not necessarily from a compact set of processes. We need to control the modulus of continuity as well.

Theorem 7.6. Let A closed in M be such that the family of one dimensional marginals of Q as Q varies over A forms a tight family of measures on X. Then

$$J_{x_0}(A) \leq - \inf_{Q\varepsilon A} H(Q) .$$

Proof: Let us denote by A_M the family of marginals of A. Given any sequence $\varepsilon_n \to 0$, there exists $K_n \subset X$ such that $q(K_n) \geq 1 - \varepsilon_n$ for $q \varepsilon A_M$. Since $x_0 \to P_{x_0}$ is weakly continuous there exists $C_n \subset D[0,1]$ such that C_n is compact there and $P_x(C_n) \geq 1 - \eta_n$ for all $x \varepsilon K_n$. Denoting by $\tilde{C}_n$ the comnplement of C_n it is easily checked that for all $x \varepsilon X$

$$E^{P_x}\{\exp[\nu X_{K_n}(\omega(0))X_{\tilde{C}_n}(\omega)\} \leq 1 + \eta_n(e^\nu - 1) .$$

From the continuous analog of lemma 4.8

$$E^{P_x}\{\exp[\int_0^t \nu X_{K_n}(\omega(s))\, X_{\tilde{C}_n}(\theta_s\omega)]\} \leq \exp[t \log(1 + \eta_n(e^\nu - 1))] .$$

Therefore allowing for an error of $\frac{1}{t}$ for periodization

$$\hat{Q}_{t,x_0}\{A \cap \{Q:Q(\hat{C}_n) \geq \frac{1}{t} + 2\,\varepsilon_n\}$$

$$\leq \exp[t \log[1 + \eta_n(e^\nu - 1)] - \varepsilon_n\nu t] .$$

Pick $\lambda > 0$, $\nu = \lambda n^2$, $\varepsilon_n = \frac{1}{n}$ and $\eta_n = \exp[-\lambda n^2]$. Then

$$\hat{Q}_{t,x_0}\{A \cap \{Q:Q(\tilde{C}_n) \geq \frac{1}{t} + \frac{2}{n}\} \leq e^{t\log 2-\lambda nt} .$$

If we let

$$A_t = \{Q:Q(\tilde{C}_n) \leq \frac{1}{t} + \frac{2}{n} \text{ for all } n \geq 1\}$$

then

$$\hat{Q}_{t,x_0}\{A \cap \tilde{A}_t\} \leq e^{t\log 2}(\frac{e^{-\lambda t}}{1-e^{-\lambda t}}) .$$

Therefore

$$\limsup_{t\to\infty} \frac{1}{t} \log \hat{Q}_{t,x_0}(A \cap \tilde{A}_t) \leq \log 2 - \lambda . \tag{7.1}$$

It is easy to check tht $A_\infty = \cap_t A_t$ is compact in M and if $G \supset A \cap A_\infty$ is open that

$A_t \cap A \subset G$ for t sufficiently large. Theorem 7.5 and 7.1 provide a proof of Theorem 7.6. If we now assume Hypothesis 2) one can obtain easily

$$E^{P_{x_0}}[\exp \int_0^t V(\omega(s))ds] \leq C$$

for every $t \geq 0$. With this estimate the proof of the upper bound for closed sets proceeds exactly like the discrete case.

Lower bound: The only essential difference with the discrete case is the ergodic theorem: If

$$\frac{dQ_\omega}{dP_{\omega(0)}}\Big|F_t^0 = \exp[\psi(t,\omega)]$$

then $\psi(t + s,\omega) = \psi(t,\omega) + \psi(s,\theta_t\omega)$ is an additive functional. To establish the ergodic theorem almost everywhere we need to show

$$E^Q \sup_{0\leq t\leq 1} |\psi(t,\omega)| < \infty .$$

This is the content of the following lemma:

Lemma 7.7. Let α,β be two probability measures on a measurable space (X,F). Let F_t $0 \leq t \leq 1$ be an increasing family of subfields with $F_1 = F$. Let $h(\beta;\alpha) < \infty$ and $\psi(\omega,t) = \log \frac{d\beta}{d\alpha}\Big|F_t$. Then

$$E^\beta[\sup_{0\leq t\leq 1} |\psi(\omega,t)|] < \infty .$$

Proof: From standard martingale inequalities

$$\beta\{\omega: \inf_{0\leq t\leq 1} \psi(\omega,t) \leq -\ell\} = \beta\{\omega: \inf_{0\leq t\leq 1} R(t,\omega) \leq e^{-\ell}\} \leq e^{-\ell} .$$

We therefore only have to show

$$E^\beta[\sup_{0\leq t\leq 1} \psi(\omega,t)] < \infty .$$

But from entropy inequalities it is sufficient to show that

$$E^\alpha[\sup_{0\leq t\leq 1} e^{\psi(\omega,t)}] < \infty$$

or

$$E^\alpha[\sup_{0\leq t\leq 1} R(t,\omega)] < \infty .$$

Since R is a martingale we need only the integrability $R(\log R)^+$ which is of course true because $h(\beta,\alpha) < \infty$. Now the lower bound is completed as before and we have

Theorem 7.8. Under hypothesis (1), (2) and (3) a large deviation principle holds with the rate function $H(Q)$. In particular $H(Q)$ is a rate function:

We can define for $\mu \in M$

$$I(\mu) = -\inf_{\substack{u>0\\ u\in D}} \int \left(\frac{Lu}{u}\right)(x)\mu(dx) .$$

Assuming that the domain is big enough we have if

$$I_h(\mu) = -\inf_{u>0} \int \log \frac{\Pi_h u}{u} (x)\mu(dx)$$

then

$$I_h(\mu) \leq h\, I(\mu)$$

and

$$\lim_{h\to 0} \frac{1}{h} I_h(\mu) = I(\mu) .$$

We can use this to prove the contraction principle:

Theorem 7.9

$$\inf_{Q: q=\mu} H(Q) = I(\mu) .$$

Proof: If we denote by Q_h the Markov chain at times that are multiples of h, with invariant measure which is obtained in lemmas 5.10 and 5.11 then the entropy of Q_h with respect to the basic π_h Markov chain with initial distribution μ is given by $I_h(\mu)$ per time gap h. Therefore on the h "grid" in the unit interval $h(Q_h, P_{\mu,h}) \leq \frac{1}{h} I_h(\mu) \leq I(\mu)$. We can fill in the gaps of the grid by the conditional distributions of bridges of span h. This leads to the measure P_μ as filled in $P_{\mu,h}$ and some measure Q_h' for filled in $Q_h \cdot h(Q_h', P_\mu) = h(Q_h, P_{\mu,h}) \leq I(\mu)$. It is easy to show from this entropy estimate the tightness of Q_h' and the limit Q is stationary, has marginal μ and $H(Q) \leq I(\mu)$. The other half of the theorem is just as trivial as the discrete case.

Finally if $p(t,x,dy)$ has a symmetric density $p(t,x,y)$ with respect to a reference measure α then

Theorem 7.10. $I(\mu) < \infty$ if and only if $\mu << \alpha$ and if $f = \frac{d\mu}{d\alpha}$ then $\sqrt{f} \in L_2(\alpha)$ is in the domain of $(-L)^{1/2}$ in $L_2(\alpha)$.

$$I(\mu) = ||(-L)^{1/2}\sqrt{f}||^2 .$$

Section 8. Application to the Problem of the Wiener Sausage

A problem that comes up in the study of density of states for Schrödinger operators with certain random potentials near the edge of the energy spectrum is the following:

Let $\beta(t)$ be d-dimensional Brownian motion starting from the origin. Let $\varepsilon > 0$ be fixed. Consider

$$C_t = \{x : x = \beta(s) \text{ for some } 0 \leq s \leq t\} ,$$

$$C_t^\varepsilon = \{\bigcup_{x \in C_t} s(x,\varepsilon)\} = \{x : |x - \beta(s)| < \varepsilon \text{ for some } 0 \leq s \leq t\} .$$

C_t is just the image of the Wiener path up to time t, and C_t^ε is the sausage arount it of radius ε.

The problem is to show that

$$\lim_{t\to\infty} \frac{1}{t^{d/(d+2)}} \log E\{e^{-v|C_t^\varepsilon|}\} = -k(d,v) \tag{8.1}$$

exists and is nonzero. Here $|C_t^\varepsilon|$ is the d-dimensional volume of the sausage C_t^ε up to time t. The actual physical problem involves the Brownian motion that is conditioned to return to the origin at time t. But for large t, one can see easily that the difference between the free Brownian motion and the conditional Brownian motion is small enough that the formula (8.1) is unaffected by it. We will study only the free Brownian motion.

We will first carry out a Brownian change of scale so that $t^{d/(d+2)}$ appears naturally

Let us replace $\beta(s)$, $0 \leq s \leq t$, by

$$t^{1/(d+2)}\beta(t^{-2/(d+2)}s) \quad \text{for } 0 \leq s \leq t^{d(d+2)} ,$$

which is again a Brownian motion. Therefore the distribution of $|C_t^\varepsilon|$ is the same as that of $t^{d(d+2)}|C^{\varepsilon t^{-1/(d+2)}}_{d/(d+2)}|$. If we let $\tau = t^{d/(d+2)}$, then

$$E\{e^{-v|C_t^\varepsilon|}\} = E\{\exp[-\tau v|C_\tau^{\varepsilon\tau^{-1/d}}|]| .$$

The problem therefore reduces to showing that

$$\lim_{t\to\infty} \frac{1}{t} \log E\{\exp[-vt|C_t^{\varepsilon t^{-1/d}}|]\} = - k(v,d) \tag{8.2}$$

exists and is nonzero.

A basic fact in what follows is the behavior of

$$P[\beta(s) \in G \text{ for } 0 \leq s \leq t]$$

where G is a smooth bounded open set containing the origin. If $L(t,\omega)$ is the random measure representing the occupation time of the Brownian motion, i.e., if

$$L(t,\omega)(A) = \frac{1}{t}\int_0^t \chi_A(\beta(s))ds ,$$

then

$$\beta(s) \in G \text{ for } 0 \leq s \leq t \Leftrightarrow \operatorname{supp} L(t,\omega) \subset G .$$

We can therefore estimate

$$P[\beta(s) \in G \text{ for } 0 \leq s \leq t] \leq P[\operatorname{supp} L(t,\omega) \quad G] .$$

Since the set $\{\mu : \operatorname{supp} \mu \quad \bar{G}\}$ is a compact set, we can estimate

$$\limsup_{t\to\infty} \frac{1}{t} \log P[\beta(s) \in G \text{ for } 0 \leq s \leq t]$$

$$\leq - \inf_{\mu:\mu(\bar{G})=1} I(\mu)$$

$$\leq - \inf_{\substack{f=0 \text{ off } \bar{G} \\ f\geq 0, \int f dx=1}} \left[\frac{1}{8}\int \frac{|\nabla f|^2}{f} dx\right]$$

$$= - \inf_{\substack{g=0 \text{ off } \bar{G} \\ \int g^2 dx=1}} \left[\frac{1}{2}\int |\nabla g|^2 dx\right] = \lambda(G) .$$

The lower estimate

$$\liminf_{t\to\infty} \frac{1}{t} \log P[\beta(s) \in G \text{ for } 0 \leq s \leq t] \geq -\lambda(G)$$

is also easy to derive. See [1] for details. We can combine the two halves in the form of a lemma:

Lemma 8.1. <u>For nice open sets G</u>

$$\lim_{t\to\infty} \frac{1}{t} \log P[\beta(s) \in G \text{ for } 0 \leq s \leq t] = -\lambda(G) .$$

<u>Lower bound</u>. Let G be a bounded open set containing the origin with smooth boundary. Let $\lambda(G)$ be the first eigenvalue of $-\frac{1}{2}\Delta$ in G with Dirichlet boundary conditions. Then according to Lemma 8.1

$$P[\beta(s) \in G \text{ for } 0 \leq s \leq t] = \exp[-\lambda(G) + o(t)] .$$

If $\beta(s) \in G$ for $0 \leq s \leq t$, then

$$|C\xi_t^{-1/d}| \leq |G| + o(1) \quad \text{as } t \to \infty .$$

Therefore

$$\liminf_{t\to\infty} \frac{1}{t} \log E\{\exp[-v|C_t^{\xi t^{-1/d}}|]\} \geq (v|G| + \lambda(G)) .$$

Taking the infimum over all such sets G, we find

$$\liminf_{t\to\infty} \frac{1}{t} \log E\{\exp[-v|C_t^{\xi t^{-1/d}}|]\} \geq -k(v,d)$$

where

(8.3) $$k(v,d) = \inf_G [v|G| + \lambda(G)] .$$

We have proved

Theorem 8.2.

$$\liminf_{t\to\infty} \frac{1}{t} \log E\{\exp[-v|C_t^{\xi t^{-1/d}}|]\} \geq -k(v,d) .$$

We now turn to the

Derivation of the upper bound. Let us replace R^d by the d-dimensional torus T_ℓ^d and consider Brownian motion on the torus. We will denote by E_ℓ expectation with respect to the Brownian motion on the torus of size ℓ. Since any set in R^d projected to T_ℓ^d has volume no larger than the original volume,

$$E\{\exp[-v|C_t^{\xi t^{-1/d}}|]\} \leq E_\ell\{\exp[-v|C_t^{\xi t^{-1/d}}|]\} .$$

We will show that

$$\lim_{\ell\to\infty} \limsup_{t\to\infty} \frac{1}{t} \log E_\ell\{\exp[-v|C_t^{\xi t^{-1/d}}|]\} \leq -k(v,d) .$$

Upper bounds on the torus. Let $\phi(x)$ be a function with support $\{x: |x| \leq \varepsilon\}$ which is nonnegative and has $\int\phi(x)dx = 1$. Then, if $\phi_t(x) = t\phi(xt^{1/d})$, then

$$C_t^{\xi t^{-1/d}} = \{x: L_t * \phi_t > 0\}$$

where L_t is the occupation distribution and * denotes convolution. We will denote by f_t,

$$f_t = L_t * \phi_t ,$$

the mollified local time. The problem we have reduces to two lemmas:

Lemma 8.3.

$$\limsup_{t\to\infty} \frac{1}{t} \log E_\ell\{\exp[-vt|x: f_t > 0|]\} \leq -\inf_f [I(f) + v|x; f > 0|]$$

$$= \inf_G [v|G| + \lambda_\ell(G)] ,$$

and

Lemma 8.4.

$$\liminf_{\ell\to\infty} \inf_G [v|G| + \lambda_\ell(G)] = \inf_G [v|G| + \lambda(G)] .$$

Of the two lemmas, the second is a standard approximation lemma involving truncation methods. We will not carry out the proof, but will only refer to [1]. We will sketch a proof of Lemma 8.3. If we consider the L_1 topology for densities on T_ℓ^d, then $|x:f > 0|$ is a lower semicontinuous functional of the density f. In view of Theorem 2.2, it is sufficient to prove the large deviation principle for f_t in the L_1 topology with a rate function I(f).

Lemma 8.5. Let ψ be any mollifier, i.e., a smooth probability density. Then

$$\limsup_{t\to\infty} \frac{1}{t} \log P[f_t * \psi \in C] \leq - \inf_{f:f*\psi\in C} I(f)$$

for any C closed in L_1.

Proof: The map $f \to f * \psi$ is continuous from M with weak topology to L_1 with norm topology. So the large deviation principle in the weak topology for L_t, which implies the large deviation principle in the weak topology for f_t, is converted into a large deviation principle for $f_t*\psi$ in the norm topology of L_1. Theorem 2.3 provides the precise proof. □

We now state without proof Lemma 8.6. We will then state and prove Lemma 8.7, which will imply Lemma 8.3 and our main result. Finally we will prove Lemma 8.6.

Lemma 8.6.

$$\limsup_{t\to\infty} \frac{1}{t} \log P[||f_t * \psi - f_t|| \geq \rho] \leq - k_\rho(\psi)$$

where $k_\rho(\psi) \to \infty$ as $\psi \to \delta_0$ for each $\rho > 0$.

Proof: The proof will be given after the proof of Lemma 8.7.

Lemma 8.7. The large deviation principle holds for f_t with the rate function I(f) in the space L_1 with norm topology.

Proof: Upper bound.

$$P[f_t \in C] \leq P[f_t * \psi \in \overline{C}^\rho] + P[|f_t - f_t * \psi| > \rho] .$$

Therefore

$$\limsup_{t\to\infty} \frac{1}{t} \log P[f_t \in C] \leq - \inf[\inf_{f:f*\psi\in C^\rho} I(f), k_\rho(\psi)] .$$

Letting $\psi \to \delta_0$ and $\rho \to 0$, we get

$$\lim_{\psi\to\delta_0} \inf_{f:f^*\psi\delta\bar{C}^\rho T\rho} I(f) = \inf_{f\varepsilon\bar{C}^\rho T\rho} I(f) \text{ and } \lim_{\rho\to 0} \inf_{f\varepsilon\bar{C}^\rho} I(f) = \inf_{f\varepsilon C} I(f) ,$$

provided C is closed in L_1.

Lower bound. For an open set G around f,

$$P[f_t \varepsilon G] \geq P[f_t * \psi \varepsilon G_1] - P[||f_t - f_t*\psi|| \geq \rho]$$

where G_1 is a smaller open set around f such that the sphere around G_1 of radius ρ is contained in G. The result is again obvious from Lemma 8.6. □

From Lemma 8.7 we obtain Lemma 8.3 by an application of Thereom 2.2. If we now combine it with the lower bound, i.e. Theorem 8.2, and take Lemma 8.4 for granted, then we have

Theorem 8.8.

$$\lim_{t\to\infty} \frac{1}{t} \log E\{\exp[-v|C_t^{\varepsilon t^{-1/d}}|]\} = -k(v,d)$$

where k(v,d) is given by (8.3).

We now turn to the

Proof of Lemma 8.6.

$$\begin{aligned} ||f_t * \psi - f_t|| &= \sup_{|g|\leq 1} \left|\int (f_t * \psi - f_t(x)g(x)dx\right| \\ &= \sup_{|g|\leq 1} \left|\int (L_t * \phi_t * \psi - L_t * \phi_t)(x)g(x)dx\right| \\ &= \sup_{|g|\leq 1} \left[\int h_t(x)L_t(dx)\right] \end{aligned}$$

where

$$h_t(x) = (g * \phi_t * \psi - g * \phi_t)(x) .$$

(We have assumed that ϕ_t and ψ are symmetric.)

The map $g \to \theta$ defined by $\theta = g*\phi_t$ is a compact map of the unit ball $|g| \leq 1$. Therefore for any $\rho > 0$ we can find a finite number $N = N(t,\rho)$ of functions $\theta_1,\cdots,\theta_N$ such that the image of the unit ball is covered by spheres around $\theta_1,\cdots,\theta_N$ of radius $\rho/2$. We can assume that $\theta_1,\cdots,\theta_N$ are all bounded by 1 as well. Then

$$||f_t * \psi - f_t|| \leq \frac{\rho}{4} + \sup_{1\leq i\leq N} \left[\int (\theta_i(x) - (\theta_i * \psi)(x))L_t(dx)\right] ,$$

$$P[||f_t * \psi - f_t|| \geq \rho] \leq N \sup_{1\leq i\leq N} P\left[\int \chi_i(x)L_t(dx) \geq \frac{\rho}{2}\right]$$

$$\leq N \sup_{1 \leq i \leq N} P\left[\int_0^t \chi_i(\beta(s))ds \geq \frac{\rho}{2} t\right]$$

$$\leq Ne^{-z\rho t/2} E_\ell\{\exp[z \int_0^t \chi_i(\beta)(s))ds]\}$$

where $\chi_i(x) = \theta_i(x) - (\theta_i * \psi)(x)$.

One can show that for any χ with $|\chi| \leq 2$

$$E_\ell\{\exp[z \int_0^t \chi(\beta(s))ds]\} \leq C_z \exp[t\lambda_\ell(z\chi)]$$

where $\lambda_\ell(z\chi)$ is the largest eigenvalue of

$$\frac{1}{2}\Delta + z\chi \quad \text{on } T_\ell^d .$$

If, for each $\rho > 0$, $N(t,\rho) \leq \exp[D_\rho t]$ for some D_ρ, then

$$\frac{1}{t} \log P[||f_t * \psi - f_t|| \geq \rho] \leq D_\rho - \frac{z\rho}{2} + \sup_{\chi:\chi=\theta-\theta*\psi, |\theta|\leq 1} \lambda(z\chi) .$$

One verifies that $\sup \lambda(z\chi) \to 0$ as $\psi \to \delta_0$ for each $z > 0$. Therefore

$$\limsup_{\psi\to\delta_0} \limsup_{t\to\infty} \frac{1}{t} \log P[||f_t * \psi - f_t|| \geq \rho] \leq D_\rho - \frac{z\rho}{2} ,$$

and by letting $z \to \infty$ we will obtain our lemma. We no need only the estimation of $N(\rho,t)$ to complete the proof of Lemma 8.6.

$$\theta(x) = t \int g(y)\phi((x - y)t^{1/d})dy ,$$

$$|\theta(x_1) - \theta(x_2)| \leq \int |\phi((x_1 - x_2)t^{1/d} + y) - \phi(y)|dy ,$$

$$\sup_{|x_1-x_2|\leq h} |\theta(x_1) - \theta(x_2)| \leq \omega(ht^{1/d})$$

where ω is the L_1 modulus of continuity of ϕ. Therefore

$$|\theta(x_1) - \theta(x_2)| \leq \eta \quad \text{if } ht^{1/d} \leq \eta' , \quad \text{i.e., if } h \leq \eta' t^{-1/d} .$$

We can divide the torus T_ℓ^d into small cubes of size $\eta' t^{-1/d}$, and we will then have $t/(\eta')^d$ cubes. In order to cover the unit ball, we need step functions that are constant on cubes, and an easy estimate provides the bound

$$N \leq [\frac{C}{\rho}]^{t/(\eta')^d} .$$

This almost completes the proof of Lemma 8.6.

Finally we need to show that

$$\inf_G [\lambda(G) + v|G|] = k(v,d) > 0 .$$

If we expand a region by a factor σ, then $\lambda(\sigma G) = (1/\sigma^2)\lambda(G)$ and $|\sigma G| = \sigma^d|G|$. Then

$$\inf_{\sigma>0}\left[\frac{\lambda(G)}{\sigma^2} + v\sigma^d|G|\right] = c(v,d)|G|^{2/(d+2)}[\lambda(G)]^{d/(d+2)}$$

where c(v,d) can be calculated explicitly. Therefore

$$k(v,d) = c(v,d) \inf_{|G|=1} [\lambda(G)]^{d/(d+2)} .$$

A rearrangement argument tells us that the infimum is attained when G is the sphere of unit volume in R^d. This calculates k(v,d) explicitly, and k(v,d) > 0. For details see [1]. □

Section 9. The Polaron Problem

A problem that comes up in statistical mechanics, known as the polaron problem, leads to the following question concerning Brownian motion. Does

$$(9.1)\qquad \lim_{t\to\infty} \frac{1}{t} \log E \left\{\exp\left[\alpha \int_0^t \int_0^t \frac{e^{-|s-\sigma|}}{|\beta(s)-\beta(\sigma)|} \, d\sigma \, ds\right]\right\} = g(\alpha)$$

exist, where $\beta(\cdot)$ is the three-dimensional tied-down Brownian motion in the interval [0,t]? And how does $g(\alpha)$ behave for large α? A conjecture by Pekar states that

$$\lim_{\alpha\to\infty} \frac{g(\alpha)}{\alpha^2} = \sup_{\substack{\phi\in L_2(R^d)\\ ||\phi||_2=1}} \left[2\iint \frac{\phi^2(x)\phi^2(y)}{|x-y|} \, dx \, dy - \frac{1}{2}\int |\nabla\phi|^2 dx\right] .$$

We will use our methods to prove the conjecture.

First we note that

$$\begin{aligned}\int_0^t \int_0^t \frac{e^{-|s-\sigma|}}{|x(\sigma)-x(s)|} \, ds \, d\sigma &= 2\int_0^t d\sigma \int_0^t \frac{e^{-(s-\sigma)}}{|x(s)-x(\sigma)|} \, ds \\ &= 2\int_0^t d\sigma \int_\sigma^\infty \frac{e^{-(s-\sigma)}}{|x(s)-x(\sigma)|} \, ds + o(t) \\ &= 2\int_0^t F(\theta_s\omega) ds + o(t)\end{aligned}$$

where θ_s is the shift and

$$F(\omega) = \int_0^\infty \frac{e^{-\sigma}}{|x(\sigma)-x(0)|} \, d\sigma .$$

By our large deviation results, one expects $g(\alpha)$ to exist in (9.1) and to be given by the variational formula

$$g(\alpha) = \sup_Q \left[2\alpha E^Q F(\omega) - H(Q)\right]$$

where $H(Q)$ is the entropy relative to Brownian motion of the stationary process Q and Q varies over all stationary processes with values in R^3. There are two technical problems here. The first is the fact that we have tied-down Brownian motion and not free Brownian motion. For t large there is very little difference, and this can be made precise. The details are in [3]. A more serious problem is the fact that Brownian motion does not satisfy the conditions for obtaining upper bounds. The lower bound, hoever, follows painlessly by our methods. To get upper bounds, one notices that if we replace Brownian motion by the Ornstein-Uhlenbeck process with generator

$$\frac{1}{2}\Delta - \varepsilon x\cdot\nabla$$

for some small $\varepsilon > 0$, the theory applies, and moreover the expectation for Brownian motion is dominated by the expectation of the OU process for every $\varepsilon > 0$. If $H_\varepsilon(Q)$ is the entropy relative to the OU process and $H(Q)$ is the entropy relative to Brownian motion, then for any Q

$$H_\varepsilon(Q) = H(Q) + \varepsilon \int ||x(0)||^2 dQ - \frac{3\varepsilon}{2} \geq H(Q) - \frac{3\varepsilon}{2} .$$

Therefore

$$\limsup_{t\to\infty} \frac{1}{t} \log E\left\{\exp\left[\alpha \int_0^t\int_0^t \frac{e^{-|\sigma-s|}}{|\beta(\sigma)-\beta(s)|}\right]\right\}$$
$$\leq \sup_Q \left[2\alpha E^Q F(\omega) - H_\varepsilon(Q)\right] \leq \sup_Q \left[2\alpha E^Q F(\omega) - H(Q)\right] + \frac{3\varepsilon}{2} .$$

By letting $\varepsilon \to 0$ we obtain that the limit (15.1) exists and g is given by

$$g(\alpha) = \sup_Q \left[2\alpha E^Q F(\omega) - H(Q)\right] . \tag{9.2}$$

Using (9.2) and Brownian scaling, one can get

$$\frac{g(\alpha)}{\alpha^2} = \sup_Q \left[\frac{2}{\alpha^2} E^Q\left\{\int_0^\infty \frac{e^{-t/\alpha^2}}{|x(t)-x(0)|} dt - H(Q)\right\}\right] , \tag{9.3}$$

and now we have to see what happens to

$$E^Q\left\{\frac{2}{\alpha^2}\int_0^\infty \frac{e^{-t/\alpha^2}}{|x(t)-x(0)|} dt\right\} \text{ as } \alpha \to \infty .$$

Writing q(t,dx,dy) for the two-dimensional distribution of x(0) and x(t) under the stationary process Q, we have

$$\lim_{\alpha\to\infty} \frac{1}{\alpha^2} \int_0^\infty e^{-t/\alpha^2} dt \int\int \frac{q(t,dx,dy)}{|x-y|} = \lim_{t\to\infty} \int\int \frac{q(t,dx,dy)}{|x-y|}$$
$$= \int\int \frac{q(dx)q(dy)}{|x-y|} .$$

This is not quite correct. However, if Q is ergodic, the independence of x(0), x(t) in an average sense, for $t \to \infty$, is enough to give the final answer. This argument is essentially correct.

There is also a serious problem of interchanging sup and limit on α. If we could carry this out we would have

$$\lim_{\alpha\to\infty} \frac{g(\alpha)}{\alpha^2} = \sup_Q \left[2\int\int \frac{q(dx)q(dy)}{|x-y|} - H(Q)\right] \tag{9.4}$$
$$= \sup_q \left[2\int\int \frac{q(dx)q(dy)}{|x-y|} - I(q)\right]$$

by the contraction principle. Since

$$I(q) = \frac{1}{8}\int \frac{|\nabla f|^2}{f} dx$$

if $q(dx) = f(x)dx$ and $I(q) = \infty$ otherwise, the variational formula in (9.4) reduces to Pekar's conjecture. Incidentally, the unbounded nature of the function $1/|x|$ causes additional technical problems that need to be handled. All this has been rigorously justified, and the details can be found in [3].

Section 10. Large deviations and laws of the iterated logarithm

Let $\ell(t,\cdot)$ be the local time of the one dimension Brownian motion defined by

$$\ell(t,y) = \int_0^t \delta(\beta(s) - y)ds$$

One knows that $\ell(t,y)$ is jointly continuous in t and y. If we define

$$\hat{\ell}(t,y) = \frac{1}{\sqrt{t}} \ell(t,\sqrt{t}y)$$

and

$$\tilde{\ell}(t,y) = \frac{1}{\sqrt{t \log\log t}} \ell(t,\sqrt{t/\log\log t}\; y)$$

then the distribution of $\hat{\ell}(t,y)$ is independent of t by Brownian scaling. One can get functional laws of the iterated logarithm $\tilde{\ell}(t,\cdot)$ by showing that the set of limit points of $\tilde{\ell}(t,\cdot)$ as $t \to \infty$ are precisely the set of subprobability densities

$p(y)$ with $\int p(y)dy \leq 1$ and $\frac{1}{8} \int \frac{[p'(y)]^2}{p(y)} dy \leq 1$. In particular we can take functionals F which are nice and obtain

$$\limsup_{t\to\infty} F(\tilde{\ell}(t,\cdot)) = \sup_C F(p(\cdot))$$

where C is the set of limit points described earlier.

If we take $F(p(\cdot)) = p(0)$ we obtain

$$\limsup_{t\to\infty} \frac{\ell(t)}{\sqrt{t \log \log t}} = \sqrt{2}\ a\ e$$

If we take $F(p()) = \inf_q [\ell : \int_{-\ell} p(y)dy = 1]$ then we obtain (taking liminf rather than limsup)

$$\liminf_{t\to\infty} \sqrt{\frac{\log\log t}{t}} \sup_{0\leq s\leq t} |\beta(s)| = \frac{\pi}{\sqrt{8}}.$$

This is the so-called "other law of iterated logarithm". There are several other examples that one can think of. For some of the details see [2].

Concluding Remarks

The large deviation theory that we have developed during these lectures depends on rather stringent assumptions on the transition probabilities $p(t,x,dy)$. These assumptions are strong enough to ensure the existence of at most one invariant probability measure for the Markov Process. If we were to drop this strong ergodicity assumption then the large deviation rate, even when they exist could start to depend on the starting point. To be more precise, if the Markov Process were to admit several invariant probability measures, the extremals among them being ergodic, then the large deviation rate for the types of sets that we had considered before, namely

$$P_x[L(t,\omega,\cdot) \in A]$$

could have the following type of behavior:

$$P_x[L(t,\omega,\cdot) \in A] = \exp[-t\ I_\alpha(A) + o(t)]$$

for almsot all x ω·r·t α. Here α is an ergodic invariant probability measure and $I_\alpha(\cdot)$ is computed in terms of a rate function depending on α. We can also start with initial distribution α and then

$$P_\alpha[L(t,\omega,\cdot) \in A] = \exp[-t\ \hat{I}_\alpha(A) + o(t)]$$

where $\hat{I}_\alpha(\cdot)$ is computed in terms of a slightly different rate function also depending on α. I_α takes care of large deviations in the evolution where $\tilde{I}_\alpha$ takes care of large deviations in the initial conditions as well. There are some interesting examples of infinite particle systems where computations have been made to illustrate such behavior. The relevant references are [6], [7] and [8].

Bibliographical Remarks:

The results outlined here appeared originally in several articles. There are now several sources available that provide a general survey of the large deviation theory along with a list of references: Varadhan [10], [11], Stroock [9] and Ellis [5] are good sources. The missing details in some of the applications of Sections 6, 8, 9, 10 can be found in [1], [2] and [3] and [4].

References

[1] Donsker, M. D. and Varadhan, S.R.S. The Asymptotics of The Wiener Sausage, Comm. Pure Appl. Math. Vo. XXVIII, 1975, 525-565.

[2] Donsker, M.D. and Varadhan S.R.S. On Laws of Iterated Logarithms for Local Times, Comm. Pure Appl. Math. Vol. XXX, 1977, 707-753.

[3] Donsker, M.D. and Varadhan, S.R.S. Asymptotics for the Polaron, Comm. Pure Appl. Math. Vol. XXXVI, 1983, 505-528.

[4] Donsker, M.D. and Varadhan, S.R.S. Large Deviations for Stationary Gaussian Processes, Comm. Math. Physics 97, 1985, 187-210.

[5] Ellis, Richard S. Entropy, Large Deviations and Statistical Mechanics, Springer-Verlag, New York, 985.

[6] Cox and Griffeath, Large Deviations for some infinite particle system occupation times. Contemp. Math. Vol. 41, 1985, 3-54.

[7] Bramson, Cox and Griffeath. Large deviations for infinite particle systems, preprint, 1985.

[8] Lee, T.Y. Large Deviation theory for the empirical density of non interacting infinite particle systems, 1986, Thesis NYU.

[9] Stroock, D.W. An introduction to the theory of large deviations Springer-Verlag, New York, 1984.

[10] Varadhan, S.R.S. Large Deviations and Applictions, SIAM, Philadelphia, 1984.

[11] Varadhan, S.R.S. Large Deviations and Applications, Expo. Math. 3, 1985, 251-272.

APPLICATIONS OF NON-COMMUTATIVE FOURIER ANALYSIS TO PROBABILITY PROBLEMS

Persi DIACONIS

P. DIACONIS : "APPLICATIONS OF NON-COMMUTATIVE FOURIER ANALYSIS TO PROBABILITY PROBLEMS"

1. Introduction.

These lectures develop a set of tools that are useful for solving probability problems connected to groups and homogeneous spaces. Here, the tools are used to solve problems like how long should a Markov chain run to get close to its stationary distribution. In particular, how many times should a deck of cards be shuffled to be close to random? The answer is, about 7 times for a 52 card deck.

A main component is the representation of the group involved as a subgroup of a matrix group. Group representations is one of the most active areas of modern mathematics. These notes may be regarded as an introduction, aimed at probabilists, to noncommutative Fourier analysis.

The first topic is a set of introductory examples -- random walk on the discrete circle Z_n , and the cube Z_2^n.

These tools are specialized to a problem on the symmetric group: if a random card and the top card are repeatedly transposed, it takes about $n \log n$ switches to mix up n cards. Analysis of this problem makes full use of the representation theory of the symmetric group. On the other hand, because of the symmetries involved (the measure is invariant under conjugation by all permutations fixing 1) the Fourier transforms are all diagonal matrices, so the analysis is not too difficult.

Highly symmetric measures often have simple Fourier transforms. In the easiest case, the measure might be bi-invariant under a subgroup $(P(ksk') = P(s))$ arising from a Gelfand pair. Then the Fourier transform is a matrix with a single non-zero entry. If the measure is constant on conjugacy classes $(P(tst^{-1}) = P(s))$ the Fourier transform is a constant times the identity matrix.

The example above introduces an interpolation between these two: there is a group G and subgroups $H \supset K$. The measure is invariant under conjugation by H and bi-invariant under K. Then the Fourier transform is diagonal, so analysis is still "simple". As shown, it is possible to explicitly compute the diagonal entries and answer any of the standard questions such as rates of convergence to stationarity, time to return, or hit a point, etc.

The next section applies the group theory to a practical problem -- generating random "objects" on a computer. The objects might be things like permutations or rotations. A general procedure -- the subgroup algorithm -- is developed, shown to specialize to the best existing algorithms in familiar cases, and shown to give new algorithms as well.

This leads to the problem of when can we take a square root of Haar measure in a non-trivial way. The answer is a bit surprising -- never in the Abelian case, and almost always in the non-Abelian case. For a finite group G, there is a non-uniform probability Q such that $Q*Q$ = uniform, if and only if G is not Abelian or the product of some two-element groups with the eight element quaternion group.

Prerequisites and background material.

The material developed here uses only the basics of representation theory. I have chosen to build these notes around J. P. Serre's (1977) little book, Linear Representations of Finite Groups. A careful reading of the first 30 or so pages of Serre will suffice. I will make constant references to Serre as we go along, so these notes may be regarded as a (hopefully) motivated way to work through Serre.

Alternative, highly readable sources for this material are Naimark and Stern (1982) or Curtis and Reiner (1982). This last is also encyclopedic, containing many further topics.

My favorite introductions to group theory are Herstein (1975), Rotman (1973), and Suzuki (1982). Again, this last is extremely readable as an introduction, and encyclopedic in scope.

Finally, I find the surveys by Mackey (1978, 1980) make tantalizing reading -- he seems to find everything related to group representations.

The present notes have some overlap with Diaconis (1982, 1987). There I develop the basic material more slowly, and present the statistical applications. Here, I present fresh material on partially invariant probabilities and treat factorization problems. The two sets of notes use common notation, and may be regarded as complementary.

Acknowledgement.

Much of the material presented here has been developed over the years with coauthors. Chief among these is Mehrdad Shahshahani who taught me group representations. David Aldous, Fan Chung, and R. L. Graham have also been active collaborators. Jean Bretagnolle and Lucien Birge arranged my trip to Paris. P. L. Hennequin put a tremendous amount of energy into making all details work out. Phillippe Bougerol, Laura Ellie, and Claude Kipnis made my stay a pleasure. Thanks to all.

2. First Examples -- A Random Walk on the Circle.

To begin, consider the group of integers modulo n, denoted Z_n, as an additive group. I usually think of Z_n as n points around a circle. The simplest random walk on Z_n can be described as a particle that hops left or right each time with probability $\frac{1}{2}$. Mathematically we consider the probability $Q(1) = Q(-1) = \frac{1}{2}$, $Q(j) = 0$ otherwise.

Repeated steps are modeled by convolution. For probabilities P and Q on Z_n

$$P*Q(j) = \sum_{i=0}^{n-1} P(j-i)Q(i) .$$

As usual, define $Q^{*n} = Q*Q^{*n-1}$.

The Fourier transform of a probability P on Z_n is familiar as the discrete Fourier transform of applied mathematics

$$\hat{P}(j) = \sum_{k=0}^{n-1} e^{\frac{2\pi ijk}{n}} P(k) . \tag{2.1}$$

Thus for simple random walk, $\hat{Q}(j) = \cos(\frac{2\pi j}{n})$.

Let U be the uniform distribution on Z_n, so $U(j) = \frac{1}{n}$ for all j. The fact that the roots of unity sum to zero gives $\hat{U}(j) = 0$ if $j \neq 0$, $\hat{U}(0) = 1$. Fourier analysis proofs of convergence to uniformity work by showing that $\hat{Q}(j)^k$ converges to zero for all $j \neq 0$ as k gets large.

It is easy to see from the definition (2.1) that transforms turn convolution into multiplication:

$$\widehat{P*Q}(j) = \hat{P}(j)\ \hat{Q}(j) .$$

Finally, we have the Fourier inversion and Plancherel theorem

(2.2) $$P(j) = \frac{1}{n}\sum_{k=0}^{n-1} e^{\frac{-2\pi ijk}{n}}\ \hat{P}(k)$$

(2.3) $$\sum_{j=0}^{n-1} P(j)Q(j) = \frac{1}{n}\sum_{k=0}^{n-1} \hat{P}(k)\hat{Q}(k) .$$

These are easy to prove. Take (2.2): since both sides are linear in P, it is enough to take $P(j) = \delta_m(j)$, a point mass at m. Then $\hat{\delta}_m(k) = e^{2\pi imk/n}$ and (2.2) follows from the roots of unity summing to zero. For (2.3), again both sides are linear in P; taking $P = \delta_m$ reduces to (2.2). Clearly (2.2) and (2.3) remain true if P and Q are any functions, not necessarily probabilities.

We now have all the tools needed to analyze simple random walk on Z_n. It is intuitively clear that as time goes on the distribution of the particle tends to uniform. To make this precise, we use the familiar total variation distance.

(2.4) $$\|Q^{*k}-U\| = \max_A|Q^{*k}(A)-U(A)| = \frac{1}{2}\sum_{j=0}^{n-1}|Q^{*k}(j)-U(j)| = \frac{1}{2}\max_{\|f\|=1}|E_k(f)=E_U(f)| .$$

THEOREM 2.1. Let n be odd. There are universal positive constants a, b, c, such that for all n and k,

(2.5) $$ae^{-bk/n^2} \le \|Q^{*k}-U\| \le ce^{-bk/n^2} .$$

Remarks (1) If n is even, the particle is at an even position after an even number of steps and so never converges. This can be seen from the transform as well, $\hat{Q}^k(n/2) = (-1)^k$ which does not converge to zero.

(2) The theorem says that to get close to uniform k must be somewhat larger than n^2. This can be seen from the central limit theorem; after all, the particle must have a good chance of getting to the side opposite zero. But in k steps, the particle is quite likely to be within $\pm\sqrt{k}$ of zero, so $k >> n^2$ is needed. This can be made into a proof (but needs the Berry-Esseen theorem).

(3) The bounds above are uniform in k and n. I think of n large, and wanting to know how large $k = k(n)$ should be to force the variation distance to be small.

The proof of Theorem 2.1 relies on the following useful lemma.

Lemma 2.1 (upper bound lemma). Let P be a probability on Z_n

$$\|P-U\|^2 \leq \frac{1}{4}\sum_{j=1}^{n-1}|\hat{P}(j)|^2 .$$

Proof: From the definition (2.4) of variation distance,

$$\|P-U\|^2 = \frac{1}{4}\Big(\sum_{j=0}^{n-1}|P(j)-U(j)|\Big)^2 \leq \frac{n}{4}\sum_{j=0}^{n-1}|P(j)-U(j)|^2$$
$$= \frac{1}{4}\sum_{j=1}^{n-1}|\hat{P}(j)|^2 .$$

The inequality is Cauchy-Schwarz. The final equality uses the Plancherel theorem (2.3) applied to $|P(j)-U(j)|$, and $\hat{U}(0) = \hat{P}(0)$, $\hat{U}(j) = 0$ otherwise.

Remark. The upper bound lemma has turned out to be a rather precise tool, despite the apparent sloppy use of the Cauchy-Schwarz inequality. This is demonstrated by the matching lower bound.

Proof of Theorem 2.1. Using the upper bound lemma we have

$$4\|Q^{*k}-U\|^2 \leq \sum_{j=1}^{n-1}\cos\Big(\frac{2\pi j}{n}\Big)^{2k} .$$

To bound the sum we use simple properties of cosine. To begin with, $\cos\frac{2\pi j}{n} = \cos\frac{2\pi(n-j)}{n}$ so the sum equals twice the sum from 1 to $(n-1)/2$. Next, for $0 \leq x \leq \pi/2$,

$$1 - \cos x = \frac{x^2}{2} - \frac{x^4}{24} + \ldots \geq \frac{x^2}{2} - \frac{x^4}{24} \geq \frac{x^2}{2}\Big(1 - \Big(\frac{\pi}{2}\Big)^2\frac{1}{12}\Big) \geq \frac{x^2}{3} .$$

Since $e^{-x} \geq 1-x$,

$$\cos x \leq 1 - \frac{x^2}{3} \leq e^{-\frac{x^2}{3}} \quad \text{for } 0 \leq x \leq \pi/2 .$$

Thus, for $u = e^{-\frac{8}{9}\frac{\pi^2 k}{n^2}}$

$$(2.6) \qquad \sum_{1\le j\le n/4} \cos\left(\frac{2\pi j}{n}\right)^{2k} \le \sum_{1\le j\le n/4} u^{j^2} \le u + \sum_{j\ge 1} u^{2j} \le 2u$$

where the last inequality assumes $k \ge n^2$. Taking $c > 1$ in (2.5) covers the other cases.

Finally, for $\frac{n}{4} < j \le \frac{n-1}{2}$, let $\ell = j - \frac{n-1}{2}$,

$$\cos\left(\frac{2\pi}{n}\left(\frac{n-1}{2} - \ell\right)\right) = \cos\left(\frac{2\pi(\ell + \frac{1}{2})}{n}\right) \qquad 0 \le \ell < \frac{n}{4}.$$

Precisely the same bound (2.6) applies with u replaced by $u^{\frac{1}{2}}$.

Combining bounds we have,

$$\|Q^{*k} - U\| \le \sqrt{2}\, e^{-\frac{2}{9}\pi^2 k/n^2}.$$

This completes the proof of the upper bound. For the lower bound, use the right-most expression in (2.4) with $f(j) = \cos(\frac{2\pi j}{n})$. It again follows from the sum of roots of unity being 1 that $E_U(f) = 0$. From the definition of Fourier transform $E_k(f) = \cos(\frac{2\pi}{n})^k$. Thus,

$$2\|Q^{*k} - U\| \ge \cos\left(\frac{2\pi}{n}\right)^k \ge \left(1 - \frac{1}{2}\left(\frac{2\pi}{n}\right)^2\right)^k \ge e^{-2\pi^2 k/n^2}$$

where $1 - x \ge e^{-2x}$, $0 \le x < \frac{3}{4}$ was used. □

Remarks. (1) Clearly with more work, slightly better constants can be had in the inequalities. In particular the constants in the exponents can be taken as equal.

(2) Notice how much information about cosine was used. One of the problems to be faced is developing such knowledge in the non-commutative case.

(3) This proof was given in such detail because it seems a model of what is coming. In particular there is a lead term $\cos(\frac{2\pi}{n})^{2k}$ that dominates. The other terms sum up to something smaller.

(4) Similar arguments work for other simple measures on Z_n, for example, a step chosen each time uniformly in $[-a,a]$. If a stays bounded, and n gets large, the rate is the same: more than n^2 steps are needed. It is unknown what is the fastest converging measure with a given number of support points. For

example let P put mass $\frac{1}{4}$ on 4 points in Z_n. What choice of points makes P^{*k} converge to uniform most rapidly? (I believe k points can be chosen so that $n^{2/(k+1)}$ steps suffice.)

(5) Even for a simple group like Z_n, there can be very delicate problems. Here is an example. In considering schemes to generate pseudo random numbers on a computer, Chung, Diaconis and Graham (1987) studied random variables satisfying

$$X_k = aX_{k-1} + \varepsilon_k (\text{mod } n)$$

where ε_k was a sequence of i.i.d. random variables on Z_n. Observe that if $a = 1$, this is essentially simple random walk on the circle. By our previous discussion, it takes somewhat more than n^2 steps to get random when $a = 1$, if ε_k is "small".

For more general values of a, the Markov chain (2.7) can be represented as a convolution of independent variables as follows

$$X_0 = 0,\quad X_1 = aX_0 + \varepsilon_1 = \varepsilon_1,\quad X_2 = a\varepsilon_1 + \varepsilon_2,\quad X_3 = a^2\varepsilon_1 + a\varepsilon_2 + \varepsilon_3,\ \ldots,$$
$$X_k = a^{k-1}\varepsilon_1 + a^{k-2}\varepsilon_2 + \ldots + \varepsilon_k (\text{mod } n)\ .$$

Thus X_k has Fourier transform $\Pi_{i=0}^{k-1}\hat{Q}(a^i k)$, with Q the law of ε_i. The upper bound lemma can be used to bound the distance to uniform.

Chung, Diaconis and Graham let ε take values $0, \pm 1$ each with probability $\frac{1}{3}$. When $a = 1$, it takes n^2 steps to get random as for simple random walk. When $a = 2$, and n is of form $2^\ell - 1$, it takes about $\log n \log\log n$. This is sharp (that many steps are needed). For almost all n, $1.02 \log n$ steps are enough to get close to random, although no explicit sequence n_i, tending to infinity, is known which gets random after $c \log n_i$ for any fixed c. The analysis involved seems fairly delicate. No serious work has been done on the problem for general values of $a \neq 2$.

3. Random Walk on Z_2^n.

The second example begins as a classical piece of mathematical physics -- the Ehrenfest's urn scheme. This involves two urns containing collectively n balls.

At a typical stage a ball is chosen at random and switched into the other urn. Years ago, someone observed this could be lifted to a random walk on a group.

Let Z_2^n be the group of binary n-tuples with coordinatewise addition modulo 2. Each coordinate stands for one of the n-balls, with a 1 representing the ball being in the left urn.

A typical step in the chain thus becomes pick a coordinate at random and change the entry to its opposite mod 2. This is random walk with

$$P(100\ldots0) = P(010\ldots0) = \ldots = P(000\ldots1) = \frac{1}{n}.$$

This walk as a periodicity problem -- after an even number of steps it is at a binary vector with an even number of ones. It is not hard to deal with even and odd numbers of steps directly. Another possibility is to consider the walk in continuous time. Let us get rid of periodicity in another way: allow the walk to stay where it is with small probability. A natural candidate is

$$Q(000\ldots0) = Q(100\ldots0) = Q(010\ldots0) = \ldots = Q(000\ldots1) = \frac{1}{n+1}. \tag{3.1}$$

To analyze repeated steps, the analog of the Fourier analysis used above is needed. This is again easy and well known. The Fourier transform of any function P is just

$$\hat{P}(x) = \sum_y (-1)^{x\cdot y} P(y),$$

where $x \in Z_2^n$, and the sum runs over the group. One way to see that this is the appropriate definition is to observe that for Z_2, the two roots of unity involved in the Fourier transform are just ± 1. In general, the appropriate "dual group" of a product is the product of the "dual groups".

The analog of the Fourier inversion theorem is

$$P(y) = \frac{1}{2^n} \sum_x (-1)^{x\cdot y} \hat{P}(x). \tag{3.2}$$

This can be proved as above by specializing to P a δ function when the result reduces to the classical formula of inclusion-exclusion (identifying $x \in Z_2^n$ with the indicator function of a subset of $\{1,2,\ldots,n\}$). From here the Plancherel theorem follows by linearity. The upper bound lemma becomes

$$\|P^{*k}-U\|^2 \leq \frac{1}{4}\sum_{y\neq 0} |\hat{P}(y)|^{2k} . \tag{3.3}$$

Using these results we will prove that a bit more than $\frac{1}{4} n \log n$ steps are needed and suffice to get close to random.

<u>THEOREM 3.1</u>. For Q defined by (3.1) on the group Z_2^n, let

$$k = \frac{1}{4}(n+1)\log n + c(n+1) .$$

Then

$$\|Q^{*k}-U\|^2 \leq \frac{1}{2}\left(e^{e^{-4c}} - 1\right) \tag{3.4}$$

$$\underline{\lim_{n}} \|Q^{*k}-U\| \geq (1-8e^{c}) . \tag{3.5}$$

<u>Proof</u>: Let $|y|$ be the number of ones in the binary vector y. The Fourier transform is $\hat{Q}(y) = \Sigma_x(-1)^{x\cdot y}\, Q(x) = 1 - \frac{2|y|}{n+1}$. Using the upper bound lemma,

$$\|Q^{*k}-U\|^2 \leq \frac{1}{4}\sum_{y\neq 0}\left(1-\frac{2|y|}{n+1}\right)^{2k} = \frac{1}{4}\sum_{j=1}^{n}\binom{n}{j}\left(1-\frac{2j}{n+1}\right)^{2k}$$

$$\leq \frac{1}{2}\sum_{j=1}^{\frac{n+1}{2}}\binom{n}{j}\left(1-\frac{2j}{n+1}\right)^{2k} \leq \frac{1}{2}\sum_{j=1}^{\infty}\frac{n^j}{j!}e^{-4jk/n+1}$$

$$= \frac{1}{2}\left(e^{e^{-4c}}-1\right) .$$

This completes the proof of the upper bound. In this example, the proof of the lower bound requires an extension of the idea used in Theorem 2.1. To explain, consider the just completed calculation for the upper bound. The dominant term is $n\left(1-\frac{2}{n+1}\right)^{2k}$, all other terms being smaller. This dominant term arises from the Fourier transform at vectors y with $|y| = 1$. The new idea is to use the "slow terms" in the upper bound lemma as a clue to defining a random variable on the group, with Fourier transform the slow terms.

Here, this heuristic suggests we define the random variable

$$Z(x) = \sum_{i=1}^{n} (-1)^{x_i} = n - 2|x| .$$

Under the uniform distribution U on Z_2^n, the coordinates x_i are i.i.d. Bernoulli with parameter $\frac{1}{2}$. It follows that

(3.6) $$E_U\{Z\} = 0 , \quad Var_U\{Z\} = n .$$

Under the measure Q^{*k}, the moments of Z are

(3.7) $$E_k(Z) = n(1 - \frac{2}{n+1})^k ,$$

(3.8) $$Var_k(Z) = n + n(n-1)(1 - \frac{4}{n+1})^k - [E_k(Z)]^2 .$$

Indeed, from the definition of Z, $E_k(Z) = \Sigma_{i=1}^n E_k((-1)^{X_i}) = \Sigma_{i=1}^n \hat{Q}_k(e_i) = n(1 - \frac{2}{n+1})^k$, where e_i is a binary vector with a 1 in the ith place and zeroes elsewhere. The proof of (3.8) is similar. Straightforward estimates show that, for fixed c, as $n \to \infty$,

(3.9) $$E_k(Z) = \sqrt{n}\, e^{-c/2} (1 + O(\frac{\log n}{n}))$$

(3.10) $$Var_k(Z) = n + O(e^{-c} \log n) .$$

Proceeding heuristically for a moment, Chebychev's inequality and (3.6) imply that under U, Z lives on $\pm 10\sqrt{n}$ to good approximation. On the other hand, (3.9), (3.10) imply that under Q^{*k}, Z is supported on $\sqrt{n}\, e^{-c/2} \pm 10\sqrt{n}$, which is disjoint from $\pm 10\sqrt{n}$ to good approximation. It follows that the variation distance tends to 1 as n tends to infinity.

Proceeding more rigorously, let $K = e^{-c/2}$ and $A = \{x \in Z_2^n, |x| \le \frac{n}{2} - \frac{K}{4}\sqrt{n}\}$. Chebychev's inequality implies

$$U(A) \le \frac{1}{(K/2)^2}$$

$$Q^{*k}(A) \ge 1 - \frac{1 + o(1)}{(K/2)^2} .$$

This yields the lower bound (3.5), from definition (2.4). □

Remarks. (1) The neat argument given here owes much to participants at the St. Flour seminar.

(2) The argument for lifting a Markov chain to a random walk is abstracted with further examples and problems given in Diaconis and Shahshahani (1987a).

(3) There is an important difference between the asymptotic behavior on the cube and circle. On the cube, convergence to uniformity exhibits a "cutoff phenomena", while convergence on the circle doesn't. To explain, you might think that the variation distance between Q^{*k} and U tends smoothly to zero as k increases. In fact, in both the circle and cube (and most other examples) the variation distance stays quite close to its maximum value of 1 as k increases, and then cuts down to a small value where it decreases to zero geometrically fast. The cutoff phenomenon happens in a neighborhood of the region of decrease.

For the n point circle, if k goes to infinity with n in such a way that k/n^2 tends to ℓ, the variation distance tends to a continuous monotone decreasing limiting function $\theta(\ell)$. (Indeed, computation shows that $\theta(\ell)$ is a "theta function".)

Next consider the cube. If k tends to infinity with n in such a way that $k/n \log n$ tends to ℓ, the variation distance tends to a step function taking value 1 if $\ell < \frac{1}{4}$, and 0 if $\ell > \frac{1}{4}$. It is not clear if the limit exists when $\ell = \frac{1}{4}$.

The point for the present discussion is that for the cube, and most other examples, the transition from 1 to 0 happens in a short time (in an interval of length n at a scale of $n \log n$). For the circle the decrease happens in an interval of length proportional to n^2, at a scale of n^2.

The cutoff phenomenon happens so often, it has become a problem to explain it. See Section 6 of Aldous and Diaconis (1986).

(4) Simple random walk on the cube has been applied to a problem arising in the analysis of algorithms related to the simplex algorithm of linear programming. Aldous (1984) uses the analysis above to show that a large class of algorithms

must take an exponential number of steps to find the minimum of most "nice" functions on Z_2^n.

(5) Here is a class of problems where similar analysis is of interest. In considering the effect of stirring food while cooking, we often appeal to an ill framed "ergodic theorem" for an explanation of why things get evenly browned. Seeking to formulate this more precisely, Richard Stanley and I consider n potato slices arranged in a ring (as around the circumference of a frying pan). At each time, a spatula of radius d is put in at a random position, and the d potatoes under the spatula are turned over in place. It is intuitively clear that the pattern of "heads and tails" formed will tend to be uniformly distributed.

A first question is, how many moves does it take, and how does it depend on d? The results surprised us. To get rid of parity, assume that n and d are relatively prime. It turns out that then the variation distance to uniform behaves exactly like simple random walk on the cube; $\frac{1}{4} n \log n + cn$ moves suffice, no matter what d is used.

The problem can be set up as a random walk on Z_2^n with underlying measure Q uniform on the n cyclic shifts of the vector beginning with d ones and ending with $n-d$ zeroes. From here, the analysis can be carried out as for Theorem 2.2. However, there is a clever way to see why the result is true without calculation. The reader is referred to Diaconis and Stanley (1986) for further discussion.

4. The General Set-up, Examples, and Problems.

Let G be a finite group. Let P and Q be probabilities on G. Define, for $s \in G$

$$P*Q(s) = \sum_t P(st^{-1})Q(t) .$$

Thus, to get s, one must first pick t from Q and then "undo" things by picking st^{-1} from P. Now $P*Q \neq Q*P$ in general.

Fourier analysis is based on the notion of a representation of G. This is a homomorphism $\rho: G \to GL(V)$, where V is a finite dimensional vector space over the complex numbers. We denote $\dim V = d_\rho$. Thus, a representation assigns matrices to group elements in such a way that $\rho(st) = \rho(s)\rho(t)$.

The <u>Fourier transform</u> of P at representation ρ is the matrix

$$\sum_s P(s)\rho(s) .$$

From the definitions, the transform satisfies:

$$\widehat{P*Q}(\rho) = \hat{P}(\rho)\hat{Q}(\rho) .$$

To go further, we need the notation of an irreducible representation. Define (ρ,V) to be <u>irreducible</u> if there are no non-trivial subspaces W in V stable under G in the sense that $\rho(s)W \subset W$ for all $s \in G$. Irreducible representations make up the building blocks of all representations, and are known (more or less explicitly) for almost any group arising in applications.

It may be useful to carry along an example. Consider S_n, the permutation group on n letters. We will denote permutations by $\pi(i)$, $1 \leq i \leq n$. We are all familiar with three representations of S_n:

<u>The trivial representation</u>. Take $V = C$, and define $\rho(\pi)v = v$ for all and $v \in V$. This is a 1-dimensional representation.

<u>The alternating representation</u>. Take $V = C$, and define $\rho(\pi)v = \text{sgn}(\pi) \cdot v$, where $\text{sgn}\,\pi$ is 1 if π can be written as an even number of transpositions and $\text{sgn}\,\pi$ is -1 otherwise. A first course in group theory proves that $\text{sgn}\,\pi\eta = \text{sgn}\,\pi\ \text{sgn}\,\eta$, so this is a representation of dimension 1.

<u>The permutation representation</u>. This assigns permutation matrices to permutations in the usual way. To define it, take V as the usual n dimensional space with basis $e_1, e_2, \ldots, e_n$, where e_i has a 1 in the ith position, and zeroes elsewhere. To define a linear map $\rho(\pi)$ from V to itself, it is sufficient to define $\rho(\pi)(e_i) = e_{\pi(i)}$ and extend by linearity. The matrix of $\rho(\pi)$ with respect to the basis e_i is the usual permutation matrix.

This representation is reducible. Indeed, if $w = e_1 + e_2 + \ldots + e_n$, the linear span W of w is an irreducible space because $\rho(\pi)$ acts by permuting the coordinates. Similarly $W^{\perp} = \{v \in V : \Sigma v_i = 0\}$ is a non-trivial invariant subspace. The linear mappings $\rho(\pi)$ send this space into iteself, and one speaks of the restriction of ρ to $W^{\perp}$. Clearly

$$V = W \oplus W^{\perp} .$$

The representation $W^{\perp}$ is irreducible (see Serre's Exercise 2.6), so this gives a decomposition of V into irreducible invariant subspaces.

For $n = 3$, there are only three distinct (non-isomorphic) irreducible representations, the trivial, alternating, and 2-dimensional representation.

It is an instructive exercise to compute the six 2×2 matrices for the 2-dimensional representation. One way to do this considers the basis $e_1 - e_2$, $e_1 - e_3$ for $W^{\perp}$. Apply $\rho(\pi)$ to these vectors and express as a linear combination of the basis vectors. For example, if π is the transposition (12), $\rho(\pi)(e_1-e_2) = -(e_1-e_2)$, $\rho(\pi)(e_1-e_3) = e_2-e_3 = (e_1-e_3) - (e_1-e_2)$, so the matrix is $\begin{pmatrix}-1 & 0\\ -1 & 1\end{pmatrix}$.

As an example of a naturally occurring probability on S_n, consider the example of random transpositions. Imagine n cards in a row on the table. To start, card one is at the left, card two is next, and card n is at the right of the row. Each time, the left and right hands touch random cards (so $L = R$ with probability $\frac{1}{n}$) and the two cards are switched. This defines a measure on S_n:

$$Q(id) = \frac{1}{n}$$

$$Q(ij) = \frac{2}{n^2}$$

$$Q(\pi) = 0 \quad \text{otherwise.}$$

Repeatedly transposing is modeled by convolution, and one can ask (and answer) the usual questions. These problems are solved using Fourier analysis in Diaconis and Shahshahani (1981).

Return to the case of a general finite group G. The uniform distribution is denoted $U(s) = 1/|G|$. As for the Abelian case, we have

Lemma 4.1. For ρ a non-trivial irreducible representation, $\hat{U}(\rho) = 0$.

Proof: Any argument for this needs Schur's lemma (Serre (1977, Proposition 4). One simple way is to note $\hat{U}(\rho) = 1/|G| \Sigma_s \rho(s)$. Now the (i,j) entry of this matrix is the inner product of the function on G defined by $\rho_{ij}(\cdot)$ and the function 1. But Serre's proposition shows that the matrix entries of distinct irreducible entries are orthonormal. □

The other basic results needed to work with Fourier analysis are the Fourier inversion and Plancherel formulas.

THEOREM 4.1. Let f be a function on G, then

(a) $f(s) = \frac{1}{|G|}\Sigma_\rho d_\rho \operatorname{Tr}(\rho(s^{-1})\hat{f}(\rho))$.

(b) Let f and g be functions on G, then

$$\Sigma_s f(s^{-1})g(s) = \frac{1}{|G|} \Sigma_\rho d_\rho \operatorname{Tr}(\hat{f}(\rho)\hat{g}(\rho)) .$$

In both formulas, the sum is over all irreducible representations ρ of G.

Proof: (a) Both sides are linear in f, so it suffices to let f run through a basis of all functions on G. For $f(s) = \delta_{st}$, $\hat{f}(\rho) = \rho(t)$. The right side equals

$$\frac{1}{|G|} \Sigma_\rho d_\rho \operatorname{Tr}(s^{-1}t) . \tag{4.1}$$

This expression is $1/|G|$ times the character of the regular representation evaluated at $s^{-1}t$; indeed Serre's Corollary 1 (pg. 18) says that each irreducible representation ρ occurs d_ρ times in the regular representation. Serre shows that the character of the regular representation is zero unless $s = t$, when the character is $|G|$. This proves (a).

For (b), again both sides are linear in f. Take $f(s) = \delta_{st}$; it must be shown that

$$g(t^{-1}) = \frac{1}{|G|} \Sigma_\rho d_\rho \operatorname{Tr}(\rho(t)\hat{g}(\rho)) .$$

This was proved in part (a). □

Remarks. (1) The Fourier inversion theorem shows how knowing $\hat{f}(\rho)$ at all the irreducible representations determines f.

(2) The Plancherel theorem says that the inner product between two functions equals the inner product between their Fourier transforms. It is surprisingly useful.

(3) Let's show how the theorems of this section specialize to the more familiar theorems of the last section. For $G = Z_n$, the irreducible representations are all 1-dimensional. There are n different representations with the jth being $\rho_j(k)v = e^{\frac{2\pi ijk}{n}} v$, $0 \le j < n$. Of course, a 1-dimensional representation has a 1 by 1 "matrix", here, multiplication by $e^{\frac{2\pi ijk}{n}}$. The Fourier inversion theorem specializes to

$$f(s) = \frac{1}{n} \sum_{j=0}^{n-1} e^{\frac{-2\pi ijs}{n}} \hat{f}(\rho_j)$$

where

$$\hat{f}(\rho_j) = \sum_{k=0}^{n-1} e^{\frac{2\pi ijk}{n}} f(k) .$$

Similarly, for $G = Z_2^n$, all representations are 1-dimensional. There is a distinct irreducible representation for each $x \in Z_2^n$; it is

$$\rho_x(y)v = (-1)^{x \cdot y} v .$$

From this, (2.8) follows as a special case of (a).

The main application of Theorem 4.1 in present circumstances is to the upper bound lemma, first used by Diaconis and Shahshahani (1981).

Lemma 4.2 (upper bound lemma). Let P be a probability on the finite group G. Let U be uniform on G, and define

$$\|P-U\| = \frac{1}{2} \Sigma_s |P(s)-U(s)| .$$

Then

$$\|P-U\|^2 \le \frac{1}{4} \Sigma_\rho^* d_\rho \|\hat{P}(\rho)\|^2 ,$$

where $\|A\|^2 = Tr(AA^*)$, and the sum is over non-trivial irreducible representations of G.

Proof: Using the Cauchy-Schwarz inequality and then the Plancherel theorem,

$$\|P-U\|^2 = \frac{1}{4}[\Sigma|P(s)-U(s)|]^2 \leq \frac{1}{4}\Sigma|P(s)-U(s)|^2$$

$$= \frac{1}{4}\Sigma_\rho d_\rho Tr[(\hat{P}(\rho)-\hat{U}(\rho))(\hat{P}(\rho)^* - \hat{U}(\rho)^*]$$

$$= \frac{1}{4}\Sigma_\rho^* d_\rho Tr(\hat{P}(\rho)\hat{P}(\rho)^*) .$$

Where the last step used $\hat{U}(\rho) = 0$ for any non-trivial irreducible and $\hat{P}(\rho) = 1$ if ρ is the trivial representation. We may assume, without loss, that $\rho(s)$ is a unitary matrix, so $\Sigma P(s^{-1})\rho(s) = \Sigma P(s)\rho(s^{-1}) = \hat{P}(\rho)^*$. □

Some Examples. We have seen three examples above, Z_n, Z_2^n, and S_n. These groups can themselves supply a potentially unlimited source of problems. To make the point, consider the following natural measure on S_n -- all will be said in the language of shuffling cards.

Example 4.1 (transpose random to top). Here the measure is supported on the identity and (12), (13), ..., (1n). A probabilistic description -- pick a card at random and transpose it with the top card. A Fourier solution is presented in the next section. It takes n log n steps to get random.

Example 4.2 (random adjacent). Here the measure lives on the identity and (i,i+1), $1 \leq i \leq n-1$. It is unknown just how many transpositions it takes to mix things up. It is easy to see that at least n^3 moves are needed -- following a single card, which essentially does simple random walk when it is hit. It's hit, on average, twice in n steps and takes n^2 such to get random. It is known (Aldous 1985) that $n^3 \log n$ steps suffice. Presumably this is the correct rate.

Example 4.3 (top in at random). Probabilistically, each time the top card is taken off and inserted in a random position. The measure lives on the identity and the n-1 cyclic permutations (21), (321), (4321), ..., (nn-1...1). While no Fourier argument is known, Diaconis and Aldous (1986) show n log n moves mix things up.

Example 4.4 (riffle shuffling). This is the most usual way of mixing cards. The cards are cut in half (according to a binomial distribution) and then interlaced in the following way: If a cards are in the left hand, b cards are in the right hand, drop a card from the left with probability $a/(a+b)$ (and from the right otherwise). This mathematical model chooses shuffles like ordinary card players. Diaconis and Aldous (1986) present arguments showing $\log n$ repetitions are needed to mix up n cards. The cut off occurs at 7 shuffles for 52 cards.

Example 4.5 (overhand shuffles). This is the next most commonly used way to mix up n cards. A mathematical description involves repeatedly flipping a coin (parameter p) and labeling the back of cards 1 through n with the outcomes. Then cut off all cards up to and including the first head, remove them from the deck and put them in a pile. Remove all up to and including the second head, and put them on the first removed pile. Continue, finally removing all remaining cards and putting them on top of the newly formed pile. This mixes the cards by repeatedly cutting off small packets and reversing the order of the packets.

Robin Peamantle has found a coupling argument that bounds the number of repeated shuffles needed between n^2 and $n^2 \log n$. Thus while 7 riffle shuffles suffice, more than 2500 overhand shuffles are required to mix a 52 card deck.

Several other natural measures on S_n arise from lifting random walks to S_n from an associated homogeneous space. See section of Diaconis and Shahshahani (1987a).

While they won't be dealt with in detail here, a number of other groups arise. For example, the groups A_n -- the set of (a,b) with a and b in Z_n and a relatively prime to n -- arises in considering random number generation schemes

$$X_{k+1} = a_k X_k + b_k (\mathrm{mod}\ n) .$$

Here A_n is the affine group mod n. As a second example, the group $GL_2(Z_n)$ -- the 2×2 invertible matrices with entries mod n -- arises in analyzing second order recurrences:

$$X_{k+1} = a_k X_k + b_k X_{k-1} (\mathrm{mod}\ n) .$$

Diaconis and Shahshahani (1986) contains further details and references.

Some Problems. These notes focus on the problem of number of steps required to achieve uniformity. Two other natural problems are: (a) Starting at x, how many steps are required to hit a set A. (b) How many steps are required to hit every point.

Both problems can be attacked by using the techniques developed here. For the first problem, let Q be a probability on the finite group G. For $s,t \in G$, let F^n_{st} be the chance that t is first hit at time n starting at s. For $|Z| < 1$, let

$$F_{st}(Z) = \sum_{n=1}^{\infty} F^n_{st} \, Z^n .$$

Straightforward Markov chain theory couples with Fourier analysis to give

THEOREM 4.2. For $|z| < 1$, $(I-z\hat{Q}(\rho))$ is invertible and

$$F_{st}(z) = \frac{\Sigma_\rho \, d_\rho \, \mathrm{Tr}[\rho(st^{-1})(I-z\hat{Q}(\rho))^{-1}]}{\Sigma \, d_\rho \, \mathrm{Tr}[I-z\hat{Q}(\rho)]^{-1}} \qquad |z| < 1 .$$

This result is developed in Flatto, Odlyzko, and Wales (1985) and Diaconis (1987, Chapter 3F). The theorem can be used to approximate the moments and first hitting distributions for simple random walk on Z_n, and for random transpositions on S_n.

Matthews (1985, 1986) has developed analytical techniques for dealing with the second problem as a random walk version of the coupon collector's problem.

Aldous (1982, 1983) has given a very insightful treatment of both problems from a purely probabilistic point of view. He shows that if the random walk converges to uniform rapidly (roughly in order $\log |G|$), then problems (a) and (b) have the same answers as they would for uniform steps. Thus problem (a) becomes, drop balls at random into $|G|$ boxes, how long is required until a ball hits a given box. The time divided by $|G|$ has an approximate exponential distribution. Problem (b) becomes the classical coupon collector's problem, so $|G| \log |G| + c|G|$ steps are required to have a good chance of hitting every point. These results show how getting bounds on rates of convergence allow other questions to be answered.

5. A Library Problem.

A. Introduction. A collection of mathematical models has been introduced to study how a row of books becomes disarranged as a reader used them. This work is developed and reviewed by

The model studied here evolves as follows. Picture n books in a line on a shelf. The books are labeled in order, 1, 2, 3, ..., n from left to right. To start, a user removes book 1 (and hopefully spends some time reading it!). Then, a second book is chosen at random. It is removed, and book 1 put in place of this just removed book. This is continued repeatedly.

After a while the line of books becomes quite mixed up. In what follows it is shown that it takes about $n \log n$ steps to lose all trace of the initial order.

This problem is presented here as an example of the potential of the analytic methods developed above. It turns out that the Fourier analysis can be carried out in closed form: the Fourier transforms are diagonal matrices with "simple" entries. The example will be used as a vehicle for teaching about the representation theory of the symmetric group. It is not easy, but every effort will be made to make it generally instructive.

Suppose the process ends by placing the last book removed back into place 1 (which has been left vacant throughout). Then, the process can be modeled as a random walk on the symmetric group that repeatedly transposes a random card with the top card. To avoid a parity problem (which is easy to deal with but distracting) let us allow the identity as a choice each time (so the book in hand is retained). This gives the following probability on S_n, writing $(1j)$ for the permutation transposing 1 and j:

$$Q(id) = 1/n$$

$$Q(1j) = 1/n \quad 2 \leq j \leq n$$

$$Q(\pi) = 0 \quad \text{elsewhere.}$$

The object of this section is to prove the following result:

THEOREM 5.1. Let $k = n \log n + cn$. Then, for $c > 1$ and all n

(5.2) $$\|Q^{*k} - U\| \leq ae^{-c} \text{ for a universal constant } a.$$

For $c < 0$, and all n

(5.3) $$\|Q^{*k} - U\| > b > 0 \text{ for a universal constant } b.$$

Remark. This result shows that in a strong sense, order $n \log n$ moves are needed and suffice to mix up the books. As usual, the analysis yields a complete spectral analysis of the transition matrix of the underlying Markov chain. Thus problems like time to first hit a fixed permutation, or time to first return can be attacked by the usual methods. Flatto, Odlyzko, and Wales (1985) carry out detailed calculation for these variants.

As explained, the proof is laid out as a tutorial on the representation theory of the symmetric group. We first describe the irreducible representations. Then, observing that the measure Q is invariant under conjugation by S_{n-1}, we argue that the Fourier transform $\hat{Q}(\rho)$ is a diagonal matrix with known entries. Using a classical character formula of Frobenius the diagonal entries are determined explicitly. Next, the upper bound lemma is used to prove the upper bound (5.2). Finally the analytic approach is used to prove the lower bound for (5.3).

B. Representations of S_n. This classical piece of mathematics was developed around the turn of the century by Frobenius and Young. The most accessible modern treatment is in James (1978). A recent encyclopedic treatment appears in James and Kerber (1981).

By a partition $\lambda = (\lambda_1, \lambda_2, \ldots, \lambda_m)$ of n we mean a sequence $\lambda_1 \geq \lambda_2 \geq \ldots \geq \lambda_m$ of positive integers with $\lambda = \lambda_1 + \ldots + \lambda_m$. There is a one to one correspondence between partitions of n and irreducible representations of S_n. To describe this correspondence, the notion of a Young diagram will be useful. This is an array of empty boxes, with λ_1 boxes in the first row, λ_2 boxes in the 2nd row etc. For example, if $n = 7$, the diagram for the partition $(4,2,1)$ is

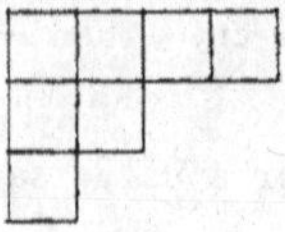

For each λ there is an irreducible representation ρ_λ. The explicit computation of ρ_λ is not required in what follows.

The facts needed will now be summarized.

(5.4) The dimension d_λ of the irreducible representation corresponding to the partition λ_n is the number of ways of placing the numbers $1, 2, \ldots, n$ into the Young diagram of λ in such a way that the entries in each row and column are decreasing.

<u>Remarks</u>. Consider the partition $(n-1,1)$ of n. Any number between 2 and n can be put into the single box in the second row and then monotonicity determines the rest. Thus $d_{n-1,1} = n-1$. Similarly $d_{n-k,k} = \binom{n}{k} - \binom{n}{k-1}$. The dimensions come up in a wide variety of combinatorial applications, and much more is known about them. Diaconis (1987, Chapter 7B) gives pointers to the literature. A proof of (5.4) is in James (1978, Section 6).

From Serre (1977, page 18), the squares of the dimensions of all irreducible representations add to the order of the group. Thus we have $d_\lambda \le \sqrt{n!}$. For an illustration of the power of different approaches to the study of dimensions, the reader might try deriving this from (5.4). From this and (5.4)

$$d_\lambda \le \binom{n}{\lambda_1}\sqrt{(n-\lambda_1)!}\ . \tag{5.5}$$

<u>Proof</u>: The first row of any allowable placement can be chosen in at most $\binom{n}{\lambda_1}$ ways. For each choice of first row, the number of ways of completing the placement of numbers is smaller than the dimension of the representation $(\lambda_2, \lambda_3, \ldots)$ of $S_{n-\lambda_1}$. This is at most $\sqrt{(n-\lambda_1)!}$.

The value of the characters of S_n are integers. While much is known about the characters, there is nothing like a useable formula presently available except in special cases. We will make heavy use of (5.6): The character of the irreducible representation of S_n corresponding to the partition λ, evaluated at a transposition τ satisfies

(5.6) $$\frac{\chi_\lambda(\tau)}{d_\lambda} = \frac{1}{n(n-1)} \sum_{j=1}^{m} [(\lambda_j - j)(\lambda_j - j + 1) - j(j-1)] .$$

This formula appears in early work of Frobenius. An accessible proof in modern notation appears in Ingram (1960).

The final fact needed about the representation of S_n is called the branching theorem. Consider ρ_λ a representation of S_n. The subgroup S_{n-1} sits in S_n as all permutations fixing 1. The representation ρ_λ can be restricted to S_{n-1}. This restriction is no longer irreducible but decomposes into irreducible representations of S_{n-1}.

(5.7) (Branching Theorem). The restriction of ρ_λ to S_{n-1} decomposes into irreducible representations of S_{n-1} associated to all partitions obtained from λ by removing a single box from the Young diagram of λ as to have a proper partition of $n-1$. In particular, no two constituents of the restriction are equivalent.

Remarks. For example, the representation $\lambda = (3,2,2,1)$ decomposes into $(2,2,2,1)$, $(3,2,1,1)$, $(3,2,2)$. The branching theorem is proved in James (1978, Section 9). We begin to draw some corollaries relevant to the basic problem.

(5.8) Let P be a probability on S_n which is invariant under conjugation by S_{n-1}: $P(sts^{-1}) = P(t)$ for $t \in S_n$, $s \in S_{n-1}$. For any partition λ of n, there is a basis for the associated representation ρ_λ, independent of P, such that the Fourier transform

$$\hat{P}(\rho_\lambda) = \Sigma\, P(t)\rho_\lambda(t)$$

is a diagonal matrix with explicitly computable entries.

Proof: Using (5.6), the representation ρ_λ restricted to S_{n-1} splits into non-equivalent pieces $\lambda^1, \lambda^2, \ldots, \lambda^j$ say with each λ^i a partition of n-1 derived from λ by removing a single box. Choose a basis such that $\rho_\lambda(s)$ is block diagonal with j blocks for $s \in S_{n-1}$. For definiteness, suppose $j = 2$. Then for $s \in S_{n-1}$,

$$\rho(s) = \begin{pmatrix} \rho^1(s) & 0 \\ 0 & \rho^2(s) \end{pmatrix}.$$

The Fourier transform $\hat{P}$ may be written

$$\hat{P} = \begin{pmatrix} \hat{P}^1 & \hat{P}^2 \\ \hat{P}^3 & \hat{P}^4 \end{pmatrix}.$$

Invariance under conjugation yields, for $s \in S_{n-1}$

$$\rho(s)\hat{P} = \hat{P}\rho(s)$$

or

$$\begin{pmatrix} \rho^1(s)\hat{P}^1 & \rho^1(s)\hat{P}^2 \\ \rho^2(s)\hat{P}^2 & \rho^2(s)\hat{P}^4 \end{pmatrix} = \begin{pmatrix} \hat{P}^1\rho^1(s) & \hat{P}^2\rho^2(s) \\ \hat{P}^3\rho^1(s) & \hat{P}^4\rho^2(s) \end{pmatrix}.$$

Since ρ^1 and ρ^2 are non-isomorphic irreducible representations, Schur's lemma (Serre (1977, pg. 7)) implies $\hat{P}^2$ and $\hat{P}^3$ are zero. It further implies $\hat{P}^1$ and $\hat{P}^4$ are constant multiples of the identity. □

Remark. (5.7) was proved by Flatto, Odlyzko, and Wales (1985). More generally, if G is a finite group H a subgroup, and the restriction of any irreducible to H is multiplicity free, then the Fourier transform of all probabilities on G that are invariant under conjugation by H are simultaneously diagonalizable. It follows that all such probabilities commute with each other under convolution.

These ideas can be pushed somewhat further to yield a larger class of probabilities where Fourier analysis becomes tractable. For example, take $G = S_n$, $H = S_k \times S_{n-k}$. Consider probabilities invariant under conjugation by S_k and bi-invariant under S_{n-k}. An argument similar to the one above shows that here too the Fourier transform is diagonal.

In the language of shuffling cards, this invariance becomes the following: consider n cards face down in a row on the table. Suppose $k \leq n/2$. Remove the left most k cards and mix them thoroughly. Remove k cards at random from the remaining $n-k$ positions. Mix them thoroughly. Place the first k cards into the position of the second group. Place the second group of k cards into the first k positions. This completes a single shuffle.

It is possible to abstract somewhat further. See Greenhalgh (1987).

Returning to the problem at hand, Hansmartin Zeuner has suggested an elegant way to calculate the Fourier transform $\hat{Q}$. The idea is to use Schur's lemma as follows: The measure $\hat{Q}$ is supported on all transpositions not in S_{n-1}. The Fourier transform for the measure supported on all transpositions is a constant times the identity (by Schur's lemma). The Fourier transform of the measure supported on all transpositions inside S_{n-1} is diagonal with explicitly computable entries. Thus the difference between linear combinations of these measures is diagonal by subtraction. Here are some details.

If τ denotes a transposition, and ρ an irreducible representation

$$(5.9) \qquad \sum_{g \in S_n} \rho(\tau) = cI \quad \text{with} \quad c = \frac{\binom{n}{2}\chi_\rho(\tau)}{d_\rho} .$$

with $\chi_\rho(\tau) = \operatorname{Tr} \rho(\tau)$ -- the character of ρ at τ, and $d_\rho = \dim \rho$.

Indeed, the matrix M at the left of (5.9) is left unchanged by conjugation $\rho(s)M\,\rho(s^{-1}) = M$ for all $s \in G$. Schur's lemma shows it must be a constant times the identity. Taking traces of $M = cI$, and using the fact that any two transpositions have the same character yields (5.9).

Remark. If instead of randomly transposing with 1, we consider the measure which transposes arbitrary random pairs, then (5.9) gives its Fourier transform as a constant times the identity. Thus Fourier analysis is straightforward. Diaconis and Shahshahani (1981) carry out details.

Return to the problem at hand. Choose a basis so $\rho(\tau)$ is block diagonal for $\tau \in S_{n-1}$, with blocks $\rho^1(\tau), \ldots, \rho^j(\tau)$ say. The sum over τ in S_{n-1} is diagonal with the first d_{ρ^1} entries equal to $\binom{n-1}{2}\chi_{\rho^1}(\tau)/d_{\rho^1}$, the next d_{ρ^2}

entries equal to $\binom{n-1}{2}\chi_{\rho^2}(\tau)/d_{\rho^2}$, etc. Subtracting gives the following:

(5.10) Let Q be defined by (5.1). Fix a partition λ of n. Suppose it decomposes into $\lambda^1, \ldots, \lambda^j$ partitions of S_{n-1}, under the branching rule. Then in a suitable basis, $\hat{Q}(\rho)$ is diagonal, with diagonal entries in blocks of length d_{λ^i}, the ith entry being $\frac{1}{n}(\lambda_i - i+1)$.

Proof: By subtraction, the transform has as diagonal entries

$$\frac{1}{n} + \frac{n-1}{2}\frac{\chi_\rho(\tau)}{d_\rho} - \frac{(n-1)(n-2)\chi_{\rho^i}(\tau)}{2n\, d_{\rho^i}},$$

Using Frobenius' formula (5.6) this last expression simplifies to what is claimed.□

Remarks. An equivalent formula is derived by Flatto, Odlyzko, and Wales (1985). The argument given generalizes to any probability on a finite group G which is invariant under conjugation by a subgroup H. What is needed is that ρ restricted to H be multiplicity free, and that if the support of Q is closed up under conjugation by G, only elements in H are added.

The upper bound lemma can now be used to yield the following

$$\text{(5.11)} \qquad \|Q^{*k}-U\|^2 \leq \frac{1}{4}\sum_\lambda^* d_\lambda \sum_{\lambda^i} d_{\lambda^i}\left(\frac{1}{n}(\lambda(i) - i+1)\right)^{2k}$$

where the outer sum is over all partitions of n except (n), the inner sum is over partitions λ^i of $n-1$ derived from λ by removing a box in row i of λ, with row i having $\lambda(i)$ boxes.

To aid understanding of the analysis that follows, consider the lead term in (5.11). This corresponds to the partition $\lambda = (n-1,1)$ of n. Then $d_\lambda = n-1$. The branching theorem (5.7) says $\lambda^1 = n-2$, $\lambda^2 = n-1$, with corresponding dimensions $d_{\lambda^1} = n-2$, $d_{\lambda^2} = 1$. The term to be bounded is thus

$$(n-1)\left\{(n-2)\left(\frac{(n-1)-1+1}{n}\right)^{2k} + 1\left(\frac{1-2+1}{n}\right)^{2k}\right\}$$

$$= (n-1)(n-2)\left(1-\frac{1}{n}\right)^{2k}.$$

This last is asymptotically

$$e^{-\frac{2k}{n} + 2 \log n} .$$

If $k = n \log n + cn$, this is e^{-2c}. It turns out that this is the slowest term, other terms being geometrically smaller so that the whole sum in the upper bound lemma is bounded by this lead term.

To understand the behavior of most terms in the sum, suppose it could be shown that $\frac{(\lambda(i)-i+1)}{n} < c < 1$, for most partitions. Then the innermost term is bounded above by c^{2k} and can be removed from the sum. The inner sum of dimensions equals d_λ, and the sum of d^2 equals $n!$ Thus k must be chosen so large as to kill $c^{2k}\, n!$. It follows that k of order $n \log n$ will do.

The remaining details are straightforward but somewhat tedious. The argument follows the lines of Diaconis and Shahshahani (1981), and Diaconis (1987) where complete details are given for the virtually identical task of bounding the measure associated to random transpositions. Further details on the upper bound are omitted here.

C. A lower bound. Two techniques are available for a lower bound: guessing at a set where the variation distance is small, and using character theory. The first approach leads to looking at the number of fixed points after k steps. If k is small, there will be many cards surely not hit while a well mixed deck has one fixed point on average. This can be carried out much as in Diaconis and Shahshahani (1981).

The second approach has shown itself to be a versatile tool and will be worked through in detail for the present example. It begins by considering the slow term in the upper bound lemma. This came from the $n-1$ dimensional representation. To aid interpretability, consider the random variable $f(\pi)$ = # fixed points of π. Under the uniform distribution U

(5.12) $$E_U(f(\pi)) = 1 , \quad \mathrm{Var}_U(f(\pi)) = 1 .$$

These results are well known from classical work on the matching problem where it is shown that under U, $f(\pi)$ has an approximate Poisson(1) distribution.

Diaconis (1987, Chapter 7, Exercise 1) shows the first n moment of $f(\pi)$ equal the first n moments of Poisson(1).

To compute the mean and variance of f under the convolution measure, consider the n-dimensional representation ρ of S_n which assigns π to its associated permutation matrix $\rho(\pi)$. Clearly

$$f(\pi) = \mathrm{Tr}\,\rho(\pi) .$$

Now ρ decomposes into two invariant irreducible pieces, the trivial and n-1 dimensional representations. Since the trace is unchanged by change of basis, choose an orthogonal basis for each invariant subspace.

$$(5.13)\quad E_k(f(\pi)) = \Sigma\, \mathrm{Tr}(\rho(\pi))P^{*k}(\pi) = \mathrm{Tr}\,\Sigma\, \rho(\pi)P^{*k}(\pi) = \mathrm{Tr}\begin{Bmatrix} 1 & 0 \\ 0 & \hat{P}^k(n-1,1) \end{Bmatrix}$$

$$= 1 + (n-2)(1-\frac{1}{n})^k$$

For $k = n\log n + cn$, $E_k = 1 + e^{-c}(1+O(\frac{1}{n}))$ which is large for $c \ll 0$.

This is not enough to derive a lower bound for the total variation distance because the function $f(\pi)$ isn't bounded. To go further, a variance is needed. The trick here is to use the fact that the character of a tensor product is the pointwise product of the character (Serre (1977, pg. 9)). We want to compute the expectation of $\{\mathrm{Tr}\,\rho(\pi)\}^2$. The tensor product of the n-dimensional representation decomposes into the direct sum of irreducibles. For the case at hand James-Kerber (1981, 2.9.16) yields

$$(5.14)\qquad [\mathrm{Tr}(\rho(\pi))]^2 = 2 + 3\chi_{n-1,1}(\pi) + \chi_{n-2,2}(\pi) + \chi_{n-2,1,1}(\pi) ,$$

with $\chi_\lambda(\pi)$ being the character of the irreducible representation determined by λ at the permutation π. Thus, the computation of the quadratic term on the left has been linearized. For the pieces involved, the branching theorem shows how the average of the characters summed over Q^{*k} split into diagonal matrices. This and the dimension formulas give the following

	λ	decomposition	
	n-1,1	n-2,1	n-1
dim	n-1	n-2	1
diag		$(1-\frac{1}{n})^k$	0
	n-2,2	n-3,2	n-2,1
dim	$\frac{n(n-3)}{2}$	$\frac{(n-1)(n-4)}{2}$	n-2
diag		$(1-\frac{2}{n})^k$	$(\frac{1}{n})^k$
	n-2,1,1	n-3,1,1	n-2,1
dim	$\frac{(n-1)(n-2)}{2}$	$\frac{(n-2)(n-3)}{2}$	n-2
diag		$(1-\frac{2}{n})^k$	$(-\frac{1}{n})^k$

From these computations,

$$E_k\{f(\pi)^2\} = 2 + 3\{(n-2)(1-\tfrac{1}{n})^k\} + \{\tfrac{(n-1)(n-4)}{2}(1-\tfrac{2}{n})^k + (n-2)(\tfrac{1}{n})^k\}$$
$$+ \{\tfrac{(n-2)(n-3)}{2}(1-\tfrac{2}{n})^k + (n-2)(\tfrac{-1}{n})^k\} .$$

For $k = n \log n + cn$, straightforward calculus shows

$$E_k(f) = 1 + e^{-c}(1+O(\tfrac{1}{n})) \qquad \mathrm{Var}_k(f) = 1 + e^{-c}(1+O(\tfrac{1}{n}))$$

with all error terms holding uniformly for $c \in [-\log n, \log n]$.

Now, Chebychev's inequality implies that

$$Q^{*k}\{\pi : |f(\pi)-E_k| > a\sqrt{\mathrm{Var}_k}\} \le \frac{1}{a^2} .$$

So, for $-c$ large, Q^{*k} is supported on $I_a^d = e^{-c} \pm ae^{-c/2}$ with probability $1/a^2$. On the other hand, the Poisson approximation to the number of fixed points of a random permutation under the uniform distribution U (or Chebychev's inequality) shows that under U, I_a has vanishingly small probability for fixed a, and $-c$ large. This implies the lower bound and completes the proof of the theorem. □

Remarks. This argument was given here to show that the arguments of Diaconis and Shahshahani (1981) apply in some generality. Chapter 3D of Diaconis (1987) provides a host of applications of the same analysis to problems in coding, Radon transforms, combinatorial group theory, and elsewhere. The extensions indicated here carry over to these problems.

It is possible to extend yet further. The measures in Diaconis and Shahshahani were constant on conjugacy classes. The measures here are invariant under conjugation by S_{n-1}. A similar analysis can be carried out for measures invariant under conjugation by S_{n-k}, and bi-invariant under S_k. The Fourier transforms are again diagonal.

Similar results seem to hold for the orthogonal group O_n, and some other cases. Greenhalgh gives conditions on a group, and subgroups insuring that suitably invariant measures have diagonal Fourier transforms: If $G \supset H \supset K$, then all probabilities on G invariant under conjugation by H and bi-invariant under K have simultaneously diagonalizable Fourier transforms if and only if any irreducible representation ρ of H which restricts to the identity matrix on K is multiplicity free when induced up to G.

6. Factorization Problems.

When can a probability P be factored as $P_1 * P_2$? Such factorizations are a crucial element of algorithms to generate pseudo-random integers, permutations, and other group valued random variables on a computer. A study of factorization illuminates the question of when is more mixing better? We will construct measures P and Q on the symmetric group which satisfy $P*P = U$, with U the uniform distribution, but $P*Q*P*Q \ldots *P*Q$ is always nonuniform. Here mixing twice with P is much better than adding additional randomization, mixing first with P, then with Q, etc.

The study of factorizations can lead to elegant characterizations. For example, when does there exist a probability P, non-uniform, such that $P*P = U$? In other words, when is there a non-trivial square root of the uniform distribution?

We will show such probabilities exist if and only if the underlying group G is not Abelian, nor the product of the eight element quaternion group and product of two element groups -- Theorem 1 below. Most of these applications involve the machinery of group representations developed in Section 4 above.

A. Example. Generating random permutations.

For generating random bridge hands, Monte Carlo investigation of rank tests in statistics, and other applications, a source of pseudo-random permutations of n objects is useful. If n is small, a useful approach is to set up a 1-1 correspondence between the integers from 1 to $n!$ and permutations and then use a source of pseudo-random integers. The factorial number system is sometimes used for this purpose, see pages 64 and 192 of Knuth (1981). For larger n, like 52, the most frequently used algorithm involves a factorization of the uniform distribution. Informally, at the ith stage a random integer J_i between i and n is chosen and i and J_i are transposed. Call the probability distribution at the ith stage P_i, it will be shown below that $P_1 * P_2, \ldots, * P_{n-1}$ is a factorization of the uniform distribution. Further discussion of this algorithm is on pp. 139-141 of Knuth (1981). The factorization has recently been applied in the theoretical problem of finding the order of a random permutation by Bovey (1980). It also forms the basis of fast algorithms for manipulating permutations. See Furst et al (1980). The following algorithm abstracts the idea to any finite group.

Subgroup algorithm. Let G be a finite group. Let $G_0 = G \supset G_1 \supset \ldots \supset G_r$ be a nested chain of subgroups, not necessarily normal. Let C_i be coset representatives for G_{i+1} in G_i, $0 \leq i < r$. Clearly G can be represented as $G \cong C_0 \times C_1 \times \ldots \times C_{r-1} \times G_r$ in the sense that each $g \in G$ has a unique representation as $g_0 g_1 \ldots g_{r-1} g_r$ with $g_i \in C_i$ and $g_r \in G_r$. Let P_i be the uniform distribution on C_i and G_r respectively. The convolution $P_1 * P_2 * \ldots * P_r$ is then a factorization of the uniform distribution on G.

Specializing to the symmetric group, consider the chain $S_n \supset S_{n-1} \supset S_{n-2} \supset \ldots \supset \{id\}$. Here S_{n-i} is represented as the set of permutations of n letters that fix the first i letters. Then, coset representatives C_i

can be chosen as the set of transpositions transposing i and letters larger than i. The subgroup algorithm suggests a class of algorithms that interpolate between the factorial number system and random transpositions: Let $S_n \supset S_{n_1} \supset \ldots \supset S_{n_r}$ with $n > n_1 > \ldots > n_r$. Here the size of the cosets C_i are allowed to get large and a variant of the factorial number systems permits choice of a random coset element from a random integer. For example, consider the chain $S_n \supset S_{n-2} \supset S_{n-4} \supset \ldots \supset \{id\}$. Coset representatives for $S_{n-2(i+1)}$ in S_{n-2i} are permutations bringing a pair of elements between $i+1$ and n into positions $i+1$ and $i+2$. These permutations may be ordered lexographically, setting up a 1-1 correspondence between them and numbers $1, 2, \ldots, (n-i)(n-i-1)$. An advantage of this method is that it requires fewer calls to the random number generator.

The subgroup algorithm can be used to generate random positions in the currently popular Rubic's cube puzzle. Diaconis and Shahshahani (1987) give a continuous version of the algorithm and show how it gives the fastest algorithm for generating random orthogonal, unitary, or symplectic matrices.

A different application of the factorization suggested by the subgroup algorithm is to computer generation of pseudo-random integers (mod N). Given two sources of pseudo-random integers X and Y, computer scientists sometimes form a new sequence $Z = X+Y \pmod N$. Knuth (1981, p. 631) contains a discussion of work by Marsaglia and others. Solomon and Brown (1980) show that this procedure brings Z closer to uniform. Marshall and Olkin (1980, p. 383) generalizes this to any finite group. It is natural to seek distributions of X and Y such that Z is exactly uniform. The subgroup algorithm does this when N is composite. For N prime, see Lemma 6.2 below. These ideas can be used to generate truly large random integers stored in mixed radix arithmetic.

B. <u>More mixing is not necessarily better.</u>

We begin by constructing a non-uniform probability P on the symmetric group S_n with the property that $P*P = U$. To do this, define a function $f(\pi)$ on S_n by defining its Fourier transform as

$$\hat{f}(\rho) = \begin{cases} 0 & \text{if } \rho \text{ is not the } n-1 \text{ dimensional representation} \\ N & \text{at the } n-1 \text{ dimensional representation} \end{cases}$$

where N is a nilpotent matrix with a single 1 in the $(1,n-1)$ position and zeroes elsewhere. Thus N^2 is the zero matrix.

Now f is defined by the inversion theorem as

$$f(\pi) = \frac{(n-1)}{n!} \rho(\pi^{-1})_{n-1,n-1} .$$

Now, for any fixed $\varepsilon \leq 1/n-1$, the measure

$$P(\pi) = U(\pi) + \varepsilon f(\pi)$$

is a probability which is not uniform but satisfies $P*P = U$. To see this last, compute Fourier transforms at any irreducible representation.

Next note that if Q is any probability on S_n, $P*Q$ has as its Fourier transform

$$1 \quad \text{at the trivial representation}$$

$$\begin{pmatrix} \hat{Q}_{n-1,1} & \cdots & \hat{Q}_{n-1,n-1} \\ 0 & \cdots & 0 \\ 0 & \cdots & 0 \end{pmatrix} \quad \text{at the } (n-1) \text{ dimensional representation}$$

$$0 \quad \text{at all other representations.}$$

It follows that if $\hat{Q}_{n-1,1} \neq 0$ (and thus for most probabilities Q)

$$(P*Q)^{*k} \neq U \quad \text{for any } k .$$

The counterexample constructed here is fairly explicit; it might be useful to have an even more explicit probabilistic description.

Later in this section these ideas will be generalized to any finite group and a converse of sorts will be proved. The next example contains some easy variations.

C. A problem of Levy.

Throughout, P is a probability on a finite group G. The probability P is decomposable if P can be decomposed as a convolution $P = P_1 * P_2$ with P_i not fixed at a point. This definition rules out the trivial decomposition $P_1 = P * \delta_g$ with δ_g a point mass at group element g and $P_2 = \delta_{g^{-1}}$. Levy (1953) gave a nice example of a measure which is not decomposable: Take the group as S_3, fix p in $(0,1)$ and let P put mass p on the identity and mass $(1-p)$ on a 3-cycle. It is not hard to show that the set S of support points of P is not of the form $S = S_1S_2$ with S_i of cardinality 2 or more, so P is not decomposable. The following result shows that if $P \ll dg$ with a positive continuous density, then P is decomposable. The idea of the proof is broadly useful.

THEOREM 6.1. Let G be compact; let $P = f dg$ with $f > \varepsilon > 0$. Then P is decomposable.

Proof: Suppose $f > \varepsilon > 0$. Let probability measures P_i be defined by

$$P_1 = \frac{1}{1+\varepsilon_1}\{f + \varepsilon_1\}dg \ , \quad P_2 = \frac{1}{1+\varepsilon_2}\{\delta_{id} + \varepsilon_2 dg\} \ ,$$

with ε_i chosen so $\varepsilon_1 + \varepsilon_2 + \varepsilon_1\varepsilon_2 = 0$; e.g., $\varepsilon_1 = -\varepsilon_2/(1+\varepsilon_2)$ and ε_2 chosen positive but so small that $P_i \geq 0$. Then

$$P_1 * P_2 = f dg + \{\varepsilon_1 + \varepsilon_2 + \varepsilon_1\varepsilon_2\}dg = f dg \ . \qquad \square$$

Remarks. In the case of finite groups, this gives an easy proof of a theorem of P. J. Cohen (1959). Cohen showed that if the density f is continuous, then the measures P_i can be chosen to have densities. Note that in our construction P_2 is not absolutely continuous with respect to dg when G is infinite. Cohen gives an example of a probability density on a compact subset of $\mathbb{R}$ which cannot be written as a convolution of two probabilities with densities. An earlier example of Levy and a review of the literature on $\mathbb{R}$ appear in Chapter 6 of Lukacs (1970). Lewis (1967) shows that the uniform distribution on $[-0,1]$ cannot be written as a convolution of two probabilities with densities. It is well known that the convolution of singular measures can have a density. See Rubin (1967) and Hewitt and Zukerman (1966) for some examples.

D. Decomposing the uniform.

Turn now to decomposing the uniform distribution U on a compact group. The subgroup algorithm gives any easy method for decomposing the uniform distribution on a finite group. Consideration of the circle group and the subgroup of kth roots of unity suggests that the result generalizes:

Lemma 6.1. Let G be a compact, Polish group with a closed subgroup H. Then, the uniform distribution is decomposable, with non-uniform factors.

Proof: Let $\pi: G \to G/H$ be the canonical map. Let Q be the image of the uniform distribution under π. Take a measurable inverse $\phi: G/H \to G$ with the property that $\pi\,\phi\{gH\} = \{gH\}$. The existence of ϕ under our hypothesis follows from Theorem 1 of Bondar (1976). Let P_1 be the image of Q under ϕ. Let P_2 be the uniform distribution on H. To prove that $P_1 * P_2$ is uniform, consider any continuous function f on G. By definition

$$\int_G f(g)\, P_1 * P_2(dg) = \int_G\int_G f(g_1 g_2)\, P_1(dg_1) P_2(dg_2) = \int_{\phi(G/H)}\int_H f(g_1 g_2)\, P_1(dg_1) P_2(dg_2)$$
$$= \int_G f(g)\, U(dg) .$$

The final equality in the display follows from Theorem 2 in Bondar. □

We next show how to decompose the uniform distribution on groups with no proper subgroups: the integers mod a prime. It is easy to see that the uniform distribution is not semi-decomposable on Z_2 or Z_3. One approach uses the fact that $1+z$ and $1+z+z^2$ are irreducible over the reals. Factorization of U would lead to a factorization of the associated polynomial. Note that Levy's example on S_3 discussed above is really Z_3 since his basic measure lives on the identity and the two 3-cycles.

Lemma 6.2. Let $p \geq 5$ be prime. Let Z_p be the integers mod p. Then the uniform distribution is decomposable into non-uniform factors.

Proof: For $i = 1, 2, \ldots, \frac{p-1}{2}$, let a_i, b_i be determined by

$$a_i + 2b_i = 1 \ , \quad a_i + 2b_i \cos\left(\frac{2\pi i^2}{p}\right) = 0 \ .$$

Noting that $\cos(\frac{2\pi i^2}{p}) \neq 1$ for i in the indicated range;

$$b_i = \left\{2\left(1 - \cos\frac{2\pi i^2}{p}\right)\right\}^{-1} \ , \quad a_i = -\cos\left(\frac{2\pi i^2}{p}\right)\Big/\left(1 - \cos\frac{2\pi i^2}{p}\right) \ .$$

Define signed measures Q_i on Z_p by

$$Q_i(0) = a_i \ , \quad Q_i(i) = Q_i(-i) = b_i \ , \quad Q_i(j) = 0 \ \text{ otherwise} \ .$$

The argument depends on the discrete Fourier transform of a measure. If P is a measure on Z_p and $k \in Z_p$, define

$$\hat{P}(k) = \frac{1}{p}\sum_{j=0}^{p-1} P(j)e^{2\pi ijk/p} \ .$$

For the uniform distribution,

$$\hat{U}(k) = \begin{cases} 1 & \text{if } k = 0 \\ 0 & \text{otherwise} \end{cases} \ .$$

It is easy to check that for $k \neq 0$, $\hat{Q}_k(\pm k) = 0$, $\hat{Q}_k(0) = 1$. Now let signed measures R_1 and R_2 be defined by

$$R_1 = \mathop{*}_{i=1}^{a} Q_i \ , \quad R_2 = \mathop{*}_{i=a+1}^{(p-1)/2} Q_i \quad \text{for fixed } 1 \leq a \leq (p-1)/2 \ .$$

Finally, for sufficiently small ε the measures $U + \varepsilon R_1$ and $U + \varepsilon R_2$ are positive measures and can be normed to be probabilities, say P_1 and P_2. We claim $U = P_1 * P_2$. Indeed, for $k \neq 0$, $\widehat{P_1 * P_2}(k) = \hat{P}_1(k)\hat{P}_2(k) = 0$. To show that the decomposition is non-trivial, it suffices to show that R_i are non-zero, $i = 1,2$. This follows from the fact that for $k \neq j$, $\hat{Q}_k(j) \neq 0$. Indeed,

$$\begin{aligned} \hat{Q}_k(j) &= a_k + 2b_k \cos\left(\frac{2\pi jk}{p}\right) \\ &= \frac{1}{1 - \cos(2\pi k^2)}\left\{-\cos\left(\frac{2\pi k^2}{p}\right) + \cos\left(\frac{2\pi jk}{p}\right)\right\} . \end{aligned}$$

This is zero if and only if $j = k$. □

Remarks. Factoring the uniform distribution on Z_p is sufficiently close to some classical factorization results to warrant discussion. A well known elementary probability problem argues that it is impossible to load two dice so that the sum is uniform. More generally, Dudewicz and Dann (1972) show that it is impossible to find probabilities P_1 and P_2 on the set $\{1,2,\ldots,n\}$ such that P_1*P_2 is the uniform distribution on $\{2,\ldots,2n\}$. A related result asks for a decomposition of the uniform distribution on the set 0, 1, 2, ..., N. Lukacs (1970, pp. 182-183) reviews the literature on this problem. He shows factorization is possible when, and only when, N is prime. The difference between the three results is this: In Lemma 6.2, and in the subgroup factorization, addition is (mod n). In the dice result, both factors must be supported on $\{1,\ldots,n\}$ while the uniform distribution is on $\{2,\ldots,n\}$. In the results reported in Lukacs, the factors are permitted to have arbitrary support.

The results above can be combined into the following.

THEOREM 6.2. The uniform distribution on a compact Polish group G is decomposable into non-uniform factors unless G is Z_2 or Z_3.

Proof: For finite groups, Lemma 6.2 and the subgroup algorithm prove the claim, since a finite group with no proper subgroups is the residues of a prime. We now argue that every infinite compact group contains a closed non-trivial subgroup. A topological group has no small subgroups (NSS) if there exists a neighborhood U of the identity such that the only subgroup in U is {id}. Clearly, a group which has small subgroups contains non-trivial closed subgroups. A famous theorem of Gleason (1952) implies that a group with NSS is a Lie group. The structure of compact Lie groups is well known; see, for example, Chapter 11 of Pontryagin (1966): If G is Abelian, then the connected component of the identity is a finite dimensional torus which certainly has non-trivial closed subgroups, hence G does. If G is not Abelian, then its maximal torus is a non-trivial closed subgroup. □

E. Square roots.

On a compact Abelian group the factorization $U = P*P$ is impossible unless P is uniform. This follows because all irreducible representations are one-dimensional and $0 = \hat{U}(\rho) = \widehat{P*P}(\rho) = \hat{P}(\rho)^2$ implies $\hat{P}(\rho) = 0$. For non-Abelian groups, things are more complex. The following relates to B above.

Example. On S_3 a square root P of U can be defined as follows: using cycle notation for permutations let

$$P(id) = \frac{1}{6},\ P(12) = \frac{1}{6},\ P(23) = \frac{1}{6} + h,\ P(31) = \frac{1}{6} - h,\ P(123) = \frac{1}{6} - h,\ P(132) = \frac{1}{6} + h\ ,$$

for any h with $0 \leq h \leq \frac{1}{6}$.

To motivate Theorem 6.3 let us explain how this example was found. We seek a probability P on S_3 such that $\hat{P}(\rho)^2 = 0$ for each non-trivial irreducible representation ρ. Let us find a function f on S_3 such that $\hat{f}(\rho) = 0$ for all irreducible ρ and then $P(\pi) = \frac{1}{6} + \varepsilon f(\pi)$, with ε chosen small enough that $P(\pi) \geq 0$ will do the job. There are three irreducible representations of S_3, the trivial representation ρ_t, the alternating representation ρ_a, and a two-dimensional representation ρ_2. If $\hat{f}(\rho_2)$ is a non-zero nilpotent matrix $\hat{f}(\rho_2) = \begin{pmatrix} 0 & * \\ 0 & 0 \end{pmatrix}$ and $\hat{f}(\rho_t) = \hat{f}(\rho_a) = 0$, then $\hat{f}(\rho) \equiv 0$. This gives five linear relations for the six numbers $f(\pi)$. The example above resulted from solving these equations. The following theorem gives a generalization.

THEOREM 6.3. Let G be a compact, non-commutative group. The following conditions are equivalent.

a) There is a probability measure $P \neq U$ such that $P*P = U$.

b) There is an irreducible representation ρ of G such that the algebra $R_\rho = \{\Sigma_{g \in G}\ \mathbb{R}\rho(g)\}$ contains nilpotent elements.

Remark. The quaternions $\pm 1, \pm i, \pm j, \pm k$ form a finite non-commutative group such that the uniform distribution does not have a non-trivial square root. This follows from Theorem 6.4 below which identifies all finite groups satisfying condition b).

The proof of Theorem 6.3 requires some notation. Throughout we assume that all irreducible representations are given by unitary matrices. If ρ is a representation, let $\tilde{\rho}(g)$ be defined as $\bar{\rho}(g) = \rho(g^{-1})'$. The following lemma is used in the proof of Theorem 6.3.

Lemma 6.3. Let μ be a bounded measure on a compact group G. Then μ is real if and only if $\hat{\mu}(\tilde{\rho}) = \overline{\hat{\mu}(\rho)}$ for every irreducible ρ.

Proof: If μ is real, then

$$\hat{\mu}(\tilde{\rho}_{ij}) = \int \bar{\rho}_{ij}(g)\, \mu(dg) = \overline{\hat{\mu}(\rho_{ij})} .$$

Conversely, suppose μ is a measure such that $\hat{\mu}(\bar{\rho}) = \overline{\hat{\mu}(\rho)}$. This means

$$0 = \int \bar{\rho}_{ij}(g)\, \mu(dg) - \int \bar{\rho}_{ij}(g)\, \bar{\mu}(dg)$$

or

$$0 = \int \rho_{ij}(g)\, \bar{\mu}(dg) - \int \rho_{ij}(g)\, \mu(dg) .$$

Since this holds for every irreducible ρ, the Peter-Weyl theorem implies that the set function $\bar{\mu} - \mu$ is zero, so μ is real. □

Proof of Theorem 6.3: If $U = P*P$, then $\hat{P}(\rho)^2 = 0$ and $\hat{P}(\rho) \neq 0$ for some ρ^* because $P \neq U$. Thus $R_{\rho*}$ has nilpotents. Conversely, let $\gamma_1 \in R_{\rho*}$ be nilpotent. If $\gamma_1^n = 0$ and n is the smallest such power, then set $\gamma = \gamma_1^{n-1}$. This is non-zero and $\gamma^2 = 0$. Define a continuous function f on G as follows. Set $\hat{f}(\rho) = 0$ if $\rho \neq \rho^*$ or $\tilde{\rho}^*$, $\hat{f}(\rho^*) = \gamma$, and if ρ^* is not equivalent to $\tilde{\rho}$, $\hat{f}(\tilde{\rho}^*) = \bar{\gamma}$. This defines a non-zero continuous function f through the Peter-Weyl theorem. Because of Lemma 6.3, f is real. Clearly, $f(\rho)^2 = 0$ for all irreducible ρ. It follows that for ε suitably small, $P = (1 + \varepsilon f(g))dg$ is a probability satisfying $P*P = U$. □

A sufficient condition for Theorem 6.3 is that G have a real representation of dimension 2 or greater: If ρ^* is an n-dimensional real representation, let $f(g) = \varepsilon\, \rho^*_{1n}(g)$. Then by the Schur orthogonality relations, for any $\rho \neq \rho^*$, $f(\rho) = 0$. Also, Schur's relations imply $\hat{f}(\rho^*)$ is an $n \times n$ matrix which is zero except that the 1, n entry is $\varepsilon \int \rho_{1n}^2\, dg > 0$. Thus $\hat{f}(\rho^*)^2 = 0$. Let a probability P be defined by $P = (1+\varepsilon f)dg$, with ε chosen so that P is positive. Then $\hat{P*P}(\rho) = \hat{U}(\rho)$ for all irreducible representations. As an example, the adjoint representation of a compact simple Lie group has a basis with respect to which it is real orthogonal. Thus, the group $SO(n)$ of proper notations for $n = 3$, and $n \geq 5$ admits a square root of U.

The next result classifies all finite groups such that the uniform distribution has a non-trivial square root.

THEOREM 6.4. The uniform distribution on a finite group G is decomposable if and only if G is not Abelian or the product of the quarternions and a finite number of two-element groups.

Proof: It was argued above that Abelian groups do not admit a non-trivial square-root of the uniform distribution. In light of Theorem 6.3, the non-Abelian groups with the property that $R_\rho(G)$ has no nilpotents must be classified. We will use a lemma of Sehgal (1975). Some notation is needed. Let Q denote the rational numbers, and let $Q(G)$, the rational group ring denote the set of formal linear combinations of elements of G with rational coefficients. A non-Abelian group in which every subgroup is normal is called Hamiltonian. Theorem 12.5.4 of Hall (1959) shows that every Hamiltonian group is of the form $G = A \times B \times H$, where A is an Abelian group of odd order, B is a product of a finite number of two-element groups, and H is the eight element group of quarternions $\{\pm 1, \pm i, \pm j, \pm k\}$. The following lemma has been abstracted from Sehgal (1975). The result also appears in Pascaud (1973).

Lemma 6.4. If $Q(G)$ has no nilpotents, then G is Hamiltonian.

Proof: Observe first that if R is any ring with unit and no nilpotents, then an idempotent $e^2 = e$ in R commutes with every element of R. Indeed, the equation $0 = [er(1-e)]^2$ implies $er(1-e) = 0$, so $er = ere$. Similarly, $re = ere = er$. Now let $R = Q(G)$, let H be a subgroup of G , and set $e = \frac{1}{|H|} \Sigma_{h\in H} h$. It follows that for each $g \in G$, $geg^{-1} = e$, and this implies that for each $h \in H$, $ghg^{-1} \in H$, so H is normal. □

Proof of Theorem 6.4: Map $Q(G)$ into $R_\rho(G)$ by mapping $g \to \rho(g)$ and extending by linearity. This is an algebra homomorphism. We thus get a map from $Q(G)$ into $\Pi_\rho R_\rho(G)$. From Proposition 10 of Serre (1977) this map is 1-1. Since no $R_\rho(G)$ has nilpotents, neither does $Q(G)$. Lemma 6.4 implies that G has the form $G = A \times B \times H$ where A is an Abelian group of odd order. If A is not zero, choose a character χ taking atleast one complex value. Let ρ be the irreducible representation of H which sends $i \to \begin{bmatrix} 0 & i \\ i & 0 \end{bmatrix}$ and $j \to \begin{bmatrix} 0 & -1 \\ 1 & 0 \end{bmatrix}$. Then $\chi \otimes \rho$ is an irreducible two-dimensional representation so $R_{\chi\otimes\rho}(G)$ consists of all 2×2 matrices, with complex entries, and so contains nilpotents. □

Theorem 6.4 is also true for compact groups, the difference being that an infinite product of two-element groups is allowed in addition to a single copy of the eight-element quarternions. The difference in difficulty of proof is substantial; a non-trivial extension problem must be solved in classifying compact groups with all closed subgroups normal. See Diaconis and Shahshahani (1986). The present section is drawn from an earlier unpublished written report jointly with Shahshahani. It shows how Fourier analysis can be used without getting bogged down in too many analytical details.

7. Other Material.

These lectures are an introduction to a rapidly growing field. The tools and problems can be used and applied in many other areas. This brief section gives some pointers to the literature.

A. Other techniques for random walk. The present write-up emphasizes the use of Fourier analysis to solve random walk problems. There are two other techniques which give precise results for this type of problem. The most widely known technique is coupling where one constructs a pair of processes evolving marginally like the random walk. One is in its stationary distribution. The second starts deterministically. If T is the first time the processes meet, the coupling inequality says

$$\|P^{*k}-U\| \leq P\{T>k\} .$$

It is sometimes possible to find tractable couplings so that the stopping time can be bounded or approximated, and for which the inequality above is roughly an equality.

The best introduction to this subject in the random walk context is Aldous (1983a). A satisfying theoretical result due to Griffeath (1975, 1978) and Pitman (1976) says that a maximal coupling always exists which achieves equality in the coupling bound. Of course, this is only useful in theory. The construction involves knowing all sorts of things about the process and is useless for practical purposes.

A second tool for bounding rates of convergence is the technique of strong uniform times. Here one constructs a stopping time T with the property that the random walk is uniformly distributed when stopped at T, even given the value of T:

$$P\{S_n=s|T=n\} = 1/|G| .$$

One then has the same inequality as above, bounding the variation distance.

Strong uniform times are introduced and applied by Aldous and Diaconis (1986, 1987). They prove that there is always a strong uniform time achieving equality for a stronger notion of distance: $\max_s(1-|G|Q^{*k}(s))$.

The value of strong uniform times comes from the possibility of finding tractable times which permit analysis. Diaconis (1987a, Chapter 4) gives many examples. Diaconis (1987b) begins to build a general theory that unifies the examples.

Coupling and strong uniform times are "pure probability" techniques. The Fourier methods are purely analytic. At present, each has success stories the other techniques can't handle. Diaconis and Aldous (1987) review the connections.

B. Statistics on groups. There are a number of practical problems where data is collected from naturally occurring processes on groups or homogeneous spaces. Such data can be analyzed directly using tests or models derived from the structure of the underlying group. Diaconis (1987) develops many techniques and examples in detail. Here is a brief "advertisement" for these ideas.

An example of such data is a collection of rankings: A panel of experts may each rank 4 wines. A community may rank candidates for leadership or alternative energy sources. A statistical regression problem may admit variables in some order; under bootstrapping, a number of orders may appear.

Each of these problems leads to data in the form of a function $f(\pi)$, for π in S_n - the permutation group on n letters. In the wine tasting example $n = 4$, and $f(\pi)$ is the number (or proportion) of experts choosing the ranking π.

To analyze such data one looks at obvious averages -- which wine has the highest (lowest) average score. Can the rankings be clustered into meaningful groups? Are the rankings usefully related to covariates such as sex, nationality, etc.

Data can come in only partially ranked form (rank your favorite k of this years n movies). This leads to data naturally regarded as living on the homogeneous space S_n/S_{n-k} with S_{n-k} the subgroup of permutations fixing $1, 2, \ldots, k$.

For a continuous example, data is sometimes collected on the orientation of objects. This may be regarded as data with values in the orthogonal group $SO(3)$. Directional data leads to points on the sphere. Higher dimensional spherical data arises in testing for normality with many small samples having different, unknown means and variances.

A variety of special purpose tools have been developed for individual problems. Three general approaches are suggested in Diaconis (1987a). These involve notions of distance (metrics), a version of spectral analysis, and a technique for building probability models. These are briefly described in the next three paragraphs.

C. <u>Metrics</u>. Data analysis can begin by defining a notion of distance between observables. For data on a group G, let (ρ,V) be a representation and $\| \ \|$ a matrix norm on $GL(V)$. Then

$$d(s,t) = \|\rho(s)-\rho(t)\|$$

defines a two-sided invariant distance on G.

Alternatively, if V has an inner product, ρ is unitary, and $v \in V$, then

$$d(s,t) = \|\rho(s)^{-1}v-\rho(t)^{-1}v\|$$

is a left invariant metric.

These approaches, specialized to the symmetric group, give standard distances used by statisticians as measure of nonparametric association. These include Spearman's "rho" and "footrule" or Hamming distance. These facts and other topics discussed in this section are developed in Chapter 6 of Diaconis (1987a).

With a metric chosen, there are procedures available for most standard tasks: For example to test if two samples can reasonably be judged to come from the same population, look at the labels of close points. If the populations are different, close points will tend to belong to the same sample. A dozen other applications are illustrated in Chapter 6 of Diaconis (1987a).

This approach can also be developed for partially ranked data, and other homogeneous spaces. Critchlow (1985) develops theory, applications and examples. His book contains tables and computer programs for popular cases.

D. Spectral analysis. Let G be a finite group. The Fourier inversion theorem (4.1) can be interpreted as showing that the matrix entries of the irreducible representations $\rho_{ij}(\cdot)$, thought of as functions on G, form a natural basis for all functions on G. Thus, if $f(s)$ is a given function, it can be Fourier expanded as

$$f(s) = \Sigma\, c(i,j\,\rho)\,\rho_{ij}(s)$$

for some coefficients $c(i,j,\rho)$ depending on f.

When $G = Z_n$, this is the usual expansion of a function into sines and cosines familiar from the discrete Fourier transform. In this case, the largest coefficients are used to suggest approximations to f as a sum of a few simple periodic functions.

In the general case, the large coefficients may reveal a simple structure to f that is not obvious from direct inspection of the numbers $\{f(s)\}$, $s \in G$.

This is a rich idea which includes the classical analysis of designed experiments (ANOVA) as a special case. There one has data $f(x)$ indexed by a finite set X. There is a group of symmetries operating transitively on X. The space of all functions $L(X)$ gives a representation of G (indeed G acts on X and so on $L(X)$). This representation decomposes into irreducible subspaces. One can project a given $f \in L(X)$ into these subspaces (which often have convenient names and interpretations like "grand mean", "row effects" or "column effects").

Here, spectral analysis consists of the projection of f into the irreducible subspaces and the approximation of f by as few projections as give a reasonable fit. This includes the Fourier expansion as a special case.

Many specific examples are computed and discussed in Chapter 7 of Diaconis (1987a) and in Diaconis (1987b).

E. Building models. Let G be a finite group. Any positive probability P on G can be written as

$$P(s) = e^{\log P(s)} = e^{\Sigma c(i,j,\rho)\rho_{ij}(s)} .$$

Conversely, one can consider log-linear models defined in terms of the matrix $\rho_{ij}(\cdot)$.

Such models have been used in special cases by social scientists interested in ranking data. If ρ is the irreducible $n-1$ dimensional representation of S_n, and Θ is an $(n-1) \times (n-1)$ matrix, the probability

$$P(\pi) = ce^{\mathrm{Tr}(\Theta\rho(\pi))}$$

defines a "first order" model which specifies that a linear weight is assigned to each object in each position and the chance that π is chosen is a linear combination of these weights (on a log scale).

When specialized to the sphere, such models give the standard exponential families: the Fisher-von Mises distribution and Bingham's distribution. It seems natural to consider other representations and high order models. These ideas are developed more fully in Beran (19), Verducci (1983) and in Chapter 9 of Diaconis (1987a).

These are not the only ways of analyzing data on groups, but they do present a host of possibilities that are worth trying before developing more specialized procedures. They offer new applications for the powerful tools developed by group theorists. They suggest fresh questions and techniques within group theory.

It is still surprising that Fourier analysis is such an effective tool when working on the line. This surprise should not constrain the broad usefulness of Fourier techniques in more general problems.

References

Aldous, D. (1982). Markov chains with almost exponential hitting times. Stochastic Proc. Appl. 13, 305-310.

Aldous, D. (1983a). Random walk on finite groups and rapidly mixing Markov chains. In Seminaire de Probabilites XVII, 243-297. Lecture Notes in Mathematics 986.

Aldous, D. (1983b). Minimization algorithms and random walk on the d-cube. Ann. Prob. 11, 403-413.

Aldous, D. and Diaconis, P. (1986). Shuffling cards and stopping times. American Mathematical Monthly 93, 333-348.

Aldous, D. and Diaconis, P. (1987). Strong uniform times and finite random walks. Advances in Appl. Math. 8, 69-97.

Beran, R. (1979). Exponential models for directional data. Ann. Statist. 7, 1162-1178.

Bovey, J. D. (1980). An approximate probability distribution for the order of elements of the symmetric group. Bull. London Math. Soc. 12, 41-46.

Chung, F., Diaconis, P. and Graham, R. L. (1987). A random walk problem involving random number generation. To appear in Ann. Prob.

Cohen, P. J. (1959). Factorization in group algebras. Duke Math. J. 26, 199-205.

Critchlow, D. (1985). METRIC METHODS FOR ANALYZING PARTIALLY RANKED DATA. Lecture Notes in Statistics No. 34. Springer-Verlag, Berlin.

Curtis, C. W. and Reiner, I. (1982). REPRESENTATION OF FINITE GROUPS AND ASSOCIATIVE ALGEBRA, 2nd edition. Wiley, New York.

Diaconis, P. (1982). Lectures on the use of group representations in probability and statistics. Typed Lecture Notes, Department of Statistics, Harvard University.

Diaconis, P. (1987). GROUP REPRESENTATIONS IN PROBABILITY AND STATISTICS. Institute of Mathematical Statistics, Hayward.

Diaconis, P. and Shahshahani, M. (1981). Generating a random permutation with random transpositions. Z. Wahrscheinlichkeitstheorie verw. Gebiete 57, 159-179.

Diaconis, P. and Shahshahani, M. (1986a). Products of random matrices as they arise in the study of random walks on groups. Contemporary Mathematics 50, 183-195.

Diaconis, P. and Shahshahani, M. (1986b). On square roots of the uniform distribution on compact groups. Proc. American Math'l Soc. 98, 341-348.

Diaconis, P. and Shahshahani, M. (1987a). The subgroup algorithm for generating uniform random variables. Prob. in Engineering and Info. Sciences 1, 15-32.

Diaconis, P. and Shahshahani, M. (1987b). Time to reach stationarity in the Bernoulli-Laplace diffusion model. SIAM J. Math'l Analysis 18, 208-218.

Diaconis, P. and Stanley, R. (1986). Mathematical aspects of cooking potatoes. Unpublished manuscript.

Dies, J. E. (1983). CHAÎNES DE MARKOV SUR LES PERMUTATIONS. Lecture Notes in Math. 1010. Springer-Verlag, New York.

Flatto, L., Odlyzko, A. M. and Wales, D. B. (1985). Random shuffles and group representations. Ann. Prob. 13, 154-178.

Furst, M., Hopcroft, J. and Luka, E. (1980). Polynomial time algorithms for permutation groups. Proc. 21st FOCS I, 36-41.

Gleason, A. (1952). Groups without small subgroups. Amer. Math. 56, 193-212.

Greenhalgh, A. (1987). Ph.D. disseration, Department of Mathematics, Stanford University.

Griffeath, D. (1975). A maximal coupling for Markov chains. Z. Wahrscheinlichkeitstheorie verw. Gebiete 31, 95-100.

Griffeath, D. (1978). Coupling methods for Markov chains. In G. C. Rota (ed.) STUDIES IN PROBABILITY AND ERGODIC THEORY, 1-43.

Hall, M. (1959). THE THEORY OF GROUPS. MacMillan, New York.

Herstein, I. N. (1975). TOPICS IN ALGEBRA, 2nd edition. Wiley, New York.

Hewitt, E. and Zukerman, H. (1966). Singular measures with absolutely continuous convolution squares. Proc. Camb. Phil. Soc. 62, 399-420.

Ingram, R. E. (1950). Some characters of the symmetric group. Proc. Amer. Math. Soc. 1, 358-369.

James, G. D. (1978). THE REPRESENTATION THEORY OF THE SYMMETRIC GROUPS. Lecture Notes in Mathematics 682. Springer-Verlag, Berlin.

James, G. and Kerber, A. (1981). THE REPRESENTATION THEORY OF THE SYMMETRIC GROUP. Addison-Wesley, Reading, Massachusetts.

Knuth, D. (1981). THE ART OF COMPUTER PROGRAMMING. Vol. II, 2nd edition. Addison-Wesley, Menlo Park, California.

Levy, P. (1953). Premiers Elements de l'Arithmetique des Substitutions Aleatoires. C.R. Acad. Sci. 237, 1488-1489.

Lewis, T. (1967). The factorization of the rectangular distribution. J. Appl. Prob. 4, 529-542.

Lukacs, E. (1970). CHARACTERISTIC FUNCTIONS, 2nd edition. Griffin, London.

Mackey, G. (1978). UNITARY GROUP REPRESENTATIONS IN PHYSICS, PROBABILITY, AND NUMBER THEORY. Benjamin/Cummings.

Mackey, G. (1980). Harmonic analysis as the exploitation of symmetry. Bull. Amer. Math. Soc. 3, 543-697.

Marshall, A. W. and Olkin, I. (1979). INEQUALITIES: THEORY OF MAJORIZATION AND ITS APPLICATIONS. Academic Press, New York.

Matthews, P. (1985). Covering problems for random walks on spheres and finite groups. Ph.D. dissertation, Department of Statistics, Stanford University.

Matthews, P. (1987). Covering problems for Brownian motion on a sphere. To appear Ann. Prob.

Naimark, M. and Stern, A. (1982). THEORY OF GROUP REPRESENTATIONS. Springer-Verlag, New York.

Pascaud, J. (1973). Anneaux de groups réduits. C.R. Acad. Sci. Paris, Sér. A 277, 719-722.

Pemantle, R. (1987). An analysis of overhand shuffles. To appear Ann. Prob.

Pitman, J. W. (1976). On the coupling of Markov chains. Z. Wahrscheinlichkeitstheorie verw. Gebiete 35, 315-322.

Pontrijagin, L. S. (1966). TOPOLOGICAL GROUPS. Gordon and Breach.

Rotman, J. (1973). THE THEORY OF GROUPS: AN INTRODUCTION, 2nd edition. Allyn and Bacon, Boston.

Rubin, H. (1967). Supports of convolutions of identical distributions. Proc. Fifth Berkeley Symp. on Mathematics, Statistics, and Probability. Vol. 2 415-422.

Sehgal, S. K. (1975). Nilpotent elements in group rings. Manuscripta Math. 15, 65-80.

Serre, J. P. (1977). LINEAR REPRESENTATIONS OF FINITE GROUPS. Springer-Verlag, New York.

Suzuki, M. (1982). GROUP THEORY, I, II. Springer-Verlag, New York.

Verducci, J. (1982). Discriminating between two probabilities on the basis of ranked preferences. Ph.D. dissertation, Department of Statistics, Stanford University.

RANDOM FIELDS AND DIFFUSION PROCESSES

Hans Föllmer

H. FOLLMER : "RANDOM FIELDS AND DIFFUSION PROCESSES"

INTRODUCTION

In these lectures, a random field P will be a probability measure on a product space S^I where S is some state space, for example $S = \{-1, +1\}$ or $S = C[0,1]$, and where I is some countable set of sites, for example the d-dimensional lattice Z^d.

We take the point of view of R.L. Dobrushin: A random field will be specified, but not necessarily uniquely determined, by a system of conditional probabilities π_V, where V is a finite subset of I, and where $\pi_V(\cdot|\eta)$ is the conditional distribution on S^V given the configural η outside of V. In that case, P will also be called a Gibbs measure with respect to the local specification (π_V). P is called a Markov field if $\pi_V(\cdot|\eta)$ only depends on the values $\eta(i)$ for the sites i in some finite "boundary" of V.

We want to give an introduction to some of the connections between Gibbs measures and infinite dimensional stochastic processes, and in particular between Markov fields and Markov processes. These connections appear on different levels:

(A) Markov fields as Markov processes

(B) Markov fields as invariant measures of Markov processes

(C) Markov processes as Markov fields.

To begin with, any Markov field on $S^I = (S^J)^Z$ with $I = Z^d$ and $J = Z^{d-1}$ can be viewed as the law of a stochastic process with state space S^J. This process may or may not be a Markov process: this is the problem of the "global" Markov property.

(B) has been a central topic in the study of time evolution, of interacting particle systems, and has been studied in various different contexts; cf., for example, [Li]. We are going to look at (B) for a class of infinite dimensional diffusion processes $X = (X_i)_{i \in I}$ of the form

$$dX_i = dW_i + b_i(X(t),t))dt \quad (i \in I)$$

where (W_i) is a collection of independent Brownian motions. Under some bounds on the interaction in the drift terms, the time reversed process is again of this form, and there is an infinite dimensional analogue to the classical duality equation

$$b + \hat{b} = \nabla \log \rho$$

which relates forward drift, backward drift and the density of the process at any given time. In the infinite dimensional case, the density is replaced by the system of conditional densities at each site, i.e., by a local specification. This leads to the description of invariant measures of the process as Gibbs measures and often as Markov fields. As to (C), note that the distribution of the infinite dimensional diffusion process above is a probability measure on $C[0,1]^I$ and can thus be regarded as a random field with state space $S = C[0,1]$ at each site $i \in I$. If we determine its local specification then we can apply random field techniques to the diffusion, and this may be useful, e.g., in view of large deviations or a central limit theorem not in the time but in the space direction.

At the Ecole d'Eté de Probabilités de Saint-Flour, it seemed natural to assume a very strong background in Stochastic Analysis, and to take a more introductory approach to random fields. Thus we begin with a self contained introduction to Gibbs measures in Ch. I, with special emphasis on Dobrushin's contraction technique and on the probabilistic limit theorems which are behind thermodynamical qunatities like energy and entropy. These topics are well known, and some excellent introductions are available, e.g. [Pr]. But we want to discuss various applications of the contraction technique, and the Shannon-McMillan theorem for the relative entropy with respect to a Gibbs measure will lead us to a more recent development, namely to the study of large deviations of the empirical field of a Gibbs measure.

In Ch. II we discuss some connections between Gibbs measures and infinite dimensional diffusion processes. In the spirit of (C), we first discuss some large deviations of the empirical field of an infinite collection of independent Brownian motions. Then we look at infinite dimensional diffusion processes from the point of view of time reversal. A first application is the description of invariant measures as Gibbs measures. But time reversal is also a useful tool for other purposes; for example, it can be used to study certain large deviations of the empirical field of the diffusion process. In the last section we apply Dobrushin's contraction technique to an infinite dimensional diffusion process, again in the spirit of (C).

Je voudrais remercier P.L. Hennequin de m'avoir invité à faire ce cours: c'était un grand plaisir pour moi de participer à l'Ecole d'Eté de Saint-Flour.

I. An introduction to random fields

A random field will be a probability measure μ on a countable product space S^I. In Dobrushin's approach, random fields are specified (but in general not uniquely determined) by a consistent system of conditional probabilities. This is analogous to the specification of a Markov process μ on S^Z by the semigroup of its transition kernels (which may or may not admit more than one entrance law, if any). But in a spatial setting, it is natural to replace conditional probabilities on the future given the past by conditional probabilities on the behavior in some finite subset $V \subset I$ given the situation outside of V. This leads to the specification of a random field by a consistent family of stochastic kernels (π_V) indexed by the finite subsets of I. In this way, the total order on the time axis is lost, and some of the usual techniques from the theory of Markov processes have to be modified or replaced accordingly.

Section 1 contains basic definitions and some general facts about the structure of the class of all random fields specified by a collection (π_V) of local conditional probabilities. In order to guarantee uniqueness, we have to introduce bounds on the interaction. In Section 2 we give a short introduction to Dobrushin's contraction technique which is based on estimates of the interaction between the different sites in I. This leads not only to a strong uniqueness theorem, but also to additional regularity properties of the unique Gibbs measure.

In Section 3 the local specification is given by a stationary interaction potential on a d-dimensional lattice, and this allows us to introduce "thermodynamical" quantities like specific energy. Our main purpose is to derive a Shannon-Mc Millan theorem for the relative entropy $h(\nu;\mu)$ of a stationary measure ν with respect to a stationary Gibbs measure μ. This is the key to Section 4 where we describe some joint work with S. Orey on large deviations of the empirical field of a Gibbs measure. In the second part of Section 4, the effect of a phase transition on large deviations is discussed in terms of "surface entropy".

1. Random fields and their local specification

In this section we introduce Dobrushin's description of random fields on S^I in terms of their local specification by a system (π_V) of conditional probabilities, indexed by the finite subsets V of I. The first part contains some basic definitions and remarks; in the second we collect some general results on the integral representation of the set $G(\pi)$ of all random fields specified by a given family (π_V).

1.1 Definitions

Let I be a countable set of sites, let S be a standard Borel space of states, and consider the product space $\Omega = S^I$ of all configurations $\omega : I \to S$. For $J \subseteq I$ we introduce the σ-field $\underline{\underline{F}}_J = \sigma(\omega(i)\ ;\ i \in J)$, and we write $\underline{\underline{F}} = \underline{\underline{F}}_I$. The restriction of ω to J will be denoted by ω_J.

Let $M(\Omega)$ denote the class of probability measures on $(\Omega, \underline{\underline{F}})$. A probability measure $\mu \in M(\Omega)$ will also be called a random field.

(1.1) Remark. $(\Omega,\underline{\underline{F}})$ is again a standard Borel space. Thus we can choose, for any $\mu \in M(\Omega)$ and any $V \subseteq I$, a regular conditional distribution of μ with respect to $\underline{\underline{F}}_{V^c}$, i.e., a stochastic kernel $\pi_V(\omega,dy)$ from $(\Omega,\underline{\underline{F}}_{V^c})$ to $(\Omega,\underline{\underline{F}})$ such that

(1.2) $$\pi_V(\omega,\cdot) = \delta_\omega \quad \text{on } \underline{\underline{F}}_{V^c}$$

and

(1.3) $$E_\mu[\varphi|\underline{\underline{F}}_{V^c}](\omega) = \int \pi_V(\omega,d\eta)\varphi(\eta) =: (\pi_V\varphi)(\omega)$$

for any $\underline{\underline{F}}$-measurable function $\varphi \geq 0$. These conditional distributions are consistent in the following sense: since $\underline{\underline{F}}_{W^c} \subseteq \underline{\underline{F}}_{V^c}$ for $W \supseteq V$, we have

$$\begin{aligned}(\pi_W\pi_V\varphi)(\omega) &= E_\mu[E_\mu[\varphi|\underline{\underline{F}}_{V^c}]|\underline{\underline{F}}_{W^c}](\omega) = E_\mu[\varphi|\underline{\underline{F}}_{W^c}](\omega)\\ &= (\pi_W\varphi)(\omega) \qquad \mu\text{-a.s.}\end{aligned}$$

for any measurable $\varphi \geq 0$, hence

$$(\pi_W \pi_V)(\omega,\cdot) = \pi_W(\omega,\cdot) \qquad \mu\text{-a.s.} \tag{1.4}$$

We are now going to prescribe the local conditional behavior of a random field by fixing a system of conditional distributions π_V for the finite subsets $V \subseteq I$. These conditional distributions will be consistent in a strict sense, i.e., without the intervention of null sets. We are then going to study the class of all random fields which are compatible with these kernels in the sense of (1.3)

(1.5) <u>Definition.</u> For each finite $V \subseteq I$, let π_V be a stochastic kernel from $(\Omega, \underline{\underline{F}}_{V^c})$ to $(\Omega, \underline{\underline{F}})$ such that

$$\pi_V(\omega,\cdot) = \delta_\omega \quad \text{on} \quad \underline{\underline{F}}_{V^c} . \tag{1.6}$$

The collection (π_V) is called a local specification if

$$\pi_W \pi_V = \pi_W \qquad \text{for} \quad V \subseteq W . \tag{1.7}$$

A random field μ is called a Gibbs measure with respect to the local specification (π_V) if, for any finite V , π_V is a conditional distribution of μ with respect to $\underline{\underline{F}}_{V^c}$ in the sense of (1.3).

From now on we fix a local specification (π_V) and denote by $G(\pi)$ the corresponding class of Gibbs measures.

(1.8) <u>Lemma.</u> A random field μ belongs to $G(\pi)$ if and only if

$$\mu \pi_V = \mu . \tag{1.9}$$

for any finite $V \subseteq I$.

<u>Proof.</u> 1) For $\mu \in G(\pi)$ and finite V we have

$$\int \varphi d\mu = \int E_\mu[\varphi | \underline{\underline{F}}_{V^c}] d\mu = \int (\pi_V \varphi) d\mu = \int \varphi d(\mu \pi_V) ,$$

hence $\mu = \mu\pi_V$.

2) Condition (1.6) implies $\pi_V(\varphi\psi)(\omega) = \psi(\omega)(\pi_V\varphi)(\omega)$ for $\underline{\underline{F}}$-measurable $\varphi \geq 0$ and $\underline{\underline{F}}_{V^c}$-measurable $\psi \geq 0$. Thus, (1.9) implies

$$E_\mu[\varphi\psi] = E_\mu[\pi_V(\varphi\psi)] = E_\mu[\psi\pi_V\varphi] ,$$

hence (1.3).

(1.10) Remark on existence. Introducing a polish topology on S , we may view Ω as a polish space, and then $M(\Omega)$ becomes a polish space with respect to the weak topology. In particular,

$$(1.11) \qquad \mu_n \to \mu \quad \Longleftrightarrow \quad \int f_k d\mu_n \to \int f_k d\mu \qquad (k = 1,2,\ldots)$$

for a suitably chosen countable family of bounded continuous functions on Ω . Now suppose that S is compact, and that (π_V) has the following Feller property:

$$(1.12) \qquad f \in C(\Omega) \quad \Rightarrow \quad \pi_V f \in C(\Omega)$$

for any finite V . Then $\mu_n \to \mu$ implies

$$\lim_n \int fd(\mu_n\pi_V) = \lim_n \int(\pi_V f)d\mu_n = \int(\pi_V f)d\mu = \int fd(\mu\pi_V)$$

for $f \in C(\Omega)$, i.e., $\mu \to \mu\pi_V$ is a continuous map on the compact convex space $M(\Omega)$. By Schauder-Tychonow, we have $\{\mu \in M(\Omega) | \mu\pi_V = \mu\} \neq \emptyset$, and since

$$G(\pi) = \bigcap_{V \text{ finite}} \{\mu \in M(\Omega) | \mu\pi_V = \mu\}$$

$$= \bigcap_n \{\mu \in M(\Omega) | \mu\pi_{V_n} = \mu\}$$

for any increasing sequence $V_n \uparrow I$, we see that $G(\pi)$ is a non-empty convex compact set. The assumption of a compact state space can be replaced by a tightness condition on the local specification; cf. [Pr].

(1.13) Remark on spatial homogeneity. Let I be the d-dimensional lattice Z^d, and let $\theta_i : \Omega \to \Omega$ be the shift map defined by $(\theta_i\omega)(k) = \omega(i+k)$. In this case, we denote by

$$M_s(\Omega) := \{\mu \in M(\Omega) | \mu \circ \theta_i = \mu \ (i \in I)\}$$

the class of all spatially homogeneous random fields. A local specification is called spatially homogeneous if

$$\pi_V(\theta_i\omega,\cdot) = \pi_{V+i}(\omega,\cdot) \circ \theta_i$$

resp.

$$(\pi_V\varphi) \circ \theta_i = \pi_{V+i}(\varphi \circ \theta_i)$$

for $i \in I$ and finite $V \subseteq I$. For any $i \in I$, $\mu \to \mu \circ \theta_i$ defines a continuous map on $G(\pi)$: in fact,

$$\begin{aligned}\int(\pi_V\varphi)d(\mu \circ \theta_i) &= \int(\pi_V\varphi) \circ \theta_i d\mu \\ &= \int\pi_{V+i}(\varphi \circ \theta_i)d\mu \\ &= \int\varphi \circ \theta_i d\mu = \int\varphi d(\mu \circ \theta_i) ,\end{aligned}$$

and this implies $\mu \circ \theta_i \in G(\pi)$ due to (1.8). If S is compact and (1.12) is satisfied, then it follows that the set

$$G_s(\pi) = G(\pi) \cap M_s(\Omega)$$

of spatially homogeneous Gibbs measures is non-empty, convex and compact; the argument is analogous to (1.10).

(1.14) Remark on the Markov property. Suppose that for each $i \in I$ there is a finite set of neighbours $N(i) \subseteq I - \{i\}$. We say that (π_V) has the local Markov property if

$$\pi_V(\omega,\cdot) = \pi_V(\eta,\cdot)$$

whenever

$$\omega = \eta \quad \text{on} \quad \partial V := \bigcup_{i \in V} N(i) - V \;.$$

For any $\mu \in G(\pi)$, this local Markov property takes the form

$$(1.15) \qquad E_\mu[\varphi | \underline{\underline{F}}_{V^c}] = E_\mu[\varphi | \underline{\underline{F}}_{\partial V}] \qquad (\varphi \geq 0 \;,\; \underline{\underline{F}}_V\text{-measurable})$$

whenever V is finite. In this case, we also say that μ is a Markov random field. In general, the local Markov property (1.15) does not imply the global Markov property, i.e., the validity of (1.15) for any (not necessarily finite) $V \subseteq I$; see, however, Section 2.2.

1.2. Integral representation

Suppose that $G(\pi) \neq \emptyset$. In general $G(\pi)$ contains more than one element, i.e., a Gibbs measure $\mu \in G(\pi)$ is not uniquely determined by its local specification. In that case one speaks of a "phase transition". If S is compact, then $G(\pi)$ is a compact convex set, and Choquet's theorem leads to an integral representation of $G(\pi)$ in terms of extremal Gibbs measures. But such an integral representation can also be obtained without any compactness assumptions, and in a more explicit form which exhibits the role of the tail field

$$\underline{\underline{A}} = \bigcap_{V \text{ finite}} \underline{\underline{F}}_{V^c}$$

of asymptotic events; cf. [Fö1], [Dy]. The point is that martingale convergence allows us to pass from (π_V) to a limiting kernel π_∞ with respect to $\underline{\underline{A}}$:

(1.16) Theorem. If $G(\pi) \neq \emptyset$ then there exists a stochastic kernel $\pi_\infty(\omega, dy)$ from $(\Omega, \underline{\underline{A}})$ to $(\Omega, \underline{\underline{F}})$ such that

$$(1.17) \qquad \pi_\infty(\omega, \cdot) \in G(\pi)$$

for any $\omega \in \Omega$, and such that, for any $\mu \in G(\pi)$,

$$E_\mu[\varphi|\underline{\underline{A}}] = \pi_\infty\varphi \qquad \mu\text{-a.s.} \tag{1.18}$$

for $\underline{\underline{F}}$-measurable $\varphi \geq 0$.

<u>Proof.</u> 1) Fix a sequence $V_n \uparrow I$, a sequence of bounded functions f_k $(k = 1,2,\ldots)$ as in (1.11), and a Gibbs measure $\mu \in G(\pi)$. By martingale convergence,

$$\begin{aligned}\lim_n \pi_{V_n} f_k(\omega) &= \lim_n E_\mu[f_k|\underline{\underline{F}}_{V_n^c}](\omega) = E_\mu[f_k|\underline{\underline{A}}](\omega)\\ &= \int f_k(\eta)\pi_\mu(\omega,d\eta) \qquad \mu\text{-a.s.}\end{aligned}$$

where π_μ denotes a conditional probability distribution for μ and $\underline{\underline{A}}$. Thus, the set

$$\Omega_o = \{\omega|\lim_n \pi_{V_n}(\omega,\cdot) \text{ exists}\} \in \underline{\underline{A}}$$

satisfies $\mu(\Omega_o) = 1$ for any $\mu \in G(\pi)$. Defining

$$\begin{aligned}\tilde{\pi}_\infty(\omega,\cdot) &:= \lim_n \pi_{V_n}(\omega,\cdot) \quad \text{if} \quad \omega \in \Omega_o\\ &:= \mu_o \qquad\qquad\qquad \text{if} \quad \omega \notin \Omega_o\end{aligned}$$

for some fixed $\mu_o \in G(\pi)$, we obtain a kernel from $(\Omega,\underline{\underline{A}})$ to $(\Omega,\underline{\underline{F}})$ which satisfies (1.18).

2) Let $\mu \in G(\pi)$. For any $A \in \underline{\underline{A}}$ and for any k ,

$$\int_A \tilde{\pi}_\infty f_k d\mu = \int_A E_\mu[f_k|\underline{\underline{A}}]d\mu = \int_A E_\mu[\pi_V f_k|\underline{\underline{A}}]d\mu = \int_A \tilde{\pi}_\infty \pi_V f_k d\mu \ .$$

This implies $\tilde{\pi}_\infty(\omega,\cdot) = (\tilde{\pi}_\infty\pi_V)(\omega,\cdot)$ μ-a.s. for any finite V , hence $\tilde{\pi}_\infty(\omega,\cdot) \in G(\pi)$ μ-a.s. due to (1.8). Since

$$\tilde{\Omega}_o = \{\omega|\tilde{\pi}_\infty(\omega,\cdot) \in G(\pi)\} \in \underline{\underline{A}} \ ,$$

the kernel defined by

$$\pi_\infty(\omega,\cdot) := \tilde{\pi}_\infty(\omega,\cdot) \quad \text{if} \quad \omega \in \tilde{\Omega}_o$$

$$:= \quad \mu_o \qquad \text{if} \quad \omega \notin \tilde{\Omega}_o$$

has properties (1.17) and (1.18).

In the language of [Dy], we have shown that $\underline{\underline{A}}$ is an "H-sufficient statistics" with respect to $G(\pi)$. Using the general construction in [Dy], we obtain the following integral representation:

(1.19) Corollary. 1) Each $\mu \in G(\pi)$ is of the form

(1.20)
$$\mu = \int_{G_e(\pi)} \nu \quad \gamma_\mu(d\nu) ,$$

with a unique probability measure γ_μ on

$$G_e(\pi) = \{\mu \in G(\pi) | \mu = 0-1 \quad \text{on} \quad \underline{\underline{A}}\} \subseteq \{\lim_n \pi_{V_n}(\omega,\cdot) | \omega \in \tilde{\Omega}_o\} .$$

2) γ_μ is the image of μ under $\omega \to \pi_\infty(\omega,\cdot)$

3) For each probability measure γ_μ on $G_e(\pi)$, the measure μ defined by (1.20) belongs to $G(\pi)$.

(1.21) Remark. 1) $G_e(\pi)$ is the set of extremal points in $G(\pi)$.

2) By 2), two measures in $G(\pi)$ coincide as soon as they coincide on $\underline{\underline{A}}$.

In the spatially homogeneous case (1.13), an analogous construction, which combines martingale convergence and the ergodic theorem, leads to an integral representation of the class $G_s(\pi)$ of spatially homogeneous Gibbs measures. Here the role of $\underline{\underline{A}}$ is taken by the σ-field $\underline{\underline{J}}$ of shift-invariant sets:

(1.22) Theorem. If $G_s(\pi) \neq \emptyset$ then there exists a stochastic kernel $\pi_s(\omega,dy)$ from $(\Omega,\underline{\underline{J}})$ to $(\Omega,\underline{\underline{F}})$ such that

(1.23)
$$\pi_s(\omega,\cdot) \in G_s(\pi)$$

for any $\omega \in \Omega$, and

(1.24) $$E_\mu[\varphi|\underline{\underline{J}}] = \pi_s\varphi \qquad \mu\text{-a.s.}$$

for any $\mu \in G_s(\pi)$.

<u>Proof.</u> By the d-dimensional ergodic theorem, by the slight extension (2.29) of the martingale convergence theorem sometimes referred to as "Hunt's Lemma", and by (1.25) below,

$$E_\mu[f|\underline{\underline{J}}] = E_\mu[E_\mu[f|\underline{\underline{J}}]|\underline{\underline{A}}]$$

$$= \lim_n |V_n|^{-1} \sum_{i \in V_n} E_\mu[f \circ \theta_i | \underline{\underline{F}}_{V_n^c}]$$

$$= \lim_n |V_n|^{-1} \sum_{i \in V_n} \int f d(\pi_{V_n}(\omega,\cdot) \circ \theta_i)$$

if this limit exists and belongs to $G_s(\pi)$. We can now proceed as in the proof of (1.16).

(1.25) <u>Lemma.</u> For any $\mu \in M_s(\Omega)$, the σ-field $\underline{\underline{J}}$ of shift-invariant sets is contained in the μ-completion of $\underline{\underline{A}}$.

<u>Proof.</u> Let f be a bounded $\underline{\underline{J}}$-measurable function, and let $W \subseteq I$ be finite. Since $f = f \circ \theta_i$ and $\mu = \mu \circ \theta_i$,

$$||f - E_\mu[f|\underline{\underline{F}}_W]|| = ||f \circ \theta_i - E_\mu[f|\underline{\underline{F}}_W] \circ \theta_i||$$

$$\equiv ||f - E_\mu[f|\underline{\underline{F}}_{W+i}]||$$

due to (1.13), where $||\cdot||$ denotes the $L^1(\mu)$-norm. Thus,

$$||f - E_\mu[f|\underline{\underline{A}}]|| \le ||f - E_\mu[f|\underline{\underline{F}}_W]||$$

$$+ ||E_\mu[f|\underline{\underline{F}}_{W+i}] - E_\mu[f|\underline{\underline{F}}_{V^c}]||$$

$$+ ||E_\mu[f|\underline{\underline{F}}_{V^c}] - E_\mu[f|\underline{\underline{A}}]|| .$$

Take $i \in I$ such that $W + i \subseteq V^c$. Then the second term on the right is dominated by

$$||E_\mu[f|\underline{\underline{F}}_{W+i}] - f|| = ||E_\mu[f|\underline{\underline{F}}_W] - f|| .$$

By martingale convergence, V and W can be chosen such that the first and the third (hence also the second) term is $\leq \epsilon$. This shows $f = E_\mu[f|\underline{\underline{A}}]$.

(1.26) <u>Corollary.</u> $G_s(\pi)$ admits the integral representation

(1.27)
$$\mu = \int_{G_{s,e}} \nu \; \gamma_{s,\mu}(d\nu)$$

where

$$G_{s,e} = \{\mu \in G_s(\pi) | \mu = 0-1 \text{ on } \underline{\underline{J}}\}$$

is the class of all ergodic measures in $G_s(\pi)$, and where $\gamma_{s,\mu}$ is the image of μ under $\omega \to \pi_s(\omega,\cdot)$.

(1.28) <u>Remark.</u> Without the additional Gibbs structure, the construction behind (1.22) and (1.26) leads to an explicit integral representation of a stationary measure $\mu \in M_s(\Omega)$ as a mixture of ergodic measures; cf. [Dy]. For this, we will use the notation

(1.29)
$$\mu = \int \mu_\omega \mu(d\omega)$$

where μ_ω denotes the appropriate ergodic version of the conditional distribution with respect to $\underline{\underline{J}}$.

2. Dobrushin's contraction technique

In this section we give a short introduction to Dobrushin's contraction technique [Do1,2]; see also [Ro] and [DP]. This technique does not only provide a powerful uniqueness theorem, it also allows us to derive a number of additional regularity properties of the unique Gibbs measure. We illustrate this point with the global Markov property, with some covariance estimates, and with the almost sure convergence of multi-parameter martingales.

2.1. Dobrushin's comparison theorem

Let μ be a Gibbs measure on $\Omega = S^I$ with local specification (π_V). We denote by $\pi_k(\cdot|\eta)$ the conditional distribution of $\omega(k)$ given $\underline{\underline{F}}_{I-\{k\}}$, so that

$$\pi_{\{k\}}(\eta,\cdot) = \pi_k(\cdot|\eta) \times \prod_{i \neq k} \delta_{\eta(i)} \qquad (k \in I) .$$

Let us now measure the influence of site i on site k by

$$(2.1) \qquad C_{ik} := \sup \{\tfrac{1}{2}\, ||\pi_k(\cdot|\omega) - \pi_k(\cdot|\eta)|| \;:\; \omega = \eta \text{ off } i\}$$

where $||\cdot||$ denotes the total variation of a signed measure on S. The matrix $C = (C_{ik})_{i,k \in I}$ will be called Dobrushin's interaction matrix; C^n denotes the n-th power of C. For any probability measure ν on Ω let us define the vector $b = (b_k)_{k \in I}$ with components

$$(2.2) \qquad b_k := \tfrac{1}{2} \int ||\pi_k(\cdot|\eta) - \nu_k(\cdot|\eta)||\, \nu(d\eta)$$

where $\pi_k(\cdot|\eta)$ is a conditional distribution of $\omega(k)$ with respect to $\underline{\underline{F}}_{I-\{k\}}$. This is a slight modification of the definition in [Do2], which will be useful for the covariance estimates below.

In order to compare μ and ν, let us introduce the class $C(\Omega)$ of functions which can be approximated uniformly by bounded measurable functions depending only on finitely many coordinates. We say that a vector $a = (a_i)_{i \in I}$ is an estimate for μ and ν if

(2.3) $$\left|\int f d\mu - \int f d\nu\right| \le \sum_i a_i \delta_i(f) \qquad (f \in C(\Omega))$$

where

$$\delta_i(f) := \sup \{ |f(\omega) - f(\eta)| : \omega = \eta \text{ off } i \}$$

denotes the oscillation of f at site $i \in I$. For example, $a_i \equiv 1$ is always an estimate since $|f(\omega) - f(\eta)| \le \sum_i \delta_i(f)$ for any $f \in C(\Omega)$. The comparison theorem will follow from a successive improvement of this initial estimate, and for this we need an additional continuity requirement:

(2.4) $$f \in C(\Omega) \Rightarrow \pi_{\{k\}} f \in C(\Omega) \quad \text{for any } k \in I .$$

(2.5) Lemma. If a is an estimate for μ and ν then the vector $aC + b$ is also an estimate.

Proof: It is enough to check (2.3), with $aC + b$ instead of a, for functions which depend only on finitely many coordinates. Therefore, it is enough to show that for any finite subset $J \subset I$ the vector a^J with components

$$a_i^J = \min(a_i, (aC + b)_i) \qquad (i \in J)$$

$$= a_i \qquad (i \notin J)$$

is an estimate for μ and ν. We prove this by induction on the cardinality of J. For $J = \emptyset$ the statement is true. Now assume that a^J is an estimate and take $K = J \cup \{k\}$; we have to show that a^K is also an estimate. For $f \in C(\Omega)$ we have

$$\begin{aligned} \left|\int f d\mu - \int f d\nu\right| &\le \int \left[\int f d\pi_k(\cdot|\eta)\right] (\mu-\nu)(d\eta) \\ &+ \int \left|\int f d\pi_k(\cdot|\eta) - \int f d\nu_k(\cdot|\eta)\right| \nu(d\eta) \\ &\le \sum_i a_i^J \delta_i\left(\int f d\pi_k(\cdot|\cdot)\right) + b_k \delta_k(f) . \end{aligned}$$

Since

$$\delta_i\left(\int f d\pi_k(\cdot|\cdot)\right) \le \delta_i(f) + C_{ik}\delta_k(f)$$

for $i \neq k$ and $=0$ for $i = k$, we obtain

$$|\int f d\mu - \int f d\nu| \leq \sum_{i \neq k} a_i^J \delta_i(f) + (a^J C + b)_k \delta_k(f) .$$

But $a^J C \leq aC$, and since a^J with $a_k^J \leq a_k$ is also an estimate, we can replace the right side by $\sum_i a_i^K \delta_i(f)$.

Applying the lemma successively, we see that for each $n \geq 1$ the vector

$$aC^{n+1} + \sum_{m=0}^{n} bC^m$$

is an estimate. Letting $n \uparrow \infty$, taking $a_i \equiv 1$ and defining the matrix

$$D := \sum_{m=0}^{\infty} C^m ,$$

we see that the vector bD is an estimate as soon as C satisfies the condition

(2.6) $$\lim_n \sum_i C_{ik}^n = 0 \qquad (k \in I)$$

Note that (2.6) is satisfied if, for example,

(2.7) $$c := \sup_k \sum_i C_{ik} < 1 ,$$

since $\sum_i C_{ik}^n \leq c^n$ by induction.

This proves the following variant of Dobrushin's comparison theorem [Do2, Th. 3]; see also [Kü3].

(2.8) <u>Comparison Theorem:</u> Under condition (2.6) we have

$$|\int f d\mu - \int f d\nu| \leq \sum_i (bD)_i \delta_i(f) \qquad \text{for any } f \in C(\Omega) .$$

2.2 Uniqueness and global Markov property

If $\nu \in G(\pi)$, i.e., if ν has the same local specification as μ , then we have $b_i = 0$ $(i \in I)$ in (2.8), and this implies $\mu = \nu$. This is Dobrushin's well-known

(2.9) Uniqueness theorem [Do1]: Under condition (2.6) there is at most one measure $\mu \in G(\pi)$.

In fact, condition (2.6) not only implies uniqueness but also conditional uniqueness in the following sense. For any $J \subseteq I$ and for any $\eta \in S^{I-J}$, define the conditional specification $\pi_V^{J,\eta}$ ($V \subseteq J$ finite) on S^J with

$$(2.10) \qquad \pi_V^{J,\eta}(\omega, \cdot) = \pi_V(\zeta, \cdot)$$

where $\zeta(i) = \eta(i)$ for $i \in I-J$ and $\zeta(i) = \omega(i)$ for $i \in J$. The corresponding Dobrushin matrix $C^{J,\eta}$ satisfies $C_{ik}^{J,\eta} \le C_{ik}$; in particular it inherits condition (2.6). For any Gibbs measure $\mu \in G(\pi)$, let $\mu_J(\cdot|\eta)$ be a conditional joint distribution of $\omega(i)$ $(i \in J)$ with respect to $\underline{\underline{F}}_{I-J}$ and μ . For μ-almost η , $\mu_J(\cdot|\eta)$ is compatible with the conditional specification $(\pi_V^{J,\eta})$, and thus coincides with the corresponding unique Gibbs measure, due to (2.9).

Suppose, in particular, that μ is a Markov random field, i.e., that μ resp. (π_V) has the local Markov property (1.15). In this case, (2.10) shows that $(\pi_V^{J,\eta})$, hence $\mu_J(\cdot|\eta)$, only depends on the values $\eta(j)$ for $j \in \partial J$. This implies the global Markov property, i.e.,

$$E_\mu[\varphi|\underline{\underline{F}}_{J^c}] = E_\mu[\varphi|\underline{\underline{F}}_{\partial J}] \qquad (\varphi \ge 0 \ \underline{\underline{F}}_J \text{ - measurable})$$

for arbitrary (not only finite) $J \subseteq I$. Consider, in particular, the case $I = Z^d$. Here the global Markov property implies that μ may be viewed as the distribution of a Markov chain on the infinite-dimensional state space $E = S^{Z^{d-1}}$. If we denote by $L(t)$ the line $\{(t,j): j \in Z^{d-1}\}$, then the transition probabilities of the chain are of the form

$$P_t(\xi,.) = \text{distribution of } \omega_{L(t)} \text{ under } \mu_{L(t-1)}(\cdot|\omega_{L(t-1)} = \xi).$$

Let us also assume that S is finite so that E may be regarded as a compact space. Then condition (2.6) implies, in addition, that this Markov chain has the Feller property:

$$\xi_n \to \xi \Rightarrow P_t(\xi_n,\cdot) \to P_t(\xi,\cdot)$$

in the weak topology for probability measures on E. In fact, any limit point of the sequence $\mu_{L(t-1)}(\cdot|\xi_n)$ is compatible with the conditional specification induced by ξ, and so it must coincide with $\mu_{L(t-1)}(\cdot|\xi)$.

The restrictions μ_t of μ to $S^{L(t)}$ form an entrance law for the chain:

$$\mu_{t-1}P_t = \mu_t \qquad (t \in Z^1) .$$

Conversely, any entrance law induces a measure $\mu \in G(\pi)$, and so we have shown:

(2.11) <u>Corollary.</u> Suppose that $I = Z^d$, that S is finite, and that (π_V) has the local Markov property and satisfies (2.6). Then the unique Gibbs measure $\mu \in G(\pi)$ may be viewed as a Markov chain on $S^{Z^{d-1}}$, which has the Feller property and admits exactly one entrance law.

(2.12) <u>Remark</u>. We refer to [Is] for counter-examples to the conjecture that any extremal Gibbs measure with respect to a Markov specification might have the global Markov property. For an "attractive" interaction there are techniques based on monotonicity which allow to show, for example, that the $+$ and $-$ states of the Ising model have the global Markov property; cf. (3.47) below.

2.3 <u>Covariance estimates</u>

Let us now use the comparison theorem (2.8) in order to obtain estimates for certain covariances.

(2.13) <u>Theorem.</u> Under condition (2.6), the covariance of any two functions f and g in $C(\Omega)$ with respect to μ satisfies

$$|\mathrm{cov}_\mu(f,g)| \le \frac{1}{4} \sum_{i,k} \delta_i(f) D_{ki} \delta_k(g).$$

<u>Proof.</u> We may assume $g > 0$ and $\int g \, d\mu = 1$. But then we can write

$$\mathrm{cov}_\mu(f,g) = \int f \, d\nu - \int f \, d\mu ,$$

where $d\nu = g \, d\mu$ is a probability measure whose conditional probabilities are given by

$$\nu_i(d\gamma|\eta) = g(\gamma) \, [\int g \, d\pi_i(\cdot|\eta)]^{-1} \, \pi_i(d\gamma|\eta) .$$

Applying variant (2.8) of Dobrushin's comparision theorem, we obtain

$$|\mathrm{cov}_\mu(f,g)| \le \sum_i (\sum_k b_k D_{ki}) \, \delta_i(f)$$

with

$$b_k \le \frac{1}{4} \delta_k(g) \int [\int g \, d\mu_k(\cdot|x)]^{-1} \, g(x) \, \mu(dx) = \frac{1}{4} \delta_k(g) ,$$

using the elementary estimate $\frac{1}{2} ||\mu-\nu|| \le \frac{1}{4} \delta(g)$.

Let us now illustrate how this estimate provides information on the rate of decay of correlation. Following L. Gross [Gr], we consider the case $I = Z^d$ and fix a translation invariant semimetric $d(\cdot,\cdot)$ on I. Let

$$|f|_o := \sum_i e^{d(i,o)} \, \delta_i(f)$$

and

$$\gamma := \sup_k \sum_i e^{d(i,k)} \, C_{ik} .$$

Note that condition $c < 1$ in (2.7) implies $\gamma < 1$ for a suitable multiple of the Euclidean metric $|i-k|$ if $C_{ik} = 0$ for large enough $|i-k|$. Let θ_i denote the shift map on $\Omega = S^I$ associated to $i \in I$.

(2.14) Corollary. If $\gamma < 1$ then

$$\sum_i |cov_\mu(f, g \circ \theta_i)| e^{d(i,0)} \le \frac{1}{4}(1-\gamma)^{-1} |f|_0 |g|_0 \tag{2.15}$$

for f and g in $C(\Omega)$.

Proof. Applying (2.13) and the triangle inequality for $d(\cdot,\cdot)$ we obtain

$$|cov(f, g \circ \theta_i)| e^{d(i,0)}$$

$$\le \frac{1}{4} \sum_{k,j} e^{d(j,k)} D_{jk} e^{d(k,o)} \delta_k(f) e^{d(i,j)} \delta_{j-i}(g) .$$

Summing over i,j,k in that order, we get (2.15).

(2.16) Remark. These covariance estimates are a slight improvement over some similar estimates in [Gr1] and [Kü3]; cf. [Fö4].

(2.17) Remark. We have introduced the contraction technique in its simplest form, based on total variation as a measure of the distance between two probability measures on S in (2.1). The technique becomes more flexible if we replace total variation by the (Kantorovic-Rubinstein-) Vasserstein metric induced by a metric $r(\cdot,\cdot)$ on S; this extension will be used in Ch. II. For two probability measures μ and ν on S define

$$R(\mu,\nu) = \sup \frac{|\int f d\mu - \int f d\nu|}{\delta(f)} \tag{2.18}$$

where the supremum is taken over all Lipschitz functions f on S with

$$\delta(f) := \sup_{s \neq t} \frac{|f(s)-f(t)|}{r(s,t)} < \infty ,$$

cf. [Do2]. Note that (2.18) reduces to $\frac{1}{2}||\mu-\nu||$ if the metric $r(\cdot,\cdot)$ is discrete. If S is polish then (2.18) admits the dual description

$$R(\mu,\nu) = \inf \int r d\gamma \tag{2.19}$$

where the infimum is taken over all measures γ on $S \times S$ with marginals μ and ν. The contraction coefficients in (2.1) and (2.2) are now replaced by

$$C_{ik} = \sup \left\{ \frac{R(\pi_k(\cdot|\omega),\pi_k(\cdot|\eta))}{r(\omega(i),\eta(i))} : w = \eta \text{ off } i \right\} \tag{2.20}$$

and

$$b_k = \int R(\pi_k(\cdot|\eta),\nu_k(\cdot|\eta))\nu(d\eta) ; \tag{2.21}$$

the matrices $C = (C_{ik})$ and $D = \sum_{n\geq 0} C^n$ are defined as before. Assume (2.6), and in the Feller assumption (2.4) replace $C(\Omega)$ by the class $L(\Omega)$ of functions f on Ω which satisfy

$$|f(\omega) - f(\eta)| \leq \sum r(\omega(i),\eta(i))\delta_i(f) , \quad \sum_i \delta_i(f) < \infty$$

where

$$\delta_i(f) = \sup_{s\neq t} \left\{ \frac{|f(s)-f(t)|}{r(\omega(i),\eta(i))} : w = \eta \text{ off } i \right\} .$$

A measure μ is called tempered if

$$\sup_i \int r(\omega(i),\eta(i))\mu(d\omega) < \infty \tag{2.22}$$

for some fixed $\eta \in \Omega$. For a tempered Gibbs measure μ and a tempered probability measure ν, we have a uniformly bounded estimate $a = (a_k)$ to start the argument in (2.5), namely $a_k = \int r(\omega(k),\eta(k))(\mu+\nu)(d\omega)$. This leads to Dobrushin's comparison theorem (2.8) for functions $f \in L(\Omega)$, and to a corresponding uniqueness theorem for tempered Gibbs measures. The covariance estimate (2.13) takes the form

(2.23) $$|\mathrm{cov}_\mu(f,g)| \le \sigma^2 \sum_{i,k} \delta_i(f) D_{ki} \delta_k(g)$$

for f and g in $L(\Omega)$, where μ is a Gibbs measure which satisfies

$$\sigma^2 := \sup_i \int r(\omega(i),\eta(i))^2 \mu(d\omega) < \infty .$$

In the translation invariant case $I = Z^d$, exponential decay of correlation follows as in (2.14): If $\gamma < 1$ then

(2.24) $$\sum_i |\mathrm{cov}(f, g\circ\theta_i)| e^{d(i,o)} \le \sigma^2 (1-\gamma)^{-1} |f|_o |g|_o$$

for functions f and g in $L(\Omega)$; cf. [Fö4] for details.

In the translation invariant case $I = Z^d$, covariance estimates of the form (2.15) resp. (2.24) are the key to a central limit theorem; cf. [DT], [Kü3]. Suppose that the local specification is spatially homogeneous so that $c_{k-i} := c_{i,k}$ only depends on the difference $i-k$. Suppose also, for simplicity, that $c_k = 0$ for large enough $|k|$, so that

$$\sum_{k\neq o} c_k |k|^d < \infty .$$

Then the covariance estimates allow us to apply a spatial central limit theorem, e.g. in the form of [Bo]:

(2.25) <u>Theorem.</u> If $c < 1$ then the distribution of

$$S_n^*(f) = |V_n|^{-1/2} \sum_{i\in Vn} [f\circ\theta_i - \int f d\mu]$$

under the unique (tempered) Gibbs measure μ converges to the centered normal law with variance

(2.26) $$\sigma^2(f) = \sum_k \mathrm{cov}_\mu(f, f\circ\theta_k) < \infty$$

for any $f \in L(\Omega)$.

2.4 Almost sure convergence of two-parameter martingales

We consider the case $I = Z^2$ and denote by $s \leq t$ the coordinate-wise ordering of Z^2. For a bounded random variable X on $(\Omega, \underline{\underline{F}})$, consider the two-parameter martingale

$$(2.27) \qquad X_t := E_\mu[X|\underline{\underline{F}}_t] \qquad (t \in Z_+^2),$$

where $\underline{\underline{F}}_t := \sigma(\omega(i); 0 \leq i \leq t)$. We are interested in almost sure convergence as $t \uparrow \infty$. By a theorem of Cairoli [Ca], we know that almost sure convergence does hold if the underlying random field μ on S^I satisfies the following independence condition:

(2.28) For each $t = (t_1, t_2) \in Z_+^2$, the two σ-fields

$$\underline{\underline{F}}^i_{t_i} := \sigma(\omega(s);\ s_i \leq t_i) \quad (i = 1,2)$$

are conditionally independent with respect to their intersection $\underline{\underline{F}}_t = \underline{\underline{F}}^1_{t_1} \cap \underline{\underline{F}}^2_{t_2}$.

In fact, condition (2.18) allows us to write

$$X_t = E_\mu[E_\mu[X|\underline{\underline{F}}^1_{t_1}]|\underline{\underline{F}}^2_{t_2}],$$

and then it is enough to apply the following two-parameter version of the martingale convergence theorem due to Blackwell and Dubins:

(2.29) Lemma [BD]. If (Y_n) converges μ-almost surely and satisfies $\sup |Y_n| \in L^1$, then

$$\lim_{n,m} E_\mu[Y_n | \underline{\underline{G}}_m]$$

exists μ-almost surely for any increasing (or decreasing) sequence of σ-fields $(\underline{\underline{G}}_m)$; the parameter n may run through any partially ordered index set.

From the point of view of random fields, condition (2.28) is very restrictive: in most cases there is some diagonal interaction between $\underline{\underline{F}}^1_{t_1}$ and $\underline{\underline{F}}^2_{t_2}$ which does not pass through $\underline{\underline{F}}_t$. If this interaction becomes too strong then almost sure convergence may break down; see, for example, the counterexample in [DP1] . Let us now sketch how Dobrushin's condition (2.7) allows us to control this effect. Define $X_t^{(0)} \equiv X$ and

$$X_t^{(N)} = E_\mu[E_\mu[X_t^{(N-1)}|\underline{\underline{F}}^1_{t_1}]|\underline{\underline{F}}^2_{t_2}] \qquad (t \in Z_+^2) .$$

Lemma (2.29), with n ranging in Z_+^2, implies almost sure convergence of $X_t^{(N)}$ for any $N \geq 1$. But if μ is a Markov random field which satisfies condition (2.7) , then we can use the global Markov property and the comparison theorem (2.8) in order to show that

$$\lim_N \sup_t |X_t^{(N)} - X_t| = 0 \quad \text{a.s.}$$

This implies almost sure convergence of the martingale (X_t) :

(2.30) Theorem. If μ is a Markov random field which satisfies condition (2.7) then bounded two-parameter martingales converge μ-almost surely.

We refer to [Fö5] for a detailed proof, and for an example which shows that it is not enough to require that μ is uniquely determined by its conditional probabilities.

2.5 Time-inhomogeneous Markov chains and annealing

Dobrushin's contraction technique for random fields may be viewed as the spatial extension of a classical contraction method for Markov chains. In [Do3], this technique has been used systematically in order to study the asymptotic behavior of time-inhomogeneous Markov chains. The annealing algorithm is an important example in this context, and its convergence may be viewed as a special case of the following general convergence theorem.

Let $P_n(x,dy)$ $(n = 1,2,\ldots)$ be a sequence of transition kernels on some state space, and define the contraction coefficient of P_n as

$$c(P_n) := \sup_{x,y} \frac{1}{2} \|P_n(x,.) - P_n(y,.)\| .$$

For two probability measures μ and ν we have

$$\|\mu P_n - \nu P_n\| \leq c(P_n) \, \|\mu-\nu\| ,$$

and this shows that

(2.31) $$\prod_n c(P_n) = 0$$

is a sufficient condition for "asymptotic loss of memory", i.e.,

(2.32) $$\lim_n \|\mu P_1 \ldots P_n - \nu P_1 \ldots P_n\| = 0$$

for two initial distributions μ and ν ; cf. [Do3].

Now suppose that each kernel P_n has a unique invariant distribution μ_n , and that

(2.33) $$\sum_n \|\mu_{n+1} - \mu_n\| < \infty .$$

This implies the existence of a unique limiting measure $\mu_\infty = \lim \mu_n$ (in total variation). Let us also assume $c(P_n) > 0$ for all n .

(2.34) Corollary. Under (2.31) and (2.33),

$$\lim_{n} ||\nu P_1 \dots P_n - \mu_\infty|| = 0$$

for any initial distribution ν .

Proof. For a fixed N ,

$$||\nu P_1 \dots P_n - \mu_\infty||$$

$$= ||(\nu P_1 \dots P_N - \mu_\infty) P_{N+1} \dots P_n + \mu_\infty P_{N+1} \dots P_n - \mu_\infty||$$

$$\leq \prod_{K=N+1}^{n} c(P_k) + ||\mu_\infty P_{N+1} \dots P_n - \mu_\infty|| .$$

But

$$\mu_\infty P_{N+1} \dots P_n - \mu_\infty = (\mu_\infty - \mu_{N+1}) P_{N+1} \dots P_n + \mu_{N+1} P_{N+2} \dots P_n - \mu_\infty$$

$$= (\mu_\infty - \mu_{N+1}) P_{N+1} \dots P_n + \sum_{k=1}^{n-N-1} (\mu_{N+k} - \mu_{N+k+1}) P_{N+k+1} \dots P_n + \mu_n - \mu_\infty ,$$

and so

$$\sup_{n \geq N} ||\mu_\infty P_{N+1} \dots P_n - \mu_\infty||$$

$$\leq \sup_{n \geq N} 2||\mu_\infty - \mu_n|| + \sum_{j > N} ||\mu_j - \mu_{j+1}||$$

converges to 0 due to (2.33).

In this form, the contraction technique has been used to establish convergence of the following "annealing algorithm"; cf. [GG] and also [Gi]. Let $E(x)$ be a function on a product space of the form S^I , where I is a finite index set and S is some finite state space at each site $i \in I$. Our purpose is to find global minima of E . In order to avoid being trapped in one of the local minima, we use a randomized search procedure. For each $\beta > 0$ consider the Gibbs measure

$$\mu_\beta(x) = Z(\beta)^{-1} \exp(-\beta E(x)) \tag{2.35}$$

and note that μ_β converges for $\beta \uparrow \infty$ to the uniform distribution μ_∞ on the set of global minima of E. For a fixed β, the local specification of μ_β satisfies

$$\pi^\beta_{\{i\}}(x,.) = \pi^\beta_i(\cdot|x) \times \prod_{j\neq i} \delta_{x(j)}$$

with

$$\pi^\beta_i(s|x) = Z_i(\beta)^{-1} \exp[-\beta E((s,x))] \tag{2.36}$$

where (s,x) is the configuration which coincides with x off i and has value s in i. Let $I = \{1,\ldots,N\}$ be some enumeration of I. Then μ_β is the unique invariant measure for the Markov chain with transition probability

$$P_\beta = \pi^\beta_{\{i\}} \cdots \pi^\beta_{\{N\}} \tag{2.37}$$

on S^I. Thus, we can expect to be close to the global minima of the function E if we let the chain P_β run for a sufficient amount of time and for large enough β. Now the idea is to choose a sequence $\beta(n) \uparrow \infty$ such that the time-inhomogeneous Markov chain with $P_n = P_{\beta(n)}$ $(n = 1,2,\ldots)$ converges to μ_∞, i.e.,

$$\lim_n ||\nu P_{\beta(1)}\cdots P_{\beta(n)} - \mu_\infty|| = 0 \tag{2.38}$$

for any initial distribution ν. The appropriate rate at which $\beta(n)$ is allowed to go to infinity is computed in view of condition (2.31):

(2.39) Theorem [GG]: There is a constant γ such that (2.38) holds for

$$\beta(n) \leq \gamma \log n . \tag{2.40}$$

Proof. For $s \in S$ and for any $x \in S^I$ we have

$$\pi^\beta_i(s|x) \geq |S|^{-1} \exp(-\beta\delta_i(E))$$

where $\delta_i(E)$ denotes the oscillation of E in the i-th coordinate. Thus,

$$\min_{x,y} P_\beta(x,y) \geq (|S|^{-1} \exp(-\beta\Delta))^N$$

with $\Delta = \max_i \delta_i(E)$. Since

$$\frac{1}{2}||\mu-\nu|| = \sum_x (\mu(x)-\nu(x))^+$$

$$= 1 - \sum_x \min(\mu(x),\nu(x))$$

$$\geq 1 - \inf_x |S^I| \min(\mu(x),\nu(x))$$

for any two probability measures μ and ν on S^I , we obtain

$$c(P_\beta) \leq 1 - \exp(-\beta N\Delta) .$$

Thus, condition (2.31) is satisfied for

$$\beta(n) \leq (N\Delta)^{-1} \log n .$$

As to condition (2.33), note that for each $x \in S^I$ the sequence $\mu_{\beta(n)}(x)$ is either increasing or decreasing. Thus,

$$\sum_n ||\mu_{(n+1)}-\mu_{(n)}|| = \sum_{x\in S} \sum_n (\mu_{\beta(n+1)}(x)-\mu_{\beta(n)}(x))^+$$

$$= \sum_{x\in S} (\mu_\infty(x)-\mu_{\beta(1)}(x))^+ < \infty .$$

(2.41) <u>Remarks.</u> 1) See, e.g., [Ha] for further refinements concerning the constant γ .

2) Let P_ν be the distribution of the Markov chain with initial distribution ν and transition kernels $P_{\beta(n)}$ $(n = 1,2,...)$. Under condition (2.40), P_ν coincides with P_μ on the tail field $\bigcap_n \sigma(X_n,X_{n+1},...)$. But this does not yet imply ergodic behavior, i.e.,

$$\lim_{n} \frac{1}{n} \sum_{i=1}^{n} f(X_i) = \int f d\mu_\infty \tag{2.42}$$

P_μ-almost surely for functions f on S^I (there are counterexamples to Theorem 1.3 in [Gi]). But, as observed by N. Gantert, (2.42) does hold if the constant used in the proof of (2.39) is divided by 2; this follows, e.g., from the laws of large numbers for time-inhomogeneous Markov chains in [IT].

3) In [GG] the annealing technique is applied to the restauration of distorted images. An image is described as a configuration $x \in S^I$, and it is viewed as the realization of some a priori distribution

$$\mu(x) = Z^{-1} \exp[E(x)]$$

on S^I ; usually, μ is assumed to have a Markov property with respect to some graph structure on I , and $E(x)$ is specified in terms of some interaction potential as in the following section. Now suppose that we observe a distorted version y of x which is generated by some probability kernel $Q(x,y)$ on S^I , and denote by $\mu(\cdot|y)$ the corresponding a posteriori distribution on S^I ; if both Q and μ have a local Markov property, then $\mu(\cdot|y)$ is again a Markov field. In any case, one can compute explicitely the function $E(\cdot|y)$ in the Gibbsian description

$$\mu(x|y) = Z(y)^{-1} \exp[E(x|y)]$$

of $\mu(\cdot|y)$. The Bayesian estimate with respect to the loss function

$$L(x,\hat{x}) = I_{\{x \neq \hat{x}\}}$$

is given by a picture $\hat{x}$ with maximal a posteriori probability $\mu(\cdot|y)$. This leads to the problem of finding the global minima of the function $E(\cdot|y)$, and here the annealing algorithm comes in.

3. Entropy, energy and the theorem of Shannon-Mc Millan

In Section 1 Gibbs measures were introduced in terms of their local specification. If the local specification is given by a stationary interaction potential on a d-dimensional lattice, then the variational principle of Lanford and Ruelle provides an alternative global characterization in terms of specific entropy and specific energy. The purpose of this section is to give a short introduction to the probabilistic limit theorems which are behind the existence of these "thermodynamical" quantities. In particular, we derive the d-dimensional Shannon-Mc Millan theorem for the specific entropy $h(\nu;\mu)$ of a stationary measure ν with respect to a Gibbs measure $\mu \in G_s(\pi)$; cf. [Fö2], [Pr]. In view of recent work on large deviations for lattice models, these results are of renewed interest. The Shannon-Mc Millan theorem is in fact the key tool in proving the lower bound (4.3) for large deviations of the empirical field of a Gibbs measure. We are going to discuss large deviations in Section 4, and so it seems reasonable to include a self-contained exposition. In view of Ch. II we admit a general state space S , otherwise we use the simplest setting of a bounded interaction on Z^d . For unbounded interactions we refer, e.g., to [Kü1], [Kü2], [Pi], [Gu], for a general amenable group I to [Te] and [Mo], for analogous results in the theory of interactive point processes to [NZ].

3.1 Specific entropy

Let μ and ν be two probability measures on some measurable space $(E,\underline{\underline{E}})$. We define the relative entropy of μ with respect to ν as

$$(3.1) \qquad H(\mu;\nu) := \int \log \varphi \, d\mu$$

if μ is absolutely continuous with respect to ν on $\underline{\underline{E}}$ with density φ , and $H(\mu;\nu) := +\infty$ else.

(3.2) Remarks. 1) By Jensen's inequality we have $H(\mu;\nu) \geq 0$, and $H(\mu;\nu) = 0 \iff \mu = \nu$ on $\underline{\underline{E}}$.

2) For $0 < \psi \in L^1(\nu)$, the inequality in 1) implies

$$H(\mu;\nu) = H(\mu;\tilde{\nu}) + \int \log \psi \, d\mu - \log \int \psi \, d\nu$$

$$\geq \int \log \psi \, d\mu - \log \int \psi \, d\nu$$

where $d\tilde{\nu} = \psi(\int \psi d\nu)^{-1} d\nu$. Thus,

(3.3) $$H(\mu;\nu) = \sup_{\psi} [\int \log \psi \, d\mu - \log \int \psi \, d\nu] .$$

If $\underline{\underline{E}}$ is the Borel σ-field with respect to some polish topology on E , then it is enough to take the supremum over bounded continuous functions which are bounded away from 0 ; cf. [Va]. This implies that $H(\cdot;\nu)$ is lower semicontinuous with respect to the weak topology.

3) The total variation $||\mu-\nu|| = \int |\varphi-1| d\nu$ can be estimated by

(3.4) $$||\nu-\mu||^2 \leq 2 \, H(\mu;\nu) .$$

This follows from the elementary inequality $3(x-1)^2 \leq f(x)g(x)$ with $f(x) = 4 + 2x$ and $g(x) = x \log x - x + 1$ (cf. [Ke]):

$$3(\int |\varphi-1| d\nu)^2 \leq (\int \sqrt{f(\varphi)} \sqrt{g(\varphi)} d\nu)^2 \leq \int f(\varphi) d\nu \int g(\varphi) d\nu$$

$$= 6 \int \varphi \log \varphi \, d\nu .$$

We are going to use the following well-known

(3.5) Lemma. Let T be a partially ordered index set, let $\underline{\underline{A}}_t (t \in T)$ be an increasing directed family of σ-fields contained in $\underline{\underline{E}}$, and put $\underline{\underline{A}}_\infty := \sigma(\bigcup_t \underline{\underline{A}}_t)$. If

(3.6) $$\sup_t H_t(\mu;\nu) < \infty ,$$

where $H_t(\mu;\nu)$ denotes the relative entropy on $\underline{\underline{A}}_t$, then μ is absolutely continuous, with respect to ν on $\underline{\underline{A}}_\infty$, and

(3.7) $$H_\infty(\mu;\nu) = \sup_t H_t(\mu;\nu) .$$

Moreover, the densities $\varphi_t := \frac{d\mu}{d\nu}|_{\underline{\underline{A}}_t}$ and $\varphi_\infty := \frac{d\mu}{d\nu}|_{\underline{\underline{A}}_\infty}$ satisfy

$$\lim_t \log \varphi_t = \log \varphi_\infty \quad \text{in} \quad L^1(\mu) .$$

<u>Proof.</u> See, for example, [NZ] or [Or2].

Let us now go back to our product space S^I. We fix a reference probability measure λ_o on S and denote by $\lambda = \prod_{i\in I} \lambda_o$ the corresponding product measure on S^I. For $\mu \in M(\Omega)$ and for finite $V \subseteq I$ let $H_V(\mu;\lambda)$ denote the relative entropy of μ with respect to ν on the σ-field $\underline{\underline{F}}_V$. From now on we assume that I is the d-dimensional lattice Z^d, and we write

$$V_n := \{t \in Z^d \mid 0 \leq t_k \leq n \quad (k=1,\dots,d)\} .$$

(3.8) <u>Definition:</u> For any $\mu \in M_s(\Omega)$, the specific entropy of μ with respect to λ is defined as

$$(3.9) \qquad h(\mu;\lambda) := \lim_n |V_n|^{-1} H_{V_n}(\mu;\lambda) = \sup_n |V_n|^{-1} H_{V_n}(\mu;\lambda) .$$

Note that (3.2) inplies that $h(\cdot;\lambda)$ is lower semicontinuous with respect to the weak topology on $M_s(\Omega)$. The existence of the limit and the equality in (3.8) follow from the fact that $H_V(\mu;\lambda)$ is superadditive in V; cf. [Pr] Th. 8.1. It is also contained in the following spatial version of the Shannon-McMillan theorem which states that there is L^1-convergence behind (3.8):

(3.10) <u>Theorem.</u> Suppose that $h(\mu|\lambda) < \infty$. Then the limit

$$(3.11) \qquad h_\mu(\omega) := \lim_n |V_n|^{-1} \log \frac{d\mu}{d\lambda}\Big|_{\underline{\underline{F}}_{V_n}}(\omega)$$

exists in $L^1(\mu)$ and satisfies

$$(3.12) \qquad h_\mu = E_\mu[H(\mu_o(\cdot|\underline{\underline{P}});\lambda_o)|\underline{\underline{J}}]$$

where $\underline{\underline{P}} := \sigma(\omega(i);i<0)$ denotes the σ-algebra of the "past" of $0 \in Z^d$ with respect to the lexicographical order on Z^d, and where

$\mu_o(\cdot|\underline{\underline{P}})$ is a conditional distribution of $\omega(0)$ with respect to $\underline{\underline{P}}$ and μ. In particular,

$$h(\mu;\lambda) = \int H(\mu_o(\cdot|\underline{\underline{P}})(\omega);\lambda_o)\mu(d\omega). \tag{3.13}$$

2) We have

$$h_\mu(\omega) = h(\mu_\omega;\lambda) \quad \mu\text{-a.s.} \tag{3.14}$$

if $\mu = \int \mu_\omega[\cdot]\mu(d\omega)$ is the ergodic decomposition of μ in (1.29).

Proof. 1) For $W \subseteq I - \{i\}$ we denote by $\mu_i(\cdot|\underline{\underline{F}}_W)(\omega)$ a conditional distribution of $\omega(i)$ with respect to $\underline{\underline{F}}_W$ and μ. Then we can write

$$|V_n|^{-1} \log \frac{d\mu}{d\lambda}\Big|_{\underline{\underline{F}}_{V_n}}(\omega) \tag{3.15}$$

$$= |V_n|^{-1} \sum_{i\in V_n} [\log \frac{d\mu_o(\cdot|\underline{\underline{F}}_{W_{n,i}})(\omega)}{d\lambda}(\omega(0))] \circ \theta_i$$

where $W_{n,i} := \{j|j<0,\ j+i \in V_n\}$. But

$$\frac{d\mu_o(\cdot|\underline{\underline{F}}_W)(\omega)}{d\lambda_o}(\omega(o)) = \frac{d\mu}{d\nu}\Big|_{\underline{\underline{F}}_{W\cup\{o\}}}$$

if we define $\nu := \mu_{I-\{o\}} \times \lambda_o$. (3.17) below shows that assumption (3.6) is satisfied in our present case. By (3.5), the $L^1(\mu)$-convergence of (3.15) is thus reduced to the convergence of

$$|V_n|^{-1} \sum_{i\in V_n} \left[\log \frac{d\mu_o(\cdot|\underline{\underline{P}})(\omega)}{d\lambda_o}(\omega(o))\right] \circ \theta_i .$$

The spatial ergodic theorem yields the existence of the limit in (3.11) with

$$h_\mu = E_\mu[\log \frac{d\mu_o(\cdot|\underline{\underline{P}})}{d\lambda_o}\Big|\underline{\underline{J}}] . \tag{3.16}$$

and this implies (3.13).

2) Integrating (3.15) we obtain

$$|V_n|^{-1} H_{V_n}(\mu;\lambda) = |V_n|^{-1} \sum_{i \in V_n} H_{W_{n,i}}(\mu;\lambda) .$$

But $H_W(\mu;\lambda) \le H_{W'}(\mu;\lambda)$ for $W' \supseteq W$, and so we get, for a fixed $W = W_{n_o,i_o}$,

$$(3.17) \qquad H_W(\mu;\lambda) = \lim_n |V_n|^{-1} \sum_{\substack{i \in V_n \\ W_{n,i} \supseteq W}} H_W(\mu;\lambda)$$

$$\le \sup_n |V_n|^{-1} H_{V_n}(\mu;\lambda) = h(\mu;\lambda) < \infty .$$

3) The same argument as in Lemma (1.25) yields $\underline{\underline{P}} \supseteq \underline{\underline{J}}$ μ-a.s. This implies $\mu_o(\cdot|\underline{\underline{P}})(\omega) = (\mu_\omega)_o(\cdot|\underline{\underline{P}})(\omega)$ μ-a.s., hence

$$h_\mu(\omega) = E_\mu[\log \frac{d\mu_o(\cdot|\underline{\underline{P}})}{d\lambda_o}|\underline{\underline{J}}](\omega) = E_\mu[\log \frac{d(\mu_\omega)_o(\cdot|\underline{\underline{P}})}{d\lambda_o}|\underline{\underline{J}}](\omega)$$

$$= \int \log \frac{d(\mu_\omega)_o(\cdot|\underline{\underline{P}})}{d\lambda_o} d\mu_\omega = h(\mu_\omega) .$$

Our next purpose is to extend the Shannon-McMillan theorem to the case where the product measure λ is replaced by a Gibbs measure whose local specification is "nice".

3.2 Specific energy

For each finite $V \subseteq I$ let U_V be a $\underline{\underline{F}}_V$-measurable function on Ω . The collection (U_V) is called an interaction potential if

$$(3.18) \qquad U_{V+i} = U_V \circ \theta_i \qquad (i \in I)$$

$$(3.19) \qquad |U| := \sum_{O \in V} ||U_V|| < \infty$$

where $||U_V||$ is the supremum norm of U_V .

For $\omega, \eta \in \Omega$ let $(\omega,\eta)_V$ denote the configuration which coincides with ω on V and with η on V^c. The quantity

(3.20) $$E_V(\omega_V|\eta) := \sum_{A\cap V\neq\emptyset} U_A((\omega,\eta)_V)$$

will be called the conditional energy of ω on V given the environment η.

(3.21) <u>Theorem.</u> For any stationary $\mu \in M_s(\Omega)$, the pointwise specific energy

(3.22) $$e_U(\omega) := \lim_n |V_n|^{-1} \sum_{A\subseteq V_n} U_A(\omega)$$

exists, μ-almost surely and in $L^1(\mu)$, and satisfies

(3.23) $$e_U(\omega) = E_\mu[\sum_{0\in A} \frac{U_A}{|A|}|\underline{\underline{J}}] = \lim_n |V_n|^{-1} E_{V_n}(\omega_{V_n}|\eta)$$

for any η. In particular, the specific energy

(3.24) $$e_U(\mu) := \lim_n |V_n|^{-1} \int E_{V_n}(\cdot|\eta)\, d\mu = \int e_U(\cdot)\, d\mu,$$

of μ exists, and

(3.25) $$e_U(\omega) = e_U(\nu_\omega) \qquad \nu\text{-a.s.}$$

if $\nu = \int \nu_\omega\, \nu(d\omega)$ is the ergodic decomposition of ν in (1.29).

<u>Proof.</u> For $g := \sum_{0\in A} \frac{U_A}{|A|}$ we have

$$\sum_{i\in V} g\circ\theta_i = \sum_{i\in V} \sum_{\substack{A:i\in A\\ A\subseteq V}} \frac{U_A}{|A|} + \sum_{i\in V} \sum_{\substack{A:i\in A\\ A\cap V^c\neq\emptyset}} \frac{U_A}{|A|}$$

$$= \sum_{A\subseteq V} \sum_{i\in V} I_A(i) \frac{U_A}{|A|} + \sum_{\substack{A\cap V^c\neq\emptyset\\ A\cap V\neq\emptyset}} \sum_{i\in V} I_A(i) \frac{U_A}{|A|}.$$

Since

$$E_V(\omega_V|\eta) = \sum_{A\subseteq V} U_A(\omega) + \sum_{\substack{A\cap V\neq\emptyset \\ A\cap V^c\neq\emptyset}} U_A((w,\eta)_V) ,$$

we obtain

$$|E_V(\omega_V|\eta) - \sum_{i\in V} (g\circ\theta_i)(\omega) | \le 2\Delta_o(V)$$

where

$$(3.26) \qquad \Delta_o(V) := \sum_{i\in V} \sum_{\substack{A:i\in A \\ A\cap V^c\neq\emptyset}} ||U_A|| \le |V - V_l||U| + |V_l| \,\gamma_l ,$$

with

$$\gamma_l := \sum_{0\in A\not\subseteq N_l(0)} ||U_A||$$

and $V_l := \{i\in V|\ \text{dist}\,(i,V^c) > l\}$. Letting first $n \uparrow \infty$ and then $l \uparrow \infty$, we see that

$$(3.27) \qquad \lim_n \frac{\Delta_o(V_n)}{|V_n|} = 0 .$$

The theorem now follows from the d-dimensional ergodic theorem.

3.3 Specific entropy with respect to a Gibbs measure

For an interaction potential (U_V) as above we introduce the local specification

$$\pi_V(\eta,\cdot) = \pi_V(\cdot|\eta) \times \prod_{j\notin V} \delta_{\eta(j)}$$

where $\pi_V(\cdot|\eta)$ is the probability measure on S^V with density

$$(3.28) \qquad \frac{d\pi_V(\cdot|\eta)}{d\lambda_V}(\xi) = Z_V(\eta)^{-1} \exp(-E_V(\xi|\eta))$$

For $l \ge 1$ we define $\partial_l V$ and γ_l as in (3.26).

(3.29) Lemma. If $\eta = \eta'$ on $\partial_1 V$ then

$$\frac{d\pi_V(\cdot|\eta)}{d\pi_V(\cdot|\eta')} \le \exp(4|V|\gamma_1)$$

Proof. Since

$$E_V(\xi|\eta) = \sum_{\substack{A\cap V\neq\emptyset \\ A\subseteq V\cup\partial_1 V}} U_A((\xi,\eta)_V) + \sum_{\substack{A\cap V\neq\emptyset \\ A\not\subseteq V\cup\partial_1 V}} U_A((\xi,\eta)_V),$$

$\eta = \eta'$ on $\partial_1 V$ implies

$$E_V(\xi|\eta') \le E_V(\xi|\eta) + 2\,\Delta_1(V)$$

with

$$\Delta_1(V) := \sum_{\substack{A\cap V\neq\emptyset \\ A\not\subseteq V\cup\partial_1 V}} ||U_A|| \le |V|\cdot\gamma_1 ,$$

hence $Z_V(\eta') \ge Z_V(\eta)\exp(-2\,\Delta_1(V))$ and vice versa. Thus,

$$\frac{d\pi_V(\cdot|\eta)}{d\pi_V(\cdot|\eta')}(\xi) = \frac{Z_V(\eta')}{Z_V(\eta)}\exp[E_V(\xi|\eta') - E_V(\xi|\eta)] \le \exp(4|V|\,\gamma_1) .$$

Now suppose that $\mu \in G_s(\pi)$ is a stationary Gibbs measure with local specification (π_V); condition (3.19) implies in fact that $G_s(\pi) \neq \emptyset$, see [Pr]. By (3.29)

$$(3.30) \qquad \lim_n |V_n|^{-1} \sup_\xi \left|\log\frac{d\mu_{V_n}}{d\lambda_{V_n}}(\xi) - \log\frac{d\mu_{V_n}(\cdot|\eta)}{d\lambda_{V_n}}(\xi)\right| = 0$$

for any η, since $\lim_l \gamma_l = 0$ due to (3.19). Thus,

$$h(\mu;\lambda) = \lim_n \frac{1}{|V_n|}\left(-\log Z_{V_n}(\eta) - \int E_{V_n}(\cdot|\eta)\,d\mu\right)$$

due to (3.12). The existence of $e_U(\mu)$ in (3.24) implies the existence of the "pressure"

(3.31) $$p_U = \lim \frac{1}{|V_n|} \log Z_{V_n}(\eta)$$

and the thermodynamical relation

(3.32) $$-p_U = h(\mu;\lambda) + e_U(\mu) .$$

We are now in a position to derive a Shannon-McMillan theorem for the specific entropy relative to a Gibbs measure $\mu \in G_s(\pi)$:

(3.33) <u>Theorem.</u> For any $\nu \in M_s(\Omega)$, the specific relative entropy with respect to μ

(3.34) $$h(\nu;\mu) := \lim |V_n|^{-1} H_{V_n}(\nu;\mu)$$

exists and satisfies

(3.35) $$h(\nu;\mu) = e_U(\nu) + h(\nu;\lambda) + p_U \geq 0 .$$

More precisely:

(3.36) $$h_{\nu;\mu}(\omega) := \lim_n |V_n|^{-1} \log \frac{d\nu_{V_n}}{d\mu_{V_n}}(\omega_{V_n})$$

exists in $L^1(\nu)$ and satisfies, ν-almost surely,

(3.37) $$h_{\nu;\mu}(\omega) = e_U(\omega) + h_\nu(\omega) + p_U = h(\nu_\omega;\mu) \geq 0 ,$$

where $\nu = \int \nu_\omega \, \nu(d\omega)$ is the ergodic decomposition of ν in (1.29). In particular,

(3.38) $$h(\nu;\mu) = \int h_{\nu;\mu}(\omega)\nu(d\omega).$$

<u>Proof.</u> (3.30) implies, for any $\eta \in \Omega$ and in $L^1(\nu)$,

$$\lim |V_n|^{-1} \log \frac{d\nu_{V_n}}{d\mu_{V_n}}(\omega_{V_n})$$

$$= \lim |V_n|^{-1} \log \frac{d\nu_{V_n}}{d\lambda_{V_n}}(\omega_{V_n}) - \lim_n |V_n|^{-1} \log \frac{d\mu_{V_n}(\cdot|\eta)}{d\lambda_{V_n}}(\omega_{V_n})$$

$$= h_\nu(\omega) \quad + \quad e_U(\omega) \quad + p_U$$

due to (3.10), (3.23) and (3.31). Integrating with respect to ν we obtain (3.34) and (3.35). If $\nu = \int \nu_\omega \, \nu(d\omega)$ is the ergodic decomposition of ν, then (3.14), (3.25), and (3.35) imply

$$h_{\nu;\mu}(\omega) = h(\nu_\omega;\lambda) + e_U(\nu_\omega) + p_U = h(\nu_\omega;\mu) \ .$$

(3.35) shows that any $\nu \in M_s(\Omega)$ satisfies

$$e_U(\nu) + h(\nu;\lambda) \geq - p_U \ ,$$

and the minimal value $-p_U$ is in fact assumed for any $\nu \in G_s(\pi)$ due to (3.32) . This is already one direction of the following variational principle due to Lanford and Ruelle [LR].

(3.39) <u>Theorem.</u> For $\nu \in M_s(\Omega)$ and $\mu \in G_s(\pi)$, the following properties are equivalent:

i) $\nu \in G_s(\pi)$

ii) $h(\nu;\mu) = 0$

iii) The function $e_U + h(\cdot;\lambda)$ assumes in ν its minimal value $-p_U$.

<u>Proof.</u> It only remains to show ii) $\Rightarrow$ i) , and here we follow [Pr].

1) Let us fix $V = V_{n_o}$. We have to show that

(3.40) $$\nu_V(\cdot|\underline{\underline{F}}_{V^c})(\omega) = \pi_V(\cdot|\omega)$$

for ν-almost all ω . Lemma (3.29) implies

(3.41) $$\lim_{W \uparrow V^c} \frac{d\mu_V(\cdot|\underline{\underline{F}}_W)(\eta)}{d\lambda_V}(\xi) = \frac{d\pi_V(\cdot|\eta)}{d\lambda_V}(\xi)$$

uniformly in ξ, η . Lemma (3.5) implies, as in the proof of (3.10), that

$$\text{(3.42)} \qquad \lim_{W \uparrow V^c} \log \frac{d\nu_V(\cdot \mid \underline{\underline{F}}_W)(\omega)}{d\lambda_V}(\omega_V) = \log \frac{d\nu_V(\cdot \mid \underline{\underline{F}}_{V^c})(\omega)}{d\lambda_V}(\omega_V)$$

in $L^1(\nu)$. Thus,

$$\text{(3.43)} \qquad \lim_{W \uparrow V^c} \int H(\nu_V(\cdot \mid \underline{\underline{F}}_W)(\omega); \mu_V(\cdot \mid \underline{\underline{F}}_W)(\omega)\nu(d\omega)$$

$$= \int H(\nu_V(\cdot \mid \underline{\underline{F}}_{V^c})(\omega); \pi_V(\cdot \mid \omega)\nu(d\omega) .$$

In view of (3.40), we have only to show that the left side of (3.43) is 0 .

2) Let $B = V \cup \partial_1 V$. For $N = (n_o+2l)k$, V_N is the union of k^d disjoint translates $B(j)$ of B , each of the form $B(j) = V(j) \cup \partial_1 V(j)$ for some translate $V(j)$ of V . We can now write

$$H_{V_N}(\nu;\mu) = H_{V_N - \bigcup_j V(j)}(\nu;\mu)$$

$$+ \sum_j \int H(\nu_{V(j)}(\cdot \mid \underline{\underline{F}}_{C(j)})(\omega); \mu_{V(j)}(\cdot \mid \underline{\underline{F}}_{C(j)})(\omega))\nu(d\omega)$$

with $C(j) = V_N - \bigcup_{m \geq j} V(m)$, hence

$$(n_o+2l)^d \, |V_N|^{-1} H_{V_N}(\nu;\mu)$$

$$\geq \frac{1}{k^d} \sum_{j=1}^{k^d} \int H(\nu_V(\cdot \mid \underline{\underline{F}}_{W(j)})(\omega); \mu_V(\cdot \mid \underline{\underline{F}}_{W(j)})(\omega))\nu(d\omega)$$

with $\partial_1 V \subseteq W(j) \subseteq V^c$. But this shows that $h(\nu;\mu) = 0$ implies that the left side of (3.43) is 0 .

(3.44) Remark. In order to prove the Shannon-McMillan theorem for the relative entropy $h(\mu;\nu)$, we have used the local specification of ν and its thermodynamical description in terms of energy. For direct approach along the lines of the proof of (3.10), we would need a canonical choice of the conditional distribution $\nu_o(\cdot|\underline{\underline{P}})(\omega)$ which does not involve the null-sets of ν, and a continuity property which would guarantee the convergence

$$(3.45) \qquad \lim_{W \uparrow \{i;i<o\}} \frac{d\nu_o(\cdot|\underline{\underline{F}}_W)}{d\lambda_o} = \frac{d\nu_o(\cdot|\underline{\underline{P}})}{d\lambda_o}$$

with respect to μ. In that case, we would obtain the formula

$$(3.46) \qquad h(\mu;\nu) = \int \mu(d\eta) H(\mu_o(\cdot|\underline{\underline{P}})(\eta);\nu_o(\cdot|\underline{\underline{P}})(\eta))$$

in analogy to (3.13). If the local specification of ν satisfies the Dobrushin condition (2.6) and if S is finite, then (3.45) follows, as in the proof of the Feller property in (2.11), and so (3.46) does hold.

(3.47) <u>Remark on the Ising model.</u> Consider the ferromagnetic Ising model with $S = \{+1,-1\}$ and

$$U_V(\omega) = \beta\omega(i)\omega(j) \quad \text{if } V = \{i,j\} \text{ and } |i-j| = 1$$
$$= 0 \quad \text{else}$$

where $\beta > 0$. The limits $\lim_n \pi_{V_n}(\eta,\cdot)$ in (1.19) with $\eta \equiv +1$ resp. $\eta \equiv -1$ define two measures μ^+ and μ^- in $G_s(\pi)$; for large enough β, these two measures are different. For any measurable bounded function which is monotone with respect to the coordinatewise partial ordering on $\Omega = S^I$, and for any $\mu \in G(\pi)$,

$$(3.48) \qquad \int f d\mu^- \le \int f d\mu \le \int f d\mu^+$$

by the FKG inequality; cf. [Pr]. In particular,

$$(3.49) \qquad |G(\pi)| = 1 \iff \mu^+ = \mu^- .$$

Both μ^+ and μ^- have the global Markov property. In fact, for any $J \subseteq I$ the conditional distribution $\mu^+(\cdot|J)(\eta)$ can be defined as the "+-state" corresponding to the conditional specification on S^{I-J} determined by $\eta_{\partial J}$ as in (2.2); cf. [Fö3], [Go]. The monotonicity argument in [Fö3] shows that we also get (3.45), hence

(3.46), for $\nu = \mu^+$ and $\mu = \mu^-$. But the variational principle (3.39) implies $h(\mu|\mu^+) = 0$ for any $\mu \in G_s(\pi)$ and in particular for $\mu = \mu^-$. Thus, (3.46) implies

$$(3.50) \qquad \mu_o^+(\cdot|\underline{\underline{P}})(\eta) = \mu_o^-(\cdot|\underline{\underline{P}})(\eta)$$

for μ-almost all η, and for any $\mu \in G_S(\pi)$. Again by monotonicity, (3.50) implies $\mu^+(\cdot|\underline{\underline{P}})(\eta) = \mu^-(\cdot|\underline{\underline{P}})(\eta)$, and so (3.48), applied to the conditional system, leads to the following

(3.51) <u>Corollary.</u> For any $\mu \in G(\pi)$ and for μ-almost all η, the conditional specification on $J = \{j|j \nless 0\}$ determined by $\eta_{\partial J}$ admits no phase transition.

Thus, the Ising model can be viewed as a Markov chain as in 2.2, with a fixed transition probability, but with different invariant measures as soon as there is a phase transition.

4. Large deviations

Let μ be a stationary Gibbs measure with marginal distribution μ_o , and suppose that μ is an extremal point in $G_s(\pi)$, hence ergodic by (1.26). The ergodic behavior of a typical configuration under μ can be described on various levels:

(1) convergence of average values for functions on S , i.e.,

$$\lim_n |V_n|^{-1} \sum_{i \in V_n} f(\omega(i)) = \int f d\mu_\omega ,$$

(2) convergence of the empirical distribution, i.e.,

$$\lim_n |V_n|^{-1} \sum_{i \in V_n} \delta_{\omega(i)} = \mu_o ,$$

(3) convergence of the empirical field, i.e.,

$$\lim_n |V_n|^{-1} \sum_{i \in V_n} \delta_{\theta_i \omega} = \mu .$$

On each level, we can look at large deviations from ergodic behavior. In the classical case where the random variables $\omega(i)$ $(i \in Z^d)$ are independent and identically distributed under μ , large deviations on level (1) are described by the theorem of Cramér and Chernoff, on level (2) by Sanov's theorem; cf., e.g., [Az]. Level (3) was introduced by Donsker and Varadhan in [DV2]. Here the situation is more subtle even in the classical case; in particular, the spatial structure of $I = Z^d$ may come in explicitely. For the one-parameter case with a "nice" stationary process μ , we refer to [DV2] and [Or2].

In this section we give an introduction to large deviations on level (3) for Gibbs measures in the multiparameter case $I = Z^d$, with special emphasis on the lower bound. In the first part, we state a general result obtained with S.Orey [FO], where the rate of convergence to 0 of a large deviation

$$\mu[|V_n|^{-1} \sum_{i \in V_n} \delta_{\omega(i)} \in A]$$

with $\mu \notin \bar{A}$ is described in terms of specific relative entropies $h(\nu;\mu)$. Here the Shannon-Mc Millan theorem (3.33) provides the technical key to the lower bound.

In the second part we discuss some joint work with M. Ort on the effect of a phase transition. Here it may happen that A contains another Gibbs measure $\nu \in G_S(\pi)$, and this will slow down the convergence to 0 . More precisely, the order of convergence becomes exponential in the surface area $|\partial V_n|$ instead of the volume $|V_n|$, and the rate of convergence will now be described in terms of "specific surface entropies" $h_\partial(\nu;\mu)$.

4.1. Large deviations for the empirical field of a Gibbs measure

Consider a stationary Gibbs measure $\mu \in G_s(\pi)$ whose local specification is given by an interaction potential as in Section 3. For finite $V \subseteq I$ define the empirical field on V as the random element of $M_s(\Omega)$ given by

$$\rho_V(\omega) := |V|^{-1} \sum_{i \in V} \delta_{\theta_i \omega} .$$

In joint work with S. Orey [FO], it is shown that the sequence

$$\mu \circ \rho_{V_n}^{-1} \qquad (n=1,2,\dots)$$

satisfies a large deviation principle where the rate function is given by the specific relative entropy $h(\cdot;\mu)$:

(4.1) Theorem [FO]. If $A \subseteq M(\Omega)$ is open then

$$\text{(4.2)} \qquad \liminf_n |V_n|^{-1} \log \mu[\rho_{V_n} \in A] \geq - \inf_{\nu \in A \cap M_s(\Omega)} h(\nu;\mu) ,$$

and if $A \subseteq M(\Omega)$ is closed then

$$\text{(4.3)} \qquad \limsup_n |V_n|^{-1} \log \mu[\rho_{V_n} \in A] \leq - \inf_{\nu \in A \cap M_s(\Omega)} h(\nu;\mu)$$

F. Comets [Co] and S. Olla [Ol] have obtained the same result with a different method: they first study the case where μ is a product measure and then pass to the case of Gibbs measures by using Varadhan's abstraction of the Laplace method. In [FO] the case of Gibbs measures is analyzed directly. Here we show only how the lower bound follows from the thermodynamical limit laws in Section 3; for the proof of the upper bound see [FO].

Proof of the lower bound: 1) It is enough to show that

$$(4.4) \qquad \liminf |V_n|^{-1} \log \mu[\rho_{V_n} \in G] \geq - h(\nu;\mu)$$

for any $\nu \in M_s(\Omega)$ and any open neighborhood G of ν. Let us first assume that ν is ergodic. It is no loss of generality to assume that G is of the form

$$G = \bigcap_{k=1}^{n} \{ \tilde{\nu} \mid |\int f_k d\tilde{\nu} - \int f_k d\nu | < \epsilon \}$$

where $f_1,\dots,f_n$ are bounded uniformly continuous functions which are $\underline{\underline{F}}_{N_p(o)}$-measurable for some $p \geq 1$. In that case,

$$(4.5) \qquad \{ \rho_{V_n} \in G \} \in \underline{\underline{F}}_{W_n}$$

where $W_n := V_n \cup \partial_p V_n$. Let φ_{W_n} denote the Radon-Nikodym density of ν with respect to μ on $\underline{\underline{F}}_{W_n}$. The Shannon-McMillan theorem (3.36) shows that

$$(4.6) \qquad \lim_n |V_n|^{-1} \log \varphi_{W_n} = \lim_n |W_n|^{-1} \log \varphi_{W_n} = h(\nu;\mu)$$

in $L^1(\nu)$. Due to (4.5) we can write

$$(4.7) \qquad \mu[\rho_{V_n} \in G] \geq \mu[\rho_{V_n} \in G \,,\, |V_n|^{-1} \log \varphi_{W_n} < h(\nu;\mu) + \epsilon]$$

$$\geq \exp(-|V_n|(h(\nu;\mu)+\epsilon)) \, \nu[\rho_{V_n} \in G \,,\, |V_n|^{-1} \log \varphi_{V_n} < h(\nu;\mu) + \epsilon].$$

Since ν was assumed to be ergodic, $\rho_{V_n} \to \nu$ ν-a.s.. This together with (4.6) implies that the last factor in (4.7) converges to 1, and so we obtain (4.4)

2) For a general $\nu \in M_s(\Omega)$ take a sequence ν_n $(n=1,2,\ldots)$ of ergodic measures as in the following lemma. Since $\nu_n \in G$ for $n \geq n_o$, the left side of (4.4) is bounded from below by $-h(\nu;\mu)$.

The following lemma is a spatial version of a construction in [Or].

(4.8) <u>Lemma:</u> For any $\nu \in M_s(\Omega)$ there is a sequence of ergodic measures ν_n $(n=1,2,\ldots)$ such that $\nu_n \to \nu$ weakly and

$$\lim_n h(\nu_n;\mu) = h(\nu;\mu) . \tag{4.9}$$

<u>Proof:</u> 1) For $n \geq 1$ denote by ν_n' the measure which coincides with ν on each σ-field $\underline{\underline{F}}_{V_n+n\cdot k}$ $(k \in Z^d)$ and makes these σ-fields independent. The measure

$$\nu_n := |V_n|^{-1} \sum_{i \in V_n} \nu_n' \circ \theta_i^{-1}$$

is stationary. To show that ν_n is ergodic let $B \in \underline{\underline{J}}$ be an invariant event. Since $\underline{\underline{J}} \subseteq \underline{\underline{A}}$ mod ν_n by (1.25), there exists $B^* \in A$ such that $B = B^*$ ν_n-a.s., hence ν_n'-a.s. But ν_n' satisfies the Kolmogorov zero-one law, and so $\nu_n'(B) = \nu_n'(B^*) \in \{0,1\}$. Since B is invariant we obtain $\nu_n(B) = \nu_n'(B) \in \{0,1\}$.

2) For an $\underline{\underline{F}}_{V_p}$-measurable bounded function φ we have

$$\int \varphi \, d(\nu_n' \circ \theta_i^{-1}) = \int \varphi \, d\nu$$

if $i \in V_{n-p}$. Since $\lim_n |V_n|^{-1} |V_n - V_{n-p}| = 0$ we get

$$\lim_n \int \varphi d\nu_n = \int \varphi d\nu . \tag{4.10}$$

(4.10) implies weak convergence, and since $h(\cdot;\lambda)$ is lower semicontinuous we obtain

(4.11) $$\liminf_n h(\nu_n;\lambda) \geq h(\nu;\lambda) .$$

3) For $i \in V_n$ there are $(N-1)^d$ disjoint translates of V_n+i contained in $V_{N\cdot n}$, and this implies

$$H_{V_{N\cdot n}}(\nu_n' \circ \theta_i ;\lambda) \leq (N-1)^d H_{V_n}(\nu;\lambda) .$$

By convexity,

$$H_{V_{N\cdot n}}(\nu_n;\lambda) \leq |V_n|^{-1} \sum_{i\in V_n} H_{V_{N\cdot n}}(\nu_n' \circ \theta_i ;\lambda)$$

$$\leq (N-1)^d H_{V_n}(\nu;\lambda) .$$

Letting $N \uparrow \infty$, we obtain

$$h(\nu;\lambda) \geq |V_n|^{-1} H_{V_n}(\nu;\lambda) \geq h(\nu_n;\lambda) .$$

This together with (4.11) implies $\lim_n h(\nu_n;\lambda) = h(\nu;\lambda)$. But

$$h(\nu_n;\mu) = h(\nu_n;\lambda) + e_U(\nu_n) + p_U ,$$

and since

$$e_U(\nu_n) = \int \sum_{0\in A} \frac{U_A}{|A|} \, d\nu_n ,$$

we have

$$\lim_n e_U(\nu_n) = e_U(\nu)$$

due to (3.19) and (4.10). This implies (4.9) .

4.2. The effect of a phase transition

Let $\mu \in G_s(\pi)$ be a stationary Gibbs measure. If A is a open set on $M(\Omega)$ such that $\mu \notin \bar{A}$, then

$$(4.12) \qquad \lim_n \mu[\rho_{V_n} \in A] = 0 \ .$$

If a phase transition $|G_s(\pi)| > 1$ occurs then we may have $A \cap G_s(\pi) \neq \emptyset$, hence

$$\inf_{\nu \in A \cap M_s(\Omega)} h(\nu;\mu) = 0$$

due to the variational principle (3.39). Thus, the lower bound (4.2) implies

$$(4.13) \qquad \lim |V_n|^{-1} \log \mu[\rho_{V_n} \in A] = 0 \ ,$$

i.e., the convergence in (4.12) is slower than exponential in the volume $|V_n|$.

Let us suppose that the local specification (π_V) has the Markov property. Then the conditional distributions of μ, $\nu \in G_s(\pi)$ on the boundary ∂V_n coincide, and so the relative entropy on $V_n \cup \partial V_n$ is given by

$$H_{V_n \cup \partial V_n}(\nu;\mu) = H_{\partial V_n}(\nu;\mu) \ .$$

But the relative entropy $H_{\partial V_n}(\nu;\mu)$ on the surface ∂V_n is of the order of $|\partial V_n|$. This suggests that in (4.13) the volume $|V_n|$ should be replaced by the surface area $|\partial V_n|$, and that the specific entropy $h(\nu:\mu)$ should be replaced by a "specific surface entropy" of the form

$$(4.14) \qquad h_\partial(\nu;\mu) = \lim_n |\partial V_n|^{-1} H_{\partial V_n}(\nu;\mu) \ .$$

Here we scetch the argument for a lower bound in terms of surface entropy; cf. [Ort] and [FOr] for more details, and [Ort] for results on the upper bound. For the study of large deviations in terms of surface area $|\partial V_n|$ see also [Scho].

Let us consider the two-dimensional Ising model (3.47) in the presence of a phase transition $\mu^+ \neq \mu^-$. Restricted to the line $L = \{t \in Z^2; t_2 = 0\}$, μ^+ and μ^- may be viewed as stationary processes on $\{+1,-1\}^Z$, and we define $h_\partial(\mu^-;\mu^+)$ as the specific relative entropy of one process with respect to the other:

(4.15) $$h_\partial(\mu^-;\mu^+) := \int H(\mu_0^-(\cdot|\underline{\underline{P}}^1)(\omega);\mu_0^+(\cdot|\underline{\underline{P}}^1)(\omega))\mu^-(d\omega)$$

where $\underline{\underline{P}}^1 := \sigma(\omega(t) ; t \in L, t_1 < 0)$ denotes the one-dimensional "past". In (4.15), $\mu_0^+(\cdot|\underline{\underline{P}})(\omega)$ is defined, for any $\omega \in \Omega$, as the "+-state" which corresponds to the conditional specification determined by the restriction of ω to $\{t \in L; t_1 < 0\}$; cf. (3.47). In proving the lower bound below, the main problem is to show that (4.15) coincides with the surface entropy (4.14), and that there is a Shannon-Mc Millan theorem on surfaces behind the existence of the limit in (4.14).

(4.16) Theorem (FOr): Suppose that $A \subseteq M_s(\Omega)$ is an open set such that $\mu^+ \notin \bar{A}$ but $\mu^- \in A$. Then

(4.17) $$\liminf_n |\partial V_n|^{-1} \log \mu^+[\rho_{V_n} \in A] \geq h_\partial(\mu^-;\mu^+)$$

Proof. 1) We proceed as in the proof of the general lower bound (4.2), with $\mu = \mu^+$ and $\nu = \mu^-$, and so we have to control the densities

$$\mu^-[\omega_{V \cup \partial V_n}]/\mu^+[\omega_{V \cup \partial V_n}] = \mu^-[\omega_{\partial V_n}]/\mu^+[\omega_{\partial V_n}] .$$

The difference is that now we need a Shannon-Mc Millan theorem for

$$|\partial V_n|^{-1} \log(\mu^-[\omega_{\partial V_n}]/\mu^+[\omega_{\partial V_n}]) ,$$

where we divide by the surface area $|\partial V_n|$ instead of the volume $|V_n|$. By martingale convergence, it is not hard to obtain

$$\lim_n |\partial V_n|^{-1} \log \mu^-[\omega_{\partial V_n}] = \int \log \mu_o^-(\cdot|\underline{\underline{P}}^1)d\mu^-$$

in $L^1(\mu^-)$. The proof of the $L^1(\mu^-)$-convergence of

$$\lim_n |\partial V_n|^{-1} \log \mu^+[\mu_{\partial V_n}] = \int \log \mu_o^+(\cdot|\underline{\underline{P}}^1)d\mu^- \tag{4.18}$$

is more subtle: this requires a combination of martingale techniques and monotonicity arguments. We argue separately on the four sides of ∂V_n . On the first side $S(1)$, we have

$$\frac{1}{n} \log \mu^+[\omega_{S(1)}] = \frac{1}{n} \sum_{t=0}^{n-1} \log \mu_o^+(\omega(0)|\omega_t) \circ \theta_t \tag{4.19}$$

where ω_t denotes the restriction of ω to $L_t := \{i \in L ; t \le i_1 < 0\}$. Let ω_t^+ be the configuration which coincides with ω_t on L_t and is $\equiv +1$ on $\{i \in Z^2; i_1 < -t\}$. By the FKG inequality (cf., e.g., [Pr]),

$$\mu_o^+(+1|\omega_t) \le \mu_o^+(+1|\omega_t^+) ,$$

hence

$$\limsup_t \mu_o^+(+1|\omega_t) \le \lim_n \mu_o^+(+1|\omega_t^+) = \mu_o^+(+1|\underline{\underline{P}}^1)(\omega) ; \tag{4.20}$$

the limit on the right exists for any ω by monotonicity. Since $\mu_o^+(+1|\underline{\underline{P}}^1)(\omega)$ is monoton increasing in ω , we obtain

$$\begin{aligned} \mu_o^+(+1|\omega_t) &= E_{\mu^+}[\mu_o^+(+1|\underline{\underline{P}}^1)|\omega_t] \\ &\ge E_{\mu^-}[\mu_o^+(+1|\underline{\underline{P}}^1)|\omega_t] , \end{aligned} \tag{4.21}$$

using the inequality (3.48) for the conditional system determined by ω_t . But by martingale convergence,

$$\lim_t E_{\mu^-}[\mu_o^+(+1|\underline{\underline{P}}^1)|\omega_t] = \mu_o^+(+1|\underline{\underline{P}}^1)(\omega) \tag{4.22}$$

μ^--almost surely. This together with (4.20) and (4.21) implies

$$(4.23) \qquad \lim_t \mu_o^+(\omega(o)|\omega_t) = \mu_o^+(\omega(o)|\underline{\underline{P}}^1)(\omega) \qquad \mu^- \text{-a.s.}$$

Applying the ergodic theorem in (4.19) and using (4.23) we obtain

$$\lim_n |\partial V_n|^{-1} \log \mu^+[\omega_{S(1)}] = \frac{1}{4} \int \log \mu_o^+(\cdot|\underline{\underline{P}})d\mu^- .$$

On the remaining sides we have to argue "around corners", and monotonicity comes in in a more complicated manner. We refer to [FOr] for the complete argument.

II. Infinite dimensional diffusions

In this chapter we consider diffusion processes of the form

$$X(t) = (X_i(t))_{i \in I} \qquad (0 \leq t \leq 1)$$

with some countable index set I, which satisfy an infinite dimensional stochastic differential equation of the form

$$dX_i = dW_i + b_i(X(t),t)dt \qquad (i \in I) ;$$

$(W_i)_{i \in I}$ is a collection of independent Brownian motions. Such a process induces a probability measure P on $(C[0,1])^I$ and may thus be viewed as a random field with state space $S = C[0,1]$ at each site $i \in I$. This point of view is first illustrated in Section 1 where we look at some large deviations of an infinite dimensional Brownian motion.

In Section 2 we introduce a class of interactive diffusion processes of the above form where the interaction in the drift terms is controlled by an entropy condition. This condition allows us to describe the time reversed process by a stochastic differential equation of the same type, and to derive an infinite dimensional analogue to the classical duality equation which relates the forward description of a diffusion process to its backward description. In particular, this leads to the well known characterization of reversible invariant measures of interacting diffusion processes as Gibbs measures. But as in the classical duality theory of Markov processes, time reversal is also an important tool in the study of transient phenomena. This point is illustrated in Section 2.3 where we look at an explicit example of a infinite-dimensional Martin boundary and at a corresponding problem of large deviations.

In Section 3 we return to the description of an infinite dimensional diffusion process as a random field P on S^I with $S = C[0,1]$. In the interactive case, we have to determine the local specification of this random field if we want to apply the random field techniques of Ch. I. Thus, we have to look at the conditional diffusion at a site $i \in I$, given full information on the trajectories $X_j(t)$ $(0 \leq t \leq 1)$ for any $j \neq i$; this is related to a "smoothing" problem in non-linear filtering theory [Pa]. As an application, we show how Dobrushin's contraction technique in Section 2 of Ch. I can be applied to an infinite dimensional diffusion process; this leads to a spatial central limit theorem. In a similar way, and in view of large deviations in the interactive case, one could apply the "thermodynamical" techniques of Section 3 and 4 of Ch. I.

Section 2 is based on joint work with A. Wakolbinger [FW], Section 3 on recent work of J.D. Deuschel [De1,2,3]. It should be clear that this introduction to infinite dimensional diffusion processes is limited to some rather special topics; some of the central issues, for example the convergence of an interacting diffusion to an invariant Gibbs measure (e.g. [HS1,2]), are not even touched upon.

1. Some large deviations of infinite dimensional Brownian motion

Let $X_o, X_1, \ldots$ be a sequence of independent Brownian motions

$$X_i(t,\omega) \qquad (0 \le t \le 1 \ , \ i = 0,1,\ldots)$$

on R^d, each with initial distribution μ_o. We can take $X_o, X_1, \ldots$ to be the coordinate maps on $\Omega = S^I$, with $S = C([0,1],R^d)$ and $I = \{0,1,\ldots\}$, under the product measure

$$P = \prod_{i \in I} P_i$$

where P_i denotes Wiener measure on S with initial distribution μ_o. For large deviations of the empirical distribution

$$\frac{1}{n} \sum_{i=0}^{n-1} \delta_{X_i(\omega)} \in M(S) \quad ,$$

described by some subset A_o of $M(S)$, the general results of Ch. I imply

$$(1.1) \qquad P\Big[\frac{1}{n}\sum_{i=o}^{n-1} \delta_{X_i} \in A_o\Big] \sim \exp\Big[-n \inf_{Q:Q_o \in A_o} h(Q;P)\Big]$$

where Q_o denotes the marginal distribution of the stationary probability measure Q on S^I. For each n, the relative entropy $H_n(Q;P)$ with respect to P on $\underline{\underline{F}}_n := \sigma(X_o,\ldots,X_n)$ satisfies

$$H_n(Q;P) = H_n(Q;\Pi_i Q_i) + \sum_{i=o}^{n} \int \log \frac{dQ_i}{dP_i}(X_i)dQ$$

$$\geq (n+1)\, H(Q_o;P_o)\ ,$$

and this implies

$$h(Q;P) \geq h(\Pi Q_i;P) = H(Q_o;P_o)$$

for any $Q \in M_S(\Omega)$ with marginal distribution Q_o on S. Thus, (1.1) reduces to Sanov's theorem

$$(1.2) \qquad P\Big[\frac{1}{n}\sum_{i=o}^{n-1} \delta_{X_i} \in A_o\Big] \sim \exp\Big[-n \inf_{Q_o \in A_o} H(Q_o;P_o)\Big]\ .$$

Roughly speaking, (1.2) can be interpreted as follows. Under the condition that a large deviation described by A_o does occur, the process $(X_o,X_1,\ldots)$ is most likely to behave like a collection of independent diffusions, each given by that measure $Q_o \in A_o$ which minimizes the entropy $H(\cdot;P_o)$ with respect to Wiener measure P_o. Therefore, one would like to describe the minimizing measure Q_o more explicitely. We are going to illustrate this point for some specific choices of A_o. The first three examples are classical, the last appears in recent work of Dawson and Gärtner [DG].

(1.3) <u>Remark on the Girsanov transformation.</u> 1) Let Q_o be a probability measure on $S = C([0,1], R^d)$. We denote by $X(t)$ $(0 \leq t \leq 1)$ the coordinate process on S and by ν_t resp. μ_t the distribution of $X(t)$ under Q_o resp. P_o. Suppose that Q_o is absolutely continuous with respect to P_o. Then the theory of the (Cameron-Martin-Maruyama-) Girsanov transformation shows that there is an R^d - valued adapted drift process b_t $(0 \leq t \leq 1)$ on S such that

$$(1.4) \qquad \int_0^1 b_t^2 \, dt < \infty \quad Q_o\text{-a.s.}$$

$$(1.5) \qquad W^b(t) := X(t) - X(0) - \int_0^t b_s \, ds$$

is a Wiener process under Q_o,

$$(1.6) \qquad \frac{dQ_o}{dP_o} = \frac{d\nu_o}{d\mu_o} \exp\left[\int_0^1 b_t dW^b(t) + \frac{1}{2}\int_0^1 b_t^2 dt\right]$$

$$= \frac{d\nu_o}{d\mu_o} \exp\left[\int_0^1 b_t dX(t) - \frac{1}{2}\int_0^1 b_t^2 dt\right],$$

cf.,e.g., [IW]. Moreover, we have $H(Q_o;P_o) < \infty$ if and only if

$$H(\nu_o;\mu_o) + E_{Q_o}\left[\int_0^1 b_t^2 dt\right] < \infty ,$$

and in that case the entropy is given by

$$(1.7) \qquad H(Q_o;P_o) = H(\nu_o;\mu_o) + \frac{1}{2} E_{Q_o}\left[\int_0^1 b_t^2 dt\right] .$$

2) If, more generally, X is a Wiener process with respect to a filtration $(\underline{\underline{F}}_t)$ on some probability space $(\Omega,\underline{\underline{F}},P_o)$ and if Q_o is a probability measure on $\underline{\underline{F}}$ with $H(Q_o;P_o) < \infty$ then there exists an adapted drift process with "finite energy"

$$E_{Q_o}\left[\int_0^1 b_t^2 dt\right] < \infty$$

such that (1.5) holds. In that case, the drift can be computed as a stochastic forward derivative in the sense of Nelson [Ne]: for almost all $t \in [0,1]$,

$$(1.8) \qquad b_t = \lim_{h\downarrow 0} \frac{1}{h} E_{Q_o}[X(t+h)-X(t)\,|\,\underline{\underline{F}}_t]$$

$$= \lim_{\alpha,\beta\downarrow 0} \frac{1}{\alpha+\beta} \int_{t-\alpha}^{t+\beta} b_s ds$$

in $L^2(Q_o)$; cf. [Fö6].

1.1. Large deviations of the average positions: Schilder's theorem

For $B \subseteq C([0,1],R^d)$ let us consider a large deviation of the form

$$(1.9) \qquad \frac{1}{n} \sum_{i=o}^{n-1} X_i \in B ;$$

since

$$P[\frac{1}{n} \sum_{i=0}^{n-1} X_i \in B] = P_o[n^{-1/2} X_o \in B] ,$$

this can also be regarded as a large deviation of a single Brownian motion with small variance. (1.9) means that in Sanov's theorem (1.2) we take $A_o = \{Q_o | \varphi_{Q_o} \in B\}$ where

$$\varphi_{Q_o}(t) := \int X(t) dQ_o \qquad (0 \le t \le 1) .$$

Let us assume that the initial distribution μ_o is concentrated on 0 , and let us fix $\varphi \in S$ with $\varphi(o) = 0$. Then the problem is reduced to finding a measure Q_o which minimizes $H(\cdot;P_o)$ under the constraint $\varphi_{Q_o} = \varphi$. The function φ is of interest only if there exists some Q_o with $\varphi_{Q_o} = \varphi$ and finite entropy $H(\cdot;P_o)$. But this implies, due to (1.5) and (1.7) ,

$$\varphi(t) = E_{Q_o}[\int_0^t b_s ds] = \int_0^t E_{Q_o}[b_s] ds ,$$

i.e., φ is absolutely continuous with derivative $\nabla\varphi(t) = E_{Q_o}[b_t]$ for almost all t . Moreover,

$$\frac{1}{2}\int_0^1 (\nabla\varphi)^2 dt \le \frac{1}{2}\int_0^1 E_{Q_o}[b_t^2]dt$$

$$= H(Q_o;P_o) < \infty ,$$

i.e. φ is an absolutely continuous function with "finite energy". For such a φ , the lower bound $\frac{1}{2}\int_0^1 (\nabla\varphi)^2 dt$ for the entropy $H(\cdot;P_o)$ is in fact attained if Q_o is the measure with $\nu_o = \delta_o$ and deterministic drift $b_t(\cdot) \equiv \nabla\varphi(t)$. In particular, Sanov's theorem (1.2) implies

(1.10) Schilder's theorem:

$$P[\frac{1}{n}\sum_{i=0}^{n-1} X_i \in B] \sim \exp[-n \inf_{\varphi\in B_o} \frac{1}{2}\int_0^1 (\nabla\varphi)^2 dt]$$

where B_o denotes the intersection of B with the class of absolutely continuous functions with finite energy; cf., e.g., [St].

1.2 Large deviations at the terminal time: h-path processes

If we are only interested in large deviations of the empirical distribution at the terminal time $t = 1$, the set A_o in (1.2) is of the form

$$A_o = \{Q_o | Q_o \circ X(1)^{-1} \in B\}$$

for some $B \subseteq M(R^d)$. This means that, for a given probability measure ν_1 on R^d, we have to find the measure Q_o with minimal entropy $H(\cdot;P_o)$ under the constraint that the marginal distribution at time 1 is given by ν_1 . But we can write

$$(1.11) \qquad H(Q_o;P_o) = H(\nu_1;\mu_1) + \int \nu_1(dy)H(Q_o^y;P_o^y) ,$$

where Q_o^y and P_o^y denote the conditional distributions of Q_o and P_o on $S = C([0,1],R^d)$, given the terminal value $X(1,\cdot) = y$.

Thus, $H(\nu_1;\mu_1)$ is a lower bound, and this lower bound is attained by

$$(1.12) \qquad Q_o := \int \nu_1(dy)\, P_o^y \quad ,$$

i.e., by the distribution of Brownian motion starting with μ_o and conditioned to have terminal distribution ν_1 at time 1. In particular,

$$(1.13) \qquad P\Big[\frac{1}{n}\sum_{i=o}^{n-1} \delta_{X_i(1)} \in B\Big] \quad \sim \quad \exp\Big[-n \inf_{\nu_1 \in B} H(\nu_1;\mu_1)\Big]$$

Of course, (1.13) is nothing else than Sanov's theorem applied to the classical case of a sequence of independent random variables with d-dimensional normal distribution. But on the higher level (1.1) we obtain additional information. For a given ν_1 with finite entropy $H(\nu_1;\mu_1)$, denote by $h(\cdot,1)$ the density of ν_1 with respect to μ_1 and by $h(x,t)$ the space-time harmonic function

$$(1.14) \qquad h(x,t) \quad := \quad \int h(y,1)\, p_{1-t}(x,y)\, dy$$

where $p_s(x,y)$ is the transition density of d-dimensional Brownian motion. Itô's formula, applied to $\log h$, shows that the measure Q_o in (1.12) is given by the density

$$\frac{dQ_o}{dP_o} = h(X(1),1)$$

$$= h(X(0),0)\, \exp\Big[\int_0^1 b_t dX - \frac{1}{2}\int_0^1 b_t^2\, dt\Big]$$

with drift

$$(1.15) \qquad b_t = \nabla \log h(X(t),t)$$

In other words, Q_o is an h-path process in the sense of J.L. Doob. The large deviation result (1.1) can now be interpreted as follows. Under a large deviation of the empirical distribution at the terminal time $t=1$, the process $(X_o, X_1, \ldots)$ behaves like a collection of independent h-path processes, where h is the space-time harmonic function induced by the measure ν_1 which has minimal entropy $H(\cdot;\mu_1)$ under the given constraint B.

1.3. Large deviations at the initial and the terminal time: Schrödinger bridges

Let us now look at large deviations of the empirical distribution both at time 0 and at time 1 ; here we assume that the initial distribution μ_o is equivalent to Lebesgue measure. This leads to the problem of finding the measure Q_o on S which minimizes $H(\cdot;P_o)$ under the constraint that the marginal distributions ν_o at time 0 and ν_1 at time 1 are fixed.

Let P_x^y denote the conditional distribution of Wiener measure on $C([0,1],R^d)$ given an initial value x at time 0 and a terminal value y at time 1 , i.e., P_x^y is the distribution of the Brownian bridge leading from x to y. Let μ and ν denote the joint distribution of $(X(0), X(1))$ under P_o and Q_o , and let us write

$$H(Q_o;P_o) = H(\nu;\mu) + \int \nu(dx,dy)\, H(Q_x^y;P_x^y) \quad .$$

This shows that $H(\cdot;P_o)$ is minimized by the measure

$$Q_o := \int \nu(dx,dy)\, P_x^y \ , \tag{1.16}$$

if ν is the measure on $R^d \times R^d$ which minimizes $H(\cdot;\mu)$ under the constraint that the marginals are given by ν_o and ν_1 . If the marginals admit any measure on the product space with finite entropy $H(\cdot;\mu)$, then a unique minimizing measure ν does exist, and its density is of the form

$$\frac{d\nu}{d\mu}(x,y) = f(x)\, g(y) \tag{1.17}$$

with $\log f \in L^1(\nu_o)$ and $\log g \in L^1(\nu_1)$; cf. [Cs] resp. the argument below.

The representation (1.17) implies the two equations

$$(1.18)\qquad \frac{d\nu_o}{d\mu_o}(x) = f(x)\int p_1(x,y)g(y)dy$$

$$\frac{d\nu_1}{d\mu_1}(y) = g(y)\int p_1(x,y)f(x)\frac{d\mu_o}{d\lambda}(x)dx\,\frac{d\mu_1}{d\lambda}(y)^{-1}\ .$$

It implies also that the associated measure Q_o in (1.16) has the Markov property. More precisely, Q_o is an h-path process with

$$h(x,t) := \int p_{1-t}(x,y)\,g(y)dy\ .$$

In fact we can write, using (1.18) and applying Itô's formula as in the previous case,

$$\frac{dQ_o}{dP_o} = f(X(0))\,g(X(1))$$

$$= \frac{d\nu_o}{d\mu_o}(X(0))\,\exp\Big[\int_o^1 b_t dX(t) - \frac{1}{2}\int_o^1 b_t^2 dt\Big]$$

where the drift process (b_t) is given by

$$(1.19)\qquad b_t = \nabla \log h(X(t),t)\ .$$

(1.20) Remark. The dual equations (1.18), with $\mu_o = \mu_1 = \lambda$, appear in a paper by E. Schrödinger [Sch1], derived in a more heuristic manner, but clearly motivated by the question of large deviations: "Imaginez que vous observez un système de particules en diffusion, qui soient en équilibre thermodynamique. Admettons qu'à un instant donné t_o vous les ayez trouvées en répartition à peu près uniforme et qu'à $t_1 > t_o$ vous ayez trouvé un écart spontané et considérable par rapport à cette uniformité. On vous demande de quelle manière cet écart s'est produit. Quelle en est la manière la plus probable?" [Sch2]. The existence of solutions f and g is studied in [Fo], [Be]; it also follows by the probabilistic argument below. The associated "Schrödinger bridge" or "reciprocal process" Q_o in (1.16) is described in detail in [Ja]. Schrödinger comments in [Sch1] on "merkwürdige Analogien zur Quantenmechanik, die mir sehr des Hindenkens wert erscheinen"; a systematic account of the present role of Schrödinger bridges in Nelson's Stochastic Mechanics can be found in [Za].

The derivation of the crucial factorization (1.17) can be based on the following general argument, which will also be used in the next section; cf. [Cs].

Let μ be a probability measure on some polish space, and let f_i (i=1,2,...) be a sequence of bounded measurable functions. We want to minimize the entropy $H(\nu;\mu)$ under the constraints

(1.21) $$\int f_i d\nu = c_i \quad (i=1,2,\ldots) .$$

Suppose that there exists some ν which satisfies (1.21) and has finite entropy $H(\nu;\mu) < \infty$. Then, and this is well known from the theory of exponential families (see, e.g., [Az]), there is a unique ν_n which minimizes $H(\cdot;\mu)$ under the first n constraints, and its density is of the form

(1.22) $$\frac{d\nu_n}{d\mu} = Z_n^{-1} \exp\left(\sum_{i=1}^{n} \lambda_i f_i \right) .$$

(1.23) <u>Lemma.</u> There is a unique measure ν_∞ which minimizes $H(\cdot;\mu)$ under the constraints (1.21). Moreover,

(1.24) $$\lim_n H(\nu_n;\mu) = H(\nu_\infty;\mu) ,$$

and

(1.25) $$\lim_n H(\nu_\infty;\nu_n) = 0 .$$

In particular, ν_n converges to ν_∞ in total variation.

<u>Proof.</u> 1) Since $H(\nu_n;\mu)$ is monotone in n with

$$\sup_n H(\nu_n;\mu) \leq H(\nu;\mu) < \infty ,$$

and since

$$(1.26)\qquad H(\nu_m;\mu) = H(\nu_m;\nu_n) + \int \log \frac{d\nu_n}{d\mu}\, d\nu_m$$

$$= H(\nu_m;\nu_n) + H(\nu_n;\mu)$$

for any $m \geq n$, we obtain

$$\lim_n \sup_{m \geq n} H(\nu_m;\nu_n) = 0 .$$

But by Ch. I (3.3),

$$(1.27)\qquad ||\nu_m - \nu_n|| \leq (2\, H(\nu_m;\nu_n))^{1/2} ,$$

and so ν_n converges in total variation to some limit measure ν_∞ . Moreover,

$$H(\nu_\infty;\mu) \geq \lim_n H(\nu_n;\mu) \geq H(\nu_\infty;\mu)$$

where the last inequality follows from lower-semicontinuity of $H(\cdot;\mu)$ (or by Fatou's lemma, if one does not want to use topological assumptions). This shows (1.24), hence (1.25) if we use (1.26) with ν_∞ instead of ν_m .

Let us now apply the lemma to the situation above. The fact that ν on $R^d \times R^d$ has marginals ν_o and ν_1 can be expressed by constraints of the form (1.21) where each f_i only depends on one of the two coordinates. Thus, the minimizing measure is given by $\nu_\infty = \lim \nu_n$ (in total variation) , and (1.22) shows that each ν_m is of the form $d\nu_n = h_n\, d\mu$ with $h_n(x,y) = f_n(x)\, g_n(y)$. Since h_n con-verges to $h_\infty = d\nu_\infty/d\mu$ in $L^1(\mu)$, we can now conclude that also h_∞ admits a factorization $h_\infty(x,y) = f(x)\, g(y)$; cf. Lemma 2.5 in [DV]. Moreover,

$$0 \leq H(\nu_\infty;\mu) = \int \log f d\nu_o + \int \log g d\nu_1 < \infty$$

implies $\log f \in L^1(\nu_o)$ and $\log g \in L^1(\nu_1)$.

1.4 Large deviations of the flow of marginal distributions

In [DG], Dawson and Gärtner discuss large deviations of the whole flow of marginal distributions

$$\frac{1}{n}\sum_{i=1}^{n}\delta_{X_i(t,\omega)} \qquad (0 \le t \le 1) .$$

This leads to the problem of minimizing the entropy $H(Q_o;P_o)$ under the constraint that all the marginal distributions $\nu_t := Q_o \circ X(t)^{-1}$ $(0 \le t \le 1)$ are fixed in advance.

Suppose that the flow ν_t $(0 \le t \le 1)$ is admissible in the sense that there exists at least one measure Q_o on $C[0,1]$ with marginals ν_t and finite entropy $H(Q_o;P_o)$; some regularity properties of such admissible flows will be discussed in Section 2.1. Then we can write

$$H(Q_o;P_o) = H(\nu_o;\mu_o) + \frac{1}{2}\int[\int_o^1 b_t^2\, dt]dQ_o$$

where (b_t) is the drift process associated to Q_o by (1.3). Introducing the Markovian drift

$$\tilde{b}_t(x) := E_{Q_o}[b_t|X_t = x]$$

and using the notation $\langle\nu_t,f\rangle = \int f d\nu_t$, we get

$$(1.28) \qquad H(Q_o;P_o) \ge H(\nu_o;\mu_o) + \frac{1}{2}\int_o^1 \langle\nu_t,\tilde{b}_t^2\rangle\, dt .$$

In order to obtain a lower bound which only involves (ν_t) , consider a smooth function f with compact support and note that Itô's formula implies

$$\langle\nu_t,f\rangle = \langle\nu_o,f\rangle + \int_o^t \langle\nu_s,\nabla f\, \tilde{b}_s\rangle ds + \frac{1}{2}\int_o^t \langle\nu_s,\Delta f\rangle ds .$$

Thus, for almost all t ,

$$\langle\dot{\nu}_t - \frac{1}{2}\Delta\nu_t,f\rangle = \langle\nu_t,\tilde{b}_t\nabla f\rangle$$

in the distributional sense, hence

$$\text{(1.29)}\qquad \frac{1}{2}\langle \nu_t, \tilde{b}_t^2\rangle = \frac{1}{2}\langle \nu_t, (\tilde{b}_t - \nabla f)^2\rangle + \langle \nu_t, \tilde{b}\nabla f\rangle - \frac{1}{2}\langle \nu_t, (\nabla f)^2\rangle \geq \langle \dot{\nu}_t - \frac{1}{2}\Delta\nu_t, f\rangle - \frac{1}{2}\langle \nu_t, (\nabla f)^2\rangle .$$

Defining the supremum of the right side over all smooth functions f with compact support as

$$||\dot{\nu}_t - \frac{1}{2}\Delta\nu_t||_{\nu_t} ,$$

we obtain the lower bound

$$\text{(1.30)}\qquad H(Q_o;P_o) \geq H(\nu_o;\mu_o) + \int_o^1 ||\dot{\nu}_t - \frac{1}{2}\Delta\nu_t||_{\nu_t} dt$$

due to (1.28) and (1.29). This lower bound is actually attained:

(1.31) Theorem ([DG]). There exists exactly one measure Q_o with marginals ν_t $(0 \leq t \leq 1)$ and

$$\text{(1.32)}\qquad H(Q_o;P_o) = H(\nu_o;\mu_o) + \frac{1}{2}\int_o^1 ||\dot{\nu}_t - \frac{1}{2}\Delta\nu_t||_{\nu_t} dt ,$$

and this measure has the Markov property.

Proof. The constraint can be written in the form (1.21) with a countable collection of bounded functions f_i $(i=1,2,\ldots)$ on $C[0,1]$, each of the form $f_i = \varphi_i(X(t_i))$. Let Q_n be the measure which minimizes $H(\cdot;P_o)$ under the constraints

$$\int f_i dQ = c_i \qquad (1 \leq i \leq n) .$$

It follows as in (1.19) that the drift process (b_t^n) associated to Q_n is of the form $b_t^n = \nabla f^n(X(t),t)$ for all but finitely many t. Now let $Q_o = \lim_n Q_n$ denote the limiting measure given by Lemma (1.23), and let (b_t) be the associated drift. (1.3) implies

$$H(Q_o;Q_n) = \frac{1}{2} E_{Q_o}[\int_o^1 (b_t - b_t^n)^2 dt] ,$$

and since $\lim H(Q_o;Q_n) = 0$ due to (1.25), we see that (b_t) is Markovian, i.e., $b_t = \tilde{b}_t(X(t))$ Q-a.s. for almost all t. It also follows that there is a sequence of smooth functions f_n with compact support such that

$$\lim_n \langle \nu_t, (\tilde{b}_t - \nabla f_n)^2 \rangle = 0 .$$

(1.29) now implies

$$\frac{1}{2} \langle \nu_t, \tilde{b}_t^2 \rangle = || \dot{\nu}_t - \frac{1}{2} \Delta \nu_t ||_{\nu_t} ,$$

hence (1.32), since

$$H(Q_o;P_o) = H(\nu_o;\mu_o) + \frac{1}{2} \int [\int_0^1 b_t^2 dt] dQ_o$$

$$= H(\nu_o;\mu_o) + \frac{1}{2} \int_0^1 \langle \nu_t, \tilde{b}_t^2 \rangle dt .$$

2. Infinite dimensional diffusions and Gibbs measures

Let P be a probability measure on $(C[0,1])^I$ and suppose that the coordinate process $X = (X_i)_{i \in I}$ satisfies a stochastic differential equation of the form

$$dX_i = dW_i + b_i(X(t),t)dt \quad (i \in I),$$

where $(W_i)_{i \in I}$ is a collection of independent Wiener processes under P. We are going to derive a dual equation

$$dX_i = d\hat{W}_i + \hat{b}_i(X(t),t)dt \quad (i \in I),$$

for the behavior of X after time reversal, i.e., under the time reversed measure $\hat{P}$. In contrast to the finite dimensional case, such a dual equation may break down if the interaction between different sites via the drift terms (b_i) becomes too strong; cf. (2.35) below. In order to exclude such effects we use a condition of locally finite entropy.

In Section 2.1 this entropy technique is first illustrated in the finite dimensional case. Here it leads to regularity results for the distributions of the coordinate process under a measure which has finite entropy with respect to Wiener measure; in particular, the density functions are a.e. partially differentiable. These smoothness properties are derived by time reversal, not by Malliavin calculus: We use the computation of drifts as stochastic forward and backward derivates and a classical integration by parts in the resulting duality equation. As a digression, we show that an integration by parts on Wiener space, i.e., a Mallavian calculus argument, only comes in at a second stage: it is needed if we want to determine not only the the density but also the backward drift in the non-Markovian case.

In Section 2.2 we apply the entropy technique to the infinite dimensional case; this is based on joint work with A. Wakolbinger [FW]. We obtain an infinite dimensional version of the classical duality equation, which relates forward drift, backward drift and the local specification of the distribution μ_t of the process at time t. In particular, this leads to a description of invariant measures as Gibbs measures. But the duality equation is also useful for quite different purposes. This is illustrated in Section 2.3 where we look at a large deviation of the empirical field of an infinite dimensional Brownian motion.

2.1. Time reversal on Wiener space

Let P be a probability measure on $C([0,1],R^d)$ with initial distribution ν_o. We say that P has finite entropy with respect to Wiener measure if

(2.1)
$$H(P;P^*) < \infty$$

where P^* denotes Wiener measure with the same initial distribution $\mu_o^* = \mu_o$. It follows by (1.3) that the drift process (b_t) associated to P has finite energy:

(2.2)
$$\frac{1}{2} E[\int_0^1 b_t^2 dt] = H(P;P^*) < \infty .$$

As we have seen in Section 1, it is of interest to clarify the regularity properties of P which are implied by the finite entropy assumption (2.1). To begin with, the distribution μ_t of $X(t)$ under P is absolutely continuous with respect to Lebesgue measure, and we denote by $\rho_t(x)$ the density function. It turns out that this function is in fact a.e. smooth in the sense that it admits a.e. a gradient $\nabla\rho_t$:

(2.3) Proposition. For almost all t, the function ρ_t is absolutely continuous on almost all straight lines which are parallel to coordinate axes, and it satisfies

$$\int_\epsilon^1 [\int (\nabla \log \rho_t)^2 (x) \rho_t(x) dx] dt < \infty$$

for any $\epsilon > 0$.

We are going to derive (2.3) as an exercise in time reversal. Let $\hat{P} = P \circ R$ denote the image of P under the pathwise time reversal R on $C([0,1],R^d)$ defined by

$$X(t,R(\omega)) = X(1-t,\omega) .$$

For the rest of this section, E and $\hat{E}$ will denote the expectation with respect to P and $\hat{P}$.

(2.4) <u>Lemma.</u> If P has finite entropy with respect to Wiener measure then $\hat{P}$ has finite entropy up to any time $t < 1$, i.e., there exists an adapted process $(\hat{b}_t)$ such that

$$\hat{W}(t) := X(t) - X(o) - \int_0^t \hat{b}_s ds \qquad (0 \le t \le 1) \tag{2.5}$$

is a Wiener process under $\hat{P}$, and

$$\hat{E}[\int_o^t \hat{b}_s^2 ds] < \infty \tag{2.6}$$

for any $t < 1$.

<u>Proof.</u> With respect to the time reversal $\hat{P}^* := P^* \circ R$ of P^*, the coordinate process $X(t)$ is a Brownian motion conditioned to have distribution μ_o at time $t = 1$. This implies that

$$W^*(t) := X(t) - X(o) - \int_o^t c_s ds$$

with $c_s := (1-s)^{-1}(X(1)-X(s))$ is a Wiener process with respect to $\hat{P}^*$ and the filtration given by $F_t^* := \underline{\underline{F}}_t V\, \sigma(X(1))$. Since

$$H(\hat{P};\hat{P}^*) = H(P;P^*) < \infty ,$$

(1.3), 2) implies that there is a process (a_t) adapted to $(\underline{\underline{F}}_t^*)$ with finite energy under $\hat{P}$ such that

$$W_t' := W_t^* - \int_o^t a_s ds \qquad (0 \le t \le 1)$$

is a Wiener process with respect to $\hat{P}$ and $(\underline{F}_t^*)$. On the other hand,

$$\sup_s \hat{E}[X_1-X_s)^2] = \sup_s E[(X_s-X_o)^2] < \infty$$

due to (2.2), and so we get

$$\hat{E}[\int_0^t (a_s+c_s)^2 ds] < \infty$$

for any $t < 1$. It follows that the process $(\hat{b}_t)$ defined as an optional version of the process

$$\hat{E}[a_t+c_t|\underline{F}_t] \qquad (0 \le t \le 1) ,$$

satisfies (2.5) and (2.6).

Lemma (2.4) shows that under $\hat{P}$ the coordinate process satisfies a stochastic differential equation of the form $dX = d\hat{W} + \hat{b}dt$. Due to (1.8) and (2.6), we can compute the dual drift $(\hat{b}_t)$ as a stochastic forward derivative under $\hat{P}$ or, equivalently, as a stochastic backward derivative under P: for almost all $t \in [0,1]$,

$$\text{(2.7)} \qquad \hat{b}_t = \lim_{h \downarrow 0} \frac{1}{h} \hat{E}[X(t+h)-X(t)|\underline{F}_t]$$

$$= -\lim_{h \downarrow 0} \frac{1}{h} E[X(1-t)-X(1-t-h)|\hat{\underline{F}}_{1-t}] \circ R$$

in $L^2(\hat{P})$, where $\hat{\underline{F}}_{1-t} := R^{-1}(\underline{F}_t)$ denotes the σ-field of events observable from time t on.

(2.8) <u>Theorem.</u> For almost all $t \in [0,1]$, ρ_t has the properties in (2.3), and its gradient $\nabla\rho_t$ satisfies the duality equation

$$\text{(2.9)} \qquad \nabla\rho_t(x) = \rho_t(x)\, E[b_t+\hat{b}_{1-t} \circ R|X_t = x]$$

for almost all x.

<u>Proof.</u> 1) Let t_0 be a point in $(0,1)$ such that (1.8) holds for $(\hat{b}_s)$ and $t = 1-t_0$, and for (b_s) and $t = t_0$. Then, for any bounded $\hat{\underline{\underline{F}}}_{t_0}$-measurable function F on $C([0,1],\mathbb{R}^d)$

$$\text{(2.10)} \qquad E[(\hat{b}_{1-t_0} \circ R)\, F] = \hat{E}[\hat{b}_{1-t_0}(F \circ R)]$$

$$= \lim \frac{1}{h} \hat{E}[(X_{1-t_0+h} - X_{1-t_0})(F \circ R)]$$

$$= -\lim \frac{1}{h} E[(X_{t_0} - X_{t_0-h})F] \, .$$

2) For a smooth function f with compact support, Itô's formula implies

$$E[(X_{t_0} - X_{t_0-h})\, f(X_{t_0})] = E[(\int_{t_0-h}^{t_0} b_s ds)\, f(X_{t_0-h})]$$

$$+ E[(X_{t_0} - X_{t_0-h})(\int_{t_0-h}^{t_0} \nabla f(X_s) dX_s + \frac{1}{2} \int_{t_0-h}^{t_0} \Delta f(X_s) ds)] \, .$$

By (2.10) we get, using some straightforward estimates based on (1.8),

$$\text{(2.11)} \qquad E[(\hat{b}_{1-t_0} \circ R)\, f(X_{t_0})] = E[b_{t_0} f(X_{t_0})] - E[\nabla f(X_{t_0})] \, .$$

Since

$$E[\nabla f(X_{t_0})] = \int \nabla f(x) \rho_{t_0}(x) dx = -\langle \nabla \rho_{t_0}, f \rangle \, ,$$

(2.11) shows that the distributional gradient of ρ_{t_0} is in fact given by the right side of (2.9). This implies that ρ_{t_0} is absolutely continuous on almost all straight lines which are parallel to coordinate axes, and that the distributional gradient coincides with the usual gradient almost everywhere, cf., e.g., [Ma] 1.1.3. The integrability condition (2.3) now follows from (2.2) and (2.6).

(2.12) Remark. Suppose that (X_t) is a Markov process under P. Then the drift is of the form $b_t = b_t(X_t)$ with some measurable function $b_t(x)$. But the Markov property is preserved under time reversal. Thus, the dual drift is given by some measurable function $\hat{b}_t(x)$, and (2.9) reduces, but here without any regularity assumptions, to the classical duality equation

$$b_t(x) + \hat{b}_{1-t}(x) = \nabla \log \rho_t(x) ; \tag{2.13}$$

cf., e.g., [Na], [Ne], [HP]. In the case of a reversible Markov process, a smoothness property of the equilibrium density is derived in [Or1] and [Fu2].

In the general non-Markovian case, the duality equation (2.9) does not yet determine the dual drift, only its projection on the present state. For a complete description, we have to replace the functionals $F = f(X_{t_0})$ used in the proof of (2.8) by smooth functionals on $C([0,1],R^d)$ which depend on the full σ-algebra $\hat{\underline{F}}_{t_0}$. In particular, the integration by parts on R^d which was used in (2.11) has to be replaced by an integration by parts on Wiener space, i.e., by a Malliavin calculus argument.

(2.14) Remark on integration by parts on Wiener space. For a bounded predictable process (u_t) put $U_t = \int_0^t u_s ds$ and $X_t^{\epsilon,U} = X_t + \epsilon U_t$ $(0 \leq t \leq 1)$. Let us say that a function $F \in L^2(P^*)$ is P^*-smooth if there is a measurable process (φ_t) such that, for any bounded predictable process (u_t),

$$DF(.,U) = \lim_{\epsilon \downarrow 0} \frac{F(X^{\epsilon,U})-F(X)}{\epsilon} = \int_0^1 u_s \varphi_s ds \tag{2.15}$$

in $L^2(P^*)$. If F is Fréchet-differentiable on $C([0,1],R^d)$ with bounded derivative $DF(\omega,dt)$, viewed as an R^d-valued measure on $[0,1]$, then (4.2) holds with $\varphi_s(\omega) = DF(\omega,[s,1])$. (2.15) is enough to use Bismut's derivation [Bi] of the basic integration by parts formula

$$(2.16) \qquad E^*[(\int_0^1 u_s dX_s)F] = E^*[DF(.,U)] = E^*[\int_0^1 u_s \varphi_s ds] .$$

If F is Fréchet-differentiable with bounded derivative $DF(.,dt)$ and G is P^*-smooth then (4.3), applied to FG, leads to

$$(2.17) \qquad E^*[FG \int_0^1 u_t dX_t] = E^*[DF(.,U)G] + E^*[FDG(.,U)] .$$

Let us now assume that the Girsanov density $G = dP/dP^*$ in (1.3) is P^*-smooth with

$$(2.18) \qquad DG(.,U) = \int_0^1 u_t \gamma_t dt .$$

If the drift b is a bounded smooth function on $C([0,1],R^d) \times [0,1]$ with bounded Fréchet derivatives $Db_t(.,ds)$, then G is indeed P^*-smooth and satisfies (2.18) with

$$(2.19) \qquad \gamma_t G^{-1} = b_t + \int_0^1 \beta_{t,s} dW_s^b$$

and $\beta_{t,s} := Db_s(.,[t,1])$; this follows, e.g., from Theorem (A.10) in [BJ] .

(2.20) <u>Theorem.</u> Suppose that $G = dP/dP^*$ is P^*-smooth and satisfies (2.18). Then the dual drift is given, for almost all $t \in (0,1)$, by

$$(2.21) \qquad \hat{b}_{1-t} = - E[b_t + a_t | \hat{\underline{F}}_t] \circ R$$

where

$$(2.22) \qquad a_t = \frac{1}{t} (W_t^b - \int_0^t \int_0^1 \beta_{s,r} dW_r^b ds) + \int_0^1 \beta_{t,s} dW_s^b .$$

<u>Proof.</u> We fix t_0 as in the proof of (2.10). Assume that F in (2.10) is Fréchet-differentiable with bounded derivative $DF(.,dt)$;

since F is $\hat{\underline{\underline{F}}}_{t_0}$-differentiable, the measures $DF(.,dt)$ are concentrated on $[t_0,1]$. Note that (2.17) translates into

$$(2.23) \qquad E[F \int_0^1 u_t dX_t] = E[DF(.,U)] + E[FG^{-1}DG(.,U)]$$

with respect to P. For $u_t = I_{[t_0-h,t_0]}(t)$ we obtain

$$(2.24) \qquad E[(X_{t_0}-X_{t_0-h})F] = h\, E[DF(.,[t_0,1])] + E[FG^{-1} \int_{t_0-h}^{t_0} \gamma_t dt] .$$

This together with (2.10) implies

$$(2.25) \qquad - E[(\hat{b}_{1-t} \circ R)F] = E[DF(.,[t,1])] + E[FG^{-1}\gamma_t]$$

for almost all $t \in (0,1)$. Using again (2.24) with $h = t_0$ we obtain

$$E[DF(.,[t,1])] = \frac{1}{t} E[X_t F] - \frac{1}{t} E[F \int_0^t G^{-1}\gamma_s ds]$$

$$= E[F \frac{1}{t}(W_t^b - \int_0^t \int_0^1 \beta_{s,r} dW_r^b ds)]$$

Thus, we can write (2.25) as

$$E[(\hat{b}_{1-t} \circ R + b_t)F] = - E[a_t F] ,$$

and this implies (2.21).

(2.26) <u>Remark.</u> In the Markovian case with a smooth drift function $b_t(x)$ we have $\beta_{t,s} = \nabla b_s(X_s) I_{[t,1]}(s)$, and (2.21) becomes

$$(2.27) \qquad \hat{b}_{1-t}(x) = -b_t(x) + \frac{1}{t} E[\int_0^t (r\nabla b_r(X_r)-1)dW_r^b | X_t = x] .$$

Due to (2.13), this can also be read as a path space formula for $\nabla \log \rho_t(x)$.

2.2. Infinite dimensional diffusions and their time reversal

Let I be a countable index set, denote by

(2.28) $$X_i(t,\omega) \quad (0 \le t \le 1 \ , \ i \in I)$$

the coordinate processes X_i $(i \in I)$ on $\Omega = C[0,1]^I$, and let $(\underline{\underline{F}}_t)$ be the canonical right-continuous filtration on Ω . Let P be a probability measure on $(\Omega, \underline{\underline{F}}_1)$ such that the process

$$X(t) = (X_i(t))_{i \in I}$$

satisfies an infinite dimensional stochastic differential equation of the following form:

(2.29) $$dX_i = dW_i + b_i(X(t),t)dt \quad (i \in I) \ ,$$

where $W = (W_i)_{i \in I}$ is a collection of independent Wiener processes, and where $b_i(.,t)$ is a function defined on a suitable state space $E \subseteq R^I$ such that

$$P[X(t) \in E \quad (0 \le t \le 1)] = 1 \ .$$

This means, in particular, that $X(t)$ $(0 \le t \le 1)$ can be viewed as an infinite dimensional Markov process with state space E .

(2.30) A class of examples. Let γ be some finite measure on I and define $E := L^2(\gamma)$. For each $i \in I$ let b_i be a function on $E \subseteq R^I$ which is locally Lipschitz in each coordinate and satisfies the following conditions:

(2.31) $$\sup |b_i(0)| < \infty \ ;$$

(2.32) There is a constant K such that

$$(x_i - y_i)(b_i(y)) \le K(x_i - y_i)^2$$

for any $i \in I$, and for $x,y \in E$ with $x = y$ off i;

(2.33) there is a continuous linear map $C = (C_{ij}) : L^1(\gamma) \to L^1(\gamma)$ such that

$$(b_i(x) - b_i(y))^2 \le \sum_j C_{ij}(x_j - y_j)^2$$

for any $i \in I$, and for $x,y \in E$ with $x_i = y_i$.

These conditions imply that, for a given collection $(W_i)_{i \in I}$ of independent Wiener processes, the stochastic differential equation (2.29) has a unique strong solution; cf. [ShSh], [DR]. If b is an E-valued Lipschitz continuous function on the Hilbert space $E = L^2(\gamma)$, then one can use general results on stochastic differential equations in Hilbert space; cf., e.g., [Mé]. But in view of applications in Statistical Mechanics as in (2.55) below, this assumption would be too restrictive. For a discussion of (2.29) in terms of stochastic differential equations in Hilbert space with "discontinuous drift" see [LR1,2].

In general, (2.29) means that $X(t)$ $(0 \le t \le 1)$ is a Markov process whose distribution P is "locally" absolutely continuous with respect to the model for an infinite dimen-sional Brownian motion considered in Section 1. This can be made precise in various ways, by a suitable extension of (1.3) to the infinite dimensional case. For our purpose, we want to make sure that the coordinate process $X(t)$ $(0 \le t \le 1)$ is a diffusion process of the form (2.29) not only under P , but also under the time reversed measure

(2.34) $$\hat{P} = P \circ R$$

where R denotes the pathwise time reversal on Ω defined by $X(t,R(\omega)) = X(1-t,\omega)$. In contrast to the finite dimensional case, this is not true in general. The point is that in infinite dimensions the present state $X(1,\omega)$ at time $t = 1$ may contain so much information that, looking backwards, the semimartingale property of the coordinate processes may be lost:

(2.35) Example. Take $I = \{0\} \cup \{(k,l)\ ;\ k,l = 1,2,\dots\}$, $b_0 \equiv 0$ and $b_i(x,t) = f_k(t)x_o$ for $i = (k,l)$ where f_k $(k=1,2,\dots)$ is a orthonormal basis in $L^2[0,1]$. For a given collection W_i $(i \in I)$ of independent Brownian motions, equation (2.29) has a unique strong solution given by $X_o = W_o$ and

$$(2.36) \qquad X_i(t) = W_i(t) + \int_0^t f_k(s)X_o(s)ds$$

for $i = (k,l)$. Let P be the corresponding measure on $C[0,1]^I$. By the law of large numbers,

$$(2.37) \qquad \lim_{n\uparrow\infty} \frac{1}{n} \sum_{l=1}^{n} X_{(k,l)}(1) = \int_0^1 f_k(s)X_o(s)ds$$

P-almost surely for any $k \geq 1$. This shows that the whole path X_o can be reconstructed if we know the state $X(1)$ at time $t=1$. In particular, X_o cannot be a semimartingale with respect to $(\underline{\underline{F}}_t)$ and $\hat{P}$.

The loss of the semimartingale property in (2.35) is due to the long range interaction between site 0 and the sites $i \neq 0$. Let us now formulate an entropy condition which will exclude this effect. For $i \in I$ we introduce the σ-fields

$$\underline{\underline{F}}_t^i := \sigma(X(0)\ ,\ X_j(s)\ ;\ j \neq i\ ,\ 0 \leq s \leq t)$$

which contain the information on the initial state $X(0)$ and on the behavior of the coordinate processes X_j $(j \neq i)$ up to time t. Let P^i denote the probability measure on $(\Omega,\underline{\underline{F}}_1)$ which coincides with P on $\underline{\underline{F}}_1^i$ and makes $X^i(t) - X^i(0)$ $(0 \leq t \leq 1)$ a Wiener process which is independent of $\underline{\underline{F}}_1^i$; this corresponds to a decoupling of the i-th coordinate process.

(2.38) Definition. We say that P has locally finite entropy if

$$(2.39) \qquad H(P;P^i) < \infty \quad (i \in I)\ ,$$

where the entropy is computed on $\underline{\underline{F}}_1$. If (2.40) holds on $\underline{\underline{F}}_t$ then we say that P has locally finite entropy up to time t.

To (2.39) corresponds the following infinite dimensional analogue of remark (1.3) on the Girsanov transformation.

(2.40) Proposition. If P has locally finite entropy then there exist adapted processes b_i $(i \in I)$ such that

$$W_i(t) := X_i(t) - X_i(0) - \int_0^t b_i(s)ds \qquad (i \in I) \tag{2.41}$$

defines a collection W_i $(i \in I)$ of independent Wiener processes under P. These drift processes satisfy the finite energy condition

$$E[\int_0^1 b_i^2(s)ds] < \infty \qquad (i \in I) \tag{2.42}$$

and the locality condition

$$E[\sum_{j \neq i} \int_0^1 (b_j(s) - E[b_j(s)|\underline{F}^i])^2 ds] < \infty \qquad (i \in I) \tag{2.43}$$

Proof. For a fixed $i \in I$, the process $X_i(t) - X_i(0)$ $(0 \le t \le 1)$ is a Wiener process under P^i, and so the representation (2.41) and the finite energy condition (2.42) follows from (1.3), 2). The point is to show that these Wiener processes are independent, and that the locality condition (2.43) is satisfied; see [FW] for the details.

(2.44) Remarks. 1) If we would assume "globally" finite entropy, i.e.,

$$H(P;P^*) < \infty \tag{2.45}$$

where P^* is the model for an infinite dimensional Brownian motion considered in Section 1, then the sum of the energy terms in (2.42) would converge; cf. [HW]. But (2.45) would be too strong for the class of examples introduces in (2.30).

2) The finite energy condition (2.42) implies that the drift processes can be computed as stochastic forward derivatives, i.e.,

(2.46) $$b_i(t) = \lim_{h \downarrow 0} \frac{1}{h} E[X_i(t+h) - X_i(t) | \underline{\underline{F}}_t]$$

in $L^2(P)$ for any $i \in I$, due to (1.3), Condition (2.43) says that the interaction is "local in the active sense": for a given $i \in I$, the drift of most coordinates $j \neq i$ is influenced only mildly by the behavior of the i-th coordinate.

3) Suppose that under P the coordinate process is the unique strong solution of the stochastic differential equation (2.29), where the drift functions satisfy the assumptions in (2.30). If, in addition,

$$|b^i(x)| \leq \text{const}(1+||x||^p)$$

for some $p \geq 1$, and if the initial distribution μ_o satisfies

$$\int ||x||^{2p} \mu_o(dx) < \infty ,$$

then P has locally finite entropy; cf. [FW].

4) The process in (2.35) is not of locally finite entropy. In fact, let $\bar{X}_o$ denote the process defined P-a.s. by the coordinates X_i $(i \neq 0)$ through equations (2.37). Since $\bar{X}_o = X_o$ P-a.s., $\bar{X}_o$ is a Brownian motion under P, hence under P^o since $P^o = P$ on $\underline{\underline{F}}^o$. But under P^o the two Brownian motions $\bar{X}_o$ and X_o are independent, and so P^o and P are singular to each other.

From now on we assume that $X(t)$ is a Markov process under P. We want to describe the structure of the time reversed process $X \circ R$ or, equivalently, the structure of the coordinate process X under $\hat{P} = P \circ R$.

By a straightforward extension of the argument in (2.4) to the infinite dimensional case, we see that locally finite entropy is essentially preserved under time reversal:

(2.47) <u>Lemma.</u> If P has locally finite entropy then $\hat{P}$ has locally finite entropy up to any time $t < 1$.

Since $\hat{P}$ is again a Markov process, (2.40) shows that X is again described by a stochastic differential equation of the form

$$(2.48) \qquad dX_i = d\hat{W}_i + \hat{b}_i(X(t),t)dt \qquad (i \in I)$$

and that the drift functions $\hat{b}_i$ $(i \in I)$ can be computed as stochastic backward derivates in the manner of (2.46) resp. (2.7). This leads to the following infinite dimensional version of the classical duality equation (2.13).

Let μ_t denote the distribution of $X(t)$ under P, and let $\mu_{t,i}(\cdot|y)$ denote the conditional distribution of $X_i(t)$ under μ_t, given the outside configuration $y = (x_j)_{j \neq i}$. Condition (2.39) implies that $\mu_{t,i}$ is in fact absolutely continuous with some density function $\rho_{t,i}(x_i|y)$, and that these conditional densities have the following regularity properties:

(2.49) <u>Theorem (FW).</u> Suppose that P has locally finite entropy. Then, for any $i \in I$, for almost all $t \in [0,1]$ and for μ_t-almost all y, the conditional density $\rho_{t,i}(\cdot|y)$ is an absolutely continuous function, and its derivative is given by

$$(2.50) \qquad \nabla_i \rho_{t,i}(x_i|y) = \rho_{t,i}(x_i|y)\,(b_i(x,t) + \hat{b}_i(x,1-t))$$

for almost all $x_i \in \mathbb{R}$.

<u>Proof.</u> We use the same argument as in (2.8), based on the computation of drifts as stochastic forward and backward derivatives. Taking $F = f(X_i(t))G$ with $G = g(X_j(t); j \in J)$ for finite $J \subseteq I - \{i\}$ in (2.10), we obtain

$$E[(b_i(X(t),t) + \hat{b}_i(X(t),1-t))\, f(X_i)\, G] = -E[\nabla_i f(X(t))\, G] \; .$$

This implies (2.5), as in the proof of (2.8).

(2.51) <u>Remark.</u> The crucial technical point is to guarantee a priori that the coordinate processes X_i $(i \in I)$ are semimartingales under the time reversed measure $\hat{P}$; then the derivation of the duality equation (2.50) is straightforward. It is clear that X_i is a semimartingale under $\hat{P}$ with respect to any "local" filtration $(\underline{F}_t^J)$ generated by X_j $(j \in J)$ for some finite J. The problem is to control the limiting behavior as $J \uparrow I$, and this was our reason to introduce the condition of locally finite entropy.

(2.49) shows that, in the infinite dimensional case, the duality equation between forward drift, backward drift and the distribution μ_t at time t does not involve the distribution itself, only its local specification in the sense of Ch. I. Consider, in particular, the time homogeneous case $b_i(x,t) = b_i(x)$ and suppose that μ is an invariant measure of the process. Tn (2.50) shows that μ is a Gibbs measure in the sense that

$$\nabla_i \log \rho_i(x_i|y) = b_i(x_i,y) + \hat{b}_i(x_i,y) .$$

(2.52) <u>Remark.</u> Any locality properties of the conditional densities $\rho_i(x_i|\cdot)$ depend on the locality properties both of b and $\hat{b}$. Suppose, for example, that b has a spatial Markov property:

$$x = y \text{ on } N(i) \Rightarrow b_i(x) = b_i(y) \tag{2.53}$$

for some finite neighbourhood $N(i)$ of i. Then the invariant measure μ is a Markov field if and only if also the dual drift $\hat{b}$ has a Markov property of the form (2.53). Note that we have always locality in the the extended sense of (2.43), both in the forward and in the backward direction.

Now suppose that the invariant measure μ resp. the stationary process μ resp. the stationary process P is reversible, i.e., $P = \hat{P}$. This implies in particular

$$b_i = \hat{b}_i \quad (i \in I) ,$$

and so μ is a Gibbs measure with respect to b, i.e.,

$$\rho_i(\cdot|y) = z_i(y)^{-1} \exp[2 \int b_i(\cdot|y)] . \tag{2.54}$$

(2.55) <u>Remark.</u> Suppose that b is given by a pair potential (U_V), i.e., b_i is of the form

$$b_i(x) = -\frac{1}{2}\nabla_i\,(U_i(x_i) + \sum_{j\neq i} U_{ij}(x_i,x_j)) .$$

Then (2.54) means that μ is a Gibbs measure with respect to (U_V) in the sense of Ch. I.3. We refer to [ShSh] for conditions on (U_V) which guarantee that b fulfills the conditions in (2.30). If $|U_i'|$ is of polynomial growth then the condition in (2.44,3) is also satisfied, and so the preceding argument does apply.

Let us now suppose that, conversely, μ is an invariant measure which satisfies (2.54). Then (2.51) implies that both P and $\hat{P}$ are governed by the same stochastic differential equation (2.29). This is equivalent to reversibility of μ resp. of P if we assume uniqueness of solutions for (2.29):

(2.56) <u>Corollary.</u> Suppose that the infinite dimensional stochastic differential equation (2.29) has a unique weak solution on $E \subseteq R^I$, and that μ is an invariant measure concentrated on E. Then μ is reversible if and only if μ is a Gibbs measure in the sense of (2.54).

(2.57) <u>Remark.</u> 1) The characterization of reversible invariant measures of an interacting particle system as Gibbs measures has been studied in various different contexts; cf., e.g., [Li], [Lan], [HS1,2], [Fr], and in our present case [DR]. A Gibbsian description of non-reversible invariant measures appears, e.g., in [Kü4].

2) Under the assumptions in (2.56), Fukushima and Stroock [FS] used the following approach. For a Gibbs measure μ, (2.54) implies

$$\int fLg\,d\mu = \int gLf\,d\mu \tag{2.58}$$

for smooth functions f and g depending on finitely many coordinates, where

$$Lf = \sum \left(\frac{1}{2}\frac{\partial^2}{\partial x_i^2} + b_i\frac{\partial}{\partial x_i}\right)f$$

is the generator associated to (2.29). The problem is to show that (2.58) implies reversibility. Under additional regularity and compactness assumptions, Fukushima's reconstruction theorem implies that there is a reversible Markov process Q associated to L resp. to the corresponding Dirichlet form [Fu1]. But under the uniqueness assumption, Q must coincide with the measure P induced by μ, and this shows that P resp. μ is reversible. For the general non-reversible case (2.49), this Dirichlet space technique does not seem to be available. This was another reason to introduce the entropy condition.

2.3. Large deviations and Martin boundary: an infinite dimensional example

Let us return to the situation in Section 1 where

$$P = \prod_{i \in I} P_i$$

is the product of Wiener measures with initinal value 0. For a large deviation of the empirical field described by a subset A of $M(\Omega)$, the results of Ch. I imply

$$(2.58') \qquad P[|V_n|^{-1} \sum_{i \in V_n} \delta_{\Theta_i \omega} \in A] \sim \exp[-\inf_{Q \in A \cap M_s(\Omega)} h(Q;P)]$$

Now suppose that, in analogy to Section 1.2, we are interested in a large deviation of the empirical field at the terminal time $t = 1$. Then (2.58') reduces to

$$(2.59) \qquad P[|V_n|^{-1} \sum_{i \in V_n} \delta_{\Theta_i \omega(1)} \in B] \sim \exp[-\inf_{\nu \in B \cap M_s(E)} h(\nu;\mu)],$$

where B is a set of probability measures on $E = \mathbb{R}^I$ and $\mu = \prod_{i \in I} N(0,1)$ is the marginal distribution of P at atime 1.

In fact, (2.59) follows also directly from I.(4.1), applied to a sequence of independent random variables with distribution $N(0,1)$. But in view of the behavior of the whole process (X_i) it is of interest, as in Section 1.2, to describe the measure $Q \in M_s(\Omega)$ which minimizes $h(Q;P)$ under the constraint that the marginal distribution at time 1 is given by $\nu \in M_s(E)$.

For $\eta = (\eta_i)_{i\in I} \in E$, let P^η denote the infinite dimensional Brownian bridge

$$P^\eta = \prod_{i\in I} P^{\eta_i}$$

where P^{η_i} is the measure on $C[0,1]$ corresponding to the one-dimensional Brownian bridge from 0 to η_i . Then

$$P^\nu = \int P^\eta \, \nu(d\eta) \tag{2.60}$$

may be viewed as infinite dimensional Brownian motion conditioned to have distribution ν at time 1 . We have

$$P^\nu_V = \int \prod_{i\in V} P^{\eta_i} \, \nu(d\eta) \quad ,$$

hence $H_V(P^\nu;P) = H_V(\nu;\mu)$ and

$$h(P^\nu;P) = h(\nu;\mu) \; .$$

Conversely, any $Q \in M_s(\Omega)$ with marginal ν at time 1 satisfies $h(Q;P) \geq h(\nu;\mu)$, and equality holds if and only if $Q = P^\nu$.

In contrast to Section 1.2, the minimizing measure $Q = P^\nu$ is now an interacting diffusion process, no longer a collection of independent diffusions, each described as an h-path process in the sense of (1.15). Let us use time reversal in order to derive the corresponding infinite dimensional stochastic differential equation. In fact, the structure of the time reversed process $\hat{Q}$ is clear: it is an infinite dimensional Brownian motion conditioned to go to $0 \in E$ at time 1 and starting with initial distribution ν . Thus, the time reversed drift $\hat{b} = (\hat{b}_i)$ is given by

$$\hat{b}_i(x,t) = -\frac{x_i}{1-t} \qquad (i \in I) \; .$$

Applying Theorem (2.49) we obtain the following

(2.61) <u>Corollary.</u> Under the minimizing measure $Q = P^{\nu}$, the process $X(t)$ $(0 \le t \le 1)$ satisfies an infite dimensional stochastic differential equation

$$(2.62) \qquad dX_i = dW_i + b_i(X(t),t)dt \qquad (i \in I)$$

as in (2.29), and the drift is given by

$$(2.63) \qquad b_i(x,t) = \frac{x_i}{t} + \nabla_i \log \rho_{i,t}(x_i|x_j(j\neq i)) \qquad (i \in I)$$

where $(\rho_{i,t})$ is the collection of conditional densities of $\nu_t = \int \nu(d\eta)P^{\eta} \circ X(t)^{-1}$.

Note that the process is not yet uniquely determined by (2.62) and (2.63) if there is a phase transition, i.e., if ν does not determine uniquely the local specification of ν_t $(0 \le t \le 1)$; cf. [FW] for an explicit example.

(2.64) <u>Remark.</u> $E = R^I$ may be regarded as the Martin boundary associated to P. If fact, (2.60) is an integral representation for the class of all probability measures Q on $C[0,1]^I$ with given conditional probabilities

$$Q[\cdot|X(s);t \le s < 1] = P[\cdot|X(s);t \le s < 1] \qquad (0 < t < 1) .$$

These conditional probabilities, defined consistently as Brownian bridges leading from $0 \in E$ to $X(t) \in E$ at time t, form a local specification in the sense of Ch. I. The Brownian bridges P^{η} $(\eta \in E)$ are the extremal points in the class of all probability measures compatible with this local specification, and the parametrizing space E may be called the Martin boundary; cf. [Dy], [Fö1]. In the finite dimensional case, each measure P^{ν} corresponds to a space-time harmonic function $h(x,t)$ as in (1.14), and (2.60) translates into an integral representation of space-time harmonic functions in terms of the classical Martin boundary R^d of space-time Brownian motion $(X(t),t)$ on $R^d \times [0,1)$. But in the infinite dimensional case this correspondence between functions and conditional processes is lost. The reason is that, other than in classical duality theory which relies on the existence of a reference measure [Dy], the measures P^{ν}

are in general not absolutely continuous with respect to P on $\underline{\underline{F}}_t = \sigma(X(s);\ s \le t)$. In particular, P^ν cannot be described as an h-path process where the space-time harmonic function $h(x,t)$ is given by the Radon-Nikodym densities

$$h(X(t),t) = \frac{dP^\nu}{dP}\Big|_{\underline{\underline{F}}_t} \qquad (0 \le t < 1) .$$

Thus, for a description of P^ν in the infinite dimensional case, we need an alternative to (1.15), and (2.61) shows that time reversal can be used for this purpose.

3. Infinite dimensional diffusions as Gibbs measures

Let P be the distribution of an infinite dimensional diffusion process

$$dX_i = dW_i + b_i(X)dt \quad (i \in I) . \tag{3.1}$$

with time-homogeneous drift. Since P is a probability measure on $C[0,1]^I$, it can be viewed as a random field with state space $S = C[0,1]$ at each site $i \in I$. In the case of an infinite dimensional Brownian motion, we have already taken this point of view in Section 1. But if we want to apply the random field techniques of Ch. I to the general case (3.1) of an interacting diffusion then it is important to derive its local specification, i.e., the conditional distribution of X_i on $C[0,1]$ given all the other coordinates $X_j \in C[0,1]$ $(j \neq i)$. This problem has been studied by J.D. Deuschel, and in this section we give an introduction to the results in [De1,2,3].

3.1 The local specification of an infinite dimensional diffusion

In order to simplify the exposition let us assume that the initial distribution is concentrated on some point $x \in R^I$. In that case, the σ-field generated by $Y = (X_j)_{j \neq i}$ can be identified with the σ-field $\underline{\underline{F}}^i$ introduced in Section 2.2. The conditional distribution $P_i(dX_i|\underline{\underline{F}}^i)$ of X_i with respect to $\underline{\underline{F}}^i$ is of the form

$$\pi_i(dX_i|Y) = c_i(Y)^{-1}(dP/dP^i)(X)\, P^*(dX_i) \tag{3.2}$$

for $X = (X_i,Y)$, where P^* denotes Wiener measure on $C[0,1]$ and where P^i is the decoupling of the i-th coordinate used in (2.38). The density dP/dP^i could be computed in the manner of (2.4); cf. the proof of (2.23) in [FW].

In order to obtain a more explicit description of $\pi_i(dX_i|Y)$, we assume that the drift is given by a smooth pair potential of finite range as in (2.55):

$$b_i(x) = \nabla_i E_i(x) \tag{3.3}$$

with

$$E_i(x) = U_i(x_i) + \sum_{j\neq i} U_{ij}(x_i,x_j)$$

where U_i and U_{ij} are smooth symmetric functions with bounded derivatives up to order 3, and $U_{ij} = 0$ $(j \notin N(i))$ for some finite neighbourhood $N(i)$ of i. For a fixed $i \in I$, we can now use the following decoupling Q^i of the i-th coordinate. Define

$$b_{ik} = b_k - \nabla_k E_i \qquad (k \in I) \tag{3.4}$$

so that $b_{ii} = 0$ and $b_{ik} = b_k$ for $k \notin N(i) \cup \{i\}$. Let Q^i denote the distribution of the infinite dimensional diffusion

$$dX_k = dW_{ik} + b_{ik}\, dt \qquad (k \in I) \tag{3.5}$$

where $(W_{ik})_{k\in I}$ is a collection of independent Brownian motions. Under Q^i, the i-th coordinate X_i is a Wiener process which is independent of $\underline{\underline{F}}^i$, and so we can write

$$\pi_i(dX_i|Y) = Z_i(Y)^{-1}(dP/dQ^i)(X)\; P^*(dX_i) \tag{3.6}$$

with $X = (X_i,Y)$ and some normalizing factor $Z_i(Y)$. In order to compute the density dP/dQ^i more explicitey, define

$$g_i(x) = \sum_{k\in N(i)\cup\{i\}} b_{ik}(x)\nabla_k E_i(x) + \frac{1}{2}\nabla_k^2 E_i(x) + \frac{1}{2}(\nabla_k E_i(x))^2 . \tag{3.7}$$

Note that g_i may be regarded as a smooth function on $R^{N^2(i)\cup\{i\}}$ where

$$N^2(i) = N(i) \cup \bigcup_{k\in N(i)} (N(k)-\{i\})$$

denotes the second order neighbourhood of i.

(3.8) Proposition. The conditional distribution $\pi_i(dX_i|Y)$ is given by the conditional density

$$Z_i(Y)^{-1} \exp[\Phi_i(X_i,Y)] \tag{3.9}$$

with respect to Wiener measure, where

$$(3.10) \qquad \Phi_i(X) = E_i(X(1)) - E_i(X(0)) - \int_0^1 g_i(X(s))ds \; .$$

Viewed as a function on $(C[0,1])^{N^2(i)\cup\{i\}}$, Φ_i is Fréchet differentiable, and the partial derivatives $D_k\Phi_i$ $(k \in N^2(i) \cup \{i\})$, viewed as signed measures on $[0,1]$, are bounded in total variation by

$$(3.11) \qquad \|D_k\Phi_i(X)\| \le 2\|\nabla_k E_i\|_\infty + \|\nabla_k g_i\|_\infty \; .$$

<u>Proof.</u> We have

$$(dP/dQ^i)(X) = \exp\Big[\sum_{k\in N(i)\cup\{i\}} \int (b_k - b_{ik})dX_k - \frac{1}{2}\int (b_k^2 - b_{ik}^2)dt\Big]$$

with

$$\frac{1}{2}(b_k^2 - b_{ik}^2) = \frac{1}{2}(\nabla_k E_i)^2 + b_{ik}\nabla_k E_i \; .$$

By Itô's formula,

$$\sum_k \int_0^1 (b_k - b_{ik})dX_k = \sum_k \int_0^1 \nabla_k E_i dX_k$$

$$= E_i(X(1)) - E_i(X(0)) - \frac{1}{2}\sum_k \int_0^1 \nabla_k^2 E_i(X(s))ds \; ,$$

and this implies (3.9). Our assumptions on $(b_k)_{k\in I}$ imply that Φ_i , viewed as a function on S^J with $S = C[0,1]$ and $J = N^2(i) \cup \{i\}$, is Fréchet differentiable. Its partial derivatives, viewed as signed measures on $[0,1]$, are of the form

$$D_k\Phi_i(X)(dt) = \nabla_k E_i(X(1))\delta_1(dt) - \nabla_k E_i(X(0))\delta_0(dt)$$

$$- \nabla_k g_i(X(t))(dt) \; ,$$

and this implies (3.11).

(3.12) Remark. 1) Since $\pi_i(\cdot|Y)$ is absolutely continuous with respect to Wiener measure P^* on $C[0,1]$, the coordinate process X_i is of the form

$$dX_i = dW_i^Y + b_i^Y(t)dt$$

with some conditional drift process $b_i^Y = b_i^Y(t)$ $(0 \le t \le 1)$ under $\pi_i(\cdot|Y)$. As in [BM], the conditional drift b_i^Y can be computed by Clark's formula [Cl]: For $\underline{\underline{G}}_t^i = \sigma(X^i(s); s \le t)$ and $\underline{\underline{F}}_t^i = \underline{\underline{F}}^i \cup \underline{\underline{G}}_t^i$, the martingale

$$\begin{aligned} M_i(t) &= E_{Q^i}[\exp \Phi_i(X) | \underline{\underline{F}}_t^i]\,(X_i,Y) \\ &= E^*[\exp \Phi_i(\cdot,Y) | \underline{\underline{G}}_t^i](X_i) \end{aligned}$$

is of the form

$$M_i(t) = M_i(0) + \int_0^t H_i^Y(s)dX_i(s)$$

with

$$H_i^Y(s) = E^*[D_i\Phi_i(\cdot,Y)(s,1] \exp \Phi_i(\cdot,Y) | \underline{\underline{G}}_s^i] .$$

The drift is now given by

(3.13) $$b_i^Y(t) = H_i^Y(t)/M_i(t) ;$$

cf., e.g., [DM]. This implies the uniform bound

(3.14) $$||b_i^Y(t)||_\infty \le ||D_i\Phi_i||_\infty := \sup_X ||D_i\Phi_i(X,\cdot)||$$

in particular,

(3.15) $$\sigma_i^2 := \sup_Y \inf_{s \in S} \int ||X_i - s||^2 \pi_i(dX_i|Y) < \infty$$

where $||\cdot||$ denotes the supremum norm on $S = C[0,1]$. Due to the special additive structure of $\Phi_i(X)$ and the Markov property of P^*,

(3.13) also implies that, for a fixed Y, the drift process b_i^Y is of the form

$$(3.16) \qquad b_i^Y(\omega,t) = b_i^Y(X_i(\omega,t),t) ,$$

i.e., the conditional diffusion is actually a Markov process.

2) As an alternative to (3.13), the conditional drift function in (3.16) can also be computed as

$$(3.17) \qquad b_i^Y(x_i,t) = \nabla_i G_i^Y(x_i,t)$$

where G_i^Y is a solution of the partial differential equation

$$(3.18) \qquad \nabla_t G + \frac{1}{2}\nabla_i^2 G + \frac{1}{2}(\nabla_i G)^2 = g_i$$

with boundary condition

$$(3.19) \qquad G(x_i,1) = E_i(x_i) .$$

In fact, Itô's formula and (3.19) imply

$$E_i(X(1)) = G_i^Y(X(0),0) + \int_0^1 \nabla_i G_i^Y(X_i(t),t)dX_i(t)$$

$$+ \int_0^1 [\frac{1}{2}\nabla_i^2 G_i^Y + \nabla_t G_i^Y](X_i(t),t)dt ,$$

and so (3.18) shows that the densitiy in (3.9) is given by

$$\exp[\int_0^1 \nabla_i G_i^Y dX_i - \frac{1}{2}\int_0^1 (\nabla_i G_i^Y)^2 dt] .$$

This implies (3.17). Bounds and smoothness properties of the conditional drift can now be obtained by means of the Feynman-Kac formula. Moreover, the representation (3.17) leads to a characterization of the conditional drift as the solution of a stochastic control problem. Cf. [De2] for details.

3.2 Applying the contraction technique

Let us now look at the system of conditional probabilities (π_k) from the point of view of Dobrushin's contraction technique in Section I.2. For two probability measures μ and ν on $S = C[0,1]$, let

$$R(\mu,\nu) = \sup \frac{|\int f d\mu - \int f d\nu|}{\delta(f)}$$

denote the Vasserstein distance as in Ch. I (2.18); the Lipschitz constant $\delta(f)$ is defined with respect to the supremum norm on $C[0,1]$. For $\eta \in \Omega = S^I$ we use the notation

$$\pi_k(\cdot|\eta) = P_k(\cdot|\underline{\underline{F}}^k)(\eta) .$$

Recall the definition of σ_k^2 in (3.15), and put

$$||D_i\Phi_k||_\infty = \sup_{\omega\in\Omega} ||D_i\Phi_k(\omega,\cdot)||$$

as in (3.14).

(3.20) <u>Theorem.</u> For ω and η in $\Omega = S^I$, the Vasserstein distance between the conditional diffusions $\pi_k(\cdot|\omega)$ and $\pi_k(\cdot|\eta)$ satisfies

$$(3.21) \qquad R(\pi_k(\cdot|\omega),\pi_k(\cdot|\eta)) \le \sum_{i\in N^2(k)} c_{ik} \, ||\omega(i)-\eta(i)||$$

with

$$(3.22) \qquad c_{ik} \le \sigma_k \, ||D_i\Phi_k||_\infty$$

<u>Proof.</u> Take $k \in I$, $i \in N(k)$ and $\omega, \eta \in S^I$ with $w = \eta$ off i. Let f be a Lipschitz function on S with Lipschitz constant $\delta(f)$.
1) For $\alpha \in [0,1]$ define $\eta_\alpha = \omega + \alpha(\eta-\omega)$. Using the notation $\mu_\alpha = \pi_k(\cdot|\eta_\alpha)$ and

$$Z_\alpha = Z_k(\eta_\alpha) = \int \exp[\Phi_k(\cdot,\eta_\alpha)dP^* ,$$

we have

$$\frac{d}{d\alpha}\log Z_\alpha = \int \frac{d}{d\alpha}\Phi_k(\cdot,\eta_\alpha)d\mu_\alpha ,$$

hence

$$\frac{d}{d\alpha}\int f d\mu_\alpha = \frac{d}{d\alpha}(Z_\alpha^{-1}\int f \exp[\Phi_k(\cdot,\eta_\alpha)]dP^*) =$$

$$= \int f\,[\frac{d}{d\alpha}\Phi_k(\cdot,\eta_\alpha) - \frac{d}{d\alpha}\log Z_\alpha]d\mu_\alpha$$

$$= \int [f-f(s)]\,[\frac{d}{d\alpha}\Phi_k(\cdot,\eta_\alpha) - \frac{d}{d\alpha}\log Z_\alpha]d\mu_\alpha$$

for any $s \in S = C[0,1]$. This implies

$$|\frac{d}{d\alpha}\int f d\mu_\alpha|^2 \le \delta(f)^2\sigma_k^2 \operatorname{var}_{\mu_\alpha}(\frac{d}{d\alpha}\Phi_k(\cdot,\eta_\alpha)) .$$

But

$$|\frac{d}{d\alpha}\Phi_k(\cdot,\eta_\alpha)| \le ||D_i\Phi_k(\cdot,\eta_\alpha)||\ ||\omega(i)-\eta(i)|| ,$$

and so we get the estimate

$$|\frac{d}{d\alpha}\int f d\mu_\alpha| \le \delta(f)\sigma_k\ ||D_i\Phi_k||_\infty||\omega(i)-\eta(i)|| . \tag{3.23}$$

2) By (3.23)

$$|\int f d\pi_k(\cdot|\omega) - \int f d\pi_k(\cdot|\eta)| = |\int_0^1 (\frac{d}{d\alpha}\int f d\mu_\alpha)d\alpha|$$

$$\le \delta(f)\sigma_k||D_i\Phi_k||_\infty||\omega(i)-\eta(i)|| .$$

This implies (3.21) and (3.22).

By (3.8), (3.15) and (3.20), the distribution P of the diffusion (3.1) can be viewed as a tempered Gibbs measure on S^I in the sense of (2.17) in Ch. I. (3.9) and (3.10) show that this random field has a spatial Markov property. Due to (3.22) and (3.11), one can derive bounds on the pair potential which guarantee the Dobrushin condition

$$\sup_k \sum_i C_{ik} < 1 ; \tag{3.24}$$

cf. [De3] for an explicit computation. This allows to proceed as in Section 2 of Ch. I. In particular, P is the unique tempered Gibbs measure compatible with the system of conditional distributions (π_i).

Let us now consider the translation invariant case, with $I = Z^d$ and a spatially homogeneous pair potential. Then we have exponential decay of correlations in the sense of (2.24) in Ch. I. For any function f in the class $L(S^I)$ the central limit theorem in I (2.25) holds, i.e., the distribution of

$$S_n^*(f) = |V_n|^{-1/2} \sum_{k \in V_n} [f \circ \theta_k - \int f dP]$$

converges in law to a centered Gaussian random variable with variance

$$\sigma^2(f) = \sum_k \operatorname{cov}(f, f \circ \theta_k) .$$

Following K. Itô [It1], we are led to weak convergence of the process $Y^{(n)}$ defined by

$$Y_t^{(n)}(\varphi) = S_n^* (\varphi \circ X(t)) \qquad (0 \le t \le 1)$$

for a suitable class S_∞ of test functions φ on R^I. Viewed as a process with values in the corresponding space S'_∞ of tempered distributions, $Y^{(n)}$ converges in law to a continuous S'_∞-valued process $Y = (Y_t)_{0 \le t \le 1}$ with variances

$$\operatorname{var}(Y_t(\varphi)) = \sum_k \operatorname{cov}(\varphi(X(t)), \varphi(X(t) \circ \theta_k)) .$$

Moreover, the evolution of this process can be described by a linear stochastic differential equation of the form

$$(3.25) \qquad Y_t(\varphi) = Y_o(\varphi) + \int_o^t Y_s(L\varphi) ds + \int_o^t dB_s(D\varphi)$$

where

$$L\varphi = \sum_k (b_k \nabla_k + \frac{1}{2} \nabla_k^2) \varphi$$

$$D\varphi = \sum_k \nabla_k \varphi \circ \Theta_{-k} ,$$

and where $B = (B_t)_{0 \le t \le 1}$ is an S'_∞-valued Brownian motion with quadratic variation

$$\mathrm{var}(B_t(\varphi)) = \int_0^t E[\varphi^2(X(s))]ds .$$

We refer to [De3] for details, and also to [De5] for an alternative approach to (3.25) which does not pass through the description of P as a random field.

References

[AH] Albeverio, S., Høegh-Kron, R.: Uniqueness and global Markov property for Euclidean fields and lattice systems. In: Quantum fields-algebras, processes (ed. Streit, L.) 303-330, Springer (1980)

[Az] Azencott, R.: Grandes deviations et applications. In: Ecole d'Eté de Probabilités de Saint-Flour VIII, Springer Lecture Notes Math. 774, 1-176 (1980)

[Be] Beurling, A.: An automorphism of product measures. Ann. Math. 72, 189-200 (1960)

[BJ] Bichteler, K., Jacod, J.: Calcul de Malliavin pour les diffusions avec sauts. Sém. Probabilités XVII, Lecture Notes Math. 986, Springer (1983)

[Bi] Bismut, J.M.: Martingales, the Malliavin Calculus and Hypolellipticity under general Hörmander's conditions. Z. Wahrscheinlichkeitstheorie verw. Geb. 56, 469-505 (1981)

[BM] Bismut, J.M., Michel, D.: Diffusions conditionelles I, II. J. Funct. Anal. 44, 174-211 (1981), 45, 274-292 (1982).

[BD] Blackwell, D. and Dubins, L.: Merging of opinions with increasing information. Ann. Math. Statist. 33, 882-886 (1962)

[Bo] Bolthausen, E.: On the central limit theorem for stationary mixing random fields. Ann. Prob. 10, 1047-1050 (1982)

[Ca] Cairoli, R.: Une inégalité pour martingales à indices multiples et ses applications. Sém. Prob. IV, Springer Lecture Notes in Math. 124, 1-27 (1970)

[Car] Carlen, E.A.: Conservative Diffusions: A constructive approach to Nelson's stochastic mechanics. Thesis, Princeton (1984)

[Cl] Clark, J.M.C.: The representation of functionals of Brownian motion by stochastic integrals. Ann. Math. Stat. 41, 1281-1295 (1970), 42, 1778 (1971)

[Co] Comets, F.: Grandes déviations pour des champs de Gibbs sur Z^d. C.R. Acad. Sc. Paris 303, sér. 1, n° 11, 511 (1986)

[Cs] Csiszàr, I.: I-Divergence Geometry of Probability Distributions and Minimization Problems. Annals Prob. 3, No. 1, 146-158 (1975)

[DG] Dawson, D.A., Gärtner, J.: Long time fluctuations of weakly interacting diffusions. Preprint (1986).

[DM] Dellacherie, C., Meyer, P.A.: Probabilités et potential, Ch. V-VIII. Hermann (1980)

[De1] Deuschel, J.D.: Lissage de diffusion à dimension infinie et leurs propriétés en tout que mesure de Gibbs. Thèse ETH Zürich 7823 (1985)

[De2] Deuschel, J.D.: Non-linear Smoothing of Infinite-dimensional Diffusion Processes. Stochastics 19, 237-261 (1986)

[De3] Deuschel, J.D.: Infinite-dimensional diffusion processes as Gibbs measures on $C[0,1]^{\mathbb{Z}^d}$. Preprint ETH Zürich (1986)

[De4] Deuschel, J.D.: Représentation du champ de fluctuation de diffusions indépendentes par le drap brownien. Sém. Prob. XXI, Springer LN Math. 1247, 428-433 (1987)

[De5] Deuschel, J.D.: A Central limit theorem for an infinite lattice system of interacting diffusion processes. Ann. Prob. (to appear)

[Do1] Dobrushin, R.L.: Description of a random field by means of conditional probabilities and the conditions governing its regularity. Theory Probab. Appl. 13, 197-224 (1968)

[Do2] Dobrushin, R.L.: Prescribing a system of random variables by conditional distributions. Theor. Probab. Appl. 15, 458-486 (1970).

[Do3] Dobrushin, R.L.: Central limit theorems for nonstationary Markov chains. Theory Probab. Appl. 1, Nr. 4, 329-383 (1956).

[DP] Dobrushin, R.L., Pecherski, E.A.: Uniqueness conditions for finitely dependent random fields. In: Coll. Math. Soc. J. Bolyai 27, Random Fields, 223-261, North-Holland (1981)

[DT] Dobrushin, R.L., Tirozzi, B.: The central limit theorem and the problem of equivalence of ensembles. Comm. Math. Phys. 54, 173-192 (1977)

[DV1] Donsker, M.D. and S.R.S. Varadhan: Asymptotic evaluation of certain Markov expectations for large time, III. Comm. Pure Appl. Math, 29, 389-461 (1976)

[DV2] Donsker, M.D. and S.R.S. Varadhan: Asymptotic evaluation of certain Markov expectations for large time IV, Comm. Pure Appl. Math. 36, 183-212 (1983)

[DR] Doss, H., Royer, G.: Processus de diffusion associés aus mesures de Gibbs. Z. Wahrscheinlichkeitstheorie 46, 125-158 (1979)

[DP1] Dubins, L., Pitman, J.W.: A divergent, two-parameter, bounded martingale. Proc. Amer. Math. Soc. 78, 414-416 (1980)

[Dy] Dynkin, E.B.: Sufficient statistics and extreme points. Ann. Probability 6, 705-730 (1978)

[El] Ellis, R.S.: Entropy, Large Deviation and Statistical Mechanics. Grundlehren 271, Springer (1975)

[Fö1] Föllmer, H.: Phase transitions and Martin boundary, in: Sém. Prob. IX, Lecture Notes in Mathematics 465, 305-318, Springer (1975)

[Fö2] Föllmer, H.: On entropy and information gain in random fields. Z. Wahrsch. verw. Geb. 26, 207-217 (1973)

[Fö3] Föllmer, H.: On the global Markov property. In: Quantum fields, algebras, processes (ed. Streit, L.) 293-302, (1980)

[Fö4] Föllmer, H.: A covariance estimate for Gibbs measures. J. Funct. Anal. 46, 387-395 (1982)

[Fö5] Föllmer, H. Almost sure convergence of multiparameter martingales for Markov random fields. Ann. Probability 12, 133-140 (1984)

[Fö6] Föllmer, H.: Time reversal on Wiener space, Proceedings of the BiBoS-Symposium "Stochastic processes - Mathematics and Physics", Lecture Notes in Mathematics 1158, 119-129 (1986)

[FO] Föllmer, H, Orey, S.: Large deviations for the empirical field of a Gibbs measure. Ann. Prob. (to appear).

[FOr] Föllmer, H., Ort, M.: Large deviations and surface entropy Astérisque, Actes du Colloque Paul Lévy (to appear).

[FW] Föllmer, H., Wakolbinger, A.: Time reversal of infinite-dimensional diffusions. Stochastic Processes Appl. 22, 59-77 (1986)

[Fo] Fortet, R.: Résolution d'un système d'équations de M. Schrödinger. J. Math. Pures Appl. IX, 83-105 (1940)

[Fr1] Fritz, J.: Generalization of McMillan's theorem to random set functions. Studia Sci. Math. Hungar. 5, 369-394 (1970)

[Fr2] Fritz, J.: Stationary measures of stochastic gradient systems, Infinite lattice models. Z. Wahrscheinlichkeitsstheorie 59, 479-490 (1982)

[Fu1] Fukushima, M.: Dirichlet forms and Markov Processes. North Holland (1980)

[Fu2] Fukushima, M.: On absolute continuity of multidimensional symmetrizable diffusion. In Functional Analysis in Markov Processes, Springer LN in Math. 923, 146-176 (1982)

[FS] Fukushima, M., Stroock, D.W.: Reversibility of solutions to martingale problems. Preprint (1980)

[GG] Geman, S. and Geman, D.: Stochastic Relaxation, Gibbs Distributions and the Bayesian Restoration of Images. IEEE Trans. on Pattern Analysis and Machine Intelligence, Vol. 6, 721-741 (1984)

[Ge] Georgii, H.O.: Canonical Gibbs measures. Springer Lecture Notes Math. 760 (1979)

[Gi] Gidas, B.: Nonstationary Markov chains and the Convergence of the Annealing Algorithm. Journal of Stat. Physics 39, 73-130 (1985)

[Go] Goldstein, S.: Remarks on the global Markov property. Comm. Math. Phys. 74, 223-234 (1980)

[Gr1] Gross, L.: Decay of correlations in classical lattice models at high temperatures. Comm. Math. Phys. 68, 9-27 (1979)

[Gr2] Gross, L.: Thermodynamics, Statistical Mechanics, and Random Fields. In: Ecole d'Eté de Probabilités de Saint-Flour X, Springer Lecture Notes in Math. 929, 101-204 (1982)

[Gu] Gurevich, B.M.: A variational Characterization of one-dimensional Countable state Gibbs random fields. Z. Wahrscheinlichkeitstheorie verw. Geb. 68, 205-242 (1984)

[Ha] Hayek, B.: A tutorial survey of theory and applications of simulated annealing. In: Proc. 24th Conference of Decision and Control, Dec. 85 (to appear)

[HP] Haussmann, U., Pardoux, E.: Time reversal of diffusion processes. In: Proc. 4th IFIP Workshop on Stochastic Differential Equations, M. Métivier and E. Pardoux Eds., Lecture Notes in Control and Information Sciences, Springer (1985)

[HS] Holley, R., Stroock, D.W.: Applications of the stochastic Ising model to the Gibbs states. Comm. math. Phys. 48, 249-265 (1976)

[HS1] Holley, R., Stroock, D.W.: Diffusions on an infinite dimensional Torus. J. Funct. Anal. 42, 29-63 (1981)

[HS2] Holley, R., Stroock, D.W.: Logarithmic Sobolev inequalities and stochastic Ising models. Preprint M.I.T. (1986)

[HW] Hitsuda, M. Watanabe, H.: On stochastic integrals with respect to an infinite number of Brownian motions and its applications. In: Stochastic Differential Equations (ed. K. Itô), Kyoto (1976)

[It1] Itô, K.: Distribution-valued processes arising from independent Brownian motions, Math. Z. 182, 17-33 (1983)

[It2] Itô, K.: Foundations of stochastic differential equations in infinite dimensional spaces. CBMS-NSF Ser. in Appl. Math. 4 SIAM, Philadelphia (1984)

[IT] Iosifescu, M., Theodorescu, R.: Random processes and learning, Grundlehren der math. Wissenschaften, Bd. 150, Springer (1969)

[IW] Ikeda, N., Watanabe, S.: Stochastic differential equations and diffusion processes. North Holland (1981)

[Ja] Jamison, B.: Reciprocal processes, Z. Wahrscheinlichkeits-theorie verw. Geb. 30, 65-86 (1974)

[Ke] Kemperman, J.H.B.: On the optimum rate of transmitting information. In: Probability and Information theory, Springer Lecture Notes in Mathematics 89, 126-169 (1967)

[Ko] Kolmogorov, A.N.: Zur Umkehrbarkeit der statistischen Naturgesetze. Math. Ann. 113, 766-772 (1937)

[Kü1] Künsch, H.: Almost sure entropy and the variational principle for random fields with unbounded state space. Z. Wahrscheinlichkeitstheorie verw. Geb. 58, 69-85 (1981)

[Kü2] Künsch, H.: Thermodynamics and Statistical Analysis of Gaussian Random fields, Z. Wahrscheinlichkeitstheorie verw. Geb. 58, 407-421 (1981)

[Kü3] Künsch, H.: Decay of correlation under Dobrushin's uniqueness condition and its application. Comm. Math. Physics 84, 207-222 (1982)

[Kü4] Künsch, H.: Non Reversible Stationary Measures for Infinite Interacting Particle Systems. Z. Wahrscheinlichkeitstheorie verw. Geb. 66, 407-424 (1984)

[La] Lanford, O.E.: Entropy and Equilibrium states in classical statistical mechanics. Springer Lecture Notes in Physics 20, 1-113 (1973)

[LR] Lanford, O.E., Ruelle, D.: Observables at infinity and states with short range correlations in statistical mechanics. Comm. Math. Phys. 13, 194-215 (1969)

[Lan] Lang, R.: Unendlich-dimensionale Wienerprozesse mit Wechselwirkung. Z. Wahrscheinlichkeitstheorie verw. Geb. 39, 277-299 (1977)

[Li] Liggett, T.M.: Interacting Particle Systems. Springer (1985)

[LR1] Leha, G., Ritter, G.: On diffusion processes and their semigroups in Hilbert space with an application to interacting stochastic systems. Ann. Prob. 12, 1077-1112 (1984)

[LR2] Leha, G. Ritter, G.: On solutions to stochastic differential equations with discontinuous drift in Hilbert space. Math. Ann. 270, 109-123 (1985.

[LS] Liptser, R.S., Shiryaev, A.N.: Statistics of Random processes I. Springer (1977)

[Ma] Maz'ja, V.G.: Sobolev spaces. Springer (1985)

[Mé] Métivier, M.: Semimartingales. de Gruyter (1982)

[Mo] Moulin Ollagnier, J.: Ergodic Theory and Statistical Mechanics, Springer Lecture Notes in Math. 1115 (1985)

[Na] Nagasawa, M.: The Adjoint Process of a Diffusion with Reflecting Barrier, Kodai Math. Sem. Report 13, 235-248 (1961)

[Ne] Nelson, E.: Dynamical theories of Brownian motion. Princeton University Press (1967)

[NZ] Nguyen, X.X., Zessin, H.: Ergodic theorems for spatial processes. Z. Wahrscheinlichkeitstheorie verw. Geb. 48, 133-158 (1979)

[Ol] Olla, S.: Large deviations for Gibbs random fields. Preprint Rutyers University (1986)

[Or1] Orey, S.: Conditions for the absolute continuity of two diffusions. Trans. Amer. Math. Soc. 193, 413-426

[Or2] Orey, S.: Large deviations in ergodic theory, Seminar on Probability, Birkhäuser (1984)

[Ort] Ort, M.: Grosse Abweichungen und Oberflächenentropie. Diss. ETHZ (to appear)

[OW] Ornstein, D, Weiss, B.: The Shannon-Mc Millan-Breimann Theorem for a class of amenable groups. Isr. JM 44, 53-60 (1983)

[Pa] Pardoux, E.: Equations du filtrage non-linéaire, de la prédiction et du lissage. Stochastics 6, 193-231 (1982)

[Pi] Pirlot, M.: A strong variational principle for continuous spin systems. J. Appl. Prob. 17, 47-58 (1980)

[Pr] Preston, C.: Random Fields. Lecture Notes in Mathematics. Springer (1976)

[Ro] Royer, G.: Etude des champs euclidéens sur un réseau. J. Math. Pures Appl. 56, 455-478 (1977)

[Ru] Ruelle, D.: Thermodynamical Formalism. Encyclopedia of Math. and Appl. 5, Addison Wesley (1978)

[Schi] Schilder, M.: Some asymptotic formulae for Wiener integrals. Trans Amer. Math. Soc., 125, 63-85 (1966)

[Sch1] Schrödinger, E.: Ueber die Umkehrung der Naturgesetze, Sitzungsberichte Preuss. Akad. Wiss. Berlin, Phys. Math. 144, 144-153 (1931)

[Sch2] Schrödinger, E.: Sur la théorie relativiste de l'électron et l'interprétation de la mécanique quantique", Ann. Inst. H. Poincaré 2, 269-310 (1932)

[ShSh] Shiga, T., Shimizu, A.: Infinite dimensional stochastic differential equations and their applications. J. Math. Kyoto Univ. 20, 395-416 (1980)

[Scho] Schonmann, R.H.: Second order large deviation estimates for ferromagnetic systems in the pase coexistence region. Commun. Math. Phys. 112, 409-422 (1987)

[St] Stroock, D.W.: An introduction to the theory of large deviations. Springer (1984)

[Su] Sucheston, L.: On one-parameter proofs of almost sure convergence of multiparameter processes. Z. Wahrsch. verw. Gebiete 63, 43-50 (1982)

[Sug] Sugita, H.: Sobolev spaces of Wiener functionals and Malliavin's calculus. J. Math. Kyoto Univ. 25-1, 31-48 (1985)

[Te] Tempelman, A.A.: Specific characteristics and variational principle for homogeneous random fields. Z. Wahrscheinlichkeitstheorie verw. Geb. 65, 341-365 (1984)

[Th] Thouvenot, J.P.: Convergence en moyenne de l'information pour l'action de Z^2. Z. Wahrscheinlichkeitstheorie verw. Geb. 24, 135-137 (1972)

[Va] Varadhan, S.R.S.: Large Deviations and Applications. SIAM Philadelphia (1984)

[Za] Zambrini, J.C.,: Schrödinger's stochastic variational dynamics. Preprint Univ. Bielefeld (1986)

WAVES IN ONE-DIMENSIONAL RANDOM MEDIA

George C. PAPANICOLAOU

G.C. PAPANICOLAOU : "WAVES IN ONE-DIMENSIONAL RANDOM MEDIA"

1. Introduction and survey.

Small amplitude wave propagation in random media has been studied for a long time by perturbation techniques when the random inhomogeneities are small. Typically one finds that the speed of propagation of a disturbance decreases somewhat. Furthermore, in a medium without dissipation the mean amplitude of the disturbance decreases with distance travelled. This is explained by noting that wave energy is converted to small scale flucuations of the field. The fluctuating part of the field intensity is in turn calculated approximately from a transport equation, a linear radiative transport equation. All this is fairly well understood [1] although a reasonably complete mathematical theory is as lacking today as it was in the early seventies when mathematical study of waves in random media begun.

The stricking thing is that this simple and intuitive picture of the effects of random inhomogeneities on wave propagation is completely false in one-dimensional situations. It is also false in a general three-dimensional setting when the random inhomogeneities are strong. This was first noted by Anderson [2] and came to be known as Anderson localization. It says that waves do not propagate; they get blocked, the disturbances stay localized. This is usually stated in terms of the spectrum of the reduced wave equation: it is discrete with probability one and the associated eigenfunctions decay exponentially. This was first proved in [3] and is now very well understood mathematically [4,5]. In three dimensions localization in a random medium was proved by Froelich and Spencer [6]. These works examine the qualitative aspects of localization. They do not look for quantities associated with the wave motion that can be calculated in some interesting asymptotic limit. There are, of course, no exact solutions available even in one dimension so one must look for interesting asymptotic limits. They usually contain the essential features of the phenomena anyway.

One can carry out formal perturbation calculations for one or three dimensional problems when the random inhomogeneities are small and there is no obvious warning that the results are wrong in the one-dimensional case [7,8]. A more careful analysis shows how to proceed safely. We looked at this issue in detail sometime ago [9]. Using the one-dimensionality of the problem we were able to exhibit quantitatively in detail the effects of localization. Our tools were limit theorems for stochastic equations which were later used in homogenization [10, Chapter 3] as well.

In these notes we take up a review of one dimensional monochromatic waves and provide an introduction to recent work on direct and inverse problems of pulse reflection. The pulse problem is formulated in section 3. Many new and unusual results have been found recently, inluding the solution of an interesting inverse problem for random media. The work in sections 3-5 is done jointly with Burridge, White and Sheng.

One-dimensional wave problems arise naturally in many applications, in optical fiber transmission, in gravity waves in shallow channels, in layered elastic media such as the earth's mantle, etc. Is there something of what follows valid in several dimensions? The answer is not immediately negative because even though the methods are very much one-dimensional, when localization does in fact take place it should still be possible to study reflection-transmission as we do here. We hope that it will turn out this way in the near future.

2. Monochromatic waves

Let $u(x)$ be the complex valued wave field at x in R^1, let k be the free space wave number and let $n(x)$ be the index of refraction of the medium. The wave field satisfies the reduced wave equation

$$u_{xx}+k^2n^2u=0 \tag{1}$$

in an interval $0<x<L$. We assume that a wave of unit amplitude is incident from the left so that

$$u=e^{ikx}+Re^{-ikx} \quad \text{for } x<0 \tag{2}$$

and

$$u=Te^{ikx} \quad \text{for } x>L \tag{3}$$

Here R and T are the complex-valued reflection and transmission coefficients, respectively. At $x=0$ and $x=L$ the wave function u and u_x are continuous. Since k and n are real it is easily seen that the total power scattered by the slab in $0<x<L$ is equal to one

$$|R|^2+|T|^2=1 \tag{4}$$

It is also easily seen that the continuity condition and (2), (3) imply that the solution of equation (1) satisfies the two point boundary conditions

$$u+\frac{1}{ik}u_x=2 \quad at \quad x=0 \tag{5}$$

$$u-\frac{1}{ik}u_x=0 \quad at \quad x=L$$

For a random slab the index of refraction $n(x)$ is a random process defined on some probability space. If $n(x)$ is positive, bounded and piecewise continuous (measurable) the solution of (1) is a well defined process. The reflection and transmission coefficients are then random variables depending on the wave number k and the slab width L. The problem is now to find qualitative or quantitative properties of R and T given some properties of the random index of refraction. In particular to find properties that require randomness and do not hold for periodic or almost periodic media, for example.

A basic qualitative property that has been known for some time [11] is the exponential decay of the transmission coefficient as the slab width L goes to infinity. Furstenberg treated discrete versions of the problem in which the analog of $n(x)$ is independent identically distributed random variables. Kotani [4] gave recently a proof of this result with minimal hypotheses. If $n(x)$ is a stationary, ergodic process that is bounded with probability one and is nondeterministic (that is it has trivial tail sigma field) then

$$\lim_{L\to\infty}\frac{1}{L}\log|T|^2=-2\gamma(k) \tag{6}$$

with probability one and $\gamma(k)$ is a positive constant. The content of the theorem is the positivity which gives the exponential decay of the transmission coefficient with the size of the slab. The decay constant $\gamma(k)$ depends on the wave number and is the (maximal) Lyapounov exponent of (1). Its reciprocal is called the localization length because it characterizes the depth of penetration (skin depth) of the wave into a random medium occupying a half space. A random half space is a perfectly reflecting medium by (4) and (6).

The positivity of the Lyapounov exponent $\gamma(k)$ is at the root of nearly every theorem dealing with qualitative properties of (1). For example the discreteness of the spectrum in an infinite random medium with bounded, stationary ergodic index of refraction (with trivial tail field) first proved by Goldsheid, Molcanov and Pastur [3] (with many more hypotheses) and in great generality by Kotani [4].

If we want to get quantitative information about scattering by a random medium we must look into interesting asymptotic limits: large or small wave number, large or small variance of $n(x)$ about its mean, etc. Let us assume that

$$n^2(x)=1+\mu(x) \tag{7}$$

where $\mu(x)$ is a zero mean stationary random process with rapidly decaying correlation functions. Let

$$\alpha=\int_0^\infty E\{\mu(x)\mu(0)\}dx \tag{8}$$

Here E denotes expectation for the stationary process μ. The parameter α has dimensions of length and can be thought of as a correlation length of the refractive index. We can now investigate the behavior of the Lyapounov exponent $\gamma(k)$ for small $k\alpha$, or large $k\alpha$ [12]. Small $k\alpha$ will be discussed here since we are aiming at the pulse problem of the next section where this case is important. Small $k\alpha$ means of course that the wave length of the incident wave is large compared to the typical 'size' of inhomogeneities in the medium. We expect the random medium to have an effective behavior independent of the detailed characterization of $n(x)$. In fact

$$\gamma(k)\approx\gamma_0 k^2 \tag{8}$$

as $k\to 0$ where $\gamma_0=\alpha/4$. Whereas $\gamma(k)$ is a complicated functional of the process μ, γ_0 depends only on α. We have a limit theorem that goes back to Rayleigh.

The simple result (8) gives us a way to assess quantitatively the transmission coefficient $T=T(k,L)$ which is a complicated functional of μ. We see in fact from (6) and (8) that if as $L\to\infty$, $k\to 0$ like $1/\sqrt{L}$ then roughly

$$|T|\approx e^{\gamma(k)L}\approx e^{\gamma_0} \tag{9}$$

This is meant to indicate only that T should have a nontrivial distribution in this limit. When the slab of random medium is large (compared to the correlation length α) then only long wavelengths of order $\sqrt{L}$ will be transmitted. Short ones are blocked. And T has a limit distribution that can be computed explicitly.

We will describe next this theorem. First it is convenient to consider the reflection coefficient R instead of the transmission coefficient. They are related by (4). Let us denote R by $R(L,k,\alpha)$ to indicate dependence on the slab width, the wave number and the correlation length. Let $\varepsilon>0$ be a small dimensionless parameter. Define

$$R^\varepsilon(L)=R(\frac{L}{\varepsilon^2},\varepsilon k,\alpha) \tag{10}$$

and consider $R^\varepsilon(L)$, $L\geq 0$ for each ε as a process on the hyperbolic disc $H=\{R \ in \ C \mid |R|=1\}$. The scaling (10) is a convenient way to express the limit described above (9). We have the following.

Theorem

Assume that $\mu(x)$ is a stationary process that has finite moments of all orders and is rapidly mixing [see [13] for definitions and sharp conditions]. Then the process R^ε converges weakly as $\varepsilon \to 0$ to Brownian motion on the hyperbolic disc H with infinitesimal variance αk^2.

If we parametrize H by polar coordinates

$$R = e^{-i\psi} \tanh \frac{\theta}{2} \tag{11}$$

then the theorem says that the limit process $R(L)$, $L>0$ is a Markov process with generator $\frac{1}{2}\alpha k^2 \Delta$ where Δ is the Laplace-Beltrami operator on H

$$\Delta = \frac{\partial^2}{\partial \theta^2} + \coth \theta \frac{\partial}{\partial \theta} + csch^2\, \theta \frac{\partial^2}{\partial \psi^2} \tag{12}$$

This result was first obtained by Gertenstein and Vasiliev [14] who realized that in a discrete medium the reflection coefficient (defined in a slightly different way) transforms by linear fractional transformations as L changes in discrete units. Thus $R(L)$ does some kind of random walk in H as L varies. The simplest diffusion approximation to a random walk is, naturally, Brownian motion. Hence the result. A more complete derivation was given in [15] (for a slightly different but very similar problem). It was noticed then, and subsequently in much greater detail in [9,16 and also 17-20], that although this theorem is indeed a diffusion approximation to a complicated random motion in H there is a reason why the limit is Brownian motion and not a more complicated process. For example, another reasonable scaling limit for the reflection coefficient is the white noise limit where L and k are fixed as the process $\mu(x)$ tends to white noise. That could be done by replacing $\mu(x)$ by $\mu^\varepsilon(x) = (1/\varepsilon)\mu(x/\varepsilon^2)$. We again have a diffusion approximation but the limit is not Brownian motion now. The scaling (10) is special for it leads to a rapid deterministic phase rotation in H that makes the limiting process isotropic.

This observation is simple and is contained in the general limit theorems for stochastic equations [13,21-25]. It enhanced profoundly our ability to calculate statistics of interesting scattering quantities (as in part II of [9] for example and in [20]) but it seemed to be just good luck. It was when we looked recently into pulse propagation, the subject of the next section, that we realized that this simplification due to rapid phase rotation leads to striking results in the time domain that are almost entirely due to this phenomenon.

3. Propagation of pulses

3.1. Introduction

In the previous section we described some problems and results that cover many aspects of one dimensional wave propagation, are well understood and have been around for some time. Recently we came across an interesting paper by Richards and Menke [26] where extensive numerical simulations of pulse reflection from an one dimensional random half-space are carried out. The questions they asked are motivated by geophysical exploration problems. They wanted to understand for example, how to distinguish multiple scattering effects by small-scale inhomogeneities from dissipation in the medium, when one has access to reflected signals, seismograms in their case. Of course the more general basic question here is: what can one say about the medium from the reflected signal if there are small-scale inhomogeneities present that one would like to ignore in an intelligent way?

In collaboration with Burridge, Sheng and White [27-33] we begun recently to analyze this problem. We shall present here a brief review of what has been done along with a discussion of the methods that are used. We will first give a precise formulation of the physical problems. Then we will discuss the results and finally, in the next section, the methods of analysis. The complete story is somewhat long and evolving in several different directions at present.

3.2. Formulation and Scaling

We consider a one-dimensional acoustic wave propagating in a random slab of material occupying the half space $x < 0$. We will analyze in detail the backscatter at $x = 0$.

Let $p(t,x)$ be the pressure and $u(t,x)$ velocity. The linear conservation laws of momentum and mass governing acoustic wave propagation are

$$\rho(x)\frac{\partial}{\partial t}u(t,x) + \frac{\partial}{\partial x}p(t,x) = 0 \tag{2.1}$$

$$\frac{1}{K(x)}\frac{\partial}{\partial t}p(t,x) + \frac{\partial}{\partial x}u(t,x) = 0$$

where ρ is density and K the bulk modulus. We define means of ρ and $\frac{1}{K}$ as

$$\rho_o = E\,[\rho] \tag{2.2}$$

$$\frac{1}{K_o} = E\,[\frac{1}{K}].$$

In the special case that ρ and K are stationary random functions of position x, ρ_o, K_o are the constant parameters of effective medium theory. That is, a pulse of long wavelength will propagate over distances that are not too large as if in a homogeneous medium with "effective" constant parameters ρ_o, K_o, and hence with propagation speed

$$c_o = \sqrt{K_o/\rho_o}\ . \tag{2.3}$$

We consider here the case where ρ_o, K_o, c_o are not constant, but vary slowly compared to the spatial scale, l_o, of a typical inhomogeneity. We may take the "microscale" l_o to be the correlation length of ρ and $\frac{1}{K}$. We introduce a "macroscale", l_o/ε^2, where $\varepsilon > 0$ is a small parameter. It is on this macroscale that ρ_o, K_o, and other statistics of ρ and K are allowed to vary. We thus write the density and bulk modulus on the macroscale in the following scaled form.

$$\rho(x) = \rho_o\left(\frac{x}{l_o}\right)\left[1 + \eta\left(\frac{x}{l_o}, \frac{x}{\varepsilon^2 l_o}\right)\right] \tag{2.4}$$

$$\frac{1}{K(x)} = \frac{1}{K_o(x/l_o)}\left[1 + \nu\left(\frac{x}{l_o}, \frac{x}{\varepsilon^2 l_o}\right)\right]$$

where the random fluctuations η and ν have mean zero and slowly varying statistics. The mean density ρ_o and the mean bulk modulus K_o are assumed to be differentiable functions of x.

Equations (2.1) are to be supplemented with boundary conditions at $x = 0$ corresponding to different ways in which the pulse is generated at the interface. In the cases analyzed below the pulse width is assumed to be on a scale intermediate between the microscale and the macroscale. That is, the pulse is broad compared to the size of the random inhomogeneities, but short compared to the non-random variations. Thus the small scale structure will introduce only random effects which the pulse is too broad to probe in detail. In contrast, the pulse is chosen to probe the non-random macroscale, from which it reflects and refracts in the manner of ray theory (geometrial optics). We will recover macroscopic variations of the medium by examination of reflections at $x = 0$.

Let typical values of ρ_o, K_o be $\bar{\rho}$, $\bar{K}$ with $\bar{c} = \sqrt{\bar{K}/\bar{\rho}}$. Then for $f(t)$ a smooth function of compact support in $[0,\infty)$ we define the incident pulse by

$$f^{\varepsilon}(t) = \frac{1}{\varepsilon^{1/2}} f\left(\frac{\bar{c}\,t}{\varepsilon\, l_o}\right). \tag{2.5}$$

This pulse, f^{ε}, will be convolved with the appropriate Green's function depending on how the wave is excited at the interface. The pre-factor $\varepsilon^{-1/2}$ is introduced to make the energy of the pulse independent of the small parameter ε.

We consider here the "matched medium" boundary condition. It is assumed that the wave is incident on the random medium occupying $x < 0$ from a homogeneous medium occupying $x > 0$ and characterized by the constant parameters $\rho_o(0)$, $K_o(0)$. One may similarly consider an unmatched medium where ρ_o, K_o are discontinuous at $x = 0$, but we do not carry this out here. To obtain the Green's function for this problem we introduce the initial-boundary condition for a left-travelling wave which strikes $x = 0$ a time $t = 0$

$$u = l_o\, \delta\,(t + \frac{x}{c_o(0)}) \tag{2.6}$$

$$p = -\, l_o\, \rho_o(0)\, c_o(0)\, \delta\,(t + \frac{x}{c_o(0)})$$

The Green's function G will then be a right-going wave in $x > 0$ and as $x \downarrow 0$

$$G = \frac{1}{2}\left[u(t,0) - \frac{p(t,0)}{(\rho_o(0)c_o(0))} \right] \tag{2.7}$$

We non-dimensionalize by setting

$$x' = x/l_o \quad p' = p/\bar{\rho}\,\bar{c}^{\,2} \tag{2.8}$$

$$t' = \bar{c}t/l_o \quad u' = u/\bar{c}$$

By inserting (2.8) into the above equations, and dropping primes, it can be shown that without loss of generality $\bar{K}$, $\bar{\rho}$, $\bar{c}$, l_o may be taken equal to unity, after K, ρ, c are replaced by their normalized forms.

We will determine the statistics of the Green's function convolved with the pulse f^ε. Let

$$G^{\varepsilon}_{t,f}(\sigma) = (G * f^{\varepsilon})(t + \varepsilon\,\sigma) \tag{2.9}$$

$$= \int_0^{t+\varepsilon\sigma} G(t + \varepsilon\sigma - s)\, f^{\varepsilon}(s)\, ds\,.$$

We consider the above expression as a stochastic process in σ, with t held fixed. That is, for each t we consider a "time window" centered at t, and of duration on the order of a pulse width, with the parameter σ measuring time within this window.

For the analysis of this problem, we Fourier transform in time, choosing a frequency scale appropriate to the pulse $f^\varepsilon(t)$. Thus, letting

$$\hat{f}(\omega) = \int_{-\infty}^{\infty} e^{i\omega t} f(t)\, dt \tag{2.10}$$

we transform (2.1) by

$$\hat{u}(\omega, x) = \int e^{i\omega t/\varepsilon} u(t,x)\, dt \tag{2.11}$$

$$\hat{p}(\omega, x) = \int e^{i\omega t/\varepsilon} p(t,x)\, dt$$

so that

$$G^\varepsilon_{i,f}(\sigma) = \frac{1}{2\pi\varepsilon^{1/2}} \int_{-\infty}^{\infty} e^{-i\omega(t+\varepsilon\sigma)/\varepsilon} \hat{f}(\omega)\, \hat{G}(\omega) d\omega . \tag{2.12}$$

In (2.12) $\hat{G}$ is the appropriate combination of $\hat{u}$, $\hat{\rho}$ obtained by Fourier transform of (2.7).

From (2.1), (2.4), (2.11), $\hat{u}$, $\hat{p}$ satisfy

$$\frac{\partial}{\partial x}\hat{p} = \frac{i\omega}{\varepsilon} \rho_o(x) [1 + \eta(x, \frac{x}{\varepsilon^2})]\, \hat{u} \tag{2.13}$$

$$\frac{\partial}{\partial x}\hat{u} = \frac{i\omega}{\varepsilon} \frac{1}{K_o(x)} [1 + \nu(x, \frac{x}{\varepsilon^2})]\, \hat{p} .$$

In the frequency domain a radiation condition as $x \to -\infty$, is required for (2.13). One way to do this is to terminate the random slab at a finite point $x = -L$, and assume the medium is not random for $x > -L$. We can later let $L \to -\infty$ but in any case the reflected signal up to a time t is not affected by how we terminate the slab at a sufficiently distant point $-L$. This is a consequence of the hyperbolicity of (2.1).

We next introduce a right going wave A and a left going wave B, with respect to the macroscopic medium. Let the travel time in the macroscopic medium be given by

$$\tau(x) = \int_x^0 \frac{ds}{c_o(s)} \quad , \; x < 0 \tag{2.14}$$

We define A, B by

$$\hat{u} = \frac{1}{(K_o\rho_o)^{1/4}} [A\, e^{-i\omega\tau/\varepsilon} + B\, e^{i\omega\tau/\varepsilon}]$$

$$\hat{p} = (K_o\, \rho_o)^{1/4} [A\, e^{-i\omega\tau/\varepsilon} - B\, e^{i\omega\tau/\varepsilon}] \tag{2.15}$$

Putting (2.14), (2.15) into (2.13) yields equations for A, B. Define the random functions $m^{\varepsilon}(x)$ and $n^{\varepsilon}(x)$ by

$$m^{\varepsilon}(x) = m(x,\ x/\varepsilon^2) = \frac{1}{2}\,[\eta(x,x/\varepsilon^2) + \nu(x,x/\varepsilon^2)] \tag{2.16}$$

$$n^{\varepsilon}(x) = n(x,\ x/\varepsilon^2) = \frac{1}{2}\,[\eta(x,x/\varepsilon^2) - \nu(x,x/\varepsilon^2)]$$

Then

$$\frac{d}{dx}\begin{bmatrix} A \\ B \end{bmatrix} = \frac{i\omega}{\varepsilon}\left(\frac{\rho_o}{K_o}\right)^{1/2}\begin{bmatrix} m^{\varepsilon} & n^{\varepsilon}e^{2i\omega\tau/\varepsilon} \\ -n^{\varepsilon}e^{-2i\omega\tau/\varepsilon} & -m^{\varepsilon} \end{bmatrix}\begin{bmatrix} A \\ B \end{bmatrix}$$

$$+\frac{1}{4}\,\frac{(K_o\,\rho_o)'}{(K_o\,\rho_o)}\begin{bmatrix} 0 & e^{2i\omega\tau/\varepsilon} \\ e^{-2i\omega\tau/\varepsilon} & 0 \end{bmatrix}\begin{bmatrix} A \\ B \end{bmatrix}. \tag{2.17}$$

We take as boundary conditions for (2.17) that there is no right-going wave at $x = -L$, and that there is a unit left-goint wave at $x = 0$.

$$A(-L) = 0 \quad B(0) = 1 \tag{2.18}$$

$$B(-L) = T \quad A(0) = R = R^{\varepsilon}(-L,\ \omega)$$

Here T is the transmission coefficient for the slab, and $R^{\varepsilon}(-L,\ \omega)$ is the reflection coefficient. From (2.6), (2.7) we see that

$$\hat{G} = R^{\varepsilon}\,(-L,\ \omega)\,. \tag{2.19}$$

We introduce the fundamental matrix solution of the linear system (2.17). That is, let $Y(x,\ -L)$ satisfy (2.17) with the initial condition that $Y(-L,\ -L) = I$ the 2 x 2 identity. From symmetries in (2.17) it is apparent that if $(a, \bar{b})^T$ is a vector solution (bar denotes complex conjugate and T transpose), then so is $(b,\ \bar{a})^T$. Thus

$$Y = \begin{bmatrix} a & b \\ \bar{b} & \bar{a} \end{bmatrix}. \tag{2.20}$$

Furthermore, since the system has trace zero, Y has determinant one. Hence

$$|\,a\,|^2 - |b\,|^2 = 1\,. \tag{2.21}$$

Now the reflection coefficient R may be expressed in terms of a, b, by writing (2.18) in terms of propagators, i.e.

$$\begin{bmatrix} a & b \\ \bar{b} & \bar{a} \end{bmatrix} \begin{bmatrix} 0 \\ T \end{bmatrix} = \begin{bmatrix} R \\ 1 \end{bmatrix}$$

and hence

$$R = \frac{b}{\bar{a}}, \qquad T = \frac{1}{\bar{a}} \tag{2.22}$$

Now from (2.17), (2.20) we have that

$$\frac{da}{dx} = \frac{i\omega}{\varepsilon}\left[\frac{\rho_o}{K_o}\right]^{1/2}\left[m^\varepsilon a + n^\varepsilon \bar{b}\, e^{2i\omega\tau/\varepsilon}\right] + \frac{1}{4}\frac{(\rho_o K_o)'}{\rho_o K_o}\,\bar{b}\, e^{2i\omega\tau/\varepsilon}$$

$$\frac{d\bar{b}}{dx} = -\frac{i\omega}{\varepsilon}\left[\frac{\rho_o}{K_o}\right]^{1/2}\left[n^\varepsilon a\, e^{-2i\omega\tau/\varepsilon} + m^\varepsilon \bar{b}\right] + \frac{1}{4}\frac{(\rho_o K_o)'}{\rho_o K_o}\, a\, e^{-2i\omega\tau/\varepsilon} \tag{2.23}$$

$$a(-L) = 1,\ b(-L) = 0 .$$

Therefore, from (2.22), (2.23) we can derive the Riccati equation for R

$$\frac{dR^\varepsilon}{dx} = \frac{i\omega}{\varepsilon}\left[\frac{\rho_o}{K_o}\right]^{1/2}\left[n^\varepsilon e^{2i\omega\tau/\varepsilon} + 2m^\varepsilon R^\varepsilon + n^\varepsilon (R^\varepsilon)^2 e^{-2i\omega\tau/\varepsilon}\right]$$

$$+\frac{1}{4}\frac{(\rho_o K_o)'}{(\rho_o K_o)}\left[e^{2i\omega\tau/\varepsilon} - (R^\varepsilon)^2 e^{-2i\omega\tau/\varepsilon}\right] \tag{2.24}$$

$$R^\varepsilon(-L) = 0 .$$

The boundary condition at $-L$ in (2.24) is for termination of the random slab by a uniform medium. If the medium is homogeneously random beyond $-L$ ($\rho_o(x)$, $K_o(x)$ constant) then we will have total reflection at $-L$ because the wave cannot penetrate the random medium to infinite depth. In fact in a statistically homogeneous random medium we have that

$$|T| \to 0 \text{ as } L \to -\infty . \tag{2.25}$$

exponentially fast which follows from Furstenberg's theorem [11,5]. Since (2.21), (2.22) imply that $|R|^2 + |T|^2 = 1$ we have

$$|R| \to 1 \text{ as } L \to -\infty . \tag{2.26}$$

It is convenient to analyze (2.24) with a **totally reflecting termination**, so that

$$R^\varepsilon = e^{-i\psi^\varepsilon} . \tag{2.27}$$

and the number of degrees of freedom is reduced by one. Putting (2.27) into (2.24) yields

$$\frac{d}{dx}\psi^\varepsilon = -\frac{\omega}{\varepsilon}\left[\frac{\rho_o(x)}{K_o(x)}\right]^{1/2}\left[2\,m^\varepsilon(x) + 2n^\varepsilon(x)\cos\left(\psi^\varepsilon + \frac{2\omega\tau(x)}{\varepsilon}\right)\right] \tag{2.28}$$

$$+\frac{1}{2}\frac{(\rho_o K_o)'}{\rho_o K_o}\sin\left(\psi^\varepsilon + \frac{2\omega\tau(x)}{\varepsilon}\right)$$

and we take ψ^ε to be asymptotically stationary as $x \to -\infty$.

To recapitulate, the asymptotically stationary solution of (2.28), evaluated at x=0 is put into (2.27) to yield the totally reflecting reflection coefficient R^ε at frequncy ω. The frequency domain Green's function is then given by (2.19) The result is then transformed back to the time domain by (2.12).

3.3. Statement of the main results

Let $G^\varepsilon_{t,f}(\sigma)$ be the reflection process observed at $x=0$ within the time window centered at t. Then $G^\varepsilon_{t,f}(\cdot)$ converges weakly as $\varepsilon \downarrow 0$ to a stationary Gaussian process with mean zero and power spectral density

$$S_t(\omega) = |\,\hat{f}(\omega)\,|^2\,\mu(t,\omega)\,, \tag{3.1}$$

The normalized power spectral density μ is computed as follows.

Let α_{nn} be the integral of the second moment of the medium properties defined by

$$\alpha_{nn}(x) = \int_0^\infty E[n(x,y)\,n(x,y+s)]\,ds\,. \tag{3.2}$$

Let $\tau(x)$ be travel time to depth x defined by (2.14), and let $\bar{x}(\tau)$ be its inverse which is depth reached up to time t in the medium without fluctuations. Define

$$\gamma(\tau) = \frac{\alpha_{nn}(\bar{x}(\tau))}{c_o(\bar{x}(\tau))} \tag{3.3}$$

Let $W^{(N)}(\tau,\,t,\,\omega)$, $N=0,1,2...$ be the solution of

$$\frac{\partial W^{(N)}}{\partial \tau} + 2N\frac{\partial W^{(N)}}{\partial t} - 2\omega^2\,\gamma(\tau)\Big\{[N+1]^2\,W^{(N+1)}$$

$$-2N^2W^{(N)} + [N-1]^2W^{(N-1)}\Big\} = 0 \tag{3.4}$$

for

$$t, \tau > 0, \quad N = 0,1,2, \cdots$$

with

$$W^{(N)} \equiv 0 \text{ for } t < 0, \ N < 0.$$

and

$$W^{(N)}(0,t,\omega) = \delta(t)\,\delta_{N,1}\,. \tag{3.5}$$

Then

$$\mu(t,\omega) = \lim_{\tau \to \infty} W^{(0)}(\tau,t,\omega)\,. \tag{3.6}$$

The system (3.4) is hyperbolic so it is not necessary to take a limit in (3.6) because $W^{(0)}$ is constant for $\tau > t/2$. Thus

$$\mu(t,\omega) = W^{(0)}(\frac{t}{2},t,\omega) \tag{3.6a}$$

For the case of a homogeneous medium [c_o, $\gamma = const = \tilde{\gamma}$] the normalized power spectral density can be computed explicitly.

$$\mu(t,\omega) = \frac{\omega^2\tilde{\gamma}}{[1+\omega^2\tilde{\gamma}t]^2} \tag{3.7}$$

Let us now consider inverse problems associated with the pulse reflection problem. Inverse problems associated with (2.1) are of little interest before the limit $\varepsilon \to 0$ because the usual procedures (c.f. [34] and refemces therein) are overwhelmed by the fluctuations. So we want to pose inverse problems after the limit, i.e. for the reflected process $G_{t,f}(\sigma)$. Perhaps the simplest such problem is this: what can we say about the slowly varying properties of the medium if we know the power spectral density $\mu(t,\omega)$ of the windows given by (3.6a)? We have the following result.

The limit

$$\lim_{\omega \to 0} \frac{1}{\omega^2}\mu(t,\omega) = \frac{2\alpha}{c_0(\overline{x}(-t/2))} \tag{3.8}$$

exists and is given by the right side of (3.8).

This is an immediate and elementary consequence of (3.4)-(3.6). Let us consider a simple application of it. Let us assume that α is known and let

$$\Theta(t)=\frac{1}{2\alpha}\lim_{\omega\to 0}\frac{1}{\omega^2}\mu(t,\omega)=\frac{1}{c_0(\bar{x}(-t/2))} \tag{3.9}$$

be the quantity that is measured. From (2.14) we see that

$$\frac{d\bar{x}(t)}{dt}=c_0(\bar{x}(t))=\frac{1}{\Theta(-2t)}$$

Thus, the distance travelled up to time t is obtained from the reflection statistics by

$$\bar{x}(t)=\int_0^t \frac{1}{\Theta(-2s)}ds \tag{3.10}$$

Since $\tau(x)$ and $\bar{x}(\tau)$ are monotone inverse functions, (3.10) also determines the travel time $\tau(x)$ and hence the mean propagation speed $c_0(x)$ as a function of distance.

This result is in striking contrast with what the usual inverse methods give when applied to (2.1). It is known for example that quantities such as impedance can only be computed as functions of the (in general unknown from normal incidence data) travel time. In the stochastic case when fluctuations are statistically homogeneous, we are able to determine the mean speed c_0 as a function of distance into the medium from the power spectra via (3.8-3.10). This appears paradoxical at first because we get more out of (the limit of) the noisy problem. We get more because we have access to more, namely power specra which are ensemble averages and contain therefore information from more than one realization. In many applied contexts one has only a single realization of the reflected signal available. Then $\mu(t,\omega)$ must be estimated statistically from this sample and so $\Theta(t)$ in (3.9) is known only approximately and depends on the realization i.e., it is random. But we know in addition that the reflected signal is a Gaussian process (in the limit $\varepsilon\to 0$). This information is now used to get sharp estimates for the travel time statistics.

It is clear that there are many interesting problems, direct and inverse, that can be posed for the pulse reflection problem. Our framework and results provide a theoretical basis for solving some of them.

4. Explicit Calculation of the Power Spectral Density.

We next calculate the power spectrum, as $\varepsilon \downarrow 0$, or the reflection process $G^{\varepsilon}_{i,f}(\sigma)$. From (2.9) we have the correlation function $C^{\varepsilon}_{i,f}$

$$C^{\varepsilon}_{i,f}\,\sigma) \equiv E[G^{\varepsilon}_{i,f}(\sigma)\,G^{\varepsilon}_{i,f}(0)] \tag{4.1}$$

$$=\frac{1}{4\pi^2\varepsilon}\int_{-\infty}^{\infty}d\omega_1\int_{-\infty}^{\infty}d\omega_2\, e^{-i\,\omega_1 t/\varepsilon}\, e^{-i\omega_1\sigma}\, e^{i\omega_2 t/\varepsilon}$$

$$\cdot \hat{f}(\omega_1)\bar{\hat{f}}(\omega_2) E[\hat{G}(\omega_1)\bar{\hat{G}}(\omega_2)] .$$

Let

$$u^\varepsilon(\omega,h) = E[\hat{G}(\omega - \frac{\varepsilon h}{2})\bar{\hat{G}}(\omega + \frac{\varepsilon h}{2})] . \tag{4.2}$$

We will show that the limit

$$u(\omega,h) = \lim_{\varepsilon \downarrow 0} u^\varepsilon(\omega,h) \tag{4.3}$$

exists, and we will characterize it in this section. Then, after the change of variables $\omega = \frac{1}{2}(\omega_1 + \omega_2)$, $h = (\omega_2 - \omega_1)/\varepsilon$ in (4.1) we obtain in the limit $\varepsilon \to 0$

$$C_{t,f}(\sigma) \equiv \lim_{\varepsilon \downarrow 0} C^\varepsilon_{t,f}(\sigma) \tag{4.4}$$

$$= \frac{1}{4\pi^2} \int_{-\infty}^{\infty}\int_{-\infty}^{\infty} e^{-i\omega\sigma} e^{iht} \, |\hat{f}(\omega)|^2 \, u(\omega,h) \, dh .$$

Let

$$\mu(t,\omega) = \frac{1}{2\pi} \int_{-\infty}^{\infty} e^{iht} \, u(\omega,h) \, dh \tag{4.5}$$

Then from (4.4), (4.5) the power spectral density, $S_t(\omega)$ is given by

$$S_t(\omega) \equiv \int_{-\infty}^{\infty} e^{i\omega\sigma} C_{t,f}(\sigma) \, d\sigma \tag{4.6}$$

$$= |\hat{f}(\omega)|^2 \mu(t,\omega) .$$

In the remainder of this section we characterize $u(\omega,h)$ and its transform $\mu(t,\omega)$. From (4.2) and (2.19) we see that $u^\varepsilon(\omega,h)$ can be computed from knowledge of the joint statistics of R_1^ε, R_2^ε the reflection coefficients corresponding, respectively, to frequencies $\omega_1 = \omega - \varepsilon h/2$, $\omega_2 = \omega + \varepsilon h/2$. Since we are in the totally reflecting case (2.27)

$$R_1 = e^{-i\psi_1^\varepsilon}, \; R_2^\varepsilon = e^{-i\psi_2^\varepsilon} ,$$

where ψ_1^ε, ψ_2^ε correspond, respectively, to the same frequencies, and each satisfies (2.28). We shall compute the joint distribution of ψ_1^ε, ψ_2^ε as ε tends to zero.

Let

$$\psi^\varepsilon = \begin{bmatrix} \psi_1^\varepsilon \\ \psi_2^\varepsilon \end{bmatrix}$$

From (2.28) we see that ψ^ε satisfies the differential equation

$$\frac{d\psi^\varepsilon}{dx} = \frac{1}{\varepsilon} F(x, \frac{x}{\varepsilon^2}, \frac{\tau(x)}{\varepsilon}, \psi^\varepsilon) + G(x, \frac{x}{\varepsilon^2}, \frac{\tau(x)}{\varepsilon}, \psi^\varepsilon) \tag{4.7}$$

where

$$F(x,y,h,\psi) = -2\omega \left[\frac{\rho_o(x)}{K_o(x)}\right]^{1/2} \begin{bmatrix} m(x,y) + n(x,y)\cos(\psi_1 + 2\omega h - h\,\tau(x)) \\ m(x,y) + n(x,y)\cos(\psi_2 + 2\omega h + h\tau(x)) \end{bmatrix} \tag{4.8}$$

and

$$G(x,y,h,\psi) = \begin{bmatrix} G(x,y,h,\psi) \\ G_2(x,y,h,\psi) \end{bmatrix}$$

$$G(x,y,h,\psi) = h\left[\frac{\rho_o}{K_o}\right]^{1/2} (m(x,y) + n(x,y)\cos(\psi_1 + 2\omega h - h\,\tau(x)))$$

$$+ \frac{1}{2}\frac{(\rho_o(x)\,K_o(x))'}{\rho_o(x)\,K_o(x)} \sin(\psi_1 + 2\omega h - h\,\tau(x))$$

$$G_2(x,y,h,\psi) = -h\left[\frac{\rho_o}{K_o}\right]^{1/2} (m(x,y) + n(x,y)\cos(\psi_2 + 2\omega h + \tau(x)))$$

$$+ \frac{1}{2}\frac{(\rho_o(x)K_o(x))'}{\rho_o(x)K_o(x)} \sin(\psi_2 + 2\omega h + h\tau(x))$$

We assume, as in Appendix A, that the randomness in equation (4.7) is generated by an ergodic Markov process $q^\varepsilon(x) = q(x, x/\varepsilon^2)$ in Euclidean space R^d of arbitrary dimension d. It is assumed that $q(x, x/\varepsilon^2)$ is a random process on the fast, x/ε^2, spatial scale, but has slowly-varying statistics on the x scale. We express this mathematically by the assumption that $q(x,y)$ is, for fixed x, a stationary ergodic Markov process in y with infinitesimal generator Q_x, depending on x. We then write $m(x, x/\varepsilon^2) = \tilde{m}(x, q(x, x/\varepsilon^2))$, etc. (to simplify notation we will drop tildes). A very wide class of processes with small scale randomness but slowly-varying statistics can be generated in this way.

The process $(q^\varepsilon, \psi^\varepsilon) \in R^{d+2}$, the solution of (4.7) together with its coefficients, is now jointly Markovian, with infinitesimal generator

$$L_x^\varepsilon = \frac{1}{\varepsilon^2} Q_x + \frac{1}{\varepsilon} F(x, \frac{x}{\varepsilon^2}, \frac{\tau(x)}{\varepsilon}, \psi)\cdot\nabla_\psi \tag{4.9}$$

$$+ G(x, \frac{x}{\varepsilon^2}, \frac{\tau(x)}{\varepsilon}, \psi)\cdot\nabla_\psi$$

From the results of Appendix A, we have that ψ^ε converges (weakly) to a process ψ which is Markovian by itself, without the necessity of incluidng q. The limit process ψ has the x-dependent infinitesimal generator L_x, where

$$L_x = \frac{4\omega^2}{c_o^2(x)}\left\{ \alpha_{mm}(x)\,[\frac{\partial}{\partial\psi_1} + \frac{\partial}{\partial\psi_2}]^2 \right. \tag{4.10}$$

$$\left. + \frac{1}{2}\alpha_{nn}(x)\,[\frac{\partial^2}{\partial\psi_1^2} + \frac{\partial^2}{\partial\psi_2^2} + 2\cos(\psi_2 - \psi_1 + 2h\,\tau(x))\frac{\partial^2}{\partial\psi_1\partial\psi_2}\right\} .$$

In (4.10), the coefficients α_{mm}, α_{nn} are defined by the averaged second moments

$$\alpha_{mm}(x) = \int_0^\infty E[m(x,q(x,y))\,m(x,q(x,y+r))]\,dr \tag{4.11}$$

$$\alpha_{nn}(x) = \int_0^\infty E[n(x,q(x,y))n(x,q(x,y+r))]\,dr .$$

In Appendix A we show briefly how these results are obtained.

The generator (4.10) is better expressed in terms of the sum and difference variables

$$\psi = \psi_2 - \psi_1,\ \tilde\psi = \frac{1}{2}(\psi_2 + \psi_1) .$$

Then

$$L_x = \frac{4\omega^2}{c_o^2(x)}\left\{ \alpha_{mm}(x)\frac{\partial^2}{\partial\tilde\psi^2} + \frac{1}{4}\alpha_{nn}(x)\left[1 + \cos(\psi + 2h\,\tau(x))\right]\frac{\partial^2}{\partial\tilde\psi^2} \right. \tag{4.12}$$

$$\left. + \alpha_{nn}(x)\left[1 - cos\,(\psi + 2h\tau(x))\right]\frac{\partial^2}{\partial\psi^2}\right\} .$$

Using (4.12) we can now formulate the equations for $u(\omega,h)$ and its transform $\mu(t,\omega)$. From (2.19), (2.27) and (4.2), (4.3) we have that $u_1(\omega,h)$ is

$$u(\omega, h) = E[e^{i\psi}] . \tag{4.13}$$

Note that the coefficients in (4.12) do not depend on $\tilde\psi$, so that ψ is Markovian by itself. The function

$u(\omega,h)$ can therefore be calculated from the solution V of the Kolmogorov backward equation

$$\frac{\partial V}{\partial x} + \frac{4\omega^2 \alpha_{nn}(x)}{c_o^2(x)} [1-\cos(\psi+2h\,\tau(x))] \frac{\partial^2 V}{\partial \psi^2} = 0, \ x < 0. \tag{4.14}$$

with the final condition

$$V\,|_{x=0} = e^{i\psi}. \tag{4.15}$$

The function u is then

$$u(\omega, h) = \lim_{x \to -\infty} V(x, \psi, \omega, h), \tag{4.16}$$

where the limit in (4.16) exists and is independent of ψ. This follows easily assuming that α_{nn}, τ and c_o are constant for $-x$ large.

Equation (4.14) may be simplified somewhat by the change of variables

$$\hat{\psi} = \psi + 2h\,\tau(x) \tag{4.17}$$

$$\hat{x} = x$$

Then upon dropping hats, it becomes

$$\frac{\partial V}{\partial x} - \frac{2h}{c_o(x)} \frac{\partial V}{\partial \psi} + \frac{4\omega^2 \alpha_{nn}(x)}{c_o^2(x)} [1-\cos\psi] \frac{\partial^2 V}{\partial \psi^2} = 0, \text{ for } x < 0. \tag{4.18}$$

To summarize, (4.18) is to be solved for V subject to the final condition (4.15). Then u is obtained from (4.16).

Equation (4.18) can be solved by Fourier series in ψ Let

$$V = \sum_{N=-\infty}^{\infty} V^{(N)} e^{iN\psi}, \ \psi \ in \ [-\pi, \pi).$$

Then (4.18) becomes the infinite-dimensional system

$$\frac{\partial V^{(N)}}{\partial x} - \frac{2ihN}{c_o(x)} V^{(N)} + \frac{2\omega^2 \alpha_{nn}(x)}{c_o^2(x)} \Big\{ [N+1]^2 V^{(N+1)} \tag{4.19}$$

$$-2N^2 V^{(N)} + [N-1]^2 V^{(N-1)} \Big\} = 0 \ \text{ for } x < 0.$$

with boundary condition

$$V^{(N)}\,|_{x=0} = \delta_{N,1} \tag{4.20}$$

We now have that $V^{(N)} \to 0$ as $x \to -\infty$ for $N \neq 0$, and $u(\omega, h)$ is given by $\lim_{x \to -\infty} V^{(o)}$.

Equivalently, we may now formulate an infinite set of coupled linear first order partial differential equations for $\mu(t, \omega)$ the Fourier transform in h of $u(\omega, h)$, equation (4.5). Let $W^{(N)}$ be the Fourier transform in h, of $V^{(N)}$. Then we obtain easily from (4.19), (4.20) that

$$\frac{\partial W^{(N)}}{\partial x} - \frac{2N}{c_o(x)} \frac{\partial W^{(N)}}{\partial t} + \frac{2\omega^2 \alpha_{nn}(x)}{c_o^2(x)} \left\{ [N+1]^2 \, W^{(N+1)} - 2N^2 W^{(N)} + [N-1]^2 \, W^{(N-1)} \right\} = 0 \qquad (4.21)$$

with

$$W^{(N)}|_{x=0} = \delta(t) \, \delta_{N,1} \qquad (4.22)$$

The normalized power spectral density $\mu(t, \omega)$ is then given by the limit of $W^{(o)}$ as $x \to -\infty$. The formulation given in section 3 follows from making the change of variables in (4.21) from depth, x, to travel-time $\tau(x)$, equation (2.14).

We will next calculate μ explicitly for the case of a statistically homogeneous medium. That is, we assume that c_o and

$$\bar{\gamma} = \frac{\alpha_{nn}(x)}{c_o(x)} \qquad (4.23)$$

do not depend on x. We will use a forward Kolmogorov equation formulation based upon (4.18) and (4.15). For the case of c_o and $\bar{\gamma}$ constant, great simplification is achieved by first making the transformation

$$z = \cot \frac{\psi}{2} \qquad (4.24)$$

Then $z \in [-\infty, \infty)$ when $\psi \in [-\pi, \pi)$. The Kolmogorov backward equation for z is obtained by change of variables in (4.18)

$$\frac{\partial V}{\partial x} + h \frac{(1+z^2)}{c_o} \frac{\partial V}{\partial z} + 2 \frac{\omega^2 \bar{\gamma}}{c_o} \frac{\partial}{\partial z} \left\{ (1+z_2) \frac{\partial V}{\partial z} \right\} = 0 \qquad (4.25)$$

The probability density associated with (4.25), $P(x, z)$, satisfies the Kolmogorov forward equation

$$c_o \frac{\partial P}{\partial x} = -\frac{h}{\partial z} \{(1+z^2)P\} + 2\omega^2 \bar{\gamma} \frac{\partial}{\partial z} \left\{ (1+z^2) \frac{\partial P}{\partial z} \right\} \qquad (4.26)$$

The invariant density $\bar{P}_h(z)$ is obtained by setting $\partial P/\partial x = 0$ in (4.26). For $h > 0$ we have

$$\bar{P}_h(z) = \frac{h}{2\pi\omega^2\tilde{\gamma}} \int_0^{\infty} \frac{e^{-h\xi/2\omega^2\tilde{\gamma}}}{[1+(\xi+z)^2]}\, d\xi \tag{4.27}$$

For $h < 0$, symmetries in (4.26) imply that

$$\bar{P}_{-h}(-z) = \bar{P}_h(z)\,. \tag{4.28}$$

Now from (4.24) we have that

$$e^{i\Psi} = \left(\frac{z+i}{z-i}\right). \tag{4.29}$$

Therefore

$$u(\omega, h) = E[e^{i\Psi}] = \int_{-\infty}^{\infty} \bar{P}_h(z)\left(\frac{z+i}{z-i}\right) dz \tag{4.30}$$

$$= \frac{h}{2\omega^2\tilde{\gamma}} \int_0^{\infty} e^{-h\xi/2\omega^2\tilde{\gamma}} \left[\frac{\xi}{\xi+2i}\right] d\xi$$

for $h > 0$.

From (4.28), (4.30) it follows that

$$u(\omega, -h) = \overline{u(\omega, h)}\,. \tag{4.31}$$

Therefore

$$\mu(t, \omega) = \frac{1}{\pi} \operatorname{Re} \int_0^{\infty} e^{iht} u(\omega, h)\, dh\,. \tag{4.32}$$

Substitution of (4.30) into (4.32) then gives, after some elementary integrations

$$\mu(t, \omega) = \frac{\omega^2\tilde{\gamma}}{[1+\omega^2\tilde{\gamma}t]^2}\,. \tag{4.33}$$

which is the result (3.7)

5. The method of functionals

5.1. The generalized reflection functional.

In this section we shall reconsider the problem of section 3.2 and we shall show that the windowed reflection process $G^{\varepsilon}_{t,f}(\sigma)$ defined by (2.12) tends to a Gaussian process. The power spectral density of this process is given by (3.1) which is defined by (3.6) or (3.6a). The calculations in section 4 are formal because a number of interchanges of limits and integration have not been justified. In this section the calculations are carried out in a different framework that involves functional processes and avoids these difficulties. For this reason we reformulate briefly the problem and reintroduce the quantities of interest in a more general form.

The equations of motion in scaled form are

$$\rho_o[1+\eta(\frac{x}{\varepsilon^2})]u_t + p_x = 0$$

$$\frac{1}{\rho_o(x)c_o^2(x)}[1+\nu(\frac{x}{\varepsilon^2})]p_t + u_x = 0 \tag{5.1}$$

for $x<0$ and $t>0$. We assume that $\rho_o(x)$ and $c_o(x)$ are identically constant in $x>0$ and that they are differentiable functions in all of R^1, bounded and positive. The random fluctuations η and ν are taken to be stationary here, to simplify the writing. They have mean zero, take values in the interval $[-\frac{1}{2},\frac{1}{2}]$, say and are Markovian. The last hypothesis is unnecessary again but simplifies the analysis. The methods of [9] can be used here also in the general mixing case.

Define

$$\tau(x)=\int_0^x \frac{1}{c_o(s)}ds\ , \tag{5.2}$$

the travel time, for all x and note that it is like (2.14) except for signs and it is increasing. Let $\xi(\tau)$ be its inverse function which is zero at τ equal to zero. Clearly

$$\dot{\tau}=\frac{1}{c_o}\ , \qquad \dot{\xi}=c_o \tag{5.3}$$

The equations (5.1) are provided by initial and boundary conditions by specifying that a pulse is incident from the right

$$u=\frac{1}{(c_o\rho_o)^{\frac{1}{2}}}\frac{1}{\sqrt{\varepsilon}}f(\frac{t+\tau}{\varepsilon})$$

$$p = -(c_o\rho_o)^{1/2} \frac{1}{\sqrt{\varepsilon}} f\left(\frac{t+\tau}{\varepsilon}\right) \tag{5.4}$$

for $t<0$ with u and p continuous at $x=0$. The pulse shape is a smooth function, rapidly decreasing at infinity. Recall that c_o and ρ_o are identically constant in $x>0$.

We introduce the change of variables

$$u(t,x) = \frac{1}{(\rho_o c_o)^{1/2}} \tilde{u}(t,\tau)$$

$$p(t,x) = (\rho_o c_o)^{1/2} \tilde{p}(t,\tau) \tag{5.5}$$

and let

$$\eta^\varepsilon(\tau) = \eta(\xi(\tau)/\varepsilon^2), \quad \nu^\varepsilon(\tau) = \nu(\xi(\tau)/\varepsilon^2) \tag{5.6}$$

$$c_o(\tau) = c_o(\xi(\tau)), \quad \zeta(\tau) = \frac{1}{\rho_o c_o} \frac{d}{d\tau}(\rho_o c_o) \tag{5.7}$$

From (5.1) we obtain the following equations for $\tilde{u}$ and $\tilde{p}$, with the tilde dropped from now on.

$$(1+\eta^\varepsilon)u_t + p_\tau + \frac{1}{2}\zeta p = 0$$

$$(1+\nu^\varepsilon)p_t + u_\tau - \frac{1}{2}\zeta u = 0 \tag{5.8}$$

for $\tau<0$ and $t>0$ with

$$u = \frac{1}{\sqrt{\varepsilon}} f\left(\frac{t+\tau}{\varepsilon}\right), \quad p = \frac{-1}{\sqrt{\varepsilon}} f\left(\frac{t+\tau}{\varepsilon}\right) \tag{5.9}$$

for $t<0$. Equations (5.8) and (5.9) along with continuity of u and p provide a well defined initial-boundary value problem.

Because of the form (5.9) of the excitation by a wave travelling to the left and striking the interface $\tau=0$ at time $t=0$, it is convenient to introduce right and left travelling wave amplitudes A and B by

$$u = A + B, \quad p = A - B \tag{5.10}$$

We also let

$$m^\varepsilon(\tau) = \frac{\eta^\varepsilon(\tau) + \nu^\varepsilon(\tau)}{2}, \quad n^\varepsilon(\tau) = \frac{\eta^\varepsilon(\tau) - \nu^\varepsilon(\tau)}{2} \tag{5.11}$$

Then $A(\tau,t)$ and $B(\tau,t)$ satisfy the system

$$A_t + A_\tau + m^\varepsilon A_t + n^\varepsilon B_t - \frac{\zeta}{2} B = 0$$

$$B_t - B_\tau + n^\varepsilon A_t + m^\varepsilon B_t + \frac{\zeta}{2} A = 0 \tag{5.12}$$

for $\tau<0$ and $t>0$, with

$$A = 0\,, \quad B = \frac{1}{\sqrt{\varepsilon}} f(\frac{t+\tau}{\varepsilon}) \tag{5.13}$$

for $t<0$.

We now proceed as in section 3.2 via Fourier transforms to obtain an expression for the reflected signal $A(0,t)$ for $t>0$. Let $\hat{f}$ be defined by (2.11) and let

$$\hat{A}(\tau,\omega) = \int e^{i\omega t/\varepsilon} A(\tau,t)dt$$

$$\hat{B}(\tau,\omega) = \int e^{i\omega t/\varepsilon} B(\tau,t)dt \tag{5.14}$$

Then $\hat{A}$ and $\hat{B}$ satisfy the ordinary differential equations

$$\frac{d}{d\tau}\begin{bmatrix}\hat{A}\\ \hat{B}\end{bmatrix} = \frac{i\omega}{\varepsilon}\begin{bmatrix}1+m^\varepsilon & n^\varepsilon \\ -n^\varepsilon & -(1+m^\varepsilon)\end{bmatrix}\begin{bmatrix}\hat{A}\\ \hat{B}\end{bmatrix} + \frac{1}{2}\zeta\begin{bmatrix}0 & 1\\ 1 & 0\end{bmatrix}\begin{bmatrix}\hat{A}\\ \hat{B}\end{bmatrix} \tag{5.15}$$

Consider now the fundamental solution matrix Y of system (5.15). It has the form (2.20), (2.21). If we let as in (2.22)

$$R(\tau,\omega) = \frac{b(\tau,\omega)}{\bar{a}(\tau,\omega)} \tag{5.16}$$

then R satisfies the Riccati equation

$$\frac{dR}{d\tau} = \frac{i\omega}{\varepsilon}[n^\varepsilon + 2(1+m^\varepsilon)R + n^\varepsilon R^2] + \frac{1}{2}\zeta[1-R^2] \tag{5.17}$$

The meaning of this equation is as follows. Suppose that for some $\tau=-L<0$ the reflection coefficient at frequency ω for the region $(-\infty,-L]$ is known. Then to obtain the reflection coefficient at frequency ω for the region $(-\infty,0]$ we solve (5.17) in $-L<\tau\leq 0$ with initial condition $R(-L,\omega)$ equal to the known value and then evaluate the result at $\tau=0$. Assuming as in section 3.2 that we are in the totally reflecting case we may look for R in the form

$$R = e^{-i\psi} \tag{5.18}$$

Then the phase $\psi(\tau,\omega)$ satisfies the equation

$$\frac{d\psi}{d\tau} = \frac{-2\omega}{\varepsilon}[1 + m^\varepsilon(\tau) + n^\varepsilon(\tau)\cos\psi] - \zeta(\tau)\sin\psi\,, \tag{5.19}$$

in $-L<\tau\leq 0$, with an initial condition at $\tau=-L$. We also introduce the **centered phase**

$$\tilde{\psi} = \psi + \frac{2\omega\tau}{\varepsilon} \tag{5.20}$$

which satisfies

$$\frac{d\tilde{\psi}}{d\tau} = \frac{-2\omega}{\varepsilon}[m^\varepsilon + n^\varepsilon\cos(\tilde{\psi} - \frac{2\omega\tau}{\varepsilon})] - \zeta\sin(\tilde{\psi} - \frac{2\omega\tau}{\varepsilon}) \tag{5.21}$$

The functional of interest is

$$R^\varepsilon_{f,t}(\tau,\sigma) = \frac{1}{2\pi\sqrt{\varepsilon}}\int_{-\infty}^{\infty} e^{-i\omega(t+\varepsilon\sigma)/\varepsilon}\hat{f}(\omega)e^{-i\psi(\tau,\omega)}d\omega \tag{5.22}$$

When this is evaluated at $\tau=0$ it is identical to (2.12) and it is the windowed reflected signal. Instead of (5.22) it is convenient to study another more general quantity, the **generalized reflection functional** which is defined as follows. Let $\lambda^N(s,\omega)$ be real test functions (C^∞ rapidly decreasing), $N=0,\pm1,\pm2,\cdots$ with $\lambda^N(s,\omega) = \lambda^{*N}(s,-\omega)$. Here and in the sequel star denotes complex conjugate. We define the distribution-valued process $R^\varepsilon(\tau)$ by

$$<R^\varepsilon(\tau),\lambda> = \frac{1}{\sqrt{\varepsilon}}\sum_{N=-\infty}^{\infty}\int_{-\infty}^{\infty} ds\int_{-\infty}^{\infty} d\omega e^{-i\omega s/\varepsilon}\lambda^N(s,\omega)e^{-iN\psi(\tau,\omega)} \tag{5.23}$$

Clearly when $\lambda^N(s,\omega) = \delta_{N,1}\delta(t-s)\frac{1}{2\pi}e^{-i\omega\sigma}\hat{f}(\omega)$, the generalized reflection functional (5.23) is the same as (5.22).

We will think of the generalized reflection functional as a distribution valued stochastic process with $\tau\leq 0$ being the "time" parameter. The objective here is to study the limit law of the process defined by the stochastic equation (5.21) as ε tends to zero, **simultaneously** for all ω so that the law of the functional (5.22) can be analyzed. From (5.19) or (5.21) we see that for each set of test functions $\{\lambda^N\}$, the law of the process $<R^\varepsilon(\tau),\lambda>$ is known. We shall study the limit of this law as ε tends to zero.

5.2. The limit theorem

We shall assume that the fluctuations $(\eta(x),\nu(x))$ in (5.1) or $(m(x),n(x))$ in (5.11) are Markov processes (or projections of a higher dimensional Markov process) in addition to being stationary. We denote this coefficient process by $q(x)$ and let S be its state space, a compact subset of R^2 for example.

We denote by Q the infinitesimal generator of this process and by $\bar{P}$ its invariant measure. The coefficients m and n have mean zero with respect to $\bar{P}$.

Let $q^{\varepsilon}(\tau)$ denote the process $(m^{\varepsilon}(\tau), n^{\varepsilon}(\tau))$ defined by (5.6). Let $R^{\varepsilon}(\tau)$ (we omit dependence on t, N and λ) be the generalized reflection functional introduced by (5.22). The pair $(q^{\varepsilon}(\tau), R^{\varepsilon}(\tau))$ is a Markov process with the latter component distribution-valued and with τ the time parameter. The Markov property is a consequnce of the corresponding one for q^{ε} and ψ^{ε}, the solution of (5.19).

We will write the infinitesimal generator of the $(q^{\varepsilon}, R^{\varepsilon})$ process acting on suitable test functions. Such functions can be taken in the following form. Let F be a smooth function from $S \times R$ to R and let $\{\lambda^N\}$ be smooth test functions on R vanishing at infinity. We shall write the infintesimal generator acting on functions of the form

$$F = F(q,R) = F(q, \langle R,\lambda \rangle) \tag{5.24}$$

Now from the definition (5.23) and the differential equation (5.19) we see that the infinitesimal generator has the form

$$lepsF = \frac{c_0(\tau)}{\varepsilon^2} Q\, F + F' \left[\frac{1}{\varepsilon} \langle R, G_1 \lambda \rangle + \langle R, G_2 \lambda \rangle \right] \tag{5.25}$$

where $F'(q,\xi) = \dfrac{\partial F}{\partial \xi}(q,\xi)$ and

$$(G_1\lambda)^N(s,\omega) = 2i\omega m N \lambda^N(s,\omega) + i\omega n \left[(N+1)\lambda^{N+1}(s,\omega) - (N-1)\lambda^{N-1}(s,\omega) \right] \tag{5.26}$$

$$(G_2(\tau)\lambda)^N(s,\omega) = 2N \frac{\partial \lambda^N(s,\omega)}{\partial s} + \frac{\zeta(\tau)}{2} \left[(N+1)\lambda^{N+1}(s,\omega) - (N-1)\lambda^{N-1}(s,\omega) \right] \tag{5.27}$$

Note that the infinitesimal generator *leps* depends on τ so the process is inhomogeneous in the τ parameter.

To analyze the limit of the process $R^{\varepsilon}(\tau)$ we shall calculate the limit form of its generator defined on various classes of test functions. Test functions of the form (5.24) are not enough however when we want to analyze $R^{\varepsilon}_{t,f}$ of (5.22) and its moments at a single fixed t. We begin with the limit of (5.25)

This is obtained in much the same way as in the elementary case described in the appendix (and in the references cited there). Let $F = F(\langle R,\lambda \rangle)$be a fixed test function independent of q, because we are interested only in the limit of the R process, and let

$$F^{\varepsilon} = F + \frac{\varepsilon F'}{c_0(\tau)} \langle R, G_1^{\chi} \lambda \rangle \tag{5.28}$$

Here

$$(G_1^{\chi}\lambda)^N = 2i\omega\chi_m N\lambda^N + i\omega\chi_n\Big[(N+1)\lambda^{N+1} + (N-1)\lambda^{N-1}\Big] \tag{5.29}$$

with $\chi_m(q)$ and $\chi_n(q)$ the unique, zero mean solutions of the equation

$$Q\chi_m + m = 0\,, \qquad Q\chi_n + n = 0 \tag{5.30}$$

Note that in the appendix we use the same construction to solve (A.11) and we have expressed the inverese of Q in the form (A.15). A direct calculation now yields

$$lepsF^{\varepsilon} = \frac{1}{c_0(\tau)}F''\Big\{<R,G_1\lambda><R,G_1^{\chi}\lambda>\Big\} + F'\Big\{\frac{1}{c_0(\tau)}<R,G_1G_1^{\chi}\lambda> + <R,G_2\lambda>\Big\} + O(\varepsilon) \tag{5.31}$$

Let E denote expectation with respect to the invariant measure $\overline{P}$ of the q process. Let

$$\mathbf{L}_\tau F = \frac{1}{c_0(\tau)}F''\Big\{<R\times R, E[G_1\times G_1^{\chi}]\lambda\times\lambda>\Big\} + F'\Big\{\frac{1}{c_0(\tau)}<R,E[G_1G_1^{\chi}]\lambda> + <R,G_2\lambda>\Big\} \tag{5.32}$$

where $\times$ denotes tensor product. This is then the form of the limit generator for functionals of the form (5.23). If F depends explicitly on τ then the generator is $(\partial_\tau + \mathbf{L}_\tau)F$ with $\mathbf{L}_\tau$ defined by (5.32).

It is useful to write the form of the generator for the functional (5.23) when ψ is replaced by $\bar{\psi}$, the centered phase given by (5.20). Let

$$\tilde{\lambda}^N(s) = \lambda^N(s - 2N\tau) \tag{5.33}$$

If $\tilde{R}^{\varepsilon}(\tau)$ is defined by (5.23) with ψ replaced by $\bar{\psi}$, then we see that

$$<\tilde{R}^{\varepsilon}(\tau),\lambda)> = <R^{\varepsilon}(\tau),\tilde{\lambda}> \tag{5.34}$$

Also, with this change we have

$$\frac{\partial F}{\partial\tau} \rightarrow \frac{\partial F}{\partial\tau} + F'<R,-2N\frac{\partial\lambda}{\partial s}>$$

Thus,

$$(\partial_\tau + \mathbf{L}_\tau)F(\tau,<\tilde{R},\lambda>) = \frac{\partial F}{\partial\tau} + F''\Big\{<\tilde{R}\times\tilde{R}, E(G_1\times G_1^{\chi})\lambda\times\lambda>\Big\} \tag{5.35}$$

$$+ F'\Big\{<\tilde{R},E(G_1G_1^{\chi})\lambda> + <\tilde{R},G_3\lambda>\Big\}$$

Here G_3 is defined by the right side of (5.27) without the term with the s derivative.

For λ^N of the form $\delta(t-s)\delta_{N,N'}\lambda(\omega)$ the generalized reflection functional has the form

$$\langle R_{N,t}(\tau),\lambda\rangle = \frac{1}{\sqrt{\varepsilon}}\int_{-\infty}^{\infty} e^{-i\omega t/\varepsilon}\lambda(\omega)e^{-iN\psi(\tau,\omega)}d\omega \tag{5.36}$$

and similarly for the centered functional $\tilde{R}_{N,t}$ where ψ is replaced by the centered phase $\tilde{\psi}$. The explicit form of the generator (5.35) in this case is as follows.

$$\begin{aligned}
(\partial_\tau + L_\tau)F = \frac{\partial F}{\partial \tau} &+ \frac{N^2}{c_0(\tau)}F''\Big\{4\alpha_{mm}\langle\tilde{R}_{N,t},i\omega\lambda\rangle\langle\tilde{R}_{N,t},i\omega\lambda\rangle \\
&+2\alpha_{mn}\langle\tilde{R}_{N,t},i\omega\lambda\rangle\langle\tilde{R}_{N-1,t+2\tau},i\omega\lambda\rangle + 2\alpha_{mn}\langle\tilde{R}_{N,t},i\omega\lambda\rangle\langle\tilde{R}_{N+1,t-2\tau},i\omega\lambda\rangle \\
&+2\alpha_{nm}\langle\tilde{R}_{N-1,t+2\tau},i\omega\lambda\rangle\langle\tilde{R}_{N,t},i\omega\lambda\rangle + \alpha_{nn}\langle\tilde{R}_{N-1,t+2\tau},i\omega\lambda\rangle\langle\tilde{R}_{N-1,t+2\tau},i\omega\lambda\rangle \\
&+\alpha_{nn}\langle\tilde{R}_{N-1,t+2\tau},i\omega\lambda\rangle\langle\tilde{R}_{N+1,t-2\tau},i\omega\lambda\rangle + 2\alpha_{nm}\langle\tilde{R}_{N+1,t-2\tau},i\omega\lambda\rangle\langle\tilde{R}_{N,t},i\omega\lambda\rangle \\
&+\alpha_{nn}\langle\tilde{R}_{N+1,t-2\tau},i\omega\lambda\rangle\langle\tilde{R}_{N-1,t+2\tau},i\omega\lambda\rangle + \alpha_{nn}\langle\tilde{R}_{N+1,t-2\tau},i\omega\lambda\rangle\langle\tilde{R}_{N+1,t-2\tau},i\omega\lambda\rangle\Big\} \\
&+\frac{N}{c_0(\tau)}F'\Big\{4N\alpha_{nm}\langle\tilde{R}_{N,t},-\omega^2\lambda\rangle + 2N\alpha_{mn}\langle\tilde{R}_{N-1,t+2\tau},-\omega^2\lambda\rangle \\
&+2N\alpha_{mn}\langle\tilde{R}_{N+1,t+2\tau},-\omega^2\lambda\rangle + 2(N-1)\alpha_{nm}\langle\tilde{R}_{N-1,t+2\tau},-\omega^2\lambda\rangle \\
&+(N-1)\alpha_{nn}\langle\tilde{R}_{N-2,t+4\tau},-\omega^2\lambda\rangle + (N-1)\alpha_{nn}\langle\tilde{R}_{N,t},-\omega^2\lambda\rangle \\
&+2(N+1)\alpha_{nm}\langle\tilde{R}_{N+1,t-2\tau},-\omega^2\lambda\rangle + (N+1)\alpha_{nn}\langle\tilde{R}_{N,t},-\omega^2\lambda\rangle \\
&+(N+1)\alpha_{nn}\langle\tilde{R}_{N+2,t-4\tau},-\omega^2\lambda\rangle\Big\} \\
&+N\frac{\zeta(\tau)}{2}F'\Big\{\langle\tilde{R}_{N-1,t+2\tau},\lambda\rangle - \langle\tilde{R}_{N+1,t-2\tau},\lambda\rangle\Big\}
\end{aligned} \tag{5.37}$$

Here the coefficients α_{mm}, α_{nm} and α_{nn} are defined by

$$\alpha_{mm} = \int_0^\infty E\{m(x)m(0)\}dx\ , \quad \alpha_{mn} = \int_0^\infty E\{m(x)n(0)\}dx$$

and

$$\alpha_{nn} = \int_0^\infty E\{n(x)n(0)\}dx \tag{5.38}$$

as in (3.2). Note that the components of $\tilde{R}_{N,t}(\tau)$ of the generalized, centered, reflection functional are coupled for different $N's$ and $t's$.

The question of existence and uniqueness of a distribution-valued process $R(\tau)$ with generator $\mathbf{L}_\tau$ as defined above and the detailed proof of the weak convergence as ε tends to zero will not be discussed here. We note only that the existence of the limit process and the compactness of the family R^ε follow from elementary estimates derived from (5.21) or (5.19). The uniqueness of the limit law is obtained by exploiting the "linearity" of the generalized reflection process which allows us to get closed equations for its moments of each order. The second order moments are considered in the next section.

In the context of the physical problem under consideration with the time t fixed, unless $\tau \le -t/2$ we must give the (right) reflection coefficient of the region $(-\infty,\tau]$ as initial condition for the equation (5.19) or (5.21) or for the full process R. We want to study the generalized reflection functional only in the case $\tau < -t/2$ here. Then the reflection functional for the region $(-\infty,0]$, which is the physically interesting quantity, does not depend on the subregion $(-\infty,\tau]$ at all. This is because signals propagate at a finite speed and (in the present scaled variables) the reflected signal observed up to time t could not have come from scattering in the region $\tau < -t/2$. This is seen more analytically in the next section.

5.3. The Wigner functional.

Let $\lambda^{NM}(s,\omega)$ be a doubly indexed array of test functions each of which has the same properties as the test functions λ^N of the previous section. We define the generalized Wigner functional of the reflected process by

$$<W^\varepsilon(\tau),\lambda> = \sum_{N,M} \int dt \int d\omega \lambda^{NM}(t,\omega) \int dh\, e^{iht} e^{-iN\psi(\tau,\omega-\varepsilon h/2)+iM\psi(\tau,\omega+\varepsilon h/2)} \tag{5.39}$$

Formally the Wigner functional is defined by

$$W^{NM}(\tau,t,\omega) = \frac{1}{2\pi}\int e^{iht} e^{-iN\psi(\tau,\omega-\varepsilon h/2)+iM\psi(\tau,\omega+\varepsilon h/2)} dh \tag{5.40}$$

but this expression makes sense only through (5.39). The reason we introduce the Wigner functional is as follows.

To study moments of $R^{\varepsilon}_{f,t}(\tau,\sigma)$ at a fixed t we must go beyond test functions of the form (5.24) which, in view of (5.23), contain integration over t. Formally, we could take $\lambda^N(s,\omega)$ to be a delta function in s but then the powerful calculus of generators is lost. Consider the expression

$$R^{\varepsilon}_{f,t}(\tau,\sigma)R^{\varepsilon}_{f,t}(\tau,0) = \frac{1}{(2\pi)^2}\frac{1}{\varepsilon}\iint d\omega_1\,\omega_2 e^{-i\omega_1(t+\varepsilon\sigma)/\varepsilon}\hat{f}(\omega_1)e^{-i\psi(\tau,\omega_1)}e^{i\omega_2 t/\varepsilon}\hat{f}^*(\omega_2)e^{i\psi(\tau,\omega_2)} \tag{5.41}$$

obtained from (5.22). Let

$$\omega = \frac{\omega_1+\omega_2}{2}, \quad h = \frac{\omega_2-\omega_1}{\varepsilon}$$

Then, after dropping some formally small terms as $\varepsilon\to 0$, we see that

$$R^{\varepsilon}_{f,t}(\tau,\sigma)R^{\varepsilon}_{f,t}(\tau,0) \approx \frac{1}{2\pi}\int e^{-i\omega\sigma}\,|\hat{f}(\omega)|^2 W^{11}(\tau,t,\omega)d\omega \tag{5.42}$$

with W^{NM} defined by (5.40). This explains how the Wigner functional enters and why its expectation is interesting. In fact, the power spectral density $\mu(t,\omega)$ in section 3 is simply the expectation of W^{11} in the limit $\varepsilon\to 0$.

In the same way that we calculated the generator of the Markov process $(q^{\varepsilon}(\tau),R^{\varepsilon}(\tau))$ in section (5.2) we can now calculate the generator of the Markov process $(q^{\varepsilon}(\tau),W^{\varepsilon}(\tau))$ on functions of the form $F(q,\langle W,\lambda\rangle)$. We have

$$\mathrm{M}^{\varepsilon}_{\tau}F = \frac{c_0(\tau)}{\varepsilon^2}QF + F'\left\{\frac{1}{\varepsilon}\langle W,H_1\lambda\rangle + \langle W,H_2\lambda\rangle\right\} \tag{5.43}$$

where

$$\begin{aligned}(H_1\lambda)^{NM}(s,\omega) &= 2i\,(N-M)\omega\lambda^{NM}(s,\omega) + m\,2i\,(N-M)\omega\lambda^{NM}(s,\omega)\\ &+ m\Big\{i\omega(N+1)\lambda^{N+1,M}(s,\omega) + i\omega(N-1)\lambda^{N-1,M}(s,\omega)\\ &- i\omega(M-1)\lambda^{N,M-1}(s,\omega) - i\omega(M+1)\lambda^{N,M+1}(s,\omega)\Big\}\end{aligned} \tag{5.44}$$

and

$$\begin{aligned}(H_2\lambda)^{NM}(s,\omega) &= (N+M)\frac{\partial\lambda^{NM}}{\partial s} + m\,(N+M)\frac{\partial\lambda^{NM}}{\partial s} +\\ &+ \frac{\zeta(\tau)}{2}\Big\{(N+1)\lambda^{N+1,M} - (N-1)\lambda^{N-1,M}\end{aligned}$$

$$- (M-1)\lambda^{N,M-1} + (M+1)\lambda^{N,M+1}\Big\}$$

$$- \frac{n}{2}\Big\{ (N+1)\frac{\partial}{\partial s}\lambda^{N+1,M} + (N-1)\frac{\partial}{\partial s}\lambda^{N-1,M}$$

$$+ (M-1)\frac{\partial}{\partial s}\lambda^{N,M-1} + (M+1)\frac{\partial}{\partial s}\lambda^{N,M+1}\Big\} \tag{5.45}$$

The generator on functions that depend on τ explicitly is $\partial_\tau + \mathbf{M}_\tau^\varepsilon$.

To calculate the limit generator we fix a function $F(<W,\lambda>)$ and let

$$F^\varepsilon = F + \frac{\varepsilon}{c_0(\tau)} F' <W, H_1^\chi \lambda> \tag{5.46}$$

Here H_1^χ is identical to H_1 of (5.44) except that m and n are replaced by χ_m and χ_n which are defined by (5.30). We also restrict attention to test functions λ^{NM} such that

$$\lambda^{NM} = 0 \quad \text{for} \quad N \neq M \tag{5.47}$$

We call these **diagonal** test functions. We now calculate

$$\mathbf{M}_\tau^\varepsilon F^\varepsilon = \frac{1}{c_0(\tau)} F'' \Big\{ <W, H_1\lambda><W, H_1^\chi \lambda> \Big\}$$

$$+ F' \Big\{ \frac{1}{c_0(\tau)} <W, H_1 H_1^\chi \lambda> + <W, H_2 \lambda> \Big\} + O(\varepsilon)$$

Taking expectation with respect to $\overline{P}$ defines the limit generator $\mathbf{M}_\tau$

$$\mathbf{M}_\tau F = \frac{1}{c_0(\tau)} F'' <W \times W, E[H_1 \times H_1^\chi] \lambda \times \lambda>$$

$$+ F' \Big\{ \frac{1}{c_0(\tau)} <W, E[H_1 H_1^\chi] \lambda> + <W, E[H_2] \lambda> \Big\} \tag{5.48}$$

with the test functions λ^{NM} diagonal (5.47).

The interesting thing about the generator (5.48) is that it preserves homogeneity (is "linear") as before but in addition if $F = <W, \lambda>$ with λ diagonal, then $\mathbf{M}_\tau F$ is linear in W with a diagonal test function. This follows from the fact that if λ is diagonal with $\lambda^{NN} = \lambda^N$ then

$$(E[H_1 H_1^\chi]\lambda)^N = \frac{4\omega^2 \alpha_{nn}}{c_0(\tau)} \Big\{ \frac{(N+1)^2}{2} \lambda^{N+1} - N^2 \lambda^N + \frac{(N-1)^2}{2} \lambda^{N-1} \Big\}$$

Here α_{nn} is defined by (3.2) or (5.38).

Let us now use this important property of the limit generator to calculate the conditional expectation of the Wigner functional at $\tau=0$, given its value W, say, at a fixed negative τ. Let $\tilde{\lambda}^N=\tilde{\lambda}^N(\tau,t,\omega)$ be the solution of the system

$$\frac{\partial\tilde{\lambda}^N}{\partial\tau}+2N\frac{\partial\tilde{\lambda}^N}{\partial s}+\frac{4\omega^2\alpha_{nn}}{c_0(\tau)}\left\{\frac{(N+1)^2}{2}\tilde{\lambda}^{N+1}-N^2\tilde{\lambda}^N+\frac{(N-1)^2}{2}\tilde{\lambda}^{N-1}\right\}=0 \tag{5.49}$$

for $\tau<0$ and $s<t$ with the terminal condition

$$\tilde{\lambda}^N(0,s,\omega)=\frac{1}{2\pi}\delta_{N1}\delta(t-s)$$

and let $\lambda^N(\tau,t,\omega)=\tilde{\lambda}^N(\tau,t,\omega)\delta(\omega)$. Then

$$E_{W,\tau}\{<W(0),\lambda(0)>\}=<W,\lambda(\tau)> \tag{5.50}$$

because for test functions that satisfy (5.49) the scalar process $<W(\tau),\lambda(\tau)>$ is a martingale. The subscripts on the expectation on the left side of (5.50) denote conditioning.

The terminal conditions in (5.49) were chosen so that

$$E_{W,\tau}\{<W(0),\lambda(0)>\}=E_{W,\tau}\{W^{11}(0,t,\omega)\} \tag{5.51}$$

From (5.42) we see that this must be the power spectral density $\mu(t,\omega)$ of the limit reflection process as described in section 3.2. For this to be true, the right hand side of (5.50) must not depend on W and τ if $\tau\le -t/2$. But this is clearly true because of the hyperbolic nature of the system (5.49). First, $\lambda^N\equiv 0$ for $N<0$. Second, $\lambda^N\equiv 0$ for $N\ge 1$ when $\tau<-t/2$ by the support properties in the variables t and τ of the solution λ^N of (5.49) (dependance on ω is parametric and the $\delta(\omega)$ is a multiplicative factor). Thus, only λ^0 is different from zero when $\tau<-t/2$ and it is in fact constant in this range. Since $W^{00}=\delta(t)$ in all cases, we see that

$$<W,\lambda(\tau)>=\tilde{\lambda}^0(\frac{-t}{2},t,\omega) \tag{5.52}$$

for all $\tau<-t/2$. Except for changes in notation, $\tilde{\lambda}^0(\frac{-t}{2},t,\omega)$ is exactly the same as $W^{(0)}(\frac{t}{2},t,\omega)$ of (3.6a).

The above calculation shows that the finite speed of propagation, which makes (5.49) hyperbolic, has very significant implications for the generalized reflection process. In the previous paragraph we saw how the calculation of the mean Wigner distribution is done directly in the right framework and this should be compared with the calculation of section 3.4 where the support properties of the quantities of interest are hidden and the interchanges of limits are unjustified (are impossible to justify in that setting).

If we calculate higher moments of the Wigner functional by using the generator M_τ of (5.48) with diagonal test functions we find that the fluctuations of $<W(\tau),\lambda>$ are zero. This is surprising at first because it says that the law of the (diagonal part) of the Wigner functional is deterministic in the $\varepsilon \to 0$ limit. However this does not mean that the law of the Wigner distribution at a fixed time is deterministic in the limit. Only the smoothed in t functional is and that is another reason why it is necessary to introduce still more general functionals and test functions to study the generalized reflection process at a fixed time t.

5.4. The Gaussian property.

If the generator of the generalized reflection functional $R^\varepsilon(\tau)$ of (5.23) has the form (5.32) (or (5.35) in the centered case) then it cannot possibly by a Gaussian process, which would make it a Gauss-Markov or Ornstein-Uhlenbeck process. So in what sense then do we have the Gaussian property for the reflected signals as we claimed in section 3.3? The answer to this question rests entirely on the finite propagation speed of the signals because then the law of the reflection process in the time interval $[0,t]$ at $\tau=0$ becomes constant, indendent of τ for $\tau<-t/2$ and independent of the conditioning at τ. In other words the generalized reflection process at $\tau=0$ becomes ergodic after a finite, negative starting value of τ which depends on the support of the test functions as functions of time. It is this ergodic law that is Gaussian.

The ergodicity of the law of the reflection functional is shown through its moments, using the finite propagation speed as with the Wigner functional. We will not give the details here because they are lenghty. Consider now the generalized reflection functional $\tilde{R}_{N,t}(\tau)$ defined by (5.36) with ψ replaced by $\tilde{\psi}$, the centered phase. We want to show that for each sequence of test functions $\{\lambda^N(\omega)\}$

$$E_{W,\tau}\{\exp[i\sum_N<\tilde{R}_{N,t}(0),\lambda^N>]\} = \exp[-\frac{1}{2}\sum_N E_{W,\tau}\{<\tilde{W}_t^{NN}(0),|\lambda^N|^2>\}] \tag{5.53}$$

with the expectations being taken relative to the limit process and with the conditioning at $\tau<-t/2$ indicated with subscripts. The expectations do not depend on the conditioning in this case as we saw above. On the right side of (5.53), $\tilde{W}^{NN}$ denotes the shifted Wigner functional at time t

$$\tilde{W}_t^{NN}(\tau,\omega) = W^{NN}(\tau,t-2N\tau,\omega) \tag{5.54}$$

The expectation of the Wigner functional on the right side of (5.53) can be computed in a manner similar to the one we used in (5.49)-(5.52). We describe briefly the procedure. The expectation

$E_{W,\tau}\{W^{N'N'}(0,t,\omega)\}$ for $\tau<-t/2$ is equal to $\tilde{\lambda}^{ON'}(\frac{-t}{2},t,\omega)$ where $\tilde{\lambda}^{NN'}(\tau,t,\omega)$ satisfies (5.49) for $\tau<0$, $s<t$ with terminal condition

$$\tilde{\lambda}^{NN'}(0,s,\omega)=\frac{1}{2\pi}\delta_{NN'}\delta(t-s) \tag{5.55}$$

Then

$$E_{W,\tau}\{<W^{N'N'}(0,t,\cdot),|\lambda^{N'}|^2>\}=\int d\omega\tilde{\lambda}^{ON'}(\frac{-t}{2},t,\omega)|\lambda^{N'}(\omega)|^2 \tag{5.56}$$

For the shifted Wigner functional appearing on the right side of (5.53) we use (5.56) with the second argument t on the right side replaced by $t-2N\tau$.

Now suppose that we fix $t>0$ and $\tau<-t/2$ and let (5.53) define a Gaussian law for $\tilde{R}_{N,t}(0)$. We want to show that this law is invariant with respect to the Markovian evolution that generates the process. We must then show that

$$E\{\mathbf{L}_\tau \exp[i\sum_N<\tilde{R}_{N,t},\lambda^N>]\}=0 \tag{5.57}$$

where E stands for expectation with respect to the Gaussian law (5.53) and $\mathbf{L}_\tau$ is given by (5.37). In addition, we will use the independence of $R_{N,t}$ (when $\tau<-t/2$) for different t's. As we noted earlier we omit the proof of this fact in the brief description given here. To prove (5.57) under the independence condition we first calculate the expectation of each term in (5.57) when (5.37) is used for $\mathbf{L}_\tau$. This is a standard calculation with characteristic functions of Gaussian integrals. An identity then results involving the expectation of the Wigner functionals and it must be true for (5.57) to hold. This identity is none other than (5.49) with the terminal condition (5.55), the time t shifted to $t-2N\tau$ and with $\tau<-t/2$ so that the $\frac{\partial}{\partial\tau}$ term drops. This then shows that the law (5.53) is invariant. A uniqueness argument finally tells us that (5.53) is the ergodic limit law.

There are many details that we have omitted here that are needed to make the above a complete proof. The brief description we have given introduces the main ideas and the framework of the functional processes.

Appendix A. Limit Theorem for a Stochastic Differential Equation.

We consider here the behavior, as $\varepsilon \downarrow 0$, of ψ^ε given by (4.7). As discussed in section 3.4, we assume that the equation is driven by a Markov process with slowly varying parameters. Thus, we let $q(x,y)$ with values in R^d be, for each fixed x, a stationary ergodic Markov process in y, with infinitesimal generator Q_x. Equation (4.7) is then of the form

$$\frac{d}{dx}\psi^\varepsilon = \frac{1}{\varepsilon}\mathbf{F}(x, q(x,\frac{x}{\varepsilon^2}), \frac{\tau(x)}{\varepsilon}, \psi^\varepsilon) \tag{A.1}$$

$$+\,\mathbf{G}(x, q(x,\frac{x}{\varepsilon^2}), \frac{\tau(x)}{\varepsilon}, \psi^\varepsilon)$$

By ergodicity $q(x, \cdot)$ has an invariant measure $\tilde{P}_x(dq)$ which satisfies

$$\int Q_x f(q)\tilde{P}_x(dq) = 0 \tag{A.2}$$

for any test function f. We define expectation with respect to $\tilde{P}_x$ by

$$E\{\cdot\} \equiv \int \cdot \, d\,\tilde{P}_x(q) \tag{A.3}$$

Since m, n have mean zero, it is apparent from (4.8) that

$$E\{\mathbf{F}\} = 0 \tag{A.4}$$

Now the infinitesimal generator of the Markov Process $q^\varepsilon, \psi^\varepsilon$ with $q^\varepsilon(x) = q(x, x/\varepsilon^2)$ is given by (4.9), so that the Kolmogorov backward equation for this process may be written as

$$\frac{\partial V^\varepsilon}{\partial x} + \frac{1}{\varepsilon^2} Q_x V^\varepsilon + \frac{1}{\varepsilon}\mathbf{F}\cdot\nabla_\psi V^\varepsilon + \mathbf{G}\cdot\nabla_\psi V^\varepsilon = 0 \quad x < 0\,. \tag{A.5}$$

We consider final conditions at $x = 0$ which do not depend on q i.e.

$$V^\varepsilon(x, q, \psi)\,|_{x=0} = H(\psi) \tag{A.6}$$

Let

$$h = \tau/\varepsilon \tag{A.7}$$

so that $\mathbf{F} = \mathbf{F}(x, q, h, \psi)$, etc. in (A.5). We will solve (A.5), (A.6) asymptotically as $\varepsilon \downarrow 0$ by the multiple scale expansion

$$V^\varepsilon = \sum_{n=0}^{\infty} \varepsilon^n V^n(x, q, h, \psi)\,|_{h=\tau(x)/\varepsilon} \tag{A.8}$$

To expand (A.5) in multiple scales x, h, we replace $\frac{\partial}{\partial x}$ by $\frac{\partial}{\partial x} + \frac{\tau'(x)}{\varepsilon} \frac{\partial}{\partial h}$, and note that $\tau'(x) = -\frac{1}{c_o(x)} < 0$. Thus (A.5) becomes

$$\frac{1}{\varepsilon^2} Q_x V^\varepsilon + \frac{1}{\varepsilon}\left\{ \mathbf{F}\cdot\nabla_\psi V^\varepsilon - \frac{1}{c_o(x)} \frac{\partial}{\partial h} V^\varepsilon \right\} + \left\{ \mathbf{G}\cdot\nabla_\psi V^\varepsilon + \frac{\partial}{\partial x} V^\varepsilon \right\} = 0 \tag{A.9}$$

Now substitution of (A.8) into (A.9) yields a hierachy of equations for $V^{(n)}$ of which the first three are

$$Q_x V^o = 0 \tag{A.10}$$

$$Q_x V^1 + \mathbf{F}\cdot\nabla_\psi V^o - \frac{1}{c_o(x)} \frac{\partial}{\partial h} V^o = 0 \tag{A.11}$$

$$Q_x V^2 + \mathbf{F}\cdot\nabla_\psi V^1 - \frac{1}{c_o(x)} \frac{\partial}{\partial h} V^1 \tag{A.12}$$

$$+ \mathbf{G}\cdot\nabla_\psi V^o + \frac{\partial}{\partial x} V^o = 0$$

From (A.10) and ergodicity of $q(x, \cdot)$ we conclude that V^o does not depend on q.

$$V^o = V^o(x, h, \psi) \tag{A.13}$$

We next take the expectation of (A.11). Since $\mathbf{F}$ has mean zero as noted in (A.4), using (A.2) we see that (A.11) implies that

$$-\frac{1}{c_o(x)} \frac{\partial}{\partial h} V^o = 0$$

whence V^o does not depend on h

$$V^o = V^o(x, \psi) \tag{A.14}$$

Now by ergodicity, Q_x has the one-dimensional null space consisting of functions which do not depend on q. Thus Q_x does not have an inverse. However, by the Fredholm alternative, which we assume to hold for the process q, Q_x has an inverse on the subspace of functions which have mean zero with respect to $\tilde{P}_x$. We defind a particular inverse Q_x^{-1} such that its range consists of functions with vanishing mean

$$-Q_x^{-1} = \int_0^\infty e^{Q_x r}\, dr\,. \tag{A.15}$$

In terms of this Q_x^{-1} we can solve (A.11) for V^1

$$V^1 = -Q_x^{-1}\{\mathbf{F}\cdot\nabla_\psi V^o\} + V^{1,o} \tag{A.16}$$

where $V^{1,o}$ does not depend on q.

We now substitute (A.16) into (A.12) and take expectations. We also average this equation with respect to h

$$<\cdot>_h = \lim_{h_o\to\infty} \frac{1}{h_o}\int_o^{h_o} \cdot \, dh \, . \tag{A.17}$$

Since $E\{\mathbf{F}\}=0$ and $<\mathbf{G}>_h = 0$, (see (4.8)), we see that V^o must satisfy

$$\frac{\partial}{\partial x}V^o + <E\{\mathbf{F}\cdot\nabla_\psi(-Q_x^{-1})\mathbf{F}\cdot\nabla_\psi\}>_h V^o = 0 \tag{A.18}$$

This is the solvability condition for (A.12).

Equation (A.18) is the limiting backward Kolmogorov equation for ψ. It has the form

$$\frac{\partial}{\partial x}V^o + \mathbf{L}_x V^o = 0\,, \quad x<0 \tag{A.19}$$

$$V^o\,|_{x=0} = H(\psi)$$

From (A.18) the limit infinitesimal generator $\mathbf{L}_x$ is given by

$$\mathbf{L}_x = \int_0^\infty dr <E\{\mathbf{F}\cdot\nabla_\psi e^{rQ_x}\mathbf{F}\cdot\nabla_\psi\}>_h \tag{A.20}$$

Using the probabilistic interpretation of the semigroup e^{rQ_x} and expectation $E\{\cdot\}$ with respect to the invariant measure $P_x(dq)$ and the averaging $<\cdot>$, we can write (A.20) in the form

$$\mathbf{L}_x = \int_0^\infty dr E\{<\mathbf{F}(x,q(x,y),h,\psi)\cdot\nabla_\psi(\mathbf{F}(x,q(x,y+r),h,\psi)\cdot\nabla_\psi\,\cdot\,)>_h\} \tag{A.21}$$

In the application of this result in section 4, the explicit form of **F** in (4.7) is used in (A.21) to obtain (4.11).

The above analysis is of course formal, focusing on the computational aspects of the limit theorems that is, on their basis in relatively simple perturbation calculations. A rigorous treatment is easily given in the present context or even in some other much more complicated situations where the ergodic properties of the process generated by Q_x are very weak. Examples can be found in [9, 13, 23] as well as in [35] and the references cited therein.

Appendix B. Numerical simulation of the transport equations.

This appendix is joint work with Mark Asch.

The forward and backward equations.

We want to solve the Kolmogorov equation (3.4) for the canonical power spectral density $\mu(t,\omega)$ as defined in (3.6). Since τ plays the role of time, and t plays the role of space, let us rewrite (3.4) as follows.

Let $W^{(N)}(t, x, \omega)$, $N = 0,1,2...$ be the solution of

$$\frac{\partial W^{(N)}}{\partial t} + 2N\frac{\partial W^{(N)}}{\partial x} - 4\omega^2\gamma(\tau)\left\{\frac{[N+1]^2}{2}W^{(N+1)}\right.$$

$$\left. - N^2W^{(N)} + \frac{[N-1]^2}{2}W^{(N-1)}\right\} = 0 \qquad \text{(B.1)}$$

with initial condition

$$W^{(N)}(0,x,\omega) = \delta(x)\,\delta_{N,1}\,. \qquad \text{(B.2)}$$

where $x,\ t > 0$, and N is an integer 'speed'. Also

$$W^{(N)} \equiv 0 \text{ for } x < 0,\ N < 0\,.$$

Define the symmetric second difference operator

$$\Delta W^{(N)} = \frac{1}{2}W^{(N+1)} - W^{(N)} + \frac{1}{2}W^{(N-1)} \quad , N = 0,1,2,...$$

and general initial conditions

$$x(0) = y, \quad N(0) = M\,.$$

Then (B.1) and (B.2) become

$$\frac{\partial W^{(N)}}{\partial t} + 2N\frac{\partial W^{(N)}}{\partial x} - 4\omega^2\gamma(\tau)\,\Delta(N^2W^{(N)}) = 0 \qquad \text{(B.3)}$$

$$W^{(N)}(0,x,\omega) = \delta(x-y)\,\delta_{N,M}\,. \qquad \text{(B.4)}$$

If we define Q as

$$Q(t)\,W^{(N)} = 4\omega^2\gamma(t)\,N^2\,\Delta W^{(N)} \quad , N \geq 0 \qquad \text{(B.5)}$$

and the usual inner-product

$$< W,V > = \sum_{N=0}^{\infty} W^{(N)} V^{(N)}$$

Then using summation by parts, we can show that

$$< Q(t)W,V > \; = \; < W,Q^*(t) >$$

where $Q^*(t)$, the adjoint of (B.5), is given by

$$Q^*(t)\, V^{(N)} = 4\omega^2\, \gamma(t)\;\; \Delta(N^2\, V^{(N)}) \tag{B.6}$$

So finally we can write (B.3) using (B.6), as

$$\text{(FE)} \qquad \frac{\partial W^{(N)}}{\partial t} + v_N \frac{\partial\, W^{(N)}}{\partial x} = Q^*(t)\, W^{(N)} \tag{B.7}$$

$$W^{(N)}(0,x,\omega;y,M) = \delta(x-y)\delta_{N,M} \tag{B.8}$$

where $v_N = 2N \quad , N = 0,1,2,\cdots$.

This equation can be solved by various numerical methods, for example by finite difference schemes. We choose to solve it using random walks, a method which avoids direct discretizaton error in the calculations. For the analysis that follows however, we must introduce the backward equation

$$\text{(BE)} \qquad \frac{\partial u^{(N)}}{\partial s} - v_N \frac{\partial\, u^{(N)}}{\partial x} = Q(s)\, u^{(N)} \quad , s < t \tag{B.9}$$

$$u^{(N)}(t,x,\omega) = f_N(x) \tag{B.10}$$

since certain expectations of the random walk will yield the solution to this last equation. Thereafter, the relationship between the forward and backward equations will be clarified.

Note that (FE) denotes a Forward Kolmogorov Equation, (BE) a Backward one, and that we have a terminal condition (t is fixed) for (BE).

The solution by random walks.

In this section we first define the Markov chain, then show that this process solves the backward equation (B.9). Finally we describe the connection between (BE) and (FE).

The process we are about to define will have three random components :

- $N(t)$ the speed at the time of switching
- τ the random time-step
- $x(t)$ the position reached after time τ at speed N,

where the first two are independent.

STEP 1 :

Let the initial speed $N(0)=M$, and let τ_1, the first time step, be an exponential random variable with law

$$P(\tau_1 > s) = \exp[-\int_0^s 4\omega^2\gamma(\sigma)M^2 d\sigma] \quad , s \geq 0 \tag{B.11}$$

Then for the first interval, $I_1=[0,\tau_1)$, the particle moves at a constant speed $N(t)=M \quad , 0 \leq t < \tau_1$.

The position, $x(t)$, of the particle is given by

$$x(t) = y + v_M t \quad , 0 \leq t < \tau_1 \tag{B.12}$$

where y is the initial position and $v_M = 2M$. Thus we get

$$x_1 = x(\tau_1) = y + v_M \tau_1 \tag{B.13}$$

STEP 2 :

Now choose with probability ½ the speed for this step as

$$N(\tau_1+) = \begin{cases} M+1 \;, & \text{with } p=0.5 \\ M-1 \;, & \text{with } p=0.5 \end{cases} \tag{B.14}$$

Let τ_2, the second time step, be an exponential random variable with law

$$P(\tau_2 > s) = \exp[-\int_{\tau_1}^{\tau_1+s} 4\omega^2\gamma(\sigma)\,(N(\tau_1+))^2 d\sigma] \quad , s \geq 0 \tag{B.15}$$

Then for the second interval, $I_2=[\tau_1,\tau_1+\tau_2)$, the particle moves at a constant speed $N(t)=N(\tau_1+) \quad , \tau_1 \leq t < \tau_1+\tau_2$ and the position, $x(t)$, of the particle is given by

$$x(t) = x(\tau_1) + v_{N(\tau_1+)}t \quad , \tau_1 \leq t < \tau_1+\tau_2 \tag{B.16}$$

Thus we get

$$x_2 = x(\tau_1+\tau_2) = x_1 + v_{N(\tau_1+)}\tau_2 \tag{B.17}$$

STEP n :

Choose with probability ½ the speed for this step as

$$N[(\tau_1+\cdots+\tau_{n-1})+] = \begin{cases} N[(\tau_1+\cdots+\tau_{n-2})+]+1 \ , & \text{with } p=0.5 \\ N[(\tau_1+\cdots+\tau_{n-2})+]-1 \ , & \text{with } p=0.5 \end{cases} \tag{B.18}$$

Let τ_n , the nth time step, be an exponential random variable with law

$$P(\tau_n > s) = \exp[-\int_{\tau_1+\cdots+\tau_{n-1}}^{\tau_1+\cdots+\tau_{n-1}+s} 4\omega^2\gamma(\sigma)\,[N(\tau_1+\cdots+\tau_{n-1})+]^2 d\sigma\,] \tag{B.19}$$

where , $s \geq 0$. Then for the nth interval, $I_n = [\,\tau_1+\cdots+\tau_{n-1}\,,\,\tau_1+\cdots+\tau_{n-1}+\tau_n\,)$, the particle moves at a constant speed

$$N(t) = N[(\tau_1+\cdots+\tau_{n-1})+] \quad , \quad \tau_1+\cdots+\tau_{n-1} \leq t < \tau_1+\cdots+\tau_n$$

and its position, $x(t)$, is given by

$$x(t) = x(\tau_{n-1}) + v_{N(\tau_1+\cdots+\tau_{n-1})+}t \quad , \tau_1+\cdots+\tau_{n-1} \leq t < \tau_1+\cdots+\tau_n \tag{B.20}$$

Thus we get

$$x_n = x(\tau_1+\cdots+\tau_n) = x_{n-1} + v_{N(\tau_1+\cdots+\tau_{n-1})+}\tau_n \tag{B.21}$$

or we could simply write

$$x(t) = x_{n-1} + \int_0^t v_N\,ds \quad , \ 0 \leq t < \tau_n$$

The slab of random material has some finite length, x_∞ . Thus the random walk for a specific particle terminates when either the speed switches to zero (the particle is "absorbed"), or the particle leaves the slab of material with $x > x_\infty$.

This concludes the definition of the random walk.

We will now show how the solutions of (B.7) and (B.9) can be obtained by computing expectations of functionals of this random walk.

Define the expectation

$$E_{s,x,M}\{ f_{N(t)}(x(t))\} = u^{(M)}(s,x) \quad , \text{ for } s < t \tag{B.22}$$

then $u^{(M)}(t,x) = f_M(x)$ for $s=t$. If we write $Q(t) = [q^{NM}(t)]$, the matrix of the coefficients of the difference operator (B.5), then

$$q^{NM} > 0 \text{ for } N \neq M \text{ and } -\sum_{\substack{N \neq M \\ N > M}} q^{NM} = q^{NN} \quad , \; N,M \geq 0.$$

Now our expectation (B.22) is computed by conditioning on the first jump, so that either $s+\tau_1 > t$, or $s < s+\tau_1 < t$. Thus using (B.19)

$$E_{s,x,M}\{ f_{N(t)}(x(t))\} = f_M(x+v_M(t-s))\, P(\tau_1 > s-t) + \tag{B.23}$$

$$\int_s^t [\, \frac{1}{2} E_{\sigma,x+v_m\sigma,M+1}\{ f_{N(\sigma)}(x(\sigma))\} +$$

$$\frac{1}{2} E_{\sigma,x+v_m\sigma,M-1}\{ f_{N(\sigma)}(x(\sigma))\}\,]\, q^{MM}(\sigma) \exp[\, -\int_s^\sigma q^{MM}(\gamma) d\gamma\,]\; d\sigma$$

But

$$f_M(x+v_M(t-s))\, P(\tau_1 > s-t) = f_M(x+v_M(t-s)) \exp[\, -\int_s^t q^{MM}(\sigma) d\sigma\,]$$

and then we can replace all the expectations in (B.23) by $u^{(N)}$, thus obtaining

$$u^{(M)}(s,x) = f_M(x+v_M(t-s)) \exp[\, -\int_s^t q^{MM}(\sigma) d\sigma\,] + \tag{B.24}$$

$$\int_s^t [\, \frac{1}{2} u^{(M+1)}(s+\sigma, x+v_M(\sigma-s)) +$$

$$\frac{1}{2} u^{(M-1)}(s+\sigma, x+v_M(\sigma-s))\,]\, q^{MM}(\sigma) \exp[\, -\int_s^\sigma q^{MM}(\gamma) d\gamma\,]\; d\sigma$$

This last expression is the *generalized* solution of (B.9), the backward equation. The terminal condition is easily verified by substituting $s=t$ in (B.24), and we get (B.10). By formal differentiation of (B.24) we obtain (B.9), and also by integrating (B.9) along its characteristics we get the equation (B.24).

The connection between the backward and forward equations.

Since we actually solve the forward equation, (B.7), let us examine the connection between the backward and forward equations. First we rewrite (B.7), (B.9) in a more general form

$$\text{(FE)} \quad \left(-\frac{\partial}{\partial\sigma} + L^*\right) v = 0 \tag{B.25}$$

$$\text{(BE)} \quad \left(\frac{\partial}{\partial\sigma} + L\right) u = 0 \tag{B.26}$$

where we have chosen $v = W^{(N)}$, $u = u^{(N)}$ as solutions of (B.25), (B.26) respectively. Integrating by parts, we can then write a general Green's formula

$$\int_s^t d\sigma \int \left[v \left(\frac{\partial}{\partial\sigma} + L\right) u - u \left(-\frac{\partial}{\partial\sigma} + L^*\right) v \right] = 0$$

This is equivalent to

$$\sum_{N=0}^{\infty} \int dy \left[u^{(N)}(t,y) W^{(N)}(t,y) - u^{(N)}(0,y) W^{(N)}(0,y) \right] = 0 \tag{B.27}$$

From (B.7), (B.9) we also have

$$u^{(N)}(t,y) = f_N(y)$$

$$u^{(N)}(0,y) = E_{0,y,N}\{f_{N(t)}(x(t))\}$$

$$W^{(N)}(0,y) = \delta(x-y)\delta_{N,M}$$

and substituting in (B.27) we obtain,

$$\sum_{N=0}^{\infty} \int dy \left[P(t,N,y \mid s,M,x) f_N(y) \right] = \sum_{N=0}^{\infty} \int dy \left[W^{(N)}(t,y) f_N(y) \right] \tag{B.28}$$

where $P(t,N,y \mid s,M,x)$ is the transitional probability of the random walk. But (B.28) must hold for all f, so we conclude that

$$W^{(N)}(t,y) = P(t,N,y \mid s,M,x) \tag{B.29}$$

In other words, the *conditional probability* (B.29) of our process is the solution to the forward equation (B.7), whereas the *expectation* (B.22) solves the backward equation (B.9).

The binning process.

In the algorithm (see next section) we bin the particles at the position where their speed switches to zero . Hence we actually calculate this conditional probability. In particular we assume at the boundary that $N(t)=1$, $x(t)=0$, thus obtaining from (B.29) for the zeroth order term

$$W^{(0)}(\frac{x}{2},x) = P(t,0,x\,|\,0,1,0)\,|_{t=x/2} \equiv \mu(x,\omega) \tag{B.30}$$

where $\mu(x,\omega)$ is the power spectral density of (3.6a) .

Numerical simulations.

Note that in this section we return to the original notation of the theory, where t is time, and τ is the travel time.

The algorithm.

The basic flow is as follows. For a given frequency, ω , and material profile, $\gamma(\tau)$, a particle starts its random walk at initial position $t=0$ with initial speed $N=1$. Determine a random time-step, $\Delta\tau$ (see below), and update the particle's position to $t=t(0)+2N\Delta\tau$. Now check whether the particle has left the material slab, $t>t_\infty$? If yes, then the particle has escaped and we start with a new particle; if no, then the particle takes the next step. For the next step, choose the new speed as the speed of the previous step ± 1 . If the new speed equals zero, then bin the particle at its present position, and start a new walk with a new particle. If the new speed is not zero, then determine a random time-step ...

The process continues until we have sufficient realizations in each bin. Finally the desired conditional probability (B.30) is calculated, smoothed by simple averaging of five adjacent values, and output. We can proceed now to a different frequency and then to different material profiles.

The random time-step

Here lies the heart of the algorithm. First set up a table of primitives of the material profile,

$$\Gamma(\tau) = \int_0^\tau \gamma(s)\,ds \quad , \; 0 \le \tau \le \tau_\infty \tag{B.31}$$

where $\tau_\infty = \dfrac{t_\infty}{2}$, and t_∞ is taken arbitrarily equal to ten. Then $\Gamma(\tau)$ is tabulated at 100 evenly spaced points on the interval $[0,\tau_\infty]$.

The general time-step, τ_n , is assumed to be an exponential random variable with law

$$P(\tau_n > s) = \exp[-\int_{\tau^*_{n-1}}^{\tau^*_{n-1}+s} 4\omega^2\gamma(\sigma)\, N_n^2 d\sigma] \quad , \ \tau^*_{n-1} \le s < \tau^*_{n-1} + \tau_n \tag{B.32}$$

where

$$\tau^*_{n-1} = \tau_1 + \cdots + \tau_{n-1}$$

and

$$N_n = N_{n-1} \pm 1 \quad \text{with probability } ½$$

is the speed for the nth step.

Then

$$u_n = P(s \le \tau_n) = 1 - \exp[-4\omega^2 N_n^2 (\Gamma(\tau^*_{n-1}+s) - \Gamma(\tau^*_{n-1}))] \tag{B.33}$$

where

$$\Gamma(\tau^*_{n-1}+s) - \Gamma(\tau^*_{n-1}) = \int_{\tau^*_{n-1}}^{\tau^*_{n-1}+s} \gamma(\sigma)\, d\sigma$$

and u_n is a uniform $[0,1]$ random variable generated by a random number generator. We now solve (B.33) for $\Gamma(\tau^*_{n-1}+s)$ and invert to obtain

$$\tau_n = \Gamma^{-1}\left(\Gamma(\tau^*_{n-1}) + \frac{1}{4N_n^2\omega^2} \ln \frac{1}{1-u_n} \right) - \tau^*_{n-1} \tag{B.34}$$

The results.

Four basically different material profiles were used in the simulations :

- uniform profile , $\gamma(\tau) = \text{constant}$
- stepped profile , $\gamma(\tau) = b + c\tanh(\tau-\tau_1)$
- single-bump profile , $\gamma(\tau) = b + c[\tanh(\tau-\tau_1) - \tanh(\tau-\tau_2)]$
- two-bump profile , $\gamma(\tau) = b + c[\tanh(\tau-\tau_1) - \tanh(\tau-\tau_2) + \tanh(\tau-\tau_3) - \tanh(\tau-\tau_4)]$

The uniform profile.

According to (3.7), with $\gamma \equiv 1$, we expect to obtain

$$\mu(t,\omega) = \frac{\omega^2}{[1+\omega^2 t]^2} \qquad \text{(B.35)}$$

In *figure* 1 we have a mesh plot of the surface $\mu(t,\omega)$. We can clearly observe the dependence of μ on both t and ω as predicted by (B.35). The frequencies range from 0.1 to 10 in increments of 0.1 .

The stepped profiles.

Figure 2 shows the step of magnitude one, and the resulting mesh plot of the surface $\mu(t,\omega)$. The step appears in the mesh plot at $t=5$, since a particle travelling at speed 1 over a distance of $\tau = 2.5$, and then being reflected, will "arrive" after five units of time at our point of measurement.

The detailed effects are depicted more clearly in *figure* 3, where the step is now of magnitude ten, and the frequency range is $0.1 \le \omega \le 0.6$. Here we can observe the effect of the uniform profile up to $t=4$, the appearance of the step around $t=5$, the greater magnitude of μ for low frequencies, and the gradual disappearance of the hump when ω is large. In physical terms, this means that only the low frequency pulses are able to penetrate the profile and "see" the step at $\tau = 2.5$.

The single-bump profiles.

In *figure* 4 the hump once again appears around $t=5$. In addition we detect a crease on the spine of the hump. This is due to the lag in detection of the increasing and decreasing arms of the material profile. The crease is more prominent in *figure* 5 . As before we observe that only the low frequqency pulses penetrate the profile.

We consider a decreasing bump in *figure* 6 . In the upper mesh plot we cannot detect the negative bump. However, if we zoom in on the low frequency range $0.08 \le \omega \le 0.25$ (the lower mesh), the negative bump appears as a trough in the power spectral density surface. Due to the initial speed of 2 , this trough appears earlier, around $t=2.5$.

The double-bump profiles.

Figure 7 shows a profile with two sharp, separated bumps. Here the frequency range is $0.1 \leq \omega \leq 0.7$. The upper mesh plot shows only a slight hint of the second bump. But, zooming in on the low frequency range $0.1 \leq \omega \leq 0.2$ (lower mesh plot), we see that both bumps are equally well detected. All this is accentuated in *figure* 8 where we can also observe the rise and fall of the humps as the frequency increases. The lower mesh details the low frequency range.

In *figure* 9 we have removed the first of the two sharp bumps of the previous profile. Now we observe the interesting fact that the detection of the second bump is not affected by the presence or absence of the first bump. This detection depends solely on the frequency of the probing pulse.

Additional profiles.

A number of other profiles were examined. We will mention a few of these. The step of *figure* 2 starting at 2 and going down to 1 gave results qualitatively similar to those of *figure* 6 , the negative bump.

If we take the step of *figure* 2 and remove the uniform part, i.e. the step starts from 0 and jumps to 1 , then we are able to detect a hump at much greater frequencies, of the order of 5 , whereas in *figure* 2 we only reach $\omega = 0.5$.

Also considered were the effects of varying the magnitude, sharpness and distance between steps and bumps. More simple profiles, varying linearly or quadratically can also be studied with the existing code. An interesting exercise is the superposition of different profiles, both adding them together and subtracting them one from the other.

Uniform Material Profile $a(\tau)=1$

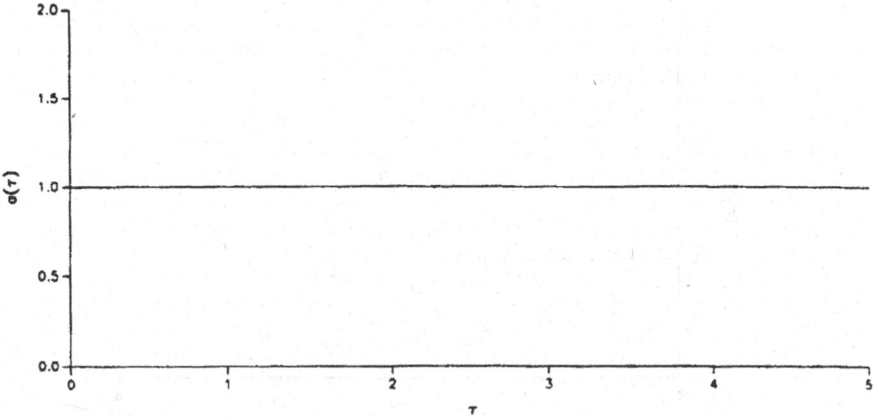

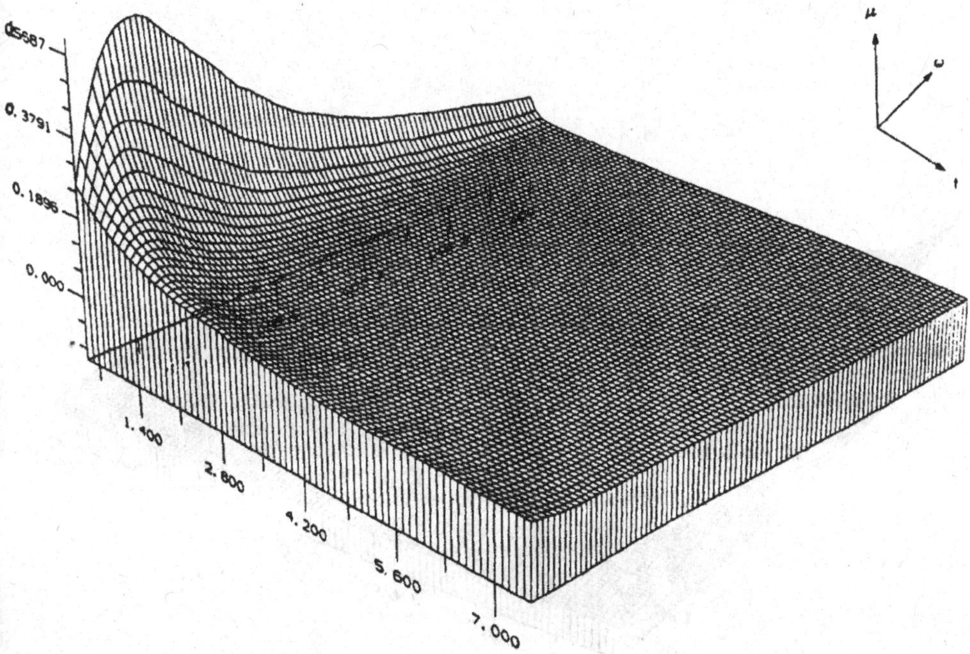

Stepped Material Profile

$a(\tau)=1.5+0.5\tanh[5(\tau-2.5)]$

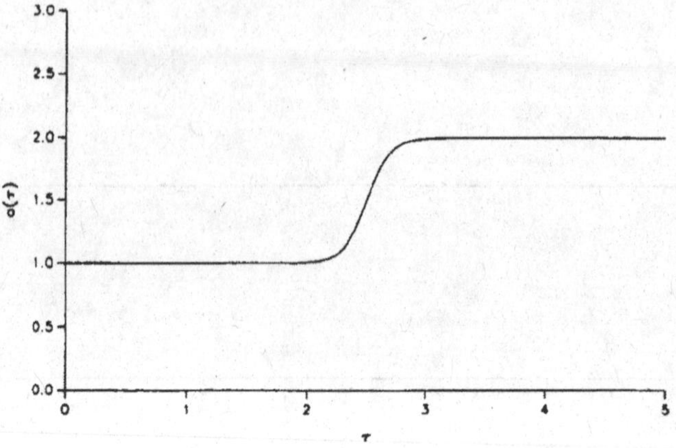

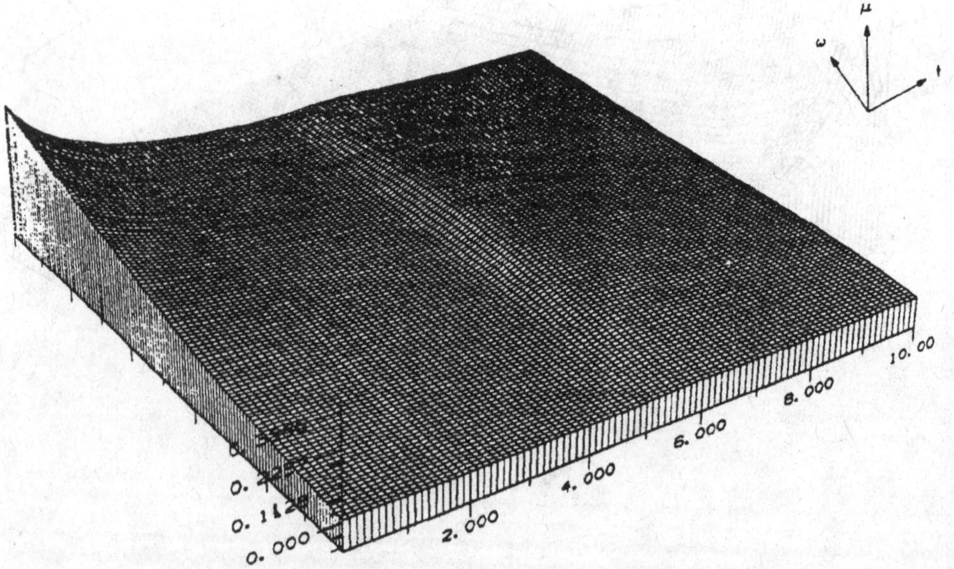

2

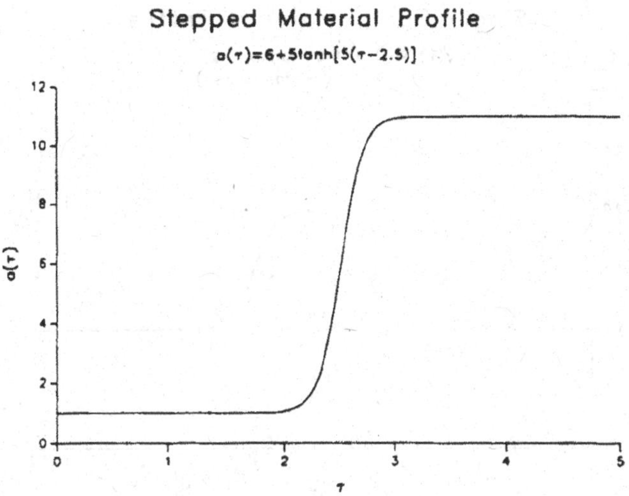
Stepped Material Profile
a(τ)=6+5tanh[5(τ−2.5)]
a(τ)
τ
0
2
4
6
8
10
12
0
1
2
3
4
5

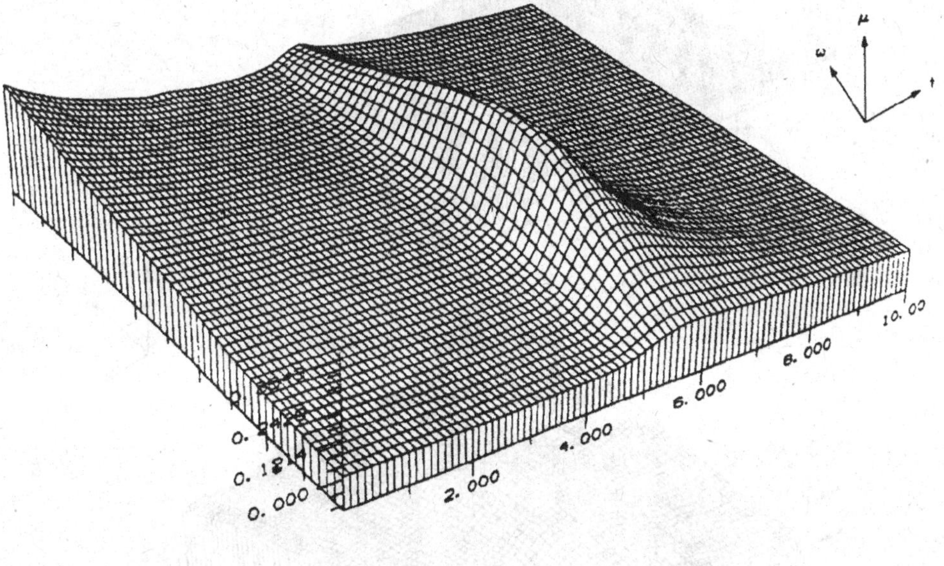
μ
ω
t
0.000
2.000
4.000
6.000
8.000
10.00

3

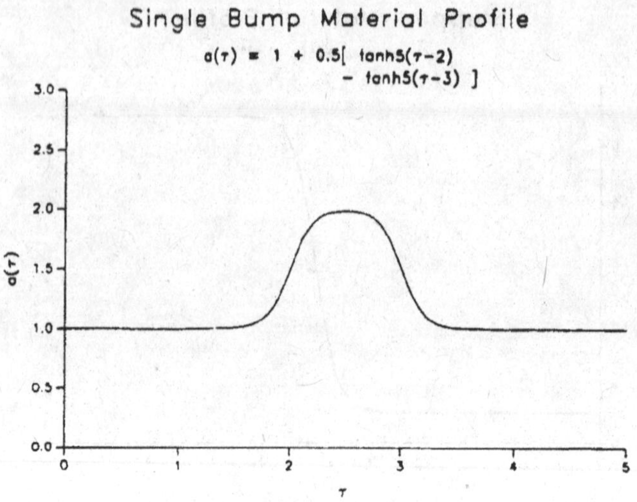
Single Bump Material Profile
a(τ) = 1 + 0.5[tanh5(τ−2) − tanh5(τ−3)]
a(τ)
τ
0.0
0.5
1.0
1.5
2.0
2.5
3.0
0
1
2
3
4
5

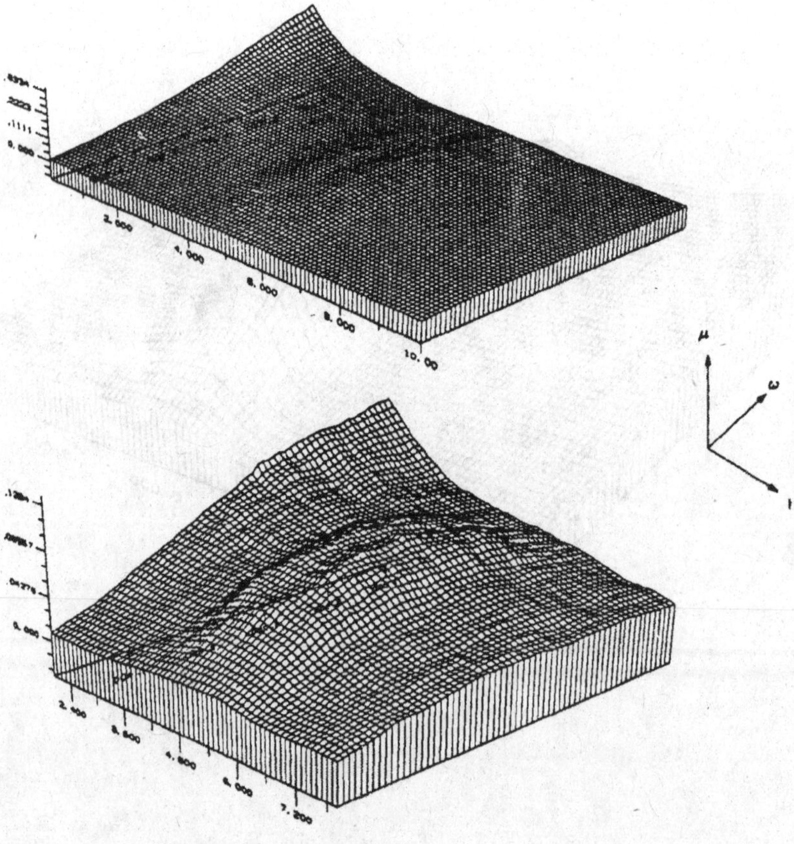
0.000
2.000
4.000
6.000
8.000
10.00
0.000
2.400
4.800
7.200
μ
ω
t

4

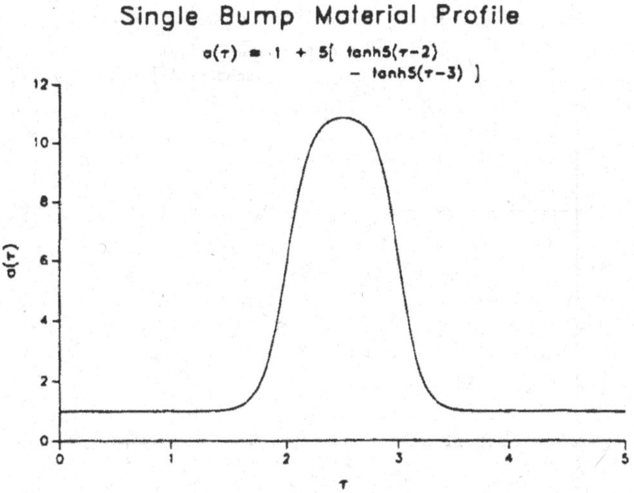
Single Bump Material Profile
a(τ) = 1 + 5[tanh5(τ−2) − tanh5(τ−3)]
a(τ)
τ
0
2
4
6
8
10
12
0
1
2
3
4
5

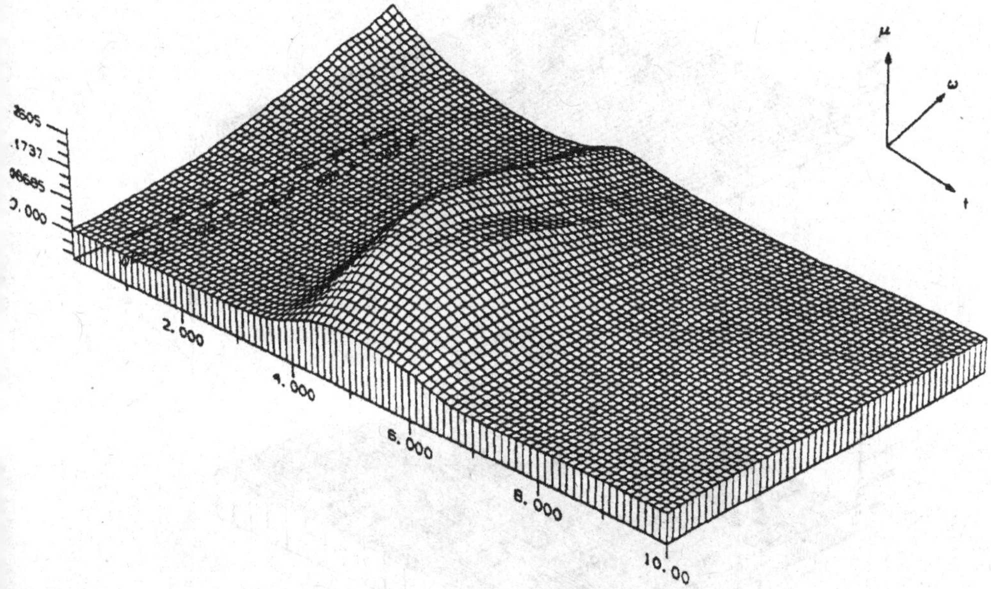
.2605
.1737
.08685
0.000
2.000
4.000
6.000
8.000
10.00
μ
ω
t

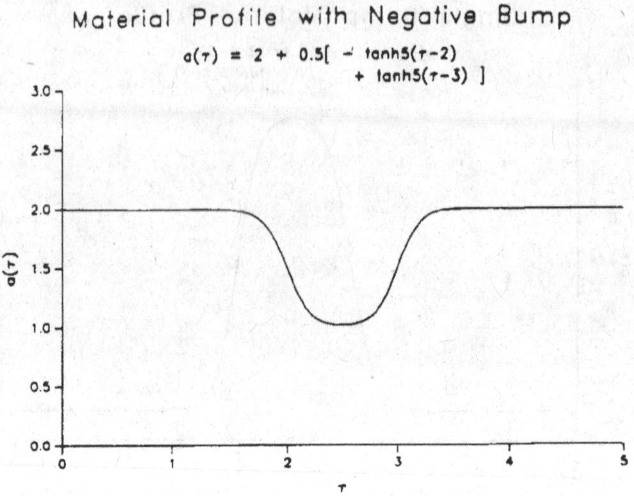
Material Profile with Negative Bump
a(τ) = 2 + 0.5[− tanh5(τ−2)
+ tanh5(τ−3)]
a(τ)
3.0
2.5
2.0
1.5
1.0
0.5
0.0
0
1
2
3
4
5
τ

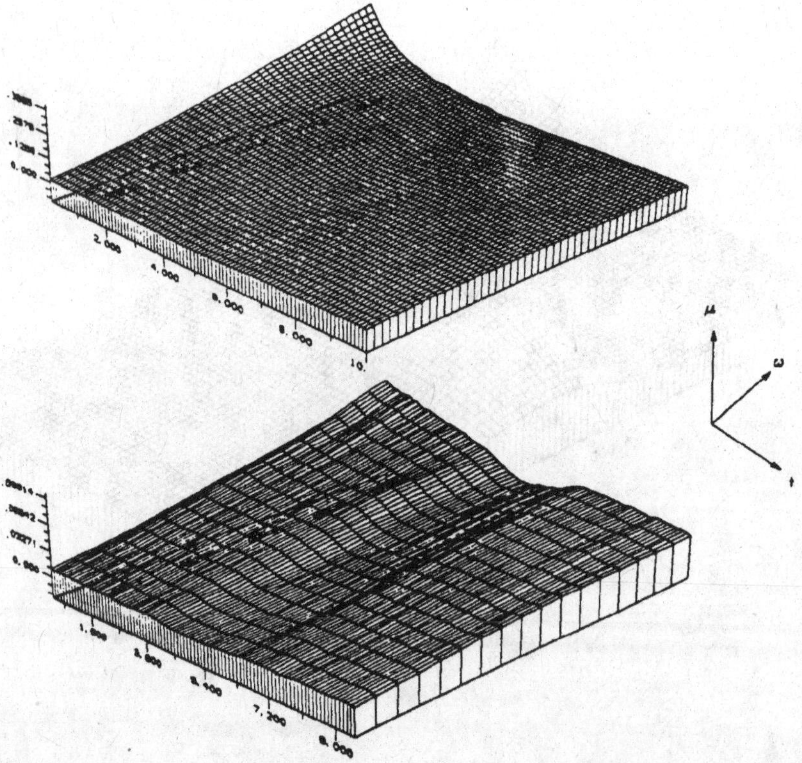
2.000
4.000
10.
μ
ω
t

6

Material Profile $a(\tau)$ with two bumps

at $\tau=1.25$, $\tau=3.75$

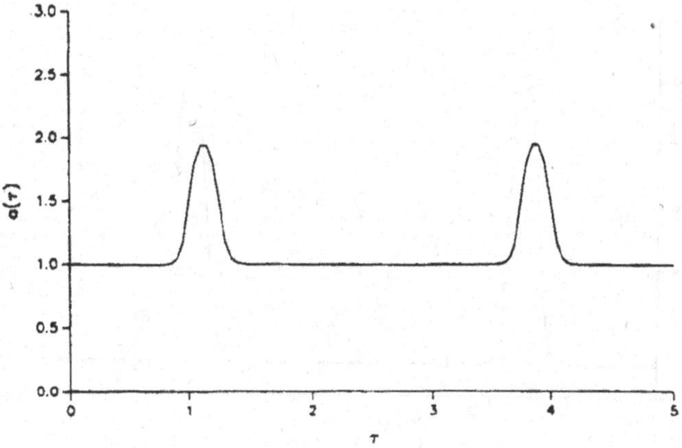

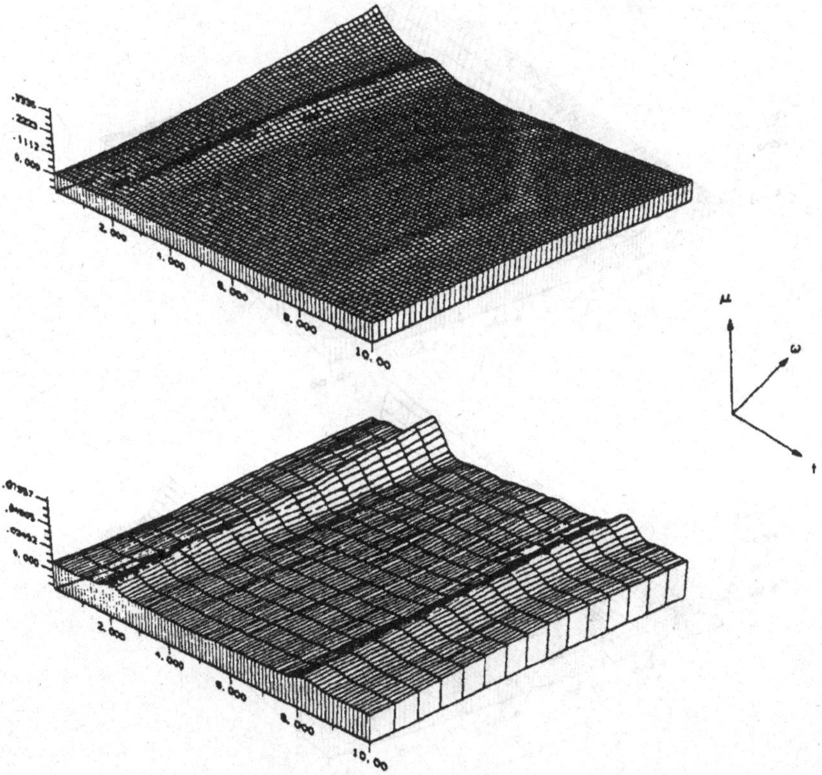

Material Profile $a(\tau)$ with two bumps

at $\tau=1.25$, $\tau=3.75$

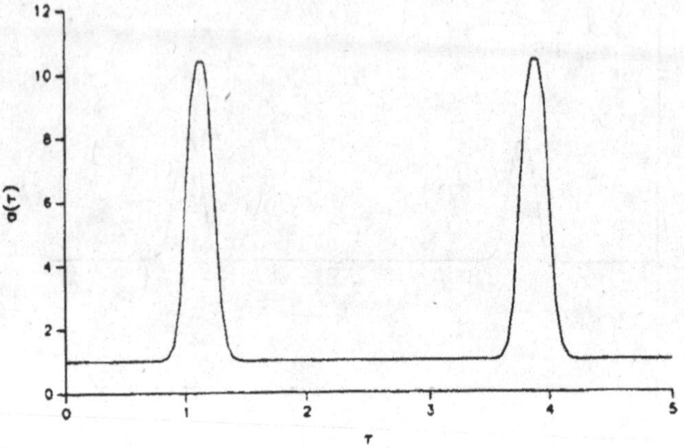

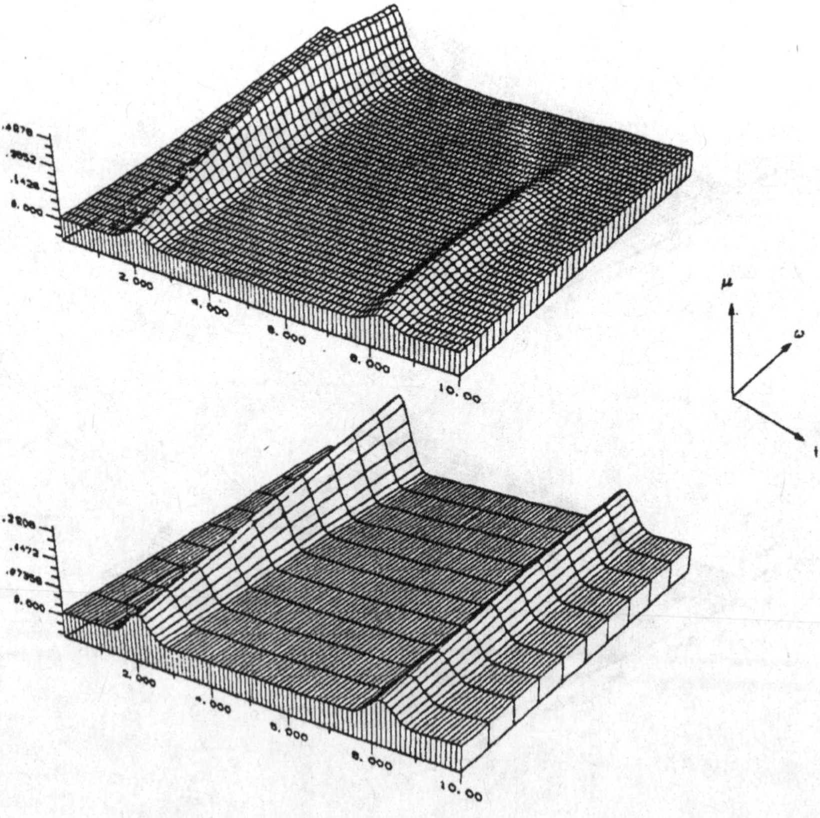

8

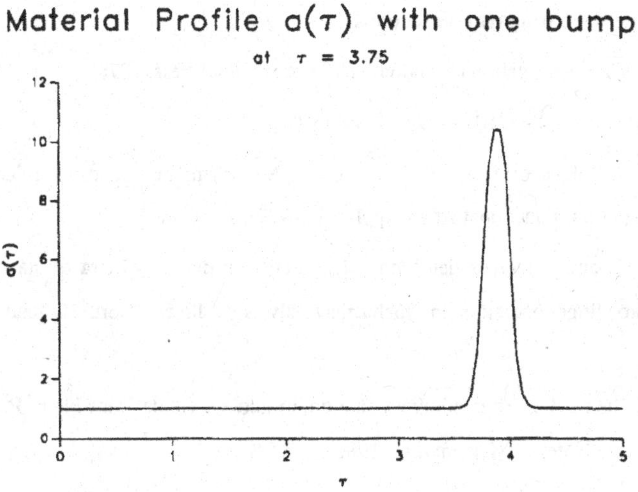
Material Profile a(τ) with one bump
at τ = 3.75
a(τ)
τ
0
2
4
6
8
10
12
0
1
2
3
4
5

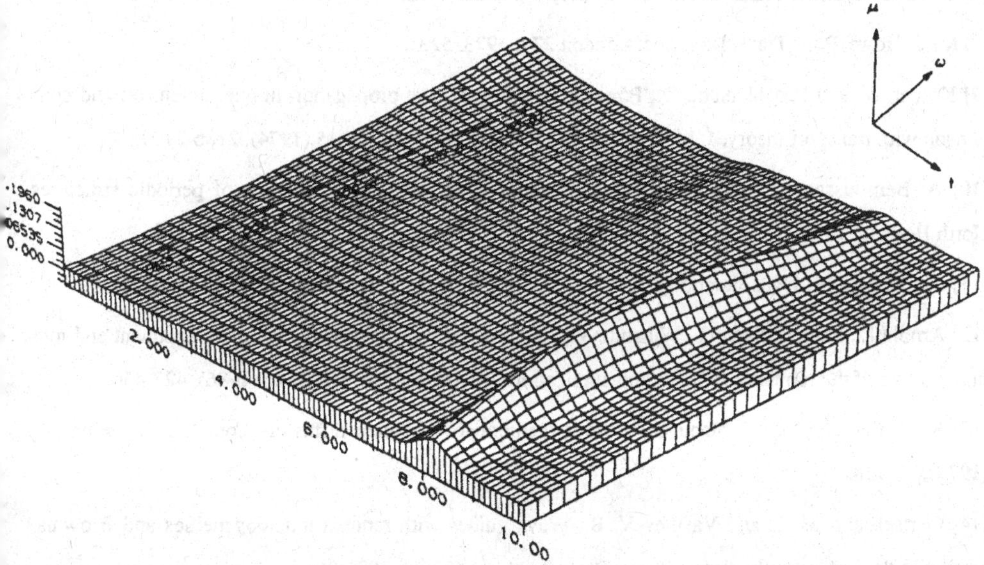
.1960
.1307
.06535
0.000
2.000
4.000
6.000
8.000
10.00
μ
ω
t

9

References for the wave propagation sections.

[1] A. Ishimaru, Wave propagation in random media, Academic Press, 1978.

[2] P. W. Anderson, Phys. Rev. 109 (1958), 1492-1505

[3] Goldsheid I. J., Molchanov S. A. and Pastur L. A., A random one-dimensional Schroedinger operator has pure point spectrum, Functional Anal. Appl., 11 (1977), 1-10.

[4] Kotani S., Lyapounov indices determine absolutely continuous spectra of stationary random one-dimensional Schroedinger operators, in Stochastic Analysis, K. Ito ed., North Holland 1984, 225-247, and preprint 1985.

[5] R. Carmona, The random Schroedinger equation, in Ecole d'ete de Saint-Flour, P. -L. Hennequin editor, Springer Lecture Notes in Mathematics, 1985.

[6] J. Froelich and T. Spencer, Absence of diffusion in the Anderson tight binding model for large disorder or low energy, 88 (1983), 151-184

[7] D. Marcuse, IEEE Trans. Microwave Theory Tech. 20, 1972, 541.

[8] M. S. Howe, Phil. Trans. Roy. Soc. London 274, 1973, 523.

[9] Kohler W. and Papanicolaou G., Power statistics for wave propagation in one dimension and comparison with transport theory, J. Math. Phys. 14 (1973), 1733-1745 and 15 (1974), 2186-2197.

[10] A. Bensoussan, J. L. Lions and G. C. Papanicolaou, Asymptotic analysis of periodic structures, North Holland, 1978.

[11] Furstenberg H., Noncommuting random products, Trans. Am. Soc., 108 (1963), 377-428.

[12] Arnold L., Papanicolaou G., Wihstutz V., Asymptotic analysis of the Lyapounov exponent and rotation number of the random oscillator and applications, SIAM J. Appl. Math., 46 (1986), 427-450.

[13] Kesten H. and Papanicolaou G., A limit theorem for turbulent diffusion, Comm. Math. Phys., 65 (1979), 97-128.

[14] Gertsenstein M. E. and Vasiliev V. B., Wave guides with random inhomogeneities and Brownian motion in the Lobachevski plane, Theory Prob. Appl., 11 (1966), 390-406.

[15] Papanicolaou G., Wave propagation in a one-dimensional random medium, SIAM J. Appl., 21 (1971), 13-18.

[16] Keller J. B., Papanicolaou G. and Weilenmann J., Heat conduction in a one-dimensional random medium, Comm. Pure Apll. Math., 32 (1978), 583-592.

[17] Gazarian Yu. L., Sov. Phys. JETP 29 (1969) 996

[18] Lang R. H., J. Math. Phys., 14 (1973), 1921

[19] Morrison J. A., J. Math. Anal. Appl., 39 (1972), 13

[20] Klyatskin V. I., Stochastic equations and waves in random media, Nauka Moscow, 1980 and Method of imbedding in the theory of random waves, Nauka, Moscow, 1986.

[21] Stratonovich R. L., Topics in the theory of noise, Elsevier, New York, 1963.

[22] Khasminskii R. Z., A limit theorem for solutions of differential equations with random right hand side, Theory Prob. Appl., 11 (1966), 390-406.

[23] Papanicolaou G., Asymptotic analysis of stochastic equations, in MAA Studies in Mathematics vol 18, M. Rosenblatt ed., MAA 1978.

[24] Kushner H. J., Approximation and weak convergence methods for random processes with applications to stochastic systems theory, MIT Press, 1984.

[25] Dawson D. A., and Papanicolaou G., A random wave process, Appl. Math. Optim., 12 (1984), 97-114.

[26] P. Richards and W. Menke, The apparent attenuation of a scattering medium, Bull. Seismol. Soc. Amer., 73 (1983), 1005-1021.

[27] R. Burridge, G. Papanicolaou and B. White, Statistics for pulse reflection from a randomly layered medium, Siam J. Appl. Math. 47 (1987), 146-168.

[28] R. Burridge, G. Papanicolaou and B. White, One dimensional wave propagation in a highly discontinuous medium, Wave Motion 10 (1988), 14-44.

[29] P. Sheng, Z.-Q. Zhang, B. White and G. Papanicolaou, Multiple scattering noise in one dimension: universality through localization lenght scales, Phys. Rev. Letters 57, number 8, 1986, 1000-1003.

[30] R. Burridge, G. Papanicolaou, P. Sheng and B. White, Direct and inverse problems for pulse reflection from inhomogeneously random halfspaces, in Advances in multiphase flow and related problems, G. Papanicolaou editor, SIAM publications, 1986, pp. 17-23.

[31] R. Burridge, G. Papanicolaou, P. Sheng and B. White, Probing a random medium with a pulse, SIAM J. Appl. Math to appear.

[32] P. Sheng, Z.-Q. Zhang, B. White and G. Papanicolaou, Wave localization characteristics in the time domain, Phys. Rev. Letters 59 (Oct. 26 1987) 1918-1921.

[33] P. Sheng, Z.-Q. Zhang, B. White and G. Papanicolaou, Minimum wave localization length in a one dimensional random medium, Phys. Rev. B 34 #7 (1986) pp. 4757-4761

[34] R. Burridge, The Gelfand-Levitan, the Marchenko and the Gopinath-Shondi integral equations of the inverse scattering theory regarded in the context of inverse impulse-response problems, Wave Motion, 2, 1980, 305-323.

[35] G. Papanicolaou, Macroscopic properties of composites, bubbly fluids, suspensions and related problems, in Les methodes de l'homogeneization:Theorie et applications en physique, CEA-EDF- INRIA series vol 57, Eyrolles, Paris, 1985, pp.229-318

Remarks on the point interaction approximation.

This is joint work with R. Figari and J. Rubinstein that appeared in vol 9 of the Springer series from the Institute for Mathematics and its Applications of the University of Minnesota in 1987.

1. Introduction.

The point interaction approximation is a way to study boundary value problems in regions with many small inclusions. For example, heat conduction in a material with many small holes that absorb heat, fluid flow in a region with many small obstacles, etc. The main idea is to replace the inclusions along with the boundary conditions on them by an inhomogeneous term in the differential equation which is then to hold evrywhere. As the name Point Interaction suggests, the effect of the inclusions is localized so the approximation is valid for small volume fractions. In this paper we shall consider the heat conduction problem as follows.

Let $u(x,t)$ be the temperature at position x in R^3 and at time $t \geq 0$. For $N=1,2,3,...$ let $w_1^N, w_2^N, ..., w_N^N$ be a sequence of points in R^3 and let D^N be the domain in R^3 defined by

$$D^N = \bigcap_{j=1}^{N} \left\{ x \mid |x-w_j^N| > \frac{\alpha}{N} \right\} \tag{1}$$

where α is a fixed positive constant. The temperature u satisfies the initial-boundary value problem

$$\frac{\partial u}{\partial t} = \Delta u \ \ in \ \ D^N, \ t>0 \tag{2}$$

$$u(x,0)=f(x) \ \text{for} \ x \ in \ D^N \ \text{and} \ u(x,t)=0 \ on \ \partial B_j^N \ \text{for} \ t>0, \ j=1,2,3,...,N \tag{3}$$

Here $f(x)$ is a smooth, positive function of compact support representing the initial temperature distribution and B_j^N is the sphere centered at w_j^N with radius α/N.

We are interested in the behavior of the solution u when N is large and the sequence of sphere centers w_j^N tends to a continuum. That is for every smooth function $\phi(x)$

$$\frac{1}{N}\sum_{j=1}^{N}\phi(w_j^N) \to \int V(x)\phi(x)dx \ \ as \ N \to \infty \tag{4}$$

where $V(x)$ is the continuum sphere center density, assumed smooth and with compact support.

A related problem which is more realistic physically is the case of spherical inclusions that melt. This means that the radii of the spheres depend on time, are denoted by $\alpha_j^N(t)/N$ and we have the additional boundary condition

$$\frac{d\alpha_j^N(t)}{dt}=\frac{-1}{N}\frac{1}{4\pi\alpha_j^N(t)/N^2}\int_{|x-w_j^N|=\alpha_j^N(t)/N}\frac{\partial u(x,t)}{\partial n}dS(x)\,,\ j=1,2,3,\cdots,N \tag{5}$$

Here n is the unit normal on the spheres pointing into the interior of the region D^N. At time zero the scaled radii are equal to $\alpha_0>0$

$$\alpha_j^N(0)=\alpha_0 \tag{6}$$

The boundary condition (5) is a simplified form of the usual one in free boundary problems: the rate of displacement of the boundary is proportional to the heat flux crossing the surface. It is simplified because the melting spheres do not change shape and their radii change in proportion to the average heat flux absorbed.

We will analyze here problem (2-4) in the continuum limit $N\to\infty$ by a relatively simple and direct method, the point interaction approximation. The radii of the spherical inclusions are already scaled in the above problems to be proportional to $1/N$. That this is appropriate scaling for a continuum limit can be seen easily by calculating the heat absorbed by a single sphere and requiring that N times this quantity be of order one as $N\to\infty$. The point interaction approximation is an intermediate step between (2-4) (or (2-6)) and the continuum limit which has features of both but is much simpler than (2-3) since the effect of the spheres is replaced by an appropriate point source term. The continuum approximation of (2-3) is the solution $\bar{u}$ of the initial value problem

$$\frac{\partial\bar{u}(x,t)}{\partial t}=\Delta\bar{u}(x,t)-4\pi\alpha V(x)\bar{u}(x,t)\ \ x\ in\ R^3\ ,\ t>0 \tag{7}$$

$$\bar{u}(x,0)=f(x)$$

Note that the volume fraction occupied by the speres goes to zero as $N\to\infty$ like N^{-2}. In the case of the melting spheres the continuum limit is given by the nonlinear diffusion equation

$$\frac{\partial\bar{u}(x,t)}{\partial t}=\Delta\bar{u}(x,t)-4\pi\alpha(x,t)V(x)\bar{u}(x,t)\ \ x\ in\ R^3\ ,\ t>0$$

$$\bar{u}(x,0)=f(x)$$

$$\frac{d\alpha(x,t)}{dt}=-\frac{\bar{u}(x,t)}{\alpha(x,t)}$$

$$\alpha(x,0)=\alpha_0$$

Note here the structure of the limit problem: it is a diffusion equation for the temperature field and an ordinary differential equation (a relaxation equation) for the continuum sphere radii. This is typical in problems where the point interaction approximation is called for as for example in waves in bubbly liquids [8].

The point interaction approximation for diffusion in a region with fixed spheres is described in section 2. A proof of its validity is given in the appendix.

Boundary value problems in regions with many small holes have been analyzed before in a variety of contexts and by several methods. Khruslov and Marchenko [1] use potential theoretic methods and give results in considerable generality regarding the possible distribution of the inclusion centers $\{w_j^N\}$, compatible with (4). Kac [2] studied (2), (3) when the points $\{w_j^N\}$ are independent identically distributed random variables over a region. He used properties of the Wiener sausage. Rauch and Taylor [3] formulated the results of Kac in a more analytic way and generalized them. Papanicolaou and Varadhan [4] studied (2-3) for nonrandom configurations of centers $\{w_j^N\}$ by probabilistic methods and obtained a strong form of convergence to the continuum limit. Ozawa [5] first considered the analysis of boundary value problems in regions with many small holes via a point interaction approximation. A study of the error in the continuum limit and a central limit theorem for it are given by Figari, Orlandi and Teta [6].

The point interaction approximation is a natural tool to analyze a variety of interesting problems in the continuum or homogenization limit. In the physical literature it goes back to Foldy's paper [7 see also 8,9] on sound propagation in a bubbly liquid and perhaps earlier. In almost all papers that followed Foldy's, the point interaction approximation is not treated as an important approximation in itself and averaging is carried out over the sphere center locations $\{w_j^N\}$. The closure problem that arises is then treated in a variety of ways depending on other parameters in the problem. In nonlinear cases, as with melting spheres and bubbles, the closure problems are much more involved. But averaging is not necessary. The continuum limit holds for deterministic sequences satisfying (4) and subject to some other conditions that hold for "most" realizations in the random case. The closure difficulties are thus avoided for many linear and nonlinear problems.

2. Point interaction approximation for diffusion in regions with many fixed inclusions.

We shall analyze the Laplace transform version of (1.2)

$$(-\Delta+\lambda)\, u^N(x) = f(x) \ , \quad x \ in \ D^N \ , \lambda > 0 \tag{1}$$

$$u^N(x) = 0 \ , \quad | x - w_j^N | = \frac{\alpha}{N}$$

Let G be the free space Green's function

$$G(x,y) = \frac{e^{-\sqrt{\lambda}\,|x-y|}}{4\pi\,|x-y|} \tag{2}$$

Using Green's theorem we may rewrite (1) in integral form

$$u^N(x) = \int_{D^N} G(x,y) f(y)\,dy - \sum_{j=1}^{N} \int_{\partial B_j^N} G(x,y) \frac{\partial u^N(y)}{\partial n}\,dS(y) \tag{3}$$

where x is in D^N and n denotes the unit outward normal to the spheres ∂B_j^N.

Now let x tend to the surface of the ith sphere in (3). Using the Dirichlet boundary condition, we rewrite (3) in the form

$$\int_{\partial B_i^N} G(x,y) \frac{\partial u^N(y)}{\partial n}\,dS(y) + \sum_{\substack{j=1 \\ j \neq i}}^{N} \int_{\partial B_j^N} G(x,y) \frac{\partial u^N(y)}{\partial n}\,dS(y) \tag{4}$$

$$= \int_{D^N} G(x,y) f(y)\,dy$$

Let

$$\frac{1}{N} Q_j^N = \int_{\partial B_j^N} \frac{\partial u^N(y)}{\partial n}\,dS(y) \quad , \; j = 1,2,\ldots,N \tag{5}$$

be the charges induced on the spheres, suitably normalized. Since $f \geq 0$, the Q_j^N are nonnegative.

The spheres B_j^N have radius of order N^{-1} so they are small. We may then consider an approximate form of (4) where we place x at the center w_j^N of the ith sphere in the first term on the left and in the sum. We may also let the y in G in the sum in (4) go to the center w_j^N. Let us denote the approximate charges by q_j^N. Then

$$\frac{1}{4\pi\alpha} q_i^N + \frac{1}{N} \sum_{\substack{j=1 \\ j \neq i}}^{N} G(w_i^N, w_j^N)\, q_j^N = \int G(w_i^N, y) f(y)\,dy \qquad i = 1,2,\ldots,N \tag{6}$$

Note that we have also extended the integration on the right to all of R^3.

System (6) is what we call the point interaction approximation (PIA). The main point is to show that, under suitable conditions on the sequence of sphere centers $\{w_j^n\}$,

$$\lim_{N \to \infty} \sup_{1 \leq j \leq N} |\, Q_j^N - q_j^N \,| = 0 \tag{7}$$

That is, the exact charges Q_j^N are well approximated by the the approximate carges q_j^N obtained from the

PIA. Once this has been shown it is easy to pass to the continuum limit in (6) using the hypothesis (1.4).

Let $q(x)$ be the solution of the integral equation

$$\frac{1}{4\pi\alpha} q(x) + \int G(x,y)V(y)q(y)dy = \int G(x,y)f(y)dy \tag{8}$$

with x in R^3. If we define

$$u(x) = \frac{q(x)}{4\pi\alpha} \tag{9}$$

we see that (8) is the integral equation version of the Laplace transform of (1.7)

$$(-\Delta + \lambda)u(x) + 4\pi\alpha V(x)u(x) = f(x) \quad , \lambda > 0 \, . \tag{10}$$

Using the regularity of the solution u of (10) and standard methods familiar from numerical analysis, along with hypothesis (1.4), we can show that

$$\lim_{N \to \infty} \sup_{1 \le j \le N} | q_j^N - q(w_j^N) | = 0 \tag{11}$$

That is, (8) is the continuum limit of the PIA (6). Combining (7) and (11) we arrive at the desired convergence of the charges

$$\lim_{N \to \infty} \sup_{1 \le j \le N} | Q_j^N - q(w_j^N) | = 0 \tag{12}$$

Once the charges have been shown to converge in the sense of (12) it is easy to show that $u^N(x)$, the solution of (1) converges to $u(x)$ the solution (10), outside a small set of points x near the surfaces of the spheres. We shall therefore concentrate here on proving (7) i.e. in proving the validity of the point interaction approximation.

Appendix.

For the proof we need some assumptions regarding the sequence of sphere centers $\{w_j^n\}$ in addition to (1.4). We will assume that

$$\inf_{i \neq j} | w_i^N - w_j^N | \geq \frac{1}{N^{1-\nu}} \quad \text{for } some \ 0 < \nu < \frac{1}{3} \tag{1}$$

$$\frac{1}{N^2} \sum_{\substack{i,j=1 \\ i \neq j}}^{N} \frac{1}{| w_i^N - w_j^N |^{3-\varepsilon}} \leq C \quad \text{for } some \quad \varepsilon > 0 \tag{2}$$

$$\frac{1}{N^3} \sum_{\substack{i,j,k=1 \\ i \neq j \neq k}}^{N} \frac{1}{| w_i^N - w_j^N |^2} \frac{1}{| w_k^N - w_i^N |^2} \leq C \tag{3}$$

These conditions are valid, for example, for sequences of independent identically distributed random sphere centers. They are valid in probability that is, there is a set of sphere center configurations that satisfy (1), (2) and (3) with probability arbitrarily close to one, uniformly in N.

The idea in introducing conditions like (1-3), contrary to what is done in [1] and [4], is that since they are valid in probability one need not worry about their form provided that they help to make the analytical part of the argument simple. This is important in nonlinear problems such as the melting spheres.

Let us now consider again the integral equation (2.4) where x is on ∂B_i^N. We average over the ith sphere

$$\begin{aligned}
&\frac{1}{4\pi(\frac{\alpha}{N})^2} \int_{\partial B_i^N} \int_{\partial B_i^N} G(x,y) \frac{\partial u^N(y)}{\partial n} dS(y) dS(x) \\
&+ \sum_{\substack{j=i \\ j \neq i}}^{N} \frac{1}{4\pi(\frac{\alpha}{N})^2} \int_{\partial B_i^N} \int_{\partial B_j^N} G(x,y) \frac{\partial u^N(y)}{\partial n} dS(y) dS(x) \\
&= \int_{\partial B_i^N} \int_{D^N} G(x,y) f(y) dy \, .
\end{aligned} \tag{4}$$

To simplify (4) we use the elementary identity.

$$\frac{1}{4\pi(\frac{\alpha}{N})^2} \int_{\partial B_i^N} G(x,y) dS(x) = \frac{1}{4\pi\,(\frac{\alpha}{N})} \frac{1 - e^{-2\sqrt{\lambda}\,\alpha/N}}{2\sqrt{\lambda}\,\alpha/N} \tag{5}$$

when y is on ∂B_i^N also. We use also the mean value theorem for any solution $h(x)$ of $(-\Delta + \lambda) h(x) = 0$

$$\frac{1}{4\pi r^2}\int_{|x|=r} h(x)dS(x) = \frac{\sinh\sqrt{\lambda}r}{\sqrt{\lambda}r}\,h(0) \tag{6}$$

Using (5) and (6), simplifying and recalling the definition (2.5) of the charges we obtain the following equations

$$\frac{\sqrt{\lambda}\,\alpha/N}{\sinh(\sqrt{\lambda}\,\alpha/N)}\;\frac{1-e^{-2\sqrt{\lambda}\alpha/N}}{2\sqrt{\lambda}\,\alpha/N}\;\frac{1}{4\pi\alpha}\,Q_i^N$$

$$+\frac{1}{N}\sum_{\substack{j=1\\ j\neq i}}^{N} G(w_i^N, w_j^N)\,Q_j^N = \int_{D^N} G(w_i^N, y)f(y)dy + \eta_i^N\,, \tag{7}$$

$$i = 1,2,\ldots,N$$

where

$$\eta_i^N = \sum_{\substack{j=1\\ j\neq i}}^{N}\int_{\partial B_j^N}\left[G(w_i^N, y) - G(w_i^N, w_j^N)\right]\frac{\partial u^N(y)}{\partial n}\,dS(y)\,. \tag{8}$$

Lemma 1.

$$\sup_{1\le i\le N}|\eta_i^N| \le \frac{C}{N^{1+\nu}}\sum_{\substack{j=1\\ j\neq i}}^{N} G(w_i^N, w_j^N)\,Q_j^N\,. \tag{9}$$

Proof: We simply note that for y on ∂B_j^N

$$\frac{e^{-\sqrt{\lambda}\alpha/N}\,|w_i^N - w_j^N|}{|w_i^N - w_j^N| + \frac{\alpha}{N}} \le \frac{G(w_i^N, y)}{G(w_i^N, w_j^N)} \le \frac{e^{\sqrt{\lambda}\alpha/N}\,|w_i^N - w_j^N|}{|w_i^N - w_j^N| - \frac{\alpha}{N}}$$

From this and hypethesis (1) the estimate (9) follows easily.

Now we return to equations (7) and use the estimate (9) along with the positivity of the Q_j^N. We see that for N large the error η_i^N can be absorbed into the sum on the left of (7) without changing signs. This and the fact that $f\in C_0^\infty(R^3)$ give us the a-priori estimate

$$0 \le Q_i^N \le C \tag{10}$$

where all constants C that appear (possibly different constants) are independent of N. We also have the estimate

$$0 \le \frac{1}{N}\sum_{\substack{j=1\\ j\neq i}}^{N} G(w_i^N, w_j^N)\,Q_j^N \le C\,. \tag{11}$$

Combining (10) and (9) we have now an estimate for the error

$$\sup_{1 \le i \le N} | \eta_i^N | \le \frac{C}{N^\nu} \qquad \nu > 0 . \tag{12}$$

If we also denote by $a_N(\lambda)$ ($\lambda > 0$ fixed) the coefficient in front of $\frac{1}{4\pi\alpha} Q_i^N$ in (7) then clearly

$$a_N \to 1 \quad \text{as} \quad N \to \infty \tag{13}$$

Put

$$f_i^N = 4\pi\alpha \int_{D^N} G(w_i^N, y) dy \quad , \quad i = 1,2,\ldots,N \tag{14}$$

and let q_i^N satisfy the PIA (2.6). We write this system in the form

$$(I + A^N) q^N = f^N \tag{15}$$

where $I = (\delta_{ij})$ is the identity matrix and

$$A_{ij}^N = \frac{4\pi\alpha}{N} G(w_i^N, w_j^N) \quad i \ne j \tag{16}$$

$$A_{ij}^N = 0 \qquad i = j$$

The vectors q^N and f^N have components q_i^N and f_i^N. Let $|| \; ||$ stand for the l^2 vector (or matrix) norm on R^N.

Lemma 2. For $\lambda > 0$ large enough.

$$\sup_N || A^N || < 1 \tag{17}$$

Proof. Clearly

$$|| A^N ||^2 \le \frac{4\pi\alpha}{N^2} \sum_{\substack{i,j=1 \\ i \ne j}}^{N} G(w_i^N, w_j^N)^2 \tag{18}$$

and this is controlled by (3) and (1.4). In addition, we can use the exponential factor in G that has the parameter λ and we can make the right side of (18) small by choosing λ large.

Thus, (15) is invertible uniformly in N and we may write

$$q^N = (I + A^N)^{-1} f^N \tag{19}$$

Clearly

$$0 \le q_i^N \le C \quad , \quad i = 1, 2, ..., N \tag{20}$$

and since the f_i^N are also uniformly bounded we have that

$$\sum_{j=i}^{N} \left[\frac{1}{\sqrt{N}} f_i^N \right]^2 \le C \tag{21}$$

Hence from (17) we deduce that

$$\frac{1}{N} \sum_{j=i}^{N} (q_i^N)^2 \le C \tag{22}$$

To estimate the difference between Q_j^N and q_j^N we need an additional estimate on the errors η_i^N.

Lemma 3.

$$\sum_{i=1}^{N} | \eta_i^N |^2 \le \frac{C}{N} \tag{23}$$

Proof. From (8) and (10) we see that

$$| \eta_i^N | \le \frac{C}{N^2} \sum_{\substack{j=1 \\ j \ne i}} \frac{1}{| w_i^N - w_j^N |^2}$$

We note that this does not follow from the estimate (9). One has to consider the η_i^N from (8) directly. From (2.4) we get

$$\sum_{i=1}^{N} | \eta_i^N |^2 \le \frac{C}{N^4} \sum_{i=1}^{N} \sum_{\substack{j=1 \\ j \ne i}}^{N} \sum_{\substack{k=1 \\ k = i}}^{N} \frac{1}{| w_j^N - w_i^N |^2} \frac{1}{| w_k^N - w_i^N |^2}$$

$$= \frac{C}{N} \left\{ \frac{1}{N^3} \sum_{\substack{i,j,k=1 \\ j \ne i \ne k}}^{N} \frac{1}{| w_i^N - w_j^N |^2} \frac{1}{| w_k^N - w_i^N |^2} \right.$$

$$\left. + \frac{1}{N^3} \sum_{\substack{j,k=1 \\ j \ne k}}^{N} \frac{1}{| w_j^N - w_k^N |^4} \right\}$$

The first sum on the right is controled by hypothesis (3). In the second sum we note that because of hypothesis (1) and (2)

$$\frac{1}{N^3} \sum_{\substack{j,k=1 \\ j \ne k}}^{N} \frac{1}{| w_i^N - w_k^N |^4} \le C \frac{(N^{1-\nu})^{1+\varepsilon}}{N} \frac{1}{N^2} \sum_{\substack{j,k=1 \\ j \ne k}}^{N} \frac{1}{| w_j^N - w_k^N |^{3-\varepsilon}} \le C$$

This estimate completes the proof of the lemma.

We can now estimate the difference $| Q_j^N - q_i^N |$ between the exact charges and the approximate charges of the PIA. In fact by subtracting (15) from (7) we see that

$$(I + A^N)(Q^N - q^N) = (1 - a^N)Q^N + \eta^N \tag{25}$$

Now $| 1 - a^N | \leq C/N$ for λ fixed and $0 \leq Q_i^N \leq C$. Thus the l^2 norm of the right side of (25) is less than a constant times $N^{-1/2}$, in view of lemma 3. From lemma 2 and this observation we conclude that

$$\sum_{j=1}^{N} (Q_j^N - q_j^N)^2 \leq \frac{C}{N} \tag{26}$$

This implies that

$$\sup_{1 \leq j \leq N} | Q_j^N - q_j^N | \leq \frac{C}{N^{1/2}} \tag{27}$$

We summarize the above in the following theorem.

Theorem

Under hypotheses (1), (2), (3) and (1.4) the charges Q_j^N of problem (2.1), defined by (2.5), and the approximate charges q_j^N satisfying the PIA (2.6) (or (15)) are uniformly close as $N \to \infty$ in the sense of (27).

As we indicated in section 2, the elementary estimate (2.11) gives us the validity of the continuum approximation (2.12) for the charges and hence for the full solution of (2.1) and (2.10).

References for the Point Interaction Section.

[1] E. I. Khruslov and V. A. Marchenko, Boundary value problems in regions with fine-grained boundaries, Naukova Dumka, Kiev, 1974.

[2] M. Kac, Probabilistic methods in some problems of scattering theory, Rocky Mountain J. Math. 4, 1974, 511-538.

[3] J. Rauch and M. Taylor, Potential and scattering theory on wildly perturbed domains, J. Funct. Anal. 18, 1975, 27-59

[4] G. Papanicolaou and S.R.S. Varadhan, Diffusion in regions with many small holes. In Stochastic Differential Systems (ed. B. Grigeliouis). Lecture Notes in Control and Information Theory 25, 190-206, Springer.

[5] S. Ozawa., On an elaboration of M. Kac's Theorem concerning eigenvalues of the Laplacian in a region with randomly distributed small obstacles, Comm. Math. Phys. 91, 1983, 473-487.

[6] R. Figari, E. Orlandi and J. Teta, The Laplacian in regions with many small obstacles: fluctuations around the limit operator 41, 1985, 465-488.

[7] L. L. Foldy, The multiple scattering of waves, Phys. Rev. 67, 1945, 107-119.

[8] R. Caflisch, M. Miksis, G. Papanicolaou and L. Ting, Effective equations for wave propagation in bubbly liquids, J. Fluid Mech. 153, 1985, 259-273 and also 160, 1-14.

[9] J. Rubinstein, NYU Dissertation, 1985.

GEOMETRIC ASPECTS OF DIFFUSIONS ON MANIFOLDS

David ELWORTHY

D. ELWORTHY : "GEOMETRIC ASPECTS OF DIFFUSIONS ON MANIFOLDS"

Introduction

INTRODUCTION

A. There were three main aspects of the theory of diffusions on manifolds presented in this course: the theory of characteristic exponents for stochastic flows; the use of the Feynman-Kac formula for the solution to the heat equation for forms to obtain geometric results, in particular on the shape of the manifold at infinity; and a technique for obtaining exact and asymptotic expansions of heat kernels. A proof of the Gauss-Bonnet-Chern theorem was given as an application of the third, and also to emphasize the fact that Malliavin calculus is not needed for probabilistic proofs of the Atiyah-Singer index theorem.

The first two aspects concern long term behaviour and the third short time behaviour.

Since the participants could not be expected to have a thorough knowledge of differential geometry, quite a lot of time was devoted to a quick course in Riemannian geometry via connections on the frame bundle. Apart from the intuitive understanding this gives to the notion of parallel translation along the paths of a diffusion, needed for the Feynman-Kac formula for forms, it allows for a global formalism and is anyway intrinsically involved in the main example considered in the discussion of characteristic exponents. Also the fermionic calculus for differential forms was described, both in order to give a proof of the Weitzenbock formula: which is a basic result needed to be able to obtain a Feynman-Kac formula, and in order to give the 'supersymmetric' proof of the Gauss-Bonnet-Chern theorem. This was taken from [33].

Probabilistically it was assumed that the participants had a reasonable understanding of stochastic differential equations in $\mathbb{R}^n$, driven by Brownian motion, and with smooth coefficients. The existence theorem for solutions of such equations on manifolds (up to an explosion time) was proved by embedding the manifold in some $\mathbb{R}^n$ and then using existence results for $\mathbb{R}^n$. See also [107].

B. Not surprisingly, perhaps, it was not possible to cover everything in these lecture notes during the 15 hours of the course. The main sections missed out were those relating to the geometry of submanifolds of $\mathbb{R}^n$: II §7, III §3; the section on moment exponents III §5; and much of V §2.

C. Acknowledgements. I would like to thank the participants for their encouragement and helpful suggestions, as well as for their help in pointing out the numerous errors in the first version of these notes. I am particularly grateful to Monique Poitier for this. From my point of view the whole summer school was extremely enjoyable as well as being very stimulating, and I am very pleased to be able to express my admiration of Professor Hennequin's efforts which made this possible. I would also like to thank Steve Rosenberg for permission to use some unpublished joint work in Chapter IV and to thank David Williams for pointing out a mistake in my original version of Chapter VI §2C.

This final camera ready version was prepared by Peta McAllister from the original version typed by Terri Moss: the difficulties they had from a combination of my writing and changing technology were considerable, and I owe them a big debt of thanks for their patience, skill and speed.

Some of the work reported in the course was supported in part by the U.K. Science and Engineering Research Council.

CHAPTER 1

STOCHASTIC DIFFERENTIAL EQUATIONS AND MANIFOLDS

§1. Some notation, running hypotheses, and basic facts about manifolds

A. We will use invariant notation as much as posssible: if U is open in a Banach space E and F is some other Banach space then the first derivative of a function $f : U \to F$ will be written

$$Df : U \to \mathbb{L}(E,F)$$

and the second

$$D^2f : U \to \mathbb{L}(E,E;F)\ , \text{etc.},$$

where $\mathbb{L}(E,F)$ and $\mathbb{L}(E,E;,F)$ are the spaces of continuous linear and bilinear maps into F respectively. Sometimes we write

$$df : U \times E \to F$$

for

$$(x,v) \to Df(x)(v)$$

with

$$(df)_x \text{ for } Df(x).$$

For continuous semi-martingales z_1, z_2 with values in finite dimensional normed spaces E^1, E^2 and for bilinear $B_t : E^1 \times E^2 \to F$ it will be more convenient to use

$$\int_0^T B_t(dz_{1,t}, dz_{2,t}) \text{ for } \Sigma_{i,j} \int_0^T B_t^{ij}\, d\langle z_1^i, z_2^j\rangle_t$$

or for some corresponding expression involving the tensor quadratic variation. Here the i,j refer to co-ordinates with respect to bases of E^1, E^2 respectively. Thus

$$D^2f(x)(dz_{1,t}, dz_{2,t}) = \Sigma_{i,j}\ \partial^2 f/\partial x^i \partial x^j\ d\langle z^i{}_1, z^j{}_2\rangle_t$$

for $E = E^i = \mathbb{R}^n$.

B. Our manifolds M will be C^∞, connected, and of finite dimension n, unless clearly otherwise. They will always be metrizable. The manifold structure is determined by some C^∞ *atlas* $\{(U_\alpha, \varphi_\alpha) : \alpha \in A\}$ where A is some index set, $\{U_\alpha : \alpha \in A\}$ is an open cover, and each φ_α is a homeomorphism of U_α onto some open subset of $\mathbb{R}^n$, such that on their domain of definition in $\mathbb{R}^n$ each coordinate change $\varphi_\alpha \circ \varphi_\beta^{-1}$ is C^∞. The pairs $(U_\alpha, \varphi_\alpha)$ are C^∞ *charts*, as are any other such which when added to the original atlas still keeps it a C^∞ atlas. For $0 \le r \le \infty$ maps $f : M \to N$ of manifolds are C^r if each $\theta_p \circ f \circ \varphi_\alpha^{-1}$ is C^r on its domain of definition when $(U_\alpha, \varphi_\alpha)$ is a chart for M and (V_p, θ_p) one for N. The map is a C^r diffeomorphism if it is a homeomorphism and both it and its inverse are C^r. The spaces $\mathbb{R}^n$ are considered as C^∞ manifolds by taking $\{(\mathbb{R}^n, \text{identity map})\}$ as atlas, and similarly for open sets U of $\mathbb{R}^n$. A subset N of $\mathbb{R}^{n+p}$ is an n-dimensional C^∞ submanifold if there is a family of C^∞ charts $\{(U_\beta, \varphi_\beta) : \beta \in B\}$ for $\mathbb{R}^m$ such that U_β covers M and $\varphi_\beta^{-1}(\mathbb{R}^n \times \{0\}) = N \cap U_\beta$ for each $\beta \in B$ (writing $\mathbb{R}^{n+p} = \mathbb{R}^n \times \mathbb{R}^p$). Then $\{(U_\beta \cap N, \varphi_\beta|U_\beta \cap N) : \beta \in B\}$ forms an atlas for N, making it a C^∞ manifold.

A C^∞ map $f : M \to \mathbb{R}^{n+p}$ is an embedding if its image $N := f(M)$ is a C^∞ submanifold of $\mathbb{R}^{n+p}$ and f gives a diffeomorphism of M onto N.

C. A tangent vector at $x \in M$ can be considered as an equivalence class of smooth curves

$$\sigma : (-\varepsilon, \varepsilon) \to M$$

some $\varepsilon > 0$ with $\sigma(0) = x$ where $\sigma_1 \sim \sigma_2$ if $d/dt\, \varphi_\alpha(\sigma_1(t))$ and $d/dt\, \varphi_\alpha(\sigma_2(t))$ agree at $t = 0$ for some (and hence all) charts $(U_\alpha, \varphi_\alpha)$ with $x \in U_\alpha$. The set of all such forms the *tangent space* T_xM to M at x. Any chart around x gives a bijection

$$T_x\varphi_\alpha : T_xM \to \mathbb{R}^n$$

which is used to give T_xM a vector space structure (independent of the choice of chart). When M is $\mathbb{R}^n$ itself, or an open subset of $\mathbb{R}^n$, the tangent space is naturally identified with $\mathbb{R}^n$ itself. A C^r map $f : M \to N$ of manifolds $r \ge 1$ determines a linear map

$$T_xf : T_xM \to T_{f(x)}N$$

(obtained by considering the curves $f \circ \sigma$ for example).

The disjoint union TM of all tangent spaces $\{T_xM : x \in M\}$ has a projection $\tau : TM \to M$ defined by $\tau^{-1}(x) = T_xM$. This is the tangent bundle : it is a C^∞ 2n-dimensional manifold with atlas

$$\{(\tau^{-1}(U_\alpha), T\varphi_\alpha) : \alpha \in A\}$$

where

$$T\varphi_\alpha(v) = (\varphi_\alpha(x), T_x\,\varphi_\alpha(v)) \in \mathbb{R}^n \times \mathbb{R}^n$$

for $v \in T_xM$, for $\{(U_\alpha, \varphi_\alpha) : \alpha \in A\}$ an atlas of M. Our C^r map $f : M \to N$ determines a C^{r-1} map $Tf : TM \to TN$ where Tf restricts to T_xf on T_xM. The assignment of Tf to f is functorial. Note that for U open in $\mathbb{R}^n$ we can identify

TU with $U \times \mathbb{R}^n$.

For $f : M \to \mathbb{R}$ write $df : TM \to \mathbb{R}$ for $v \to T_{\tau(v)}f$.

D. When $f : M \to \mathbb{R}^{n+p}$ is an embedding each $T_x f : T_x M \to \mathbb{R}^{n+p}$ is injective, and $Tf : TM \to \mathbb{R}^{n+p} \times \mathbb{R}^{n+p}$, namely $v \to (x, T_x f(v)$ for $v \in T_x M$, is an embedding. This is used to identify $T_x M$ with its image in $\mathbb{R}^{n+p}$.

Now suppose M is a closed submanifold of $\mathbb{R}^{n+p}$ (i.e. a closed subset, not necessarily compact): the identity map is an embedding and we make the above identifications. There is the *normal bundle* $\nu(M)$ to M:

$$\nu(M) = \{(x,v) \in M \times \mathbb{R}^{n+p} : v \perp T_x M\}.$$

Then $\nu(M)$ has the obvious projection onto M with fibres $\{\nu_x(M) : x \in M\}$ linear spaces of dimension p. By considering the map

$$P : M \times \mathbb{R}^{n+p} \to T_x M$$
$$P(x,v) = P_x(v) \in T_x M$$

where P_x is the orthogonal projection of $\mathbb{R}^{n+p}$ onto $T_x M$, use of the implicit function theorem shows that $\nu(M)$ is a submanifold of $\mathbb{R}^{n+p} \times \mathbb{R}^{n+p}$.

Next we will construct a *tubular neighbourhood* of M in $\mathbb{R}^{n+p}$: this will be used to save worrying about details in the proof of the existence of solutions of S.D.E. The exponential map of $\mathbb{R}^{n+p}$ is the map

$$\exp : T\mathbb{R}^{n+p} = \mathbb{R}^{n+p} \times \mathbb{R}^{n+p} \to \mathbb{R}^{n+p}$$
$$\exp(x,v) = x + v.$$

Restrict it to $\nu(M)$ to get $\psi : \nu(M) \to \mathbb{R}_{n+p}$, say. Differentiate at $(x,0)$ to get $T_{(x,0)}\psi : T_{(x,0)}\, \nu(M) \to \mathbb{R}^{n+p}$. Now $T_{(x,0)}\, \nu(M)$ consists of the direct sum of $T_x M$ (the "horizontal" part) and the tangent space to $\nu_x(M)$ at 0 which can be identified with $\nu_x(M)$ itself since it is a linear space. With this identification

$$T_{(x,0)}\psi(v,w) = v + w$$

so that $T_{(x,0)}\psi$ is a linear isomorphism. The inverse function theorem (applied in charts) implies then that ψ restricts to a C^∞ diffeomorphism of a neighbourhood of $(x,0)$ in $\nu(M)$ onto an open neighbourhood of x in $\mathbb{R}^{n+p}$. Piecing these together for different x in M we see there is an open neighbourhood of the zero section $\{(x,0) \in \nu(M) : x \in M\}$, which is mapped diffeomorphically onto an open neighbourhood of M in $\mathbb{R}^{n+p}$. In particular (using a C^∞ partition of unity on M) there is a smooth function $a : M \to (0, \infty)$ such that ψ gives a diffeomorphism $\psi : \{(x,v) \in \nu(M) : |v| < a(x)\} \to N_a(M)$ where $N_a(M) = \bigcup_{x \in M} \{y \in \mathbb{R}^{n+p} \text{ s.t. } |y-x| < a(x)\}$. This will be called the *tubular neighbourhood of* M *radius* a. Note that on $N_a(M)$ the map giving the distance squared from M

$$y \to d(y,M)^2$$

is C^∞ since $d(y,M) = |$projection on the second factor of $\psi^{-1}(y)|$. Examples to think of are the spheres S^n in $\mathbb{R}^{n+1}$ and the surface of revolution $z = (x^2 + y^2)^{-1}$, for x,y positive, in $\mathbb{R}^3$.

§2. Stochastic differential equations on M

A. For a probability space $(\Omega,\mathcal{F},\mathbb{P})$ consider a filtration $\{\mathcal{F}_t : t \geq 0\}$, assuming for simplicity that each $\mathcal{F}_t$ contains all sets of measure zero in $\mathcal{F}$.

Let $Y : M \times \mathbb{R}^m \to TM$ be C^2 with each $Y(x)(-) := Y(x,-)$ linear from $\mathbb{R}^m$ into T_xM. For $t_0 \geq 0$ let $\{z_t, t \geq t_0\}$ be a continuous semi-martingale on $\mathbb{R}^m$ and let

$$u : \Omega \to M$$

be $\mathcal{F}_{t_0}$-measurable. By a *solution to the stochastic differential equation*

$$dx_t = Y(x_t) \circ dz_t \qquad t \geq t_0 \tag{1}$$

with $\quad x_{t_0} = u$

we mean a sample continuous, adapted process,

$$x_t : \Omega \to M \qquad t_0 \leq t < \xi$$

where $\xi : \Omega \to [0,\infty]$ is a stopping time such that for any C^3 map $f : M \to \mathbb{R}$ and stopping time T with $t_0 \leq T < \xi$ there is the Ito formula

$$f(x_T) = f(u) + \int_{t_0}^{T} T_{x_s} f(Y(x_s) \circ dz_s) \quad \text{a.s.} \tag{2}$$

where $\circ$ denotes Stratonovich integral.

B. The pair (Y,z) will be called a stochastic dynamical system (S.D.S.). Write $[t_0,\xi) \times \Omega$ for $\{(t,\omega) \in [t_0,\infty) \times \Omega : t < \xi(\omega)\}$. When $M = \mathbb{R}^n$ this is equivalent to the usual definition and there is the basic existence and uniqueness theorem:

<u>Existence and Uniqueness Theorem for $\mathbb{R}^n$</u>

Given z,Y, and u as above for $M = \mathbb{R}^n$ *there is a unique maximal solution to* (1)

$$x_t : \Omega \to M \qquad t_0 < t < \zeta$$

defined up to an explosion time ζ *which is almost surely positive. On* $\{\zeta < \infty\}$, *as* $t \uparrow \zeta(\omega)$ *so* $x_t(\omega) \to \infty$ *in* $\mathbb{R}^n$ a.s.

Uniqueness holds in the sense that if $\{y_t : t_0 \leq t < \xi\}$ *is any other solution starting from* u *at time* t_0 *then* $\xi \leq \zeta$ *almost surely and* $x \mid [t_0,\xi) \times \Omega = y$ *almost surely.* Indeed this is easily deduced from the usual case of an Ito equation with globally Lipschitz coefficients: for $R = 1,2,\ldots$ take a C^∞ map φ_R :

$\mathbb{R}^n \to \mathbb{R}(>0)$ identically one on the ball $B(0;R)$ about 0 radius R and with compact support. Set $Y_R = \varphi_R Y$. In its Ito form

$$dx_t = Y_R(x_t) \circ dz_t$$

has Lipschitz coefficients and has a solution x^R, say, defined for all $t \geq t_0$, starting at u at time t_0. If S_R is its first exit time from $B(0;R)$ then $x_{R'}$ agrees with x_R up to time S_R if $R' > R$, and so a limiting solution x to (1) can be constructed as required.

C. <u>Existence and uniqueness theorem for M:</u> We will now see why the same holds for a general manifold M. The straightforward proofs of this tend to be either tedious or unsatisfying, so we will try to avoid tedium by some geometrical constructions.

First of all for there to be uniqueness there clearly has to be an ample supply of C^3 functions f. Since we will have to use that fact we may as well use a strong form of it, namely: *Whitney's embedding Theorem. There is a* C^∞ *embedding* $\varphi : M \to \mathbb{R}^{n+p}$ *of* M *onto a closed submanifold of* $\mathbb{R}^{n+p}$ *where* $p = n + 1$.

The exact codimension p will not be important, and the proof then, for some p, is a straighforward argument using partitions of unity and the implicit function theorem.

Taking such an embedding φ, identify M with its image, so that we can consider it as a submanifold of $\mathbb{R}^{n+p}$. It is easy, using C^∞ partitions of unity to extend Y to a C^2 map $\mathrm{Y} : \mathbb{R}^{n+p} \times \mathbb{R}^m \to \mathbb{R}^{n+p} \times \mathbb{R}^{n+p}$. However we will do a more explicit extension below. Since a C^2 function on $\mathbb{R}^{n+p}$ restricts to a C^2 function on M a solution to (1) is a solution to

$$dx_t = \mathrm{Y}(x_t) \circ dz_t \qquad (\bar{1})$$

in $\mathbb{R}^{n+p}$ (strictly speaking, if x_t satisfies $(\bar{1})$ then $\varphi(x_t)$ satisfies (1) since $f \equiv \bar{f} \circ \varphi$ is C^3). Thus uniqueness holds for (1) since it does for $(\bar{1})$. Conversely any C^3 map $f : M \to \mathbb{R}$ can be extended to a C^3 map $\bar{f} : \mathbb{R}^{n+p} \to \mathbb{R}$, so a solution to $(\bar{1})$ which lies on M for all time must solve (1). To prove existence of a maximal solution it is enough therefore to show that if $\{x_t : t_0 \leq t < \xi\}$ is the maximal solution to (1) with $x_{t_0} = u$ for $u : \Omega \to M$ then x_t lies in M almost surely for all t. In fact it is enough to do that for just one extension Y. In particular take a tubular neighbourhood of M radius $a : M \to \mathbb{R}(>0)$ as described in §1D; choose $R > 0$ and let

$a_R = \inf\{a(x) : x \in M \cap B(0;R+1)\} > 0.$

Choose smooth $\lambda : \mathbb{R}^{n+p} \to \mathbb{R}(\geq 0)$ with support in $B(0;\mathbb{R}+1)$ and identically one on $B(0,R)$, and smooth $\mu : [0,\infty) \to \mathbb{R}(\geq 0)$ with $\mu(x) = 1$ for $|x| \leq \frac{1}{2} a_R{}^2$ and, $\mu(x) = 0$ for $|x| > a_R{}^2$.

Let $\pi : N_a(M) \to M$ map a point to the nearest point of M to it. Using the identification ψ of $N_a(M)$ with part of $\nu(M)$ this is seen to be C^∞. Define $Y_R : \mathbb{R}^{n+p} \times \mathbb{R}^m \to \mathbb{R}^{n+p}$ by

$$Y_R(x)(e) = 0 \text{ for } x \notin N_a(M), \ e \in \mathbb{R}^m$$

$$Y_R(x)(e) = \lambda(x)\mu(d(x,M)^2)\, Y(\pi(x))(e) \text{ for } x \in N_a(M), \ e \in \mathbb{R}^m .$$

Define $f : \mathbb{R}^{n+p} \to \mathbb{R} \geq 0$ by $f(x) = \lambda(x)\mu(d(x,M)^2)$.

Inside $B(0;R)$ the map f is constant on the level sets of $d(-,M)$ while $Y_R(x)(e)$ is tangent to these sets. Therefore

$$Df(x)\,(Y_R(x)e) = 0 \qquad x \in B(0,R), \ e \in \mathbb{R}^m$$

and so any solution $\{x_t : t_0 \leq t < S\}$ to (I) starting on M satisfies $f(x_t) = 0$ almost surely until it first exits from $B(0;R)$. It therefore stays on M almost surely until this time. Since R was arbitrary we are done:

The uniqueness and existence up to an explosion time result of §2B *holds exactly as stated when* $\mathbb{R}^n$ *is replaced by a manifold* M.

D. It is now easy to see:

If N *is a closed submanifold of* M *and* $Y(x)e$ *lies in* T_xN *for all points* x *of* N *then any solution to* (1) *which starts on* N *almost surely stays on* N *for its lifetime.*

Indeed we now know there is a solution to the restriction of (1) to N which exists until it goes out to infinity on N, or for all time. This solves (1) on M and is certainly maximal: it must therefore be the unique maximal solution to (1).

E. Usually our equations will be of the form

$$dx_t = X(x_t) \circ dB_t + A(x_t)dt \qquad (1)'$$

where $X : M \times \mathbb{R}^m \to TM$ is as Y was before, $\{B_t : t \geq 0\}$ is a Brownian motion on $\mathbb{R}^m$, and A is a vector field on M, so $A(x) \in T_xM$ for each $x \in M$. This fits into the scheme of (1) with $z_t = (B_t,t) \in \mathbb{R}^m \times \mathbb{R} = \mathbb{R}^{m+1}$ and $Y(x)(e_1,r) = X(x)(e_1) + rA(x)$ when $(e_1,r) \in \mathbb{R}^m \times \mathbb{R}$. However in this case it is of course only necessary to assume that X is C^2 and A is C^1.

For an orthonormal base $e_1,\dots,e_m$ of $\mathbb{R}^m$ write B_t as $\sum_{r=1}^{m} B^r{}_t\, e_r$, so the

$\{B^r{}_t : t \geq 0\}$, $r = 1,2,\ldots,m$ are independent Brownian motions on $\mathbb{R}$. Let $X^1,\ldots,X^m$ be the vector fields

$$X^p(x) = X(x)(e_p) \qquad p = 1,\ldots,m.$$

Then (2) becomes

$$dx_t = \sum_p X^p(x_t) \circ dB^p{}_t + A(x_t)dt \tag{2}$$

§3. An Ito formula

A. We will need (2) in Ito form. One version using covariant derivatives will be given below in § 2 of Chapter II. However it will be useful to have a form which does not depend on a choice of connection for M e.g. when we need to consider equations on principal bundles it would be a nuisance to have to describe some way of covariant differentiation of vector fields on principal bundles.

For equation (1) and each $e \in \mathbb{R}^m$ let

$$(t,x) \to S(t,x)e$$

be the flow of the vector field $Y_e := Y(-)(e)$ on M, defined on some neighbourhood of $\{0\} \times M$ in $\mathbb{R} \times M$.

A vector field, e.g. Y_e, acts on $f : M \to \mathbb{R}$ to give

$$Y_e f : M \to \mathbb{R}$$

by $\quad Y_e f(x) = df(Y_e(x)).$

Thus, for our C^2 function f

$$d/dt\, f(S(t,x)e) = df(Y_e(S(t,x)e))$$
$$= Y_e f\,(S(t,x)e) \tag{3}$$

and

$$d^2/dt^2\, f(S(t,x)e) = Y_e\, Y_e f(S(t,x)e). \tag{4}$$

At $t = 0$ this gives linear and bilinear maps which we write

$$d/dt\, f\circ S(t,x)\,\Big|_{t=0} \in \mathbb{L}(\mathbb{R}^m;\mathbb{R})$$

and

$$d^2/dt^2\, f\circ S(t,x)\,\Big|_{t=0} \in \mathbb{L}(\mathbb{R}^m, \mathbb{R}^m;\mathbb{R})$$

<u>Proposition 3A</u> (Global Ito formula) *For a solution* $\{x_t : 0 \leq t < \xi\}$ *of* (1), *if* $f : M \to \mathbb{R}$ *is* C^2 *and* T *is a stopping time less than* ξ *then, using Ito integrals, almost surely:*

$$f(x_T) = f(x_0) + \int_0^T d/dt\, f \circ S(t, x_r)|_{t=0} (dz_r) + \tfrac{1}{2}\int_0^T d^2/dt^2\, f \circ S(t, x_r)|_{t=0} (dz_r, dz_r) \qquad (5)$$

In another form: if $e_1, \dots, e_m$ *is a base for* $\mathbb{R}^m$ *and*

$$z_t = \sum_p z^p{}_t\, e_p, \text{ with } Y^p = Y(-)e_p$$

then

$$f(x_T) = f(x_0) + \int_0^T \sum_p Y^p f(x_r) dz^p{}_r + \tfrac{1}{2} \int_0^T \sum_p Y^p Y^q f(x_r)\, d\langle z^p, z^q \rangle_r \qquad (6)$$

Proof. If f is C^4 the Stratonovich correction term for (2) is

$$\tfrac{1}{2} \int_0^T \sum_{p,q} dv^p{}_t\, dz^q{}_t \quad \text{where } v^p{}_t = df(Y^p(x_t)) = Y^p f(x_t),$$

and (6), and hence (5), follows by calculating $dv^p{}_t$ using (2) applied to $Y^p(t)$. For f only C^2 the easiest way to proceed is to embed M in some $\mathbb{R}^{n+p}$, extend f to some C^2 function $\bar{f}$, extend Y to $\bar{Y}$ as before and write

$$dy_t = \bar{Y}(y_t) \circ dz_t$$

in Ito form. Then apply the usual Ito formula to $\bar{f}$ to obtain (6) after restriction. //

B. For equation (2)'

$$dx_t = X(x_t) \circ dB_t + A(x_t)dt$$

form (6) becomes

$$f(x_T) = f(x_0) + \int_0^T df(X(x_r)) \circ dB_r + \int_0^T \mathcal{A} f(x_r) dr \qquad (6)'$$

where

$$\mathcal{A} f = \tfrac{1}{2} \sum_{p=1}^m X^p X^p f + Af.$$

Using the results for $\mathbb{R}^{n+p}$ after embedding M in $\mathbb{R}^{n+p}$ we see *the solutions of* (2)' *form a Markov process with differential generator* $\mathcal{A}$.

C. A sample continuous stochastic process $(y_t : 0 \le t < S)$ on M is a *semi-martingale* if $(f(y_t) : 0 \le t < \xi)$ is a semi-martingale in the usual sense whenever $f : M \to \mathbb{R}$ is C^2, (Schwartz [90]). The above formulae show that *our solutions* x_t *to* (1) *are semi-martingales.* There is also the converse result, observed by Schwarz in [90], every continuous semi-martingale y on M is the solution of some equation like (1): indeed given y take some embedding $\varphi : M \to \mathbb{R}^{n+p}$, some p. Let $z_t = \varphi(y_t)$. Then z is a semi-martingale, and if $P : M \times \mathbb{R}^{n+p} \to TM$ is the orthogonal projection map, as in §10, then y is a solution to

$$dx_t = P(x_t) \circ dz_t \qquad (7)$$

One easy way to see this is to use the projection $\pi : N_a(M) \to M$ of a tubular neighbourhood as in §1D: then, for $x_t = y_t$, equation (7) is the differential form of the equation $\pi(z_t) = z_t$.

D. For a C^1 map $f : M \to N$ of manifolds, stochastic dynamical systems (X,z) on M and (Y,z) on N are said to be *f-related* if $T_x f(X(x)e) = Y(f(x))e$ for $x \in M$ and $e \in \mathbb{R}^m$. The corresponding result for O.D.E. together with equation (5) immediately shows that *if* f *is* C^2 and $\{x_t : 0 \le t < \xi\}$ *is a solution to* $dx_t = X(x_t) \circ dz_t$, then $\{f(x_t) : 0 \le t < \xi\}$ is a solution to $dy_t = Y(y_t) \circ dz_t$ on N. For a C^3 map f it is immediate from the definition.

§4. Solution flows

A. Flows of stochastic dynamical systems were discussed in Kunita's Stochastic Flow course in 1982 , [67], and I do not want to go into a detailed discussion of their existence and properties. However I would like to describe briefly a method which gives the main properties of these flows rather quickly, and also mention some rather annoying gaps in our knowledge. The first problem is to find nice versions of the map

$$(t,x,\omega) \to F_t(x,\omega) \in M \qquad \omega \in \Omega$$

which assigns to $x \in M$ the solution to the S.D.E. starting at x at time 0. In fact throughout this section we assume $z_t = (B_t,t) \in \mathbb{R}^{m+1}$ so that we are really dealing with (1)'.

B. For M compact, or more generally for Y of compact support, or for $M = \mathbb{R}^n$ with Y having all derivatives bounded it is not difficult to show that Totoki's extension of Kolmogorov's theorem can be used to obtain a version of F such that

(i) For all $x \in M$, $\{F_t(x,-) : t \ge 0\}$ solves $dx_t = Y(x_t) \circ dz_t$ with $F_0(x,-) = x$.

(ii) Each map $[0,\infty) \times M \to M$ given by

$$(t,x) \to F_t(x,\omega) \text{ is continuous.}$$

This was the method used by Blagovescenskii and Freidlin, for example see

[43], [59], [9], [67], [78]. The extension of Kolmogorov's theorem used is:

Let (M,d) *be a complete metric space. Suppose for* $x : [0,1] \times \ldots \times [0,1]$ (p-times) $\to \mathcal{L}^0(\Omega,\mathcal{F};M)$ *there exists* $\alpha, \beta, \gamma > 0$ *such that for all* $\delta > 0$ *and* s,t *in* $[0,1]^p$

$$\mathbb{P}\{d(x_s,x_t) > \delta\} \le \beta\delta^{-\alpha} |s-t|^{p+\gamma}$$

then x *has a sample continuous version.*

The necessary estimates are most easily obtained by embedding M in some $\mathbb{R}^{n+p}$ and extending Y as before.

C. For M compact it is possible to obtain differentiability, diffeomorphism, and composition results by considering an induced stochastic differential equation on the Hilbert manifold of H^s diffeomorphisms of M for $s > \frac{1}{2}n + 3$: the solution of this equation starting at the identity map being a version of F_t, see [43]. Rather than discuss the Hilbert manifold structure of these groups of diffeomorphisms it is possible to embed M in some $\mathbb{R}^{n+p}$ and extend the S.D.S. over $\mathbb{R}^{n+p}$ as before, to have compact support. A flow for the extended system will restrict to one for the system of M. For the extended system we consider the space of diffeomorphisms of class H^s of $\mathbb{R}^{n+p}$ which are the identity outside of a fixed bounded domain U, containing the support of the extended system. We will describe this rather briefly, see [240] for details.

Suppose therefore there is the system on $\mathbb{R}^n$

$$dx_t = Y(x_t) \circ dz_t$$

where Y has compact support in U, for U open, bounded, and with smooth boundary.

For $s > n/2$ set

$H^s{}_U(\mathbb{R}^n; \mathbb{R}^p) = \{f \in H^s(\mathbb{R}^n;\mathbb{R}^p)$ with supp$f \subset U\}$ where $H^s(\mathbb{R}^n;\mathbb{R}^p)$ is the completion of the space $C^\infty{}_0(\mathbb{R}^n; \mathbb{R}^p)$ of C^∞ functions with compact support under $\langle\ \ \rangle_s$ where

$$\langle f,g\rangle_s = \sum_{|\alpha|\le s} \int_{\mathbb{R}^n} \langle D^\alpha f(x), D^\alpha g(x)\rangle dx$$

where the sum is over multi-indices $\alpha = (\alpha_1,\ldots,\alpha_n)$ with $D^\alpha = \partial^{|\alpha|}/(\partial x_1{}^{\alpha_1} \ldots \partial x_n{}^{\alpha_n})$. Because $s \ge n/2$ the evaluation map $x \to f(x)$ is continuous on H^s and so $H^s{}_U$ is a well defined closed subspace of H^s.

For $s > n/2 + 1$ set

$\mathcal{D}^s{}_U = \{f : \mathbb{R}^n \to \mathbb{R}^n$ s.t. f is a C^1 diffeomorphism and $Id - f \in H^s{}_U(\mathbb{R}^n;\mathbb{R}^n)\}$ where

Id refers to the identity map. Since diffeomorphisms are open in the C^1 topology $\mathfrak{D}^s{}_U$ is an open subset of the affine subspace $1d + H^s{}_U(\mathbb{R}^n;\mathbb{R}^n)$. It is therefore a Hilbert manifold with chart $f \to f - 1d$. Moreover

1. $\mathfrak{D}^s{}_U$ is a topological group under composition
2. For $h \in \mathfrak{D}^s{}_U$ right multiplication

$$R_h : \mathfrak{D}^s{}_U \to \mathfrak{D}^s{}_U$$
$$f \to f \circ h$$

is C^∞.

3. For $k = 0,1,2,\ldots$ composition

$$\varphi_k : \mathfrak{D}^{s+k}{}_U \times \mathfrak{D}^s{}_U \to \mathfrak{D}^s{}_U$$
$$(f,h) \to f \circ h$$

is C^k.

4. For $k = 1,2,\ldots$ inversion $h \to h^{-1}$ considered as a map

$$\mathfrak{I}_k : \mathfrak{D}^{s+k}{}_U \to \mathfrak{D}^s{}_U$$

is C^k.

Now define the right invariant stochastic dynamical system

$$dh_t = \Upsilon(h_t) \circ dz_t$$

on $\mathfrak{D}^s{}_U$, always assuming $s > n/2 + 1$, by taking

$$\Upsilon(h) : \mathbb{R}^m \to T_h \mathfrak{D}^s{}_U \simeq H^s{}_U \qquad h \in \mathfrak{D}^s{}_U$$

to be

$$\Upsilon(h)(e) = \varphi_2(Y(-)(e),h)$$

treating $x \to Y(x)(e)$ as in $H_U{}^{s+2}$. Thus Υ is C^2 by (3) above. We see

$$\Upsilon(1d)(e) = Y(-)(e)$$

and

$$\Upsilon(h)(e) = DR_h(1d)(\Upsilon(Id)e) \qquad h \in \mathfrak{D}^s{}_U$$

so that Y is right invariant.

Proposition 4C. *The solution* h_t *to* $dh_t = \Upsilon(h_t) \circ dz_t$ *starting from* 1d *exists for all time, and* $F_t(-,\omega) \equiv h_t(\omega)(-)$ *gives a flow for our equation on* $\mathbb{R}^n$ *lying in* $\mathfrak{D}^s{}_U$*. In particular there is a* C^∞ *version of this flow, such that* $t \to F_t(-,\omega)$ *is almost surely continuous into the* C^∞ *topology.*

Proof: A maximal solution certainly exists up to some predictable stopping time ξ say, with $\xi > 0$ almost surely: the theory for equations of this type on open subsets of a Hilbert space goes through just as in finite dimensions, (provided one always uses uniform estimates, i.e. uses basis free notation). Choose a predictable stopping time T with $0 \le T \le \xi$. By the right invariance of the system, for $h' \in \mathfrak{D}^s{}_U$

$$(s,\omega) \to h_s(\omega).h'$$

is a solution starting at h', so

$$(s,\omega) \to h_{s-T(\omega)}(\theta_T(\omega))h_{T(\omega)}(\omega)$$

where θ_T is the shift, is equivalent to $h_s(\omega)$ for $s > T(\omega)$. Thus

$$\xi(\omega) \ge T(\omega) + T(\theta_T(\omega)).$$

Iterating this

$$\xi(\omega) \ge \sum_{k=0}^{\infty} T\,(\theta_{kT}(\omega)).$$

Since $\{T\circ\theta_{kT}\}_{k=0}^{\infty}$ is an i.i.d. sequence we see $\xi = \infty$ almost surely. To see that $h_t(\omega)(x_0)$ is a solution to the equation on $\mathbb{R}^m$ starting at x_0 observe that the two equations are f-related for f the evaluation map $f(h) = h(x_0)$, and use the result of §3D. Finally since $s > n/2 + 1$ was otherwise arbitrary we see that almost surely $s \to h_s(\omega)$ lies in the space of continuous maps into

$\bigcap_{s=1}^{\infty} \mathfrak{D}^s{}_U$ with its induced topology: but this is just a subset of the space of C^∞ diffeomorphisms, with its relative topology. //

By embedding we deduce the existence of a C^∞ flow of diffeomorphisms $F_t(-,\omega) : M \to M$ continuous in t into the C^∞ topology when M is compact.

Results like those of Bismut and of Kunita on the stochastic differential equations for the inverses of the flows and compositions are easily obtained via the infinite dimensional Ito formula, see [24].

D. For non-compact manifolds M we can embed M and approximate the extended S.D.S. by equations with compactly supported coefficients as in §1F. This gives [66], [24a] Kunita's results on the existence of partial flows:

There is an explosion time map $\zeta : M \times \Omega \to (0,\infty]$ *and a partially defined flow* F *for* $dx_t = Y(x_t) \circ dz_t$ *such that if*

$$M(t)(\omega) = \{x \in M : t < \zeta\,(x,\omega)\} \qquad \omega \in \Omega$$

then for all $\omega \in \Omega$

(i) $M(t)(\omega)$ *is open in* M i.e. $\zeta(-,\omega)$ *is* l.s.c.

(ii) $F_t(\omega) : M(t)(\omega) \to M$ *is defined and is a* C^∞ *diffeomorphism onto an open subset of* $\mathbb{R}^n$.

(iii) *For each* x *in* M, $\zeta(x)$ *is a stopping time and* $\{F_t(x,-) : 0 \le t < \zeta(x)\}$ is *a maximal solution. Moreover if* K *is compact in* M, *if* $\zeta(K)(\omega) = \inf \{\zeta(x)(\omega) : x \in K\}$ *then on* $\zeta(K)(\omega) < \infty$, *for any* $x_0 \in M$, $\sup_{x \in K} d(x_0, F_t(x,w)) \to \infty$ *almost surely as* $t \uparrow \zeta(K)(\omega)$.

(iv) *the map* $s \to F_s(-)(\omega)$ *is continuous from* $[0,t]$ *into the* C^∞ *topology of functions on* $M(t)(\omega)$.

When we can choose $\xi \equiv \infty$ the system is said to be *strongly complete,* or *strictly conservative.* The standard example of a system which is complete but not strongly complete is

$$dx_t = dB_t$$

on $M = \mathbb{R}^2 - \{0\}$. It is complete since Brownian motions starting outside of 0 in $\mathbb{R}^2$ almost surely never hit 0. However the flow would have to be

$$F_t(x,\omega) = x + B_t(\omega)$$

so that

$$\xi(x,\omega) = \inf\{t : B_t(\omega) = -x\}.$$

Using the fact that the solution to a complete system exists for all time even when starting with a given random variable $x_0 : \Omega \to M$ (independent of the driving motion of course) it follows [27] that given completeness

$$\mu \otimes \mathbb{P}\{(x,\omega) : \xi(x,\omega) < \infty\} = 0$$

for all Borel measures μ on M. In particular for almost all $\omega \in \Omega$ each open set $M(t)(\omega)$ has full measure in M.

Even when the system is strongly complete the maps $F_t(-,\omega)$ may not be surjective: the standard example is the 2-dimensional Bessel process on $M = (0,\infty)$, see [43], [44], [59]. It seems difficult to decide when strong completeness holds. It depends on Y, not just on the generator $\mathcal{A}$: as observed by Carverhill the Ito equation

$$dz_t = (z_t/|z_t|)\, dB_t$$

on $\mathbb{C} - \{0\}$ is strongly complete (using complex multiplication for B a Brownian motion on $\mathbb{C} \simeq \mathbb{R}^2$), yet its solutions are just Brownian motions on $\mathbb{R}^2 - \{0\}$. A general problem is therefore to find conditions on $\mathcal{A}$ and M which ensure that there exists a choice of Y, giving a diffusion with generator $\mathcal{A}$, which is strongly complete. Anticipating some concepts from the next chapter: for a complete Riemannian manifold M does a lower bound on the Ricci curvature ensure strong completeness of the canonical SDS on

OM? What conditions on M ensure that it admits a strongly complete SDS with $\mathcal{A} = \frac{1}{2}\Delta$? More generally is strong completeness of Y implied by the existence of a uniform cover for Y in the sense of [43], [24a] ?

E. Assume now that the system is complete. Even if it is not strongly complete we can formally differentiate solutions in the space directions, to get "derivatives in probability" rather than almost sure derivatives, [43]. This was done for $M = \mathbb{R}^n$ in [57] to get L^2-derivatives given strong conditions on Y. These agree with the almost sure derivatives in the derivatives of the flow $F_t(-,\omega)$ where the latter exists, and we will write them here as $T_{x_0}F_t(-,\omega)$: $T_{x_0}M \to T_{x_t(\omega)}M$, as if the flow did exist.

For $v \in T_{x_0}M$ set $v_t = T_{x_0}F_t(v,-)$. Then, either by embedding in some $\mathbb{R}^{n+p}$ or by arguing directly, $\{v_t : t \geq 0\}$ satisfies an equation

$$dv_t = \delta Y(v_t) \circ dz_t$$

on TM, where now $\delta Y : TM \times \mathbb{R}^m \to T(TM)$ and is given by

$$\delta Y(v)e = \alpha\, TY_e(v)$$

for $Y_e = Y(-)e : M \to TM$ and $\alpha : TTM \to TTM$ the involution which over a chart which represents TU as $U \times \mathbb{R}^n$ and so TTU as $(U \times \mathbb{R}^n) \times (\mathbb{R}^n \times \mathbb{R}^n)$ is given by $(x,u,v,\omega) \to (x,v,u,\omega)$.

Thus in these co-ordinates

$$\delta Y((x,u))e = (x,u,Y(x)e,DY_e(x)(u)).$$

CHAPTER II

SOME DIFFERENTIAL GEOMETRY FOR PRINCIPAL BUNDLES AND CONSTRUCTIONS OF BROWNIAN MOTION

§1. Connections on principal bundles and covariant differentiation

A. A Lie group G is a C^∞ manifold with a group structure such that the maps

$$G \times G \to G \qquad \text{and} \qquad G \to G$$
$$(g_1, g_2) \to g_1 g_2 \qquad\qquad g \to g^{-1}$$

are C^∞. Standard examples include the circle S^1, the three sphere S^3 (the multiplicative group of quaternions with unit norm), orthogonal groups O(n) which have SO(n) as connected component of the identity, and non-compact groups $(\mathbb{R}^n, +)$ and GL(n).

A *right action* of G on a manifold M is a C^∞ map $M \times G \to M$ usually written $(x,g) \to x.g$ such that $x.1 = x$ (for 1 the identity element), and $(x.g_1).g_2 = x.(g_1.g_2)$. Examples are the action of G on itself by right multiplication. The natural action of GL(n) on $\mathbb{R}^n$ is a *left action*, defined similarly. Note that $x \to x.g$ is a diffeomorphism of M, which we will write as $R_g : M \to M$, with $L_g : M \to M$ for a left action.

Let g be the tangent space to G at 1. Then the left action gives a diffeomorphism (trivialization of TG)

$$Y : G \times g \to TG$$
$$Y(g)(v) = T_1 L_g(v).$$

Thus a semi-martingale z on g gives an S.D.E. on G.

The map $v \to Y_v := Y(-)(v)$ gives a bijection between g and the space of left-invariant vector fields on G. For any two vector fields X^1, X^2 on a manifold M there is another vector field $[X^1, X^2]$, the *Lie bracket*, determined by $[X^1, X^2]f = X^1X^2f - X^2X^1f$ for $f : M \to \mathbb{R}$ a C^2 function. This gives a Lie algebra structure. For a Lie group the bracket of two left invariant vectors remains left invariant, so there is an induced Lie bracket on g s.t.

$$Y_{[v_1,v_2]} = [Y_{v_1}, Y_{v_2}].$$

Let $\{\sigma(t) : t \in \mathbb{R}\}$ be the curve in G with $\sigma(0) = 1$ and $d\sigma/dt = Y_v(\sigma(t))$. It is a 1-parameter subgroup and usually written $\sigma(t) = \mathrm{Exp}(tv)$. This determines

$$\mathrm{Exp} : g \to G$$

by $v \to \mathrm{Exp}\, v$, not to be confused with other exponential maps (e.g. in §1D and §2B below).

B. A principal G-bundle over M is a map $\pi : B \to M$ of differentiable manifolds which is surjective, where B has a right G-action s.t. $\pi(b.g) = \pi(b)$ all $b \in B$, $g \in G$, and such that there is an open cover $\{U_\alpha : \alpha \in A\}$ of M for which there exist C^∞ diffeomorphisms (local trivializations)

$$\theta_\alpha : \pi^{-1}(U_\alpha) \to U_\alpha \times G$$

of the form

$$\theta_\alpha(b) = (\pi(b), \theta_{\alpha,x}(b)) \qquad x = \pi(b)$$

with

$$\theta_{\alpha,x}(b.g) = \theta_{\alpha,x}(b).g.$$

The simplest example is the product bundle $M \times G$ with obvious right action. An important example is the full *linear frame bundle* of M, $\pi : GLM \to M$, where GLM consists of all linear isomorphisms $u : \mathbb{R}^n \to T_xM$ for some $x \in M$, with π mapping u to the relevant x. The right action is just composition $u.g(e) = u(ge)$ for $e \in \mathbb{R}^n$. Each element u is called a *frame* since it can be identified with the base $(u_1, \ldots, u_n)$ of T_xM where $u_p = u(e_p)$ for $e_1, \ldots, e_n$ the standard base for $\mathbb{R}^n$.

Given principal bundles $\pi_i : B_i \to M$ with groups G_i for $i = 1,2$ a C^∞ homomorphism is a C^∞ map $h : B_1 \to B_2$ such that $h(b_1.g_1) = h(b_1).h^0(g_1)$ for some smooth group homomorphism $h^0 : G_1 \to G_2$. Thus every principal bundle is by definition locally isomorphic to the trivial (:= product) bundle, (with h^0 = identity map).

C. For a principal G-bundle $\pi : B \to M$ the tangent space TB has a naturally defined subset : the *vertical tangent bundle,* or *bundle along the fibres,*

$$VTB = \{v \in TB : T\pi(v) = 0\}.$$

A *connection on* B is an assignment of a complementary "horizontal" tangent bundle, HTB, invariant under the action of G. One way to do this is to take a g-valued 1-form $\tilde{\omega}$ i.e.

$$\tilde{\omega} : TB \to g$$

is smooth, and each restriction $\tilde{\omega}_b : T_bB \to g$, $b \in B$ is linear, with

(i) $\tilde{\omega} \circ TR_g = ad(g^{-1}) \circ \tilde{\omega} \qquad g \in G$ (8)

where $ad(g^{-1}) : g \to g$ is the *adjoint action*, namely the derivative at 1 of the map $G \to G$, $a \to g^{-1}ag$, and

(ii) $\tilde{\omega}(A^*(b)) = A \qquad b \in B, A \in g$ (9)

where A^* is the (vertical) vector field on B defined by

$$A^*(b) = d/dt\,(b.\exp tA)\,|_{t=0}.$$

Such an $\tilde{\omega}$ is called a *connection form.* Given such,one can define a horizontal tangent bundle by

$$HTB = \{v \in TB : \tilde{\omega}(v) = 0\}.$$

Then

(a) $T_bB = HT_bB \oplus VT_bB \qquad$ each $b \in B$

and

(b) $TR_g(HT_bB) = HT_{b.g}B \qquad b \in B, g \in G.$

Conversely given HTB satisfying (a) and (b) and smooth there exists a connection form inducing it. It is easy to construct connections by partitions of unity: but without additional structure there is no canonical choice.

For each trivialization $(U_\alpha, \theta_\alpha)$ there is a local section

$$s_\alpha : U_\alpha \to \pi^{-1}(U_\alpha) \subset B$$

$$s_\alpha(x) = \theta^{-1}{}_{\alpha,x}(1)$$

This can be used to pull back a connection form $\tilde{\omega}$ to a g-valued 1-form $\tilde{\omega}_\alpha = s_\alpha{}^*(\tilde{\omega})$ given by the composition $\tilde{\omega}_\alpha = \tilde{\omega} \circ Ts_\alpha : TU_\alpha \to g$. For a connection on GL(M) the components of this will give the *Christoffel symbols:* indeed a chart $(U_\alpha, \varphi_\alpha)$ for M determines a trivialization which maps u to $(T_x\,\varphi_\alpha) \circ u : \mathbb{R}^n \to \mathbb{R}^n$ for a frame u at x. Then $s_\alpha(x) = (T_x\varphi_\alpha)^{-1} : \mathbb{R}^n \to T_xM$. Define

$$\Gamma : \varphi_\alpha(U_\alpha) \to \mathbb{L}(\mathbb{R}^n; g) = \mathbb{L}(\mathbb{R}^n; L(\mathbb{R}^n;\mathbb{R}^n))$$

by

$$\Gamma(\varphi_\alpha(x))v = s_\alpha{}^*(\tilde{\omega})(T_x\varphi_\alpha{}^{-1}(v)) \qquad v \in \mathbb{R}^n \tag{10}$$

giving the classical Christoffel symbols $\Gamma^i{}_{jk}$

$$\Gamma^i{}_{jk}(y) = \langle\Gamma(y)(e_j)(e_k),e_i\rangle \tag{11}$$

where $e_1,\ldots,e_n$ is the standard base for $\mathbb{R}^n$. The connection is *torsion free* if $\Gamma(y)(v_1)(v_2) = \Gamma(y)\,(v_2)(v_1)$ or equivalently if $\Gamma^i{}_{jk} = \Gamma^i{}_{kj}$.

§2. Horizontal lifts, Covariant derivatives, geodesics and a second form of the Ito formula.

A. A connection for $\pi : B \to M$ determines a horizontal lifting map

$$H_b : T_{\pi(b)}M \to T_bB$$

which is the inverse of the restriction of $T_b\pi$ to HT_bB, a linear isomorphism. It also gives a way of horizontally lifting smooth curves in M to curves in B; a smooth curve σ in B is *horizontal* if $\dot\sigma(t) \in HTB$ for all t or equivalently if $\tilde{\omega}(\dot\sigma(t)) = 0$ all t:

For a piecewise C^1 *curve* $\sigma : [0,T) \to M$ and $b_0 \in \pi^{-1}(\sigma(0))$ *there exists a unique horizontal curve* $\sigma^\sim$ *in* B *with* $\sigma^\sim(0) = b_0$ *and* $\pi(\sigma^\sim(t)) = \sigma(t)$ *for all* t (i.e. $\sigma^\sim$ is a *lift* of σ).

In fact given a trivialization (U_α,θ_α), while $\sigma(t)$ is in U_α, for $\sigma^\sim$ to be a lift $\theta_\alpha(\sigma^\sim(t))$ has to have the form

$$\theta_\alpha(\sigma^\sim(t)) = (\sigma(t),g(t)) \in U_\alpha \times G$$

and then it will be horizontal if and only if

$$d/dt\ g(t) = -\ TR_{g(t)}\,(\tilde{\omega}_\alpha(\sigma(t))(\dot\sigma(t))) \tag{12}$$

so that local existence and uniqueness follow immediately from standard O.D.E. theory on G. To see how (12) arises note that the axioms for $\tilde{\omega}$ imply that for (v,A) in $T_{(x,a)}\,(U_\alpha \times G)$

$$\tilde{\omega}\circ(T\theta_\alpha)^{-1}(v,A) = ad(a^{-1})\tilde{\omega}_\alpha(x)v + TL_{a^{-1}}A \tag{13}$$

Note that by uniqueness and invariance, if $g \in G$ the horizontal lift of σ starting at $b_0.g$ is just $t \to \sigma^\sim(t)g$.

The only real difficulty in extending this construction to lifts of semi-martingales y_t (e.g. by treating (12) as a Stratonovich S.D.E. when $\sigma(t)$ is replaced by y_t) is to make sure the lifted process does not explode before the original one does see [77], [34], and [43] p. 175. However we will not need this.

B. For simplicity restrict attention now to an *affine connection* for M i.e. a connection on GLM. Given a piecewise C^1 curve $\sigma : [0,T) \to M$ and a vector $v_0 \in T_{\sigma(0)}M$ define the *parallel translate*

$$//_t(v_0) \in T_{\sigma(t)}M \qquad 0 \le t \le T$$

of v_0 along σ by

$$//_t(v_0) = \sigma^\sim(t) b_0^{-1}(v_0) \tag{14}$$

where $\sigma^\sim(t)$ is the horizontal lift of σ through $b_0 \in \pi^{-1}(\sigma(0))$, the result being independent of the choice of b_0.

Then $//_t : T_{\sigma(0)}M \to T_{\sigma(t)}M$ is a linear isomorphism.

If we now have *vector field* W *along* σ i.e. $W : [0,T] \to TM$ with $W(t) \in T_{\sigma(t)}M$ for each t, define its covariant derivative along σ by

$$DW/\partial t = //_t \, d/dt \, (//_t^{-1} W(t)) \tag{15}$$

Thus W is *parallel along* σ i.e. $W(t) = //_t W(0)$ iff $DW/\partial t \equiv 0$.

Over a chart $(U_\alpha, \varphi_\alpha)$ for M, with induced trivialization of $\pi^{-1}(U_\alpha)$, using the same notation as for (12)

$$T_{\sigma(t)} \varphi_\alpha (DW/\partial t) = g(t) \, d/dt \, (g(t)^{-1} v(t))$$

where $v(t) = T_{\sigma(t)} \varphi_\alpha (W(t))$, and so by (12) the local representative $T_{\sigma(t)} \varphi_\alpha (DW/\partial t)$ is given by

$$dv/dt + \Gamma(\sigma_\alpha(t))(\dot\sigma_\alpha(t))(v(t)) \tag{16}$$

for $\sigma_\alpha(t) = \phi_\alpha(\sigma(t))$ or, if $\partial/\partial x^1, \dots, \partial/\partial x^n$ denote the vector fields over U_α given by $T_x\varphi_\alpha(\partial/\partial x^i) = e_i$, the i-th element of the standard base of $\mathbb{R}^n$ and if $W(x) = \Sigma\, W^i(x)\, \partial/\partial x^i$, and

$$DW/\partial t = \Sigma\, (DW/\partial t)^i \, \partial/\partial x^i \quad \text{etc.,}$$

then

$$(DW/\partial t)^i = dW^i/dt + \Gamma^i{}_{jk}(\sigma_\alpha(t))(\dot\sigma(t)^j) W^k(t) \tag{17}$$

summing repeated indices.

By definition a curve σ in M is a *geodesic* if its velocity field $\dot\sigma$ is parallel along σ i.e.

$$D/\partial t \, \dot\sigma \equiv 0.$$

Substitution of this into (17) gives the classical local equations. The existence theory for such equations shows that for each $v_0 \in T_{x_0}M$ there exists a unique geodesic $\{\gamma(t) : 0 \le t < t_0\}$ for some $t_0 > 0$ with $\gamma(0) = x_0$ and $\dot{\gamma}(0) = v_0$. If we can take $t_0 = \infty$ for all choices of v_0 so geodesics can be extended for all time the connection is said to be *(geodesically) complete*: (note that no metric is involved so far). The geodesic γ above is often written $\gamma(t) = \exp_{x_0} tv_0$ and there is the *exponential map* defined on some domain $\mathfrak{D}$ of TM

$$\exp : \mathfrak{D} \to M \times M$$

$$\exp v = (x, \exp_x v)$$

when $v \in T_xM$. A use of the inverse function theorem shows that there is an open neighbourhood $\mathfrak{D}_0$ of the *zero section* $Z[M]$ = image of $Z : M \to TM$ given by $Z(x) = 0 \in T_xM$, such that exp maps $\mathfrak{D}_0$ diffeomorphically onto an open neighbourhood of the diagonal in $M \times M$. In particular each $\exp_x : T_xM \to M$ is a local diffeomorphism near the origin. The inverse determines a chart (U, φ) around x by $\varphi = \exp_x^{-1} : U \to T_xM \cong \mathbb{R}^n$. These are *normal* (or *geodesic*, or *exponential*) coordinates about x. If γ is a geodesic in M from x then its local representative in this chart, $\{\varphi(\gamma(t)) : 0 \le t < t_0\}$, say, is just the 1/2 ray segment $\{tv : 0 \le t < t_0\}$, where $v = \dot{\gamma}(0)$. In particular we see from (15) and (16) that *for a torsion free connection, at the centre of normal coordinates the Christoffel symbols (for that coordinate system) vanish.*

C. Let $\mathbb{L}(TM;TM) = \bigcup_{x \in M} \mathbb{L}(T_xM;T_xM)$. It has a natural C^∞ manifold structure with charts induced by the charts of M, and a smooth projection onto M, as do the other tensor bundles e.g. the cotangent bundle $T^*M = \bigcup \{\mathbb{L}(T_xM;\mathbb{R})\}$, the exterior bundles $\Lambda^p TM$, and the bundles of p-linear maps $\mathbb{L}(TM,\dots,TM;\mathbb{R}) \approx \otimes^p T^*M$.

Note that a frame u at x determines an isomorphism

$$\rho_{\mathbb{L}}(u) ; \mathbb{L}(\mathbb{R}^n;\mathbb{R}^n) \to \mathbb{L}(T_xM; T_xM)$$

$$\rho_{\mathbb{L}}(u)(T) = uTu^{-1}$$

and similarly for the other bundles mentioned:

$$\rho^*(u) : \mathbb{R}^{n*} \to T^*_xM \text{ given by } \rho^*(u)(\ell) = \ell \circ u^{-1},$$

and also

$$\rho_\wedge(u)(v_1 \wedge \dots \wedge v_p) = (uv_1) \wedge \dots \wedge (uv_p)$$

and

$$\rho_\otimes(u)T = T(u^{-1}(-),\dots,u^{-1}(-)).$$

These equations also determine representations of $GL(n)$ on $\mathbb{L}(\mathbb{R}^n,\mathbb{R}^n)$, $\wedge TM$, etc. which will also be denoted by $\rho_\mathbb{L}$, $\rho_\wedge$, etc.

D. A vector field A on M determines a map

$$A^\sim : GLM \to \mathbb{R}^n$$

by

$$A^\sim(u) = u^{-1} A(\pi(u))$$

Similarly a section B of $\mathbb{L}(TM;TM)$, i.e. a map $B:M \to \mathbb{L}(TM;TM)$ such that $B(x) \in \mathbb{L}(T_xM;T_xM)$ each x, gives

$$B^\sim : GLM \to \mathbb{L}(\mathbb{R}^n; \mathbb{R}^n)$$

by

$$B^\sim(u) = \rho_\mathbb{L}(u)^{-1} B \tag{18}$$

etc.

The *covariant derivative* ∇A of A is the section of $\mathbb{L}(TM;TM)$ defined by

$$\nabla A(x)(v) = u\, dA^\sim(v^\sim) \tag{19}$$

where $v^\sim$ is the horizontal lift $H_u v$ of v to HT_uGLM, for $v \in T_xM$. This is often written $\nabla A(v)$ or $\nabla_v A$.

Covariant derivatives of other tensor fields e.g. Sections B of $\mathbb{L}(TM;TM)$ are defined similarly: ∇B is the section of $\mathbb{L}(TM;\mathbb{L}(TM;TM))$ given by

$$\nabla B(x)(v) = \rho_\mathbb{L}(u)\, dB^\sim(v^\sim) \in \mathbb{L}(T_xM;T_xM) \tag{20}$$

for $v^\sim$ as before. In particular the higher order covariant derivatives are defined this way, e.g.

$$\nabla^2 A = \nabla(\nabla A)$$

is a section of

$$\mathbb{L}(TM;\mathbb{L}(TM;TM)) \cong \mathbb{L}(TM,TM;TM).$$

For a chart $(U_\alpha, \varphi_\alpha)$ for M around a point x, using the induced trivialization of $\pi^{-1}(U_\alpha)$ our tensor field C, say, when lifted looks like a map $\tilde{C}$ on $U_\alpha \times GL(n)$ given by $\tilde{C}'(x,g) = \rho(g)^{-1} C'(x)$ where C' is C in our coordinate system, and ρ is the relevant representation, e.g. $\rho(g) = g$ for vector fields, $\rho = \rho_\mathbb{L}$ etc. In these coordinates $v^\sim = (v, -\Gamma(x)(v))$ so $\nabla C(x)(v)$ is given by

$$dC'(v) + d_I\rho(\Gamma(x)(v))C'(x) \tag{21}$$

where $d_I\rho$ means the differential of ρ at the identity e.g. $d_I\rho^*(\Gamma(x)(v) = -(\Gamma(x)v)^*$.

In particular if our vector field A is given over U_α by

$A = \Sigma A^i \partial/\partial x^i$, etc.

then, summing repeated suffices, if $x_\alpha = \varphi_\alpha(x)$

$$[\nabla A(v)]^i = dA^i(v) + \Gamma^i{}_{jk}(x_\alpha)(v^j)(A^k(x))$$
$$= (\partial A^i/\partial x^j)\, v^j + \Gamma^i{}_{jk}(x_\alpha)(v^j)(A^k(x)) \qquad (22)$$

(where formally $\partial A^i/\partial x^j : U_\alpha \to \mathbb{R}$ means the result of acting on A^i by the vector field $\partial/\partial x^j$: in practice everything is transported to the open set $\varphi_\alpha(U_\alpha)$ of $\mathbb{R}^n$ in order to do the computations so that $\partial A^i/\partial x^j$ is computed as "$\partial/\partial x^j\, A^i(\varphi_\alpha(x^1,\ldots,x^n))$" in the sense of elementary calculus).

Comparing (17) and (22) one sees that if V is a vector field taking value v at the point x, and if σ is an integral curve of V, so $\dot\sigma(t) = V(\sigma(t))$, with $\sigma(0) = x$ then

$$DA/\partial t|_{t=0} = \nabla A(v) = \nabla_v A \qquad (23)$$

Note that if V is a vector field we can form a new vector field $\nabla_V A$ or $\nabla A(V)$ by

$$\nabla_V A(x) := \nabla A(V(x))$$

we see from (22), or by working at the centre of normal coordinates that *for a torsion free connection*

$$\nabla_V A - \nabla_A V = [V,A] \qquad (24)$$

D. Covariant differentiation behaves similarly to ordinary differentiation. For example if α is a 1-form (i.e. a section of T*M) and A is a vector field then for $v \in T_xM$

$$d(\alpha(A(\cdot)))(v) = \nabla_v\alpha(A(x)) + \alpha(\nabla_v A(x)) \qquad (25)$$

One way to see this is to write $\alpha(A(\pi(u))) = (\alpha_x \circ u) \circ u^{-1}A(\pi(u))$ for $u \in GLM$, $x = \pi(u)$. Then differentiate both sides in the direction $H_u(v)$.

E. Using the notation of §3A of Chapter I

$$d^2/dt^2\, f(S(t,x)e) = d/dt\, df(Y_e(S(t,x)e)) = d(df(Y_e(-))(d/dt\, S(t,x)e)$$

which at t = 0

$$= \nabla(df)(Y_e(x),(Y_e(x)) + df(\nabla Y_e(Y_e(x)))$$

by (25). Thus for any affine connection, (5) can be written

$$f(x_T) = f(x_s) + \int_s^T df(X(x_r)dz_r)$$

$$+ \frac{1}{2}\int_s^T \{\nabla(df)(Y(x_r)dz_r)(Y(x_r)dz_r) + df(\nabla Y(Y(x)dzr)dzr)\} \qquad (26)$$

which for equation (2)', gives the generator $\mathcal{A}$ in the form

$$\mathcal{A}f = \frac{1}{2}\sum_{p=1}^{m} \{\nabla(df)(X^p(x))(X^p(x)) + df(\nabla X^p(X^p(x)))\} + df(A(x)). \qquad (27)$$

Non-degeneracy of the S.D.S. (2') i.e. surjectivity of each $X(x) : \mathbb{R}^m \to T_xM$ is equivalent to ellipticity of $\mathcal{A}f$: the symbol of $\mathcal{A}$ is just $\Sigma\, X^p \otimes X^p$ as a section of $TM \otimes TM$, or $X(\cdot) \circ X(-)^*$ as a section of $\mathbb{L}(T^*M;TM)$).

§3. Riemannian metrics and the Laplace-Beltrami operator

A. A Riemannian metric on M assigns an inner product $\langle\ ,\ \rangle_x$ to each tangent space T_xM of M, depending smoothly on x. Over a chart $(U_\alpha,\varphi_\alpha)$ if $u = u^i\, \partial/\partial x^i$, $v = v^i\, \partial/\partial x^i$ are tangent vectors then define the $n \times n$-matrix $G(x) = [g_{ij}(x)]_{i,j}$ by

$$\langle u,v\rangle_x = g_{ij}(x)u^i v^j \qquad (28)$$

The inner product determines a metric d in the usual sense on M, compatible with its topology, by letting d(x,y) be the infimum of the lengths of all piecewise C^1 curves from x to y, where the length $\ell(\sigma)$ is

$$\ell(\sigma) = \int_a^b |\dot\sigma(t)|_{\sigma(t)}\, dt$$

for $|u|_x = \langle u,u\rangle^{\frac{1}{2}}_x$ as usual, and σ is defined on [a,b]. The Riemannian manifold (i.e. M together with $\{\langle\ ,\ \rangle_x : x \in M\}$) is (metrically) *complete* if it is complete in this metric.

For a submanifold M of $\mathbb{R}^m$ the standard inner product of $\mathbb{R}^m$ restricts to an inner product $\langle\ ,\ \rangle_x$ on each T_xM considered as a subset of $\mathbb{R}^m$, thereby determining a Riemannian structure on M. It is a highly non-trivial result, the Nash embedding theorem, that for every Riemannian metric on a manifold M there is an embedding into some $\mathbb{R}^m$ such that the induced metric agrees with the given one i.e. an *isometric embedding.* (In general a smooth map $f : M \to N$ of Riemannian manifolds is *isometric* if $\langle T_xf(u), T_xf(v)\rangle_{f(x)} = \langle u,v\rangle_x$ for all $x \in M$ and $u, v \in T_xM$; it is *an isometry* if it is also a diffeomorphism of M onto N. Thus an isometric map need not preserve distance.)

B. Given such a metric one can consider *orthonormal frames:* these are isomorphisms

$$u : \mathbb{R}^n \to T_xM$$

preserving the inner products, $\langle u(e),u(e')\rangle_x = \langle e,e'\rangle_{\mathbb{R}^n}$. The space OM of such frames is a subset of GLM, and keeping $\pi : OM \to M$ to denote the projection it forms a principal bundle with group $O(n)$: it is a subbundle of GLM in the obvious sense.

A connection on OM is called a Riemannian connection. $\tilde{\omega}$ will take values in the Lie algebra *o(n)* of O(n) which can be identified with the space of skew-symmetric $n \times n$-matrices. It can be extended over all of GLM by the action of GL(n) on GLM, insisting on condition (i) for a connection form (or (b) for the corresponding horizontal subspaces). Thus it determines a connection on GLM and so local coordinates have associated Christoffel symbols, which can be used to compute covariant derivatives.

An important point is that for this induced connection on GLM, given a curve σ in M, the horizontal lift $\sigma^\sim$ of σ to GLM starting from an orthonormal frame stays in OM and is the same as the horizontal lift for the original connection on OM. An immediate consequence (from the definitions, equations (14) and (15)) is that parallel translation preserves inner products:

$$\langle //_t v, //_t v'\rangle_{\sigma(t)} = \langle v,v'\rangle_{\sigma(o)} \qquad (29)$$

for $v, v' \in T_{\sigma(o)}M$, and for vector fields W, W' along σ

$$d/dt \langle W(t),W'(t)\rangle_{\sigma(t)} = \langle DW/\partial t, W'(t)\rangle_{\sigma(t)} + \langle W(t), DW'/\partial t\rangle_{\sigma(t)}. \qquad (30)$$

Consequently, by (23), if W_1, W_2 are vector fields and $v \in T_xM$ then

$$d\langle W_1(-),W_2(-)\rangle_{(-)}(v) = \langle \nabla W_1(v), W_2(x)\rangle_x + \langle W_1(x), \nabla W_2(v)\rangle_x \qquad (31)$$

C. The metric gives an identification of T_xM with its dual $T_x{}^*M$ by

$v \to v^\# = \langle v,-\rangle_x$. In local coordinates $(U_\alpha,\varphi_\alpha)$ let $\varphi_\alpha(y) = (x^1(y),\dots,x^n(y))$ for $y \in U_\alpha$ then $\{d_y x^1,\dots,d_y x^n\}$ form the dual basis to $\{\partial/\partial x^1,\dots,\partial/\partial x^n\}$, (strictly speaking evaluated at y). If $v = v^i\, \partial/\partial x^i$ at y then $v^* = v_i\, d_y x^i$ where

$$v_i = g_{ij}(y) v^j \tag{32}$$

Write $\ell \to \ell^\#$ for the inverse of this isomorphism also.

By choosing the vector field A such that for given $x \in M$ and $v \in T_xM$, $A(x) = v$ and $\nabla A(x) \equiv 0$, equation (25) shows that for a 1-form α

$$(\nabla_v \alpha)^\# = \nabla_v \alpha^\#. \tag{33}$$

Similarly ∇_v commutes with the 'raising and lowering of indices' on other tensor fields.

The *gradient*, grad f, or ∇f, of a C^1 function $f : M \to \mathbb{R}$ is the vector field $(df)^\#$ so

$$\langle \nabla f(x), v\rangle_x = df(v) \tag{34}$$

all $v \in T_xM$. In local coordinates $\nabla f(x) = \nabla f(x)^i\, \partial/\partial x^i$ where

$$\nabla f(x)^i = g^{ij}(x)\, \partial f/\partial x^j \tag{35}$$

where $[g^{ij}(x)]_{i,j}$ is the inverse matrix $G(x)^{-1}$ to $[g_{ij}(x)]$.

D. There will be many Riemannian connections for a given metric. However it turns out that there is a unique one which is also torsion free. This is called the *Levi-Civita connection.* It can be defined in terms of the Christoffel symbols by

$$\Gamma^\ell_{ij} = \sum_k \tfrac{1}{2} g^{k\ell}(\partial/\partial x^i\, g_{jk} + \partial/\partial x^j\, g_{ik} - \partial/\partial x^k\, g_{ij}) \tag{36}$$

It is this connection which is usually refered to when considering covariant derivatives etc. for Riemannian manifolds.

E. A Riemannian metric determines a measure on M, temporarily to be denoted by μ, such that if $(U_\alpha, \varphi_\alpha)$ is a chart then the push forward μ_α of $\mu | U_\alpha$ by φ_α is equivalent to Lebesgue measure on the open set $\varphi_\alpha(U_\alpha)$ of $\mathbb{R}^n$ with $\mu_\alpha(dx) = \sqrt{\det G(\varphi_\alpha^{-1}(x))}\, \lambda(dx)$ where λ is Lebesgue measure and G is the local representative of the metric. We shall usually just write dx for $\mu(dx)$ or $\lambda(dx)$ and write $g_\alpha(x)$ or $g(x)$ for $\det G(x)$. Note $\sqrt{g(x)} = |\det T_y\varphi_\alpha^{-1}|$ for $y = \varphi_\alpha(x)$, where 'det' refers to the determinant obtained by using $\langle\, ,\rangle_x$ and

$\langle\, ,\rangle_{\mathbb{R}^n}$.

For a C^1 vector field A on M, the *divergence*, div $A : M \to \mathbb{R}$, is given by

$$\operatorname{div} A(x) = d/dt \det T_x F_t \big|_{t=0} \tag{37}$$

where

$(t,x) \to F_t(x) \in M$

is the solution flow of A, on its domain of definition in $\mathbb{R} \times M$. It represents the rate of change of volume by the flow. It is given by

$$\operatorname{div} A(x) = \operatorname{trace} \nabla A(x). \tag{38}$$

From (37), using the change of variable formula for Lebesgue measure one gets the *divergence theorem*

$$\int_M \operatorname{div} A(x)\, dx = 0 \tag{39}$$

for M compact, and more generally. Since, by (22), if $f : M \to \mathbb{R}$,

$$\operatorname{div} fA(x) = \langle \nabla f(x), A(x)\rangle_x + f(x) \operatorname{div} A(x) \tag{40}$$

we see from this that div and $-\nabla$ are formal adjoints.

The *Laplace-Beltrami* operator Δ on C^2 functions $f : M \to \mathbb{R}$ is defined by

$$\Delta f = \operatorname{div} \nabla f$$

or equivalently

$$\Delta f = \operatorname{trace} \nabla df = \Sigma\ \nabla(df)(e_i)(e_i)$$

where $e_1,\dots,e_n$ are orthonormal. It determines a self-adjoint operator Δ on $L^2(M;\mathbb{R})$, [52], [91]. In local coordinates it has the formula

$$\Delta f(x) = g^{ij}(x)\, \partial^2 f/\partial x^i \partial x^j - g^{ij}(x)\, \Gamma^k{}_{ij}(x)\, \partial f/\partial x^k \tag{41a}$$

and

$$\Delta f(x) = g(x)^{-\frac{1}{2}}\, \partial/\partial x^i \,\{g(x)^{\frac{1}{2}}\, g^{ij}(x)\, \partial f/\partial x^j\} \tag{41b}$$

which are easily seen using (21) and (38) for (41a), and (35) and (37) for (41b).

§4. Brownian motion on M and the stochastic development

A Let M be a Riemannian manifold with its Levi-Civita connection. By a *Brownian motion* on M we mean a sample continuous process $\{x_t : 0 \le t < \xi\}$, defined up to a stopping time, which is Markov with infinitesimal generator $\frac{1}{2}\Delta$.

From the Ito formula (27) a solution of

$$dx_t = X(x_t) \circ dB_t + A(x_t)dt$$

is a Brownian motion if and only if

(i) $X(x) : \mathbb{R}^m \to T_xM$ is a projection onto T_xM for each x in M i.e. $X(x) \circ X(x)^* =$ identity; and

(ii) $A(x) = -\frac{1}{2} \sum_p \nabla X^p(X^p(x))$

When (i) but not (ii) holds we say that $\{x_t : 0 \le t < \xi\}$ is a *Brownian motion with drift*. The *drift* is the vector field $x \to A(x) + \frac{1}{2} \sum_p \nabla X^p(X^p(x))$.

B. Note, again from (27) for an arbitrary affine connection, that in general the generator $\mathcal{A}$ for our solution is elliptic if and only if each X(x) is surjective (in which case the S.D.E. is said to be *non-degenerate*). In this case each X(x) induces an inner product on T_xM , the quotient inner product, and so determines a Riemannian metric on M. Thus the *solutions to a non-degenerate* S.D.E. *are Brownian motions with drift for some (uniquely defined) metric on* M, and equivalently *any elliptic* $\mathcal{A}$ *can be written as* $\frac{1}{2}\Delta + B$ *for some first order operator* (i.e. *vector field*) B. Even working on $\mathbb{R}^n$, if one wishes to deal with elliptic generators $\mathcal{A}$, the differential geometry of the associated metric will not in general be trivial and can play an important role.

C. Although there always exist coefficients X and A satisfying (i) and (ii) there is no natural choice which can be applied to general Riemannian manifolds. However there is a canonical S.D.E. on the orthonormal frame bundle OM to M, and it turns out that the solutions to this project down to give Brownian motions on M. The construction, due to Eells and Elworthy, is as follows:

Define $X : OM \times \mathbb{R}^n \to TOM$ by

$$X(u)e = H_u(u(e)). \tag{42}$$

For given u_0 in OM let $\{u_t : 0 \le t < \zeta\}$ be a maximal solution to

$$du_t = X(u_t) \circ dB_t \tag{43}$$

where $\{B_t : 0 \le t < \infty\}$ is Brownian motion on $\mathbb{R}^n$, so now m = n. Set $x_t = \pi(u_t)$:

<u>Theorem 4C</u> $\{x_t : 0 \le t < \zeta\}$ *is a Brownian motion on* M, *defined up to its explosion time.*

<u>Proof</u> Suppose $g : M \to \mathbb{R}$ is C^2. Set $f = g \circ \pi : OM \to \mathbb{R}$. With the notation of the Ito formula (5), §3A, for $e \in \mathbb{R}^n$ and $u \in OM$ set

$$\gamma_t(u,e) = \pi(S(t,u)e). \tag{44}$$

Then $S(-,u)e$ is the horizontal lift of $\gamma_{-}(u,e)$ through u. Since

$$\dot\gamma_t(u,e) = T\pi(X(S(t,u)e)(e)) = (S(t,u)e)(e) \in T_{\gamma_t(u,e)}M$$

$\dot\gamma_t$ is parallel along γ_t and so $t \to \gamma_t(u,e)$ *is a geodesic in* M, see §2B; also $\dot\gamma_0 = u(e)$. Thus

$$d/dt\, f(S(t,u)e)|_{t=0} = d/dt\, g(\gamma_t(u,e))|_{t=0} = dg(u(e))$$

and

$$d^2/dt^2\, f(S(t,u)e)|_{t=0} = d/dt\,(dg(\dot\gamma_t(u,e)))|_{t=0}$$
$$= \nabla\, dg(\dot\gamma_t(u,e))|_{t=0} + dg(D/\partial t\, \dot\gamma_t(u,e)|_{t=0})$$
$$= \nabla\, dg(u(e))$$

since $D/\partial t\, \dot\gamma_t \equiv 0$.

Thus by (5)

$$g(x_T) = g(x_0) + \int_0^T dg(u_t \circ dB_t) + \tfrac{1}{2}\int_0^T \Delta g(x_t)dt \tag{45}.$$

That x_t is a Brownian motion can now be deduced from the martingale problem method, [92]. Alternatively, to show that x_t has the Markov property first prove that the distributions of $\{x_t : 0 \le t < \xi\}$ do not depend on the point $u_0 \in \pi^{-1}(x_0)$ in OM: having done this the Markov property is easily deduced from that of the solutions of our S.D.E. on OM, (e.g. see [43], §5C of Chapter IX). To see the lack of dependence up to distribution on u_0 observe that if $u'_0 \in \pi^{-1}(x_0)$ then there exists $g \in O(n)$ with $u'_0 = u_0.g$. By the equivariance under O(n) of the horizontal tangent spaces, condition (b) of §1C, $\{u_t.g : 0 \le t < \zeta\}$ satisfies

$$du'_t = X(u'_t) \circ dB'_t$$

where $B'_t = g^{-1}(B_t)$. By the orthogonal invariance of the distributions of Brownian motion, B'_t is again a Brownian motion and $u_t.g$ has the same distributions as the solution u'_t of $du'_t = X(u'_t) \circ dB_t$ from u'_0. Since $\pi(u_t.g) =$

$\pi(u_t)$ the required invariance for x_t follows.

The maximality of $\{x_t : 0 \le t < \zeta\}$ follows from that of $\{u_t : 0 \le t < \zeta\}$: if $u_t \to \infty$ in OM as $t \to \zeta(\omega)$ so does $\pi(u_t)$ since each $\pi^{-1}(x)$ is compact. //

D. Since each $X(u)e \in T_u OM$, the considerations of §2A suggest calling the solution $\{u_t : 0 \le t < \zeta\}$ the *horizontal lift* of the Brownian motion $\{x_t : 0 \le t < \zeta\}$ from u_0: in fact it is easy to see that it locally satisfies the local equations (12) considered as a Stratonovich equation with x_t replacing σ_t.

We can then define *parallel translation* along the sample paths of our Brownian motion

$$//_t(\omega) : T_{x_0}M \to T_{x_t(\omega)}M$$

by

$$//_t(\omega)v_0 = u_t(\omega)u_0^{-1}(v_0) \qquad \omega \in \Omega \qquad (46).$$

Also if $\{W_t : 0 \le t < \zeta\}$ is a vector field along $\{x_t : 0 \le t < \zeta\}$ i.e.

$$W_t(\omega) \in T_{x_t(\omega)}M \qquad \omega \in \Omega$$

then its covariant derivative along the Brownian paths is the vector field along $\{x_t : 0 \le t < \zeta\}$ given by

$$DW_t/\partial t = //_t \; d/dt\,(//_t^{-1} W_t) \qquad (47)$$

E. We shall not give the details but any Brownian motion on M from a point x_0 can be considered as obtained in the way described: as the *stochastic development* of a Brownian motion on $\mathbb{R}^n$. To do this we have to *anti-develop* our given Brownian motion on M. A simple way to do this is to express x_t as the solution of some S.D.E. $dx_t = Y(x_t) \circ dz_t$, as described in §3C of Chapter I. Take a horizontal lift $Y^\sim$ of Y to OM so $Y^\sim(u)e = H_u(Y(\pi(u))e)$ for $(u,e) \in OM \times \mathbb{R}^m$. Solve

$$du_t = Y^\sim(u_t) \circ dz_t \qquad (48)$$

from a given $u_0 \in \pi^{-1}(x_0)$. This is a candidate for a horizontal lift of $\{x_t : 0 \le t < \zeta\}$ (assuming it does live as long as x_t, which it does [34], [43]). For the solution set

$$B_t = \int_0^T u_t^{-1}(Y(x_t) \circ dz_t) \in \mathbb{R}^n \qquad (49)$$

One then has to use the martingale characterization of Brownian motion to show that B_t is a Brownian motion (or at least part of one). It is rather straightforward to see that $\{x_t : 0 \le t < \zeta\}$ is its stochastic development.

Note: we have used Y so that (48) and (49) can be used rather than discussing equations like the stochastic versions of (12): however, even for general semi-martingales z, the above construction gives an 'anti-development' with resultant $\mathbb{R}^n$-valued process (B_t in our case) which is independent of the choice of Y. In fact the horizontal lift $\{u_t : 0 \le t \le \zeta\}$ given by (48) is independent of Y as is most easily seen by the uniqueness of solutions to the local equations (12). The horizontality can most easily be expressed by

$$\int_0^T \tilde{\omega}\,(\circ\, du_t) = 0 \tag{50}$$

for any stopping time $\{0 \le T < \zeta\}$ where for an $\mathbb{R}^p$-valued 1-form $\theta : TN \to \mathbb{R}^p$ on some manifold N, and a continuous semimartingale $\{y_t : 0 \le t < \zeta\}$ in N we define

$$\int_0^T \theta(\circ\, dy_t) = \int_0^T \theta(Y(y_t) \circ dz_t) \tag{51}$$

where T is a stopping time less than ζ and $dy_t = Y(y_t) \circ dz_t$ for some Y and z. This has to be shown to be independent of the choice of Y and z: but this is easily done either by working with local expressions or by embedding N in some $\mathbb{R}^q$ and extending θ and any suitable Y over $\mathbb{R}^q$.

Given the horizontal lift, parallel translation along sample paths, and covariant differentiation along the paths can be defined by (46) and (47). The integral (51) of a 1-form θ on M along the paths of x_t can be written

$$\int_0^T \theta\,(\circ\, dx_t) = \int_0^T \theta(u_t \circ dB_t) \tag{52}$$

Note that this anti-development can be carried out for semi-martingales on M given any affine connection on M, as can the development itself: however a Riemannian connection was needed to construct Brownian motions since we needed the invariance of the distributions of B_t under the group $O(n)$. See [70], [76] for other situations.

F. The classical *Cartan development* maps a smooth path $\{\sigma(t) : 0 \le t < \infty\}$ on $\mathbb{R}^n$ (or $T_{x_0}M$), starting at the origin, to a path in M starting at x_0. Mathematically it is just as described above with B_t replaced by $\sigma(t)$ in (43) to yield the deterministic equation

$$du/dt = X(u(t))\,(d\sigma/dt) \tag{53}$$

with $u(0) = u_0$ a given frame at x_0. The resulting path $x(t)$ on M, defined for all time, it turns out, if M is complete, is that which is obtained by the classical mechanical procedure of "placing M on a copy of $\mathbb{R}^n$ (by u_0), then rolling M along $\{\sigma(t) : 0 \le t < \infty\}$ without slipping and taking $x(t)$ to be the point of contact of M with $\mathbb{R}^n$ at time t". The frame $u(t) : \mathbb{R}^n \to T_{x(t)}M$ represents the way M is resting on $\mathbb{R}^n$ at time t. When trying to visualize, or sketch, the situation it becomes clear that the natural objects to use are affine frames, as in [64], rather than the linear frames which we have used.

§5. Examples: Spheres and hyperbolic spaces

A. A connected Riemannian manifold is *orientable* if its orthonormal frame bundle OM has two components. This is always true for M simply connected. An orientation is then the choice of one of these, to be called SOM: it will be a principal SO(n)-bundle whose elements can be called *oriented frames.* For example there is a natural choice when $M = S^n$ given by the inclusion of the tangent spaces T_xS^n into $\mathbb{R}^{n+1}$ and the natural orientation of $\mathbb{R}^{n+1}$.

B. The natural left action of $SO(n+1)$ on $\mathbb{R}^{n+1}$ restricts to one on S^n. If $N = (1,0,\ldots,0)$ is the 'north pole' of S^n (with abuse of geography) the map of $SO(n+1)$ to S^n

$$\begin{array}{c} p \\ g \to g.N \end{array} \tag{54}$$

induces a diffeomorphism of the quotient space (with a natural differential structure which we do not need to examine)

$$\beta : SO(n+1)/SO(n) \to S^n.$$

In fact we will show that the oriented frame bundle SOS^n can be identified with $SO(n+1)$. However first we need some remarks about $SO(n+1)$ itself:

The Lie algebra $\underline{so}(n+1)$ will be identified with the space of skew-symmetric real matrices with $\underline{so}(n)$ contained in it as

$$\underline{so}(n) = \left\{\begin{pmatrix}0 & 0\\ 0 & B\end{pmatrix} : B \text{ is } n \times n \text{ and } B^* = -B\right\}$$

There is then the vector space direct sum

$$\underline{so}(n+1) = \underline{so}(n) + \underline{m} \tag{55}$$

where

$$\underline{m} = \{\underline{\xi} : \xi \in \mathbb{R}^n\}$$

for $\underline{\xi} = \begin{pmatrix}0 & -\xi^t\\ \xi & 0\end{pmatrix}$

with ξ written as a column.

Define

$$X : SO(n+1) \times \mathbb{R}^n \to TSO(n+1)$$

by

$$X(A)\xi = T_1 L_A(\underline{\xi}) \tag{56}$$

so $X(-)\xi$ is the left invariant vector field on $SO(n+1)$ corresponding to the element $\underline{\xi}$ of $\underline{m}$. Next define

$$\alpha : SO(n+1) \to SOS^n$$

by

$$\begin{aligned}\alpha(A)(\xi) &= T_A\, p(X(A)\xi) \qquad \xi \in \mathbb{R}^n\\ &= T_1(p \circ L_A)\underline{\xi}\end{aligned} \tag{57}$$

To see this is an orthonormal frame it is only necessary to check that

$$T_1(p \circ L_A) : \underline{m} \to T_{A.N}S^n$$

is an isometry where $\underline{m}$ is given the inner product induced by $\xi \to \underline{\xi}$. This is easy to check when A is the identity, and follows for general A by noting that $SO(n+1)$ acts by *isometries* on S^n (i.e. the derivative $A_* : T_xS^n \to T_{A.x}S^n$ of $x \to Ax$ preserves the Riemannian metric for each $x \in S^n$). Similarly each $\alpha(A)$ has the correct orientation.

Finally observe that α is equivariant for the right $SO(n)$ action, i.e. $\alpha(Ag) = \alpha(A)g$:

In fact if R_g denotes right multiplication by g

$$p \circ L_A = (p \circ R_g{}^{-1}) \circ LA = p \circ L_A \circ R_g{}^{-1}$$

for $g \in SO(n)$, so

$$\begin{aligned}\alpha(Ag)\xi &= T_1(p \circ L_{Ag})\xi \\ &= T_1(p \circ L_A R_g^{-1} L_g)\xi \\ &= T_1(p \circ L_A)(\mathrm{ad}\,g)^{-1}\xi \\ &= T_1(p \circ L_A)(\underline{g\xi}) \qquad (58)\end{aligned}$$

(by an elementary computation). Thus $p : SO(n+1) \to S^n$ has a principal bundle structure isomorphic to $\pi : SOS^n \to S^n$ by α.

There is a canonical connection on $p : SO(n+1) \to S^n$ defined by

$$H_A\, SO(n+1) = T_1\, L_A[\underline{m}] \qquad A \in SO(n+1) \qquad (59)$$

To see this is a connection note that for $g \in SO(n)$

$$\begin{aligned}T_A R_g[H_A\, SO(n+1)] &= T_1\, L_{Ag}(\mathrm{ad}\,g)^{-1}\,[\underline{m}] \\ &= T_1\, L_{Ag}(\underline{m}) = H_{Ag}\, SO(n+1)\end{aligned}$$

since $\underline{m}$ is invariant under adg for g in SO(n). The connection form $\tilde{\omega}_0 : TSO(n+1) \to \mathbb{R}^n$ is given by $\tilde{\omega}_0(T_1 L_A(\underline{B} + \underline{\xi})) = \underline{B}$ for $A \in SO(n+1)$, $\underline{B} \in \underline{so}(n)$ and $\underline{\xi} \in \underline{m}$.

An important property is that it is invariant under the automorphism σ_0 of $SO(n+1)$ given by $\sigma_0 = SAS$ where $S = \begin{pmatrix}-1 & 0\\ 0 & I_n\end{pmatrix}$

This connection induces a connection on $\pi : SOS^n \to S^n$ via α, i.e. a Riemannian connection on S^n. In fact this is the Levi-Civita connection, a fact which is proved using the invariance under σ_0 just mentioned: see [64] page 303.

C. This situation works much more generally. Another important example is obtained by taking the Lorentz group $O(1,n)$ of linear transformations of $\mathbb{R}^{n+1}$ which preserve the quadratic form $\langle Sx,x\rangle$ for S as in §B above. This has 4 components: let G be its identity component,

$$G = \{A \in O(1,n) : \det A = 1 \text{ and } A_{11} \ge 1\}. \qquad (60)$$

Then the Lie algebra $\underline{g}$ of G is just $\underline{o}(n,1)$ where

$$\underline{o}(n,1) = \{A \in \mathbb{L}(\mathbb{R}^{n+1};\mathbb{R}^{n+1}) : A^tS + S^tA = 0\}. \qquad (61)$$

Let

$$M = \{x \in \mathbb{R}^{n+1} : \langle Sx,x\rangle = -1 \text{ and } x^1 \ge 1\}. \qquad (62)$$

Then the form $v \to \langle Sv,v\rangle$ induces a Riemannian metric on M by restricting it to each T_xM and G acts on the left on M, preserving this structure with

$p : G \to M$

given by $p(A) = A.N$ for $N = (1,0,\dots,0)$, inducing a diffeomorphism of G/H with M for

$$H = \{A \in G : A.N = N\} = \left\{\begin{pmatrix}1 & 0\\ 0 & B\end{pmatrix} : B \in SO(n)\right\}.$$

We can identify H with SO(n).

As before there is commutative diagram

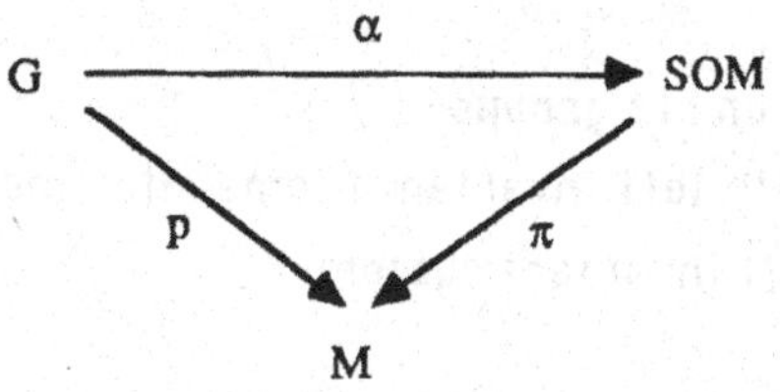

with α equivariant under the right action of SO(n), so that p can be identified with the oriented frame bundle π. For G there is the splitting

$\underline{o}(1,n) = \underline{so}(n) + \underline{m}^*$

for

$$\underline{m}^* = \{\xi^* : \xi \in \mathbb{R}^n\}, \text{ where } \xi^* = \begin{pmatrix}0 & \xi^t\\ \xi & 0\end{pmatrix} \tag{63}$$

Again there is a connection $\tilde{\omega}_0$ which induces the Levi-Civita connection for M, see [64], p. 303, and an involution σ_0.

The manifold M is *n-dimensional hyperbolic space* H^n, see Exercise (ii) at the end of §8 below and §4G of Chapter III.

The general situation where this works is that of a Riemannian *symmetric space* see [64], [65].

D. In these situations the stochastic development construction is equivalent to solving $dg_t = X(g_t) \circ dB_t$ for the X defined on G, (e.g. on SO(n+1) for S^n), and projecting by p to get Brownian motion on M. Thus for our solution with $g_0 = 1$, $\{g_t . N : t \geq 0\}$ is a Brownian motion on M starting at N.

Remark:

It is shown in [43] p. 257 that when M is a Riemannian symmetric space, so we can identify M with G/H as above, Brownian motions on G themselves project to Brownian motions on M, so if $\{g_t : t \geq o\}$ is a B.M. on G from 1 then $\{g_t.N : t \geq N\}$ is one on M starting at N, where $N \in M$ is arbitrary.

Similarly, for example by the discussion of the S(t,u)e in the proof of the Ito formula, Theorem 4C, the geodesics of M from a point N are given by $\{(\mathrm{Exp}\,tA).N : t \in \mathbb{R}\}$ where Exp is the exponential map of the group G, see §1A, and A lies in a certain n-dimensional subspace of the Lie algebra of G : in particular for $M = S^n$ we need only take $A \in \underline{m}$, and for $M = H^n$ we can take $A \in \underline{m}^*$.

§6. Left invariant S.D.S. on Lie groups

A. For G a Lie group with left invariant Riemannian metric (i.e. each L_g is an isometry) there is the left invariant system

$$dg_t = X(g_t) \circ dB_t \tag{64}$$

where n = m and $X(1) : \mathbb{R}^n \to T_1G$ is some isometry and $X(g)e = L_g(X(1)e)$ for $e \in \mathbb{R}^n$ and $g \in G$. The solutions will be Brownian motions if $\Sigma_p \nabla X^p(X^p(x)) = 0$; in particular if each $\nabla X^p(X^p(x)) = 0$. The latter means precisely that the integral curves of $d\sigma/dt = X^p(\sigma_t)$ are geodesics (just differentiate this equation along its solution). For a Lie group with both left and right invariant metric the inversing map $g \to g^{-1}$ is an isometry so that if $\{\gamma(t) : t \in \mathbb{R}\}$ is a geodesic so is $\{\gamma(t)^{-1} : t \in \mathbb{R}\}$, and therefore $\gamma(t)^{-1} = \gamma(-t)$. From this one can deduce that the geodesics are precisely the one-parameter groups, i.e. the solutions of $\dot\sigma(t) = X(\sigma(t))e$ for some e (e.g. see [79]). Thus in this bi-invariant case the solutions to (64) are Brownian motions.

The compact groups admit bi-invariant metrics, and conversely every G with a bi-invariant metric is a product $G' \times \mathbb{R}^k$ with G' compact.

B. Given a bi-invariant metric on G we have just persuaded ourselves that $\nabla A(A(x)) = 0$ for all left invariant vector fields. Therefore if B and C are both left invariant, by taking A = B + C we obtain, using (24),

$$\nabla B(C(x)) = \tfrac{1}{2}[C,B](x) \tag{65}$$

§7. The Second Fundamental form and gradient S.D.S. for an embedded submanifold.

A. Suppose now that M is a submanifold of $\mathbb{R}^n$ with induced Riemannian metric. There is then a natural S.D.S. (X,B) on M where $\{B_t : t \geq 0\}$ is Brownian motion on $\mathbb{R}^m$ and X is just the orthogonal projection map P of §1D, Chapter I, (as in the S.D.E. of equation (7)) so $X(x) : \mathbb{R}^m \to T_xM$ is the orthogonal projection.

Suppose $f : M \to \mathbb{R}$ is C^1. Let $f_0 : \mathbb{R}^m \to \mathbb{R}$ be some smooth extension. Using ∇_0 for the gradient operator on functions on $\mathbb{R}^m$ we have $df(v) = df_0(v)$ for $v \in T_xM$ and so $\nabla f(x) = X(x)(\nabla_0 f_0(x))$ for $x \in M$. Thus if $\varphi : M \to \mathbb{R}^m$ denotes the inclusion, writing $\varphi(x) = (\varphi^1(x),\dots,\varphi^m(x))$ we see

$$X^p(x) = \nabla\varphi^p(x) \qquad x \in M, p = 1,\dots,m. \tag{66}$$

For this reason our S.D.S. is often called the *gradient Brownian system* for the submanifold (or for the embedding φ). We will show that its solutions are Brownian motions on M. For this we need $\Sigma_p \nabla X^p(X^p(x)) = 0$ all $x \in M$, and so we will first examine how the covariant derivative for M is related to differentiation in $\mathbb{R}^m$.

B. Suppose Z is a vector field on M. Take some smooth extension which we will write $Z_0 : \mathbb{R}^m \to \mathbb{R}^m$. For $\nu_xM = (T_xM)^\perp$ in $\mathbb{R}^m$, as in §1D of Chapter I, there is a symmetric bilinear map

$$\alpha_x : T_xM \times T_xM \to \nu_xM$$

called the *second fundamental form* of M at x , such that Gauss's formula holds: for $v \in T_xM$

$$DZ_0(x)(v) = \nabla Z(v) + \alpha_x(Z(x),v) \tag{67}$$

One way to prove this is to *define* $\nabla Z(v)$ to be the tangential component of $DZ_0(x)(v)$ and write $\nu_x(Z_0,v)$ for its normal component. Then one can verify that $(Z,v) \to \nabla Z(v)$ satisfy the conditions which ensure it is the covariant differentiation operator for the Levi-Civita connection on M and furthermore show that $\nu_x(Z_0,v)$ has the given form for a symmetric α_x, e.g. see [65], pp. 10-13.

From this there is the bilinear map for each x in M:

$$A_x : T_xM \times \nu_xM \to T_xM$$

defined by

$$\langle A_x(u,\xi),v\rangle = \langle \alpha_x(u,v),\xi\rangle. \tag{68}$$

If $\xi : M \to \mathbb{R}^m$ is C^1 with $\xi(x) \in \nu_xM$ for all $x \in M$ and ξ_0 is a C^1 extension then Weingarten's formula gives

$$D\xi_0(x)(v) = -A_x(v,\xi(x)) + \text{a normal component} \tag{69}$$

In fact for $x \in M$

$$\langle Z_0(x), \xi_0(x)\rangle = 0$$

for Z and Z_0 as before. Therefore if $v \in T_xM$

$$\langle DZ_0(x)v, \xi_0(x)\rangle + \langle Z_0(x), D\xi_0(x)v\rangle = 0$$

i.e.

$$\langle \alpha_x(Z_0(x),v),\xi_0(x)\rangle + \langle Z_0(x), \text{tangential component of } D\xi_0(x)v\rangle = 0$$ proving (69).

C. The following goes back to Ito's work published in 1950:

Proposition 7C. *The solutions of the gradient Brownian system for a submanifold* M *of* $\mathbb{R}^m$ *are Brownian motions on* M.

Proof For the constant vector fields $E^p(x) = (\delta^{1p},\ldots,\delta^{mp})$, $p = 1$ to m on $\mathbb{R}^m$, $E^p(x) = X(x)E^p(x) + Q(x)E^p(x)$ for $Q(x) = \text{Id} - X(x)$. Therefore differentiating and taking the tangential component, for $v \in T_xM$

$$0 = \nabla X^p(v) - A_x(v,Q(x)E^p(x)). \tag{70}$$

Thus if we choose our orthonormal base $e_1,\ldots,e_m$ of $\mathbb{R}^m$ so that $e_1,\ldots,e_n$ are tangent to M at x

$$\nabla X^p(v) = 0 \qquad p = 1 \text{ to } n \tag{71}$$

while

$$\nabla X^p(X^p(x)) = 0 \qquad p = n+1,\ldots,m \tag{71'}$$

because $X^p(x) = 0$ for such p.

Thus

$$\sum_p \nabla X^p(X^p(x)) = 0 \tag{72}$$

as required. //

Note that if $X_0 : \mathbb{R}^m \to \mathbb{L}(\mathbb{R}^m,\mathbb{R}^m)$ extends X then the equation

$$dx_t = X_0(x_t) \circ dB_t$$

whose solutions lie on M when starting on M, has Ito form

$$dx_t = X_0(x_t)dB_t + \tfrac{1}{2} \sum_p DX_0{}^p(x_t)(X_0{}^p)(x_t))dt$$

and for $x \in M$

$$\sum_p DX_0{}^p(x)(X_0{}^p(x)) = \sum_p \alpha_x(X_0{}^p(x), X_0{}^p(x)) = \text{trace } \alpha_x$$

by (72) and (67). The standard example of this is the equation

$$dx_t = dB_t - \langle x_t, dB_t\rangle |x_t|^{-2} x_t - \tfrac{1}{2}(m-1)|x_t|^{-2} x_t dt \tag{73}$$

which gives Brownian motion on the sphere $S^{m-1}(r)$ of radius r if $x_0 \in S^{m-1}(r)$. For variations, extensions, and further references consult [95].

Note that from (70)

$$\operatorname{div} X^p(x) = \operatorname{trace} A_x(-, Q(x)E^p(x)) = \Sigma_{q=1}^m \langle \alpha_x(E^q(x), E^q(x)), Q(x)E^p(x)\rangle$$

$$= \Sigma_{q=1}^m \langle \operatorname{trace} \alpha_x, Q(x)E^p(x)\rangle \tag{74}$$

D. The *mean curvature normal* at x is $1/n$ trace $\alpha_x \in \nu_x M$. For a hypersurface $\nu_x M$ is a 1-dimensional subspace of $\mathbb{R}^m$ and a choice of orientation in νM (e.g. outward normal) gives the *mean curvature* as a real valued function on M.

Similarly for a hypersurface the second fundamental form can be treated as a real valued bilinear form on the tangent spaces to M. At a point x its eigenvalues are called the *principal curvatures* at x, and its eigenfunctions are the *principal directions at* x.

§8. Curvature and the derivative flow

A. Given intervals I and J of $\mathbb{R}$ and a piecewise C^1 map $u: I \times J \to M$, there are vector fields $\partial u/\partial s$ and $\partial u/\partial t$ over U (i.e. the derivatives with respect to the first and second variables respectively). Taking normal coordinates centred at a point $u(s,t)$ it is immediate from (17) that

$$D/\partial t\, \partial u/\partial s = D/\partial s\, \partial u/\partial t \tag{75}$$

If $F_t: M \times \Omega \to M$, $t \geq 0$, is a smooth flow for our S.D.S. and $v \in T_xM$, choose $\sigma: [-1,1] \to \mathbb{R}$ with $\sigma(0) = x$ and $\dot\sigma(0) = v$. Then

$$T_xF_t(v) = \partial/\partial s\, F_t(\sigma(s)) \qquad \text{a.s.} \tag{76}$$

Defining parallel translation via a horizontal lift as in §§4D, E, we can covariantly differentiate (76) in t using the analogue of (47) as definition, to get

$$Dv_t = D\partial/\partial s\, F_t(\sigma(s))$$

for $v_t = T_x F_t(v): \Omega \to T_{x_t}M$, and so by the analogue of (75), proved in exactly the same way,

$$Dv_t = D/\partial s\, Y(F_t(\sigma(s)) \circ dz_t$$

i.e.

$$Dv_t = \nabla Y(v_t) \circ dz_t \tag{77}$$

For later use, and as an exercise, we will find an equation for $|v_t|^2$ assuming now that we have a Riemannian metric and are using its Levi-Civita connection. Certainly there is the Stratonovich equation

$$d|v_t|^2 = 2\langle v_t, \circ Dv_t\rangle_{x_t} = 2\langle v_t, \nabla Y(v_t) \circ dz_t\rangle_{x_t} \tag{78}$$

obtained, for example, by parallel translation back to x. A safe way to get the Ito form is to use the Ito formulae (5): let S(t,x)(e) denote the flow of Y(-)(e) and set $\delta S(t,v) = T_x(S(t,x)(e))(v)$ for $v \in T_xM$, so

$$D/\partial t\ \delta S(t,v) = \nabla Y(\delta S(t,v))(e).$$

Therefore

$$\begin{aligned} d/dt\ |\delta S(t,v)|^2 &= 2\langle \delta S(t,v), \nabla Y(\delta S(t,v))e\rangle \\ &= 2\langle v, \nabla Y(v)e\rangle \quad \text{at } t = 0 \end{aligned}$$

and, at t = 0,

$$\begin{aligned} d^2/dt^2\ |\delta S(t,v)|^2 &= 2\langle \nabla Y(v)e, \nabla Y(v)e\rangle + 2\langle v, \nabla Y(\nabla Y(v)e)e\rangle \\ &\quad + 2\langle v, \nabla^2(Y(x)e, v)e\rangle \end{aligned} \tag{79}$$

where, for Z a vector field and $u \in T_xM$,

$$\nabla^2 Z(u,-) = \nabla_u(\nabla Z) : T_xM \to T_xM \tag{80}$$

Thus

$$\begin{aligned} d|v_t|^2 &= 2\langle v_t, \nabla Y(v_t)dz_t\rangle + \langle \nabla Y(v_t)dz_t, \nabla Y(v_t)dz_t\rangle \\ &\quad + \langle v_t, \nabla Y(\nabla Y(v_t)dz_t)dz_t\rangle + \langle v_t, \nabla^2 Y(Y(x_t)dz_t, v_t)dz_t\rangle \end{aligned} \tag{81}$$

(To be convinced of the applicability of equation (5) observe that v_t is actually a solution of the S.D.E. on TM, $dv_t = \delta Y(v_t) \circ dz_t$, given in §4E, Chapter I.) We shall simplify (81) in §8C below.

B. The curvature R of an affine connection is a section of the bundle $\mathbb{L}(TM, TM; \mathbb{L}(TM;TM))$ so it can be considered as a map

$$R : TM \oplus TM \to \mathbb{L}(TM;TM)$$

where $TM \oplus TM = \bigcup\{T_xM \oplus T_xM : x \in M\}$, such that

$$u,v,w \to R(u,v)w$$

is tri-linear in $u,v,w \in T_xM$. It can be defined by

$$R(u,v)w = \nabla^2 W(u,v) - \nabla^2 W(v,u) \tag{82}$$

where W is a vector field such that W(x) = w at the given point x of M. We will see below in §9C that this definition gives a result independent of the choice of such W. If U, V, W are all vector fields there is the new one R(U,V)W given by

$$x \to R(U(x), V(x))W(x)$$

so

$$R(U,V)W = \nabla_U \nabla_V W - \nabla_V \nabla_U W - \nabla_{[U,V]}W \tag{83}$$

for a torsion free connection (using (24) and §1D, Chapter II).

From (82) R is anti-symmetric in its first two variables. For the Levi-Civita connection of a Riemannian metric it turns out that $\langle R(u_1,v_1)u_2,v_2\rangle_x$ is antisymmetric in u_1, v_1 and in u_2, v_2 and satisfies

$$\langle R(u_1,v_1)u_2,v_2\rangle_x = \langle R(u_2,v_2)u_1,v_1\rangle_x \tag{84}$$

This is an automatic consequence of the skew-symmetry and the relation

$$R(u,v)w + R(v,w)u + R(w,u)v = 0 \tag{85}$$

which can be proved by choosing vector fields U, V, W with U(x) = u, V(x) = v, W(x) = w and which commute i.e. [U,V] = [V,W] = [W,U] = 0. For details see e.g. [64] or [79] (where a different sign is used!).

For the Levi-Civita connection the *Riemannian curvature tensor* is defined by

$$\begin{aligned} R(v_1, v_2, v_3, v_4) &= \langle R(v_1, v_2)v_4, v_3\rangle_x \\ &= -\langle R(v_1,v_2)v_3,v_4\rangle_x \end{aligned} \tag{86}$$

for $v_i \in T_xM$, (note the sign difference from [33]) and the *sectional curvature* $K_P(x)$ of a plane P in T_xM is

$$K_P(x) = R(v_1, v_2, v_1, v_2) \tag{87}$$

where $\{v_1, v_2\}$ is an orthonormal base for P.

When M is a surface, i.e. n = 2, there is a unique P at x, namely $P = T_xM$, and then $K_P(x)$ coincides with the classical *Gaussian curvature* of M at x. For M a submanifold of $\mathbb{R}^3$ with induced metric it was Gauss's famous theorem aegregium which showed

$$K_P(x) = \lambda_1 \lambda_2 \tag{87}$$

where λ_1, λ_2 are the principal curvatures at x (§7D), see [65] for example.

Continuing with the Riemannian case there is the Ricci curvature

$$Ric : TN \oplus TN \to \mathbb{R}$$

$$\begin{aligned} Ric(v_1,v_2) &= \text{trace}\,[v \to R(v,v_1)v_2] \\ &= \Sigma_j \langle R(e_j,v_1)v_2,e_j\rangle \end{aligned} \tag{88}$$

for $e_1,\dots,e_n$ an o.n. base in T_xM. Clearly it is symmetric. Observe that Ric(v,v) is the sum of the sectional curvatures of any family of (n-1) mutually orthogonal planes P in T_xM containing v, provided $|v| = 1$. Thus bounds on the Ricci curvature are weaker assumptions than bounds on the sectional curvatures.

The Ricci curvature plays a very important role in the study of diffusions. An important result of differential geometry is that if *the Ricci curvature is bounded below* i.e. if there exists a constant C with

$$Ric(u,u) \geq C|u|^2 \quad \text{for all } u \in TM$$

then for any $x_0 \in M$ *if* $r(x) = d(x,x_0)$ *then* Δr *is bounded above uniformly at all points where* r *is differentiable* [101], [58]. From the Ito formula applied to r as if it were C^2 the result of S.-T. Yau that a *complete Riemannian manifold with Ricci curvature bounded below is stochastically complete* (i.e. its Brownian motion does not explode), is no surprise [43], p. 242: see [38] for an analytic proof, and [61] for a probabilistic version with an Ito formula for $r(x_t)$ involving a local time at the points where it is not differentiable.

Taking the trace of the Ricci curvature at x gives the *scalar curvature,* a real valued function $K : M \to \mathbb{R}$

$$K(x) = \Sigma_{i=1}^n Ric(e_i,e_i) \tag{89}$$

Note that for a surface K(x) is twice the Gaussian curvature

$$K(x) = Ric(e_1,e_1) + Ric(e_2,e_2) = 2Ric(e_1,e_1)$$
$$= 2K_P(x) \tag{90}$$

A space is said to have *constant curvature* if all its sectional curvatures $K_P(x)$ are the same (it suffices to know they are independent of P for each $x \in M$ by a result of Schur, see [64]). If so

$$R(u,v)w = k\,(\langle w,v\rangle u - \langle w,u\rangle v) \tag{91}$$

where $k = K_P(x)$. For such a space of constant curvature k we see

$$Ric(u,v) = (n-1)k\,\langle u,v\rangle_x \tag{92}$$

and

$$K(x) = n(n-1)k \tag{93}$$

C. We can now get improved formulae for the norms of the derivative flow. Suppose the original S.D.E. was $dx_t = X(x_t) \circ dB_t + A(x_t)dt$ with generator $\mathcal{A}$ =

$\frac{1}{2}\Delta + Z$, where Z is a vector field and M is Riemannian. Write A = W + Z. Then, for each $x \in M$, by §4A

$$\tfrac{1}{2} \ \Sigma_p \ \nabla X^p(X^p(x)) + W(x) = 0$$

Differentiating this in the direction of v, for $v \in T_{x}M$

$$\Sigma_p \ \nabla^2 X^p(v, X^p(x)) + \Sigma_p \ \nabla X^p(\nabla X^p(v)) + 2\nabla W(v) = 0$$

whence, by (82) and the definition (88) of Ric

$$\Sigma_p \langle \nabla^2 X^p(X^p(x), v), v\rangle + \mathrm{Ric}(v,v) + \langle \Sigma_p \nabla X^p(\nabla X^p(v)) + 2\nabla W(v), v\rangle = 0 \qquad (94)$$

Substituting in (81):

$$d|v_t|^2 = 2\langle v_t, \nabla X(v_t)dB_t\rangle + 2\langle v_t, \nabla Z(v_t)\rangle dt$$
$$- \mathrm{Ric}(v_t,v_t)dt + \Sigma_p \langle \nabla X^p(v_t), \nabla X^p(v_t)\rangle dt \qquad (95)$$

In particular we see $|v_t| \in L^2(\Omega, \mathcal{F}, \mathbb{P})$ provided the quadratic forms

$v \to 2\langle v, \nabla Z(v)\rangle - \mathrm{Ric}\,(v,v)$ and

$v \to |\nabla X^p(v)|^2$, $p = 1,2,\ldots,m$

for $v \in T_xM$ are bounded above uniformly over M. Equation (95) and the growth of $|v_t|$ will be examined in detail in special cases in Chapter III.

§9. Curvature and torsion forms

A. When trying to find a useful analogue of (95) for the canonical S.D.S. on the frame bundle of a Riemannian manifold we shall need the *curvature form*, so we will describe it here and use it to prove some of the basic results about curvature which were stated in §8.

First suppose we have a principal G-bundle $\pi : B \to M$ with a connection form $\tilde{\omega}$. This is a $\mathfrak{g}$-valued 1-form on B. As with a real valued one-form it has exterior derivative

$$d\tilde{\omega} : TB \oplus TB \to \mathfrak{g}$$
$$(v_1, v_2) \to d\tilde{\omega}(v_1, v_2)$$

which is antisymmetric, satisfying

$$d\tilde{\omega}(U(b), V(b)) = U(\tilde{\omega}(V))(b) - V(\tilde{\omega}(U))(b) - \tilde{\omega}([U,V](b)) \qquad (96)$$

for U,V vector fields, where $\tilde{\omega}(V)$ is the $\mathfrak{g}$-valued function on B, $b \to \tilde{\omega}(V(b))$, etc; see the discussion of differential forms in Chapter IV.

<u>N.B.</u> Here we are departing from the convention of Kobayashi and Nomizu

[64], which would have a factor of $\frac{1}{2}$ multiplying the right hand side of (96).

The curvature form is the 2-form Ω given by

$$\Omega : TB \oplus TB \to g$$

$$\Omega(v_1,v_2) = d\tilde{\omega}(hv_1,hv_2) \qquad (97)$$

for v_1, $v_2 \in T_bB$ where $h : TB \to HTB$ is the projection onto the horizontal subspace. By the invariance property (8) of $\tilde{\omega}$, i.e. $\tilde{\omega} \circ TR_g = \mathrm{ad}(g^{-1}) \circ \tilde{\omega}$, we have

$$\Omega(T_b R_g(v_1), T_b R_g(v_2)) = \mathrm{ad}(g^{-1})\Omega(v_1,v_2) \qquad (98)$$

for $b \in B$, $g \in G$, and v_1, $v_2 \in T_bB$, because of the invariance of exterior differentiation under diffeomorphisms: in our case

$$d(\tilde{\omega} \circ TR_g)(v_1,v_2) = d\tilde{\omega}(TR_g(v_1),TR_g(v_2)).$$

B. To see the importance of Ω suppose V_1, V_2 are horizontal vector fields on B so $V_i(b) \in HT_bB$ for each b. Then $\tilde{\omega}(V_i(b)) = 0$ for all b and so by (96):

$$\Omega(V_1(b), V_2(b)) = - \tilde{\omega}([V_1,V_2](b)) \qquad (99)$$

Thus if Ω vanishes identically the Lie bracket of horizontal vectors is horizontal: but this is precisely the classical necessary and sufficient condition of Frobenius' theorem for the integrability of $\{HT_bB : b \in B\}$ i.e. for the existence of a submanifold through each b of B with HT_bB as its tangent space. (In particular if this happens for a Levi-Civita connection the canonical S.D.S. will be far from hypo-elliptic).

C. Now suppose we have an affine connection, so $B = GLM$, or OM for the Riemannian case. Then there is the *canonical 1-form*

$$\theta : TB \to \mathbb{R}^n$$

$$\theta(v) = u^{-1}(T\pi(v)) \qquad v \in T_uB \qquad (100)$$

From this we obtain the torsion form

$$\Theta(v_1,v_2) = d\theta(hv_1,hv_2) \qquad v_i \in T_uB. \qquad (101)$$

It satisfies

$$\Theta(TR_g v_1, TR_g v_2) = g^{-1}\, \Theta(v_1,v_2) \qquad (102)$$

Because of (98) and (102) there exist

$$T : TM \oplus TM \to TM$$

bilinear and skew symmetric from $T_xM \oplus T_xM \to T_xM$ and

$$R : TM \oplus TM \to \mathbb{L}(TM;TM)$$

again bilinear and skew symmetric, defined by

$$T(v_1,v_2) = u\ \Theta(H_u v_1, H_u v_2) \tag{103}$$

and

$$R(v_1,v_2)v_3 = u\,\Omega\,(H_u v_1, H_u v_2)u^{-1}(v_3) \tag{104}$$

for $v_i \in T_xM$, $u \in \pi^{-1}(x)$, and H_u the horizontal lift operator.

These are the *torsion* and *curvature tensors* of our connection. We must show that this definition of R coincides with the one in §8:

<u>**Proposition 9C.**</u>

For T, R *defined by* (104), *if* V,W *are vector fields on* M *and* v_1, v_2 *tangent vectors at* x *to* M

(i) *If* $T \equiv 0$ *then*

$$\nabla^2 V(v_1,v_2) - \nabla^2 V(v_2,v_1) = R(v_1,v_2)V(x)$$

(ii) $[V,W](x) = \nabla W(V(x)) - \nabla V(W(x)) - T(V(x),W(x))$ (104a)

In particular the connection is torsion free iff $T \equiv 0$ *or equivalently* $\Theta \equiv 0$.

<u>**Proof**</u>

(i) Choose $u_0 \in \pi^{-1}(x)$ and define the vector field V_i on B by

$$V_i(u) = H_u(uu_0^{-1}v_i) \qquad i = 1,2.$$

If $T \equiv 0$ then

$$\begin{aligned} 0 = \Theta(V_1,V_2) &= d\theta(V_1,V_2) \\ &= V_1\theta(V_2) - V_2\theta(V_1) - \theta([V_1,V_2]) \\ &= -\,\theta([V_1,V_2]) \end{aligned}$$

since $\theta(V_i)$ is constant for each i. Thus $T\pi([V_1,V_2]) = 0$ and so $[V_1,V_2]$ is vertical. It follows that

$$[V_1,V_2](u_0) = A^*(u_0)$$

for

$$A = \tilde{\omega}([V_1,V_2](u_0)) = -\,\Omega(V_1(x),\ V_2(x))$$

(by equation (9) and (99)).

Now by the definition, (19), for $\tilde{V}(u) = u^{-1}V(\pi(u))$, etc.

$$\begin{aligned} \nabla^2 V(v_1,v_2) &= u_0\ d(\widetilde{\nabla V})(V_1(u_0))(u_0^{-1}v_2) \\ &= u_0\ V_1(\widetilde{\nabla V}(-)(u_0^{-1}v_2)) \\ &= u_0 V_1 V_2(\tilde{V})(u_0) \end{aligned} \tag{105}$$

Thus $\nabla^2 V(v_1,v_2) - \nabla^2 V(v_2,v_1) = u_0[V_1,V_2]\tilde{V}(u_0)$

$$= u_0\ A^*\ \tilde{V}(u_0)$$

$= u_0 \, d/dt \, \tilde{V}(u_0 \exp tA) |_{t=0}$

$= u_0 \, d/dt \exp(-tA) u_0^{-1} V(x) |_{t=0}$

$= -u_0 A u_0^{-1} V(x)$

$= u_0 \underline{\Omega}(v_1, v_2) u_0^{-1} V(x)$

as required.

Part (ii) can be proved similarly. //

D. **Example**

Let us compute the curvature of the sphere S^n using the identification of SOS^n with $SO(n+1)$ as in §5. Using the notation of §5 to compute $\underline{\Omega}$, because of left invariance it is enough to compute it at the point 1. For this let $\xi, \eta \in \mathbb{R}^n$ with corresponding $\underline{\xi}$ and $\underline{\eta}$ in $\underline{m}$. Take the left invariant vector fields U, V on SO(n+1) which are $\underline{\xi}$ and $\underline{\eta}$ at 1. Then

$[U,V](1) = [\underline{\xi}, \underline{\eta}] = \underline{B} \in \underline{so}(n)$

which confirms that [U,V] is vertical (cf. the proof of Proposition 9C). Therefore, by definition of the connection form $\tilde{\omega}_0$ and equation (99), if $\underline{\Omega}$ is the curvature form

$\underline{\Omega}(\underline{\xi}, \underline{\eta}) = -\underline{B}.$

Now if $v \in \mathbb{R}^n$, (identified with $(0,v)$ in $\mathbb{R}^{n+1}$) we see

$$[\underline{\xi}, \underline{\eta}]v = -\langle \eta, v \rangle \xi + \langle \xi, v \rangle \eta \qquad (106)$$

Also $T_1\pi : T_1 SO(n+1) \to T_N S^n \subset \mathbb{R}^{n+1}$ is just

$A \to (0, \alpha)$

where $\underline{\alpha}$ is the component of A in $\underline{m}$, since π maps g to g.N. Thus, if $v_i \in T_N S^n = \{0\} \times \mathbb{R}^n \subset \mathbb{R}^{n+1}$, for i = 1,2,3, the horizontal lift $H_1(v_i) = \underline{v}_i$ and

$R(v_1, v_2)v_3 = \underline{\Omega}(\underline{v}_i, \underline{v}_i)v_3$

$= \langle v_2, v_3 \rangle v_1 - \langle v_1, v_3 \rangle v_2$

which shows that S^n *has constant curvature* + 1, (c.f. equation (91)).

Exercises

(i) Check that the torsion form Θ vanishes identically, so that we have a complete proof that $\tilde{\omega}_0$ gives the Levi-Civita connection.

(ii) Do the same for hyperbolic space H^n, showing it has constant curvature -1; (the difference is in the use of $\underline{\xi}^*$ given by (63) instead of $\underline{\xi}$)

§10. The derivative of the canonical flow

A. Let M be a Riemannian manifold with Levi-Civita connection. Suppose, for simplicity of notation, that its canonical S.D.S. (X,B) on OM is strongly complete with flow $\{F_t(u,\omega) : t \geq 0, u \in OM, \omega \in \Omega\}$. To get equations like those of §8A for T_uF_t we would need an affine connection for OM rather than just M. There are various candidates, e.g. see [77], but the computations get rather complicated and we will go by a direct method as in [43] (which in fact boils down to using the 'canonical flat connection' for OM).

The derivative flow lives on TOM. However there is a canonical trivialization of this tangent bundle: i.e. a diffeomorphism

$$\psi : TOM \to M \times \mathbb{R}^n \times \underline{o}(n)$$

which restricts to a linear isomorphism $\psi_u : T_uOM \to \{u\} \times \mathbb{R}^n \times \underline{o}(n)$ for each $u \in OM$. This is given for $V \in T_uOM$ by

$$\psi(V) = (u, \theta(V), \tilde{\omega}(V)) \tag{107}$$

for θ, $\tilde{\omega}$ the fundamental form and connection form. Using this, for $V_0 \in T_{u_0}OM$ we will describe $TF_t(V_0)$ in terms of $\mathbb{R}^n$ - and $\underline{o}(n)$-valued processes $\xi_t = \theta(TF_t(V_0))$, $A_t = \tilde{\omega}(TF_t(V_0))$. This is similar to a procedure used in the Malliavin calculus [17], (although here we have differentiation with respect to the initial point, and there a 'differentiation' with respect to the basic noise $\{B_t : t \geq 0\}$ is being considered as in [42a])..

We will use the analogous notation to §8A so S(t,u)e gives the flow on OM of $u_t = X(u_t)(e)$ for $e \in \mathbb{R}^n$, and $\delta S(t,V)$ gives its derivative flow on TOM.

B. A vector field J along a geodesic γ in M is a *Jacobi field* if it satisfies

$$D^2J/\partial t^2 + R(J, d\gamma/dt)\, d\gamma/dt = 0 \tag{108}$$

They arise as infinitesimal variations of geodesics, [79], (see the proof below) and as a method of computing derivatives of the exponential map.

Lemma 10B. *For* $V \in T_uOM$ *and fixed* e *in* $\mathbb{R}^n$ set $J(t,V) = T\pi(\delta S(t,V))$. *Then* $J(-,V)$ *is a Jacobi field with*

$$J(0,V) = T\pi(V)$$

and

$$D/\partial t\, J(t,V)\big|_{t=0} = u\, \tilde{\omega}(V)(e). \tag{109}$$

Proof First recall from §4C, equation (44), that $t \to \pi(S(t,u)e) = \gamma_t(u,e)$ is a geodesic: so $J(-,V)$ is a vector field along a geodesic. Next take a horizontal path σ in OM with $\sigma(0) = u$ and $\dot\sigma(0) = hV$, and the path g in O(n) given by

$$g(s) = \exp s\, \tilde\omega(V).$$

Set $\rho(s) = \sigma(s).g(s)$. Then $\dot\rho(0) = V$ so that

$$J(t,V) = \partial/\partial s\, (\pi S(t,\rho(s))e)|_{s=0}$$
$$= \partial/\partial s\, \gamma_t(\rho(s),e)|_{s=0}.$$

Therefore

$$D/\partial t\, J(t,V) = D/\partial s\;\; \partial/\partial t\, \gamma_t(\rho(s),e)|_{s=0} \tag{110}$$

and by (82)

$$D^2/\partial t^2\, J(t,V) = D/\partial s\, D/\partial t\, \dot\gamma_t(\rho(s)),e)|_{s=0}$$
$$-R(J(t,V), \dot\gamma_t(u,e))\, \dot\gamma_t(u,e)$$

which shows J is a Jacobi field since $D/\partial t\, \partial/\partial t\, \gamma_t(\rho(s),e) = 0$ because γ is a geodesic.

Clearly $J(0,V) = T\pi(V)$ as claimed, and also by (110) and the proof of Theorem 4C:

$$D/\partial t\, J|_{t=0} = D/\partial s\, \rho(s)e|_{s=0} = D/\partial s\;\; \sigma(s).g(s)e|_{s=0}$$
$$= \sigma(s)\, d/ds\, g(s)e|_{s=0} = u\;\; \tilde\omega(V)e. \quad //$$

C. From the lemma we have

$$d/dt\, \theta(\delta S(t,V)) = d/dt\, (S(t,u)e)^{-1}\, T\pi(\delta S(t,V))$$
$$= (S(t,u)e)^{-1}\, D/\partial t\, J(t,V)$$
$$= \tilde\omega(V)e \tag{111}$$

at $t = 0$, and, at $t = 0$,

$$d^2/dt^2\, \theta(\delta S(t,V)) = u^{-1}\, D^2/\partial t^2\, J(t,V)$$
$$= -u^{-1}\, R(T\pi V, ue)ue \tag{112}$$

Now take any affine connection for the manifold OM which is torsion free (e.g. the Levi-connection for the metric on OM induced by ψ). For this

$$d/dt\, \tilde\omega\, (\delta S(t,V)) = \nabla\tilde\omega(X(S(t,u)e)e)(\delta S(t,V)) + \tilde\omega(D/\partial t\delta S(t,V)) \tag{113}$$

However, by §8A,

$$D/\partial t\, \delta S(t,V) = \nabla X(\delta S(t,V))e$$

and also since $X(u)e$ is horizontal, $\tilde\omega(X(u)e) = 0$ for all $u \in OM$ whence

$$\nabla\tilde\omega(V)(X(u)e) + \tilde\omega(\nabla X(V)e) = 0 \tag{114}$$

for all $V \in T_uOM$. Substituting in (113)

$$d/dt\ \tilde{\omega}(\delta S(t,V)) = \nabla\tilde{\omega}(X(S(t,u)e)e)(\delta S(t,V)) - \nabla\tilde{\omega}(\delta S(t,V))(X(S(t,u)e)e)$$

$$= d\tilde{\omega}(X(S(t,u)e)e,\ \delta S(t,V)) \tag{115}$$

by (96) (choosing suitable U,V which commute at S(t,u)e).

To proceed further we need the first of the following, and we state the second in passing; they are valid for any affine connection, and the first for any connection on a principal bundle:

Structure Equations: For $V_1,V_2 \in T_uOM$

(i) $d\tilde{\omega}(V_1,V_2) = -[\tilde{\omega}(V_1),\ \tilde{\omega}(V_2)] + \underline{\Omega}(V_1,V_2)$

(ii) $d\theta(V_1,V_2) = -(\tilde{\omega}(V_1)\theta(V_2) - \tilde{\omega}(V_2)\theta(V_1)) + \Theta(V_1,V_2)$ (117)

__Proof__ We prove only (i), for (ii) see [64] Theorem 2.4, Chapter III, (the $\frac{1}{2}$'s in [64] come from the different convention for exterior differentiation used there).

For (i): if both V_1, V_2 are horizontal the result is clear by definition of $\underline{\Omega}$. If both are vertical we can suppose $V_1 = A^*(u)$, $V_2 = B^*(u)$ at the given point u, for $A, B \in \underline{o}(n)$. Then

$$d\tilde{\omega}(A^*,B^*) = A^*\tilde{\omega}(B^*) - B^*\tilde{\omega}(A^*) - \tilde{\omega}([A^*,B^*])$$

$$= -[A,B] = -[\omega(A^*),\ \omega(B^*)]$$

since $\tilde{\omega}(B^*)$ and $\omega(A^*)$ are constant and $[A^*,B^*] = [A,B]^*$. This gives (i) in this case because $\underline{\Omega}(A^*,B^*) = 0$.

To complete the proof it is enough to suppose V_1 is horizontal, $V_1 = V(u)$ for some horizontal vector field V, say, and $V_2 = B^*(u)$ some $B \in \underline{o}(n)$. Again

$$d\tilde{\omega}(V_1,V_2) = V\ \tilde{\omega}(B^*(u)) - B^*\omega(V_1) - \tilde{\omega}([V,B^*](u))$$

$$= -\tilde{\omega}([V,B^*](u)).$$

Thus (i) is equivalent to $[V,B^*]$ being horizontal. However this is clear since $[V,B^*](u) = d/dt\ TW_t(V(u.\exp(-tB)))|_{t=0}$ where $W_tu = u.\exp(tB)$ and horizontality is preserved under right translation. //

Applying the first structure equation to (115):

$$d/dt\ \tilde{\omega}(\delta S(t,V)) = \underline{\Omega}(X(S(t,u)e)e,\ \delta S(t,V)) \tag{118}$$

$$= (S(t,u)e)^{-1}R((S(t,u)e)e,\ J(t,V))S(t,u)e \tag{119}$$

and so, at $t = 0$

$$d^2/dt^2\ \tilde{\omega}(\delta S(t,V)) = u^{-1}\ D/\partial t\ R((S(t,u)e)e,\ J(t,V))|_{t=0}\ u$$

$$= u^{-1}\ \nabla R(ue)(ue,\ u\theta(V))u$$

$$+\ u^{-1}\ R(ue,\ u\ \tilde{\omega}(V)e)u \qquad (120)$$

since $D/t\ (S(t,u)e)e = 0$.

From these we have our equations, in Stratonovich form by (111) and (119):

$$d\xi_t = A_t \circ dB_t \qquad (121a)$$

$$dA_t = u_t^{-1}\ R(u_t \circ dB_t,\ u_t\ \xi_t)u_t \qquad (121b)$$

and in Ito form using (112) and (120), for $e_1,\dots,e_n$ an orthonormal base for $\mathbb{R}^n$:

$$d\xi_t = A_t\ dB_t - \tfrac{1}{2}\ u_t^{-1}\ \mathrm{Ric}(u_t\ \xi_t,-)^{*} \qquad (122a)$$

$$dA_t = u_t^{-1}\ R(u_t\ dB_t,u_t\xi_t)u_t + \tfrac{1}{2}\Big(u_t^{-1}\ \Sigma_i\ \nabla R(u_te_i)(u_te_i,u_t\xi_t)u_t$$

$$+\ u_t^{-1}\ \Sigma_i\ R(u_te_i,\ u_tA_te_i)u_t\Big)dt \qquad (122b)$$

(It is shown in [43] p. 168 that the covariant derivative of R can be replaced by a term in the covariant derivative of the Ricci tensor since for an orthonormal base $f_1,\dots,f_n$ for T_xM

$$\langle(\Sigma_i\ \nabla R(f_i)(f_i,w))v_1,v_2\rangle_x = \nabla \mathrm{Ric}(v_2)(v_1,w) - \nabla \mathrm{Ric}(v_1)(v_2,w) \qquad (123)$$

for all v_1,v_2, w in T_xM.)

Note that if dB_t is replaced by dt in (121a,b) we obtain the Jacobi field equation for $u_t\ \xi_t$. The fields $\{T\pi(TF_t(V)) : t \geq 0\}$ along the Brownian motion $\{\pi\ F_t(u) : t \geq 0\}$ have been called stochastic Jacobi fields, [71].

CHAPTER III: CHARACTERISTIC EXPONENTS FOR STOCHASTIC FLOWS

§1. The Lyapunov Spectrum

A. Suppose throughout this section that M is a compact connected Riemannian manifold with smooth S.D.S.

$$dx_t = X(x_t) \circ dB_t + A(x_t)dt$$

differential generator $\mathcal{A}$. Let $F_t : M \times \Omega \to M$, $t \geq 0$ denote its flow so for each $\omega \in \Omega$ we have a C^∞ diffeomorphism $F_t(-) : M \to M$, derivative $TF_t(-,\omega) : TM \to TM$. Let $(\Omega, \mathcal{F}, \mathbb{P})$ be the classical Wiener space of paths starting at 0 in $\mathbb{R}^m$ with $B_t(\omega) = \omega(t)$, and let $\theta_t : \Omega \to \Omega$ be the shift:

$$\theta_t(\omega)(s) = \omega(t+s) - \omega(t) \tag{124}$$

Then $\mathbb{P}$ is invariant under θ_t for $t \geq 0$.

A Borel probability measure ρ on M is *invariant* for our S.D.S. if

$$\mathbb{E}\, \rho \circ F_t(-,\omega)^{-1} = \rho \qquad t \geq 0 \tag{125}$$

Since M is compact there exists an invariant measure (e.g. see [102], XIII §4). The invariance of ρ depends only on $\mathcal{A}$, not on the choice of S.D.S. with $\mathcal{A}$ as generator. When $\mathcal{A}$ is elliptic then ρ is a smooth measure i.e. $\rho(dx) = \lambda(x)dx$ for some smooth λ where dx refers to the Riemannian volume element: it is also unique. This is because λ is a solution to the adjoint operator equation (e.g. see [59]).

Define

$$\Phi_t : M \times \Omega \to M \times \Omega$$

by

$$\Phi_t(x,\omega) = (F_t(x,\omega), \theta_t\omega).$$

Then for each $s,t \geq 0$

$$\Phi_t\, \Phi_s = \Phi_{t+s} \quad \text{a.s.}$$

since

$$F_t(F_s(x,\omega), \theta_s\omega) = F_{s+t}(x,\omega) \qquad \text{a.s.}$$

Also if ρ is invariant for the S.D.S., then $\rho \otimes \mathbb{P}$ is invariant for Φ_t since if $f : M \times \Omega \to \mathbb{R}$ is integrable

$$\iint f(F_t(x,\omega),\theta_t\omega)\rho(dx)\,\mathbb{P}(d\omega)$$
$$= \iiint f(F_t(x,\omega_1),\,\theta_t\omega_2)\rho(dx)\,\mathbb{P}(d\omega_1)\mathbb{P}(d\omega_2)$$
(because $\theta_t\omega$ is independent of $F_t(x,\omega)$)
$$= \iint f(x,\omega)\,\rho(dx)\,\mathbb{P}(d\omega).$$

Say that ρ is *ergodic* if $\rho \otimes \mathbb{P}$ is ergodic for $\{\Phi_t : t \geq 0\}$ i.e. if the only measurable sets in $M \times \Omega$ which are invariant under $\{\Phi_t : t \geq 0\}$ have $\rho \otimes \mathbb{P}$-measure 1 or 0. This agrees with the definition in [102]. An ergodic decomposition for any invariant ρ is given in [102].

B. In this chapter we shall be mainly concerned with looking at special examples of the following version by Carverhill [20] of Ruelle's ergodic theory of dynamical systems:

__Theorem 1B__ *Let ρ be an invariant probability measure for $\mathcal{A}$. Then there is a set $\Gamma \subset M \times \Omega$ of full $\rho \otimes \mathbb{P}$-measure such that for each $(x,\omega) \in \Gamma$ there exist numbers*

$$\lambda^{(r)}(x) < \ldots < \lambda^{(1)}(x)$$

and an associated filtration by linear subspaces of T_xM

$$0 = V^{(r+1)}(x,\omega) \subset V^{(r)}(x,\omega) \subset \ldots \subset V^{(1)}(x,\omega) = T_xM$$

such that if

$$v \in V^{(j)}(x,\omega) - V^{(j+1)}(x,\omega)$$

then

$$\lim_{t\to\infty} 1/t \log \|TF_t(v,\omega)\| = \lambda^{(j)}(x) \tag{126}$$

where $\| \ \|$ denotes the norm using the Riemannian metric of M. Moreover for $(x,\omega) \in \Gamma$ the multiplicities $m_j(x) := \dim V^{(j)}(x,\omega) - \dim V^{(j+1)}(x,\omega)$ do not depend on ω and if

$$\lambda_\Sigma(x) = \sum_j m_j(x)\,\lambda^j(x) \tag{127}$$

then

$$\lambda_\Sigma(x) = \lim\ 1/t\ \log \det T_xF_t(-,\omega). \tag{128}$$

(strictly speaking we should write $|\det T_xF_t(-,\omega)|$ here and below or use some other convention to ensure it is continuous in t with value 1 at $t = 0$).

__Proof__

Following [20] embed M in $\mathbb{R}^{n+p}$ for some p and extend X, A just as in §2C of

Chapter I to give X, A on $\mathbb{R}^{n+p}$ with compact support. Let $F_t(x,\omega)$ refer to the flow of this system, and Φ_t to the flow on $\mathbb{R}^{n+p} \times \Omega$; ρ remains an invariant measure on $\mathbb{R}^{n+p}$, concentrated on M.

Fix some time $T > 0$ and for $n = 0,1,2,\ldots$ set

$$G_n(x,\omega) = DF_T(\Phi_{nT}(x,\omega)) : \mathbb{R}^{n+p} \to \mathbb{R}^{n+p}$$

(where the D refers to differentiation in $\mathbb{R}^{n+p}$). Set

$$G^n(x,\omega) = G_{n-1}(x,\omega) \circ \ldots \circ G_0(x,\omega)$$

so

$$G^n(x,\omega) = DF_{nT}(x,\omega),$$

by the chain rule.

Fairly standard estimates show that both

$$\log^+ \|DF_T(x,\omega)\| \in L^1(M \times \Omega;\, \rho \otimes \mathbb{P}) \tag{129}$$

and

$$\log^+ \|DF_T(x,\omega)^{-1}\| \in L^1(M \times \Omega;\, \rho \otimes \mathbb{P}) \tag{130}$$

We can therefore apply the Oseledec multiplicative ergodic theorem, as in Ruelle [87], to $\{G_n(x,\omega)\}_{n=0}^{\infty}$ and obtain the theorem in a discrete time version, for X, A, and with the λ^j and m_j possibly depending on (x,ω).

To deduce the continuous time version from this as in [87] the integrability of $\sup_{0\le t\le T} \log \|DF_t(x,\omega)\|$ and of

$$\sup_{0\le t\le T} \log \|D(F_T(-,\omega) \circ F_t(-,\omega)^{-1})F_t(x,\omega)\|$$

are used. Having done this the filtrations for F_t on M are obtained by intersecting those for F with each T_xM. To show that the λ^j and m_j do not depend on ω as described, [24], use the fact that this is certainly true if ρ is ergodic (in which case they can be taken independent of x for suitable choice of Γ), and then the fact that, even if not ergodic, ρ can be decomposed into ergodic measures concentrated on disjoint subsets of M, [102]. //

C. When the system is non-degenerate ρ is ergodic, and unique, and Γ can be chosen so that the *exponents* λ^j and multiplicities m_j, and consequently the *mean exponent* $1/n\ \lambda_\Sigma$, are all independent of x.

D. When ρ is ergodic and an exponent λ^j is negative there is a *stable manifold theorem* due to Carverhill [20]. Its proof comes from Ruelle's in [87] for deterministic systems using the embedding method and regularity estimate, for T > 0,

$$\mathbb{E} \sup_{0 \le t \le T} \|F_t(-,\omega)\|_{C^2} < \infty$$

<u>Theorem 1D</u> (Stable manifold theorem) *For ergodic* ρ *with* $\lambda^{(j)} < 0$ *for some* j *the set* Γ *of Theorem* 1B *can be chosen so that for each* $(x,\omega) \in \Gamma$ *the set* $\mathcal{V}^{(j)}(x,\omega)$ *given by*

$$\mathcal{V}^{(j)}(x,\omega) = \{y: \limsup_{t\to\infty} 1/t \log d(F_t(x,\omega),F_t(y,\omega)) \le \lambda^{(j)}\} \qquad (131)$$

is the image of $V^{(j)}(x,\omega)$ *under a smooth immersion tangent to the identity at* x. //

This means there is a smooth $f : V^{(j)}(x,\omega) \to M$ which has $T_v f$ injective for each v in $V^{(j)}(x,\omega)$ and has $T_0 f = 1d$, with $\mathcal{V}^{(j)}(x,\omega)$ as its image. It will be locally a diffeomorphism onto its image but not necessarily globally.

If $\lambda^{(1)} < 0$ we say the system is *stable*. In this case $\mathcal{V}^{(1)}(x,\omega)$ will be an open subset of M.

<u>N.B.</u> When M is compact any two Riemannian metrics $\{\langle -,- \rangle_x : x \in M\}$ $\{\langle -,- \rangle'_x : x \in M\}$ are equivalent: there exists c > 0 with $c^{-1}\langle u,u\rangle'_x \le \langle u,u\rangle_x \le c\langle u,u\rangle'_x$ for $u \in T_xM$. Consequently the Lyapunov spectrum and stable manifolds are independent of the choice of metrics. This would not be true if we allowed M to be non-compact e.g. as in [26], [27].

E. The first examples to be studied were the 'noisy North-South flow on S^1' (i.e. in stereographic projection so the North pole corresponds to ∞ and the South pole to 0 the equation is

$$dx_t = \varepsilon\, Y(x_t) \circ dB_t - dt$$

where $\{B_t : t \ge 0\}$ is one-dimensional and Y corresponds to the unit vector field on S^1), with variations [21], and gradient Brownian flows, canonical flows on the frame bundle, and some stochastic mechanical flows, [22]. The latter are usually flows on the non-compact space $\mathbb{R}^n$ of the form

$$dx_t = dB_t + \nabla \log \psi \, dt \tag{132}$$

where ψ is a smooth L^2 function, with $\nabla\psi$ in L^2, which is positive (at least for the 'ground state' flow when it is the leading eigenfunction of Schrodinger operator $-\frac{1}{2}\Delta + V$, for V a real valued function). However the estimates still work to allow the exponents to be defined, [26], [27]; and something can be said even in some cases when ψ is time dependent (the quasi-periodic case, for ψ_0 a linear combination of eigenfunctions of our operator). For spaces of constant curvature, or more generally for Riemannian symmetric spaces, the frame bundle can be identified with $p : G \to M \approx G/H$ for some Lie group G as in §5 of Chapter II, and the study of the asymptotic behaviour of the canonical flow in this situation was begun by Malliavin and Malliavin in 1974/75 [71], [73]. We shall look at gradient flows and canonical flows, and for the latter look in detail at the situation for the hyperbolic plane. However we will mainly look at it directly rather than use the group theoretic approach because the latter is special to the case of symmetric spaces and so not likely to be much help in obtaining results for spaces with non-constant curvature. The group theoretic approach is carried through in detail in [10], where there is also a very nice analysis of gradient Brownian flows on spheres.

For more recent work see [63a], [63b].

§2. Mean exponents

A. The mean exponents λ_Σ are considerably easier to study than the actual exponents themselves. Since they are given by (128) :

$$\lambda_\Sigma(x) = \lim_{t\to\infty} 1/t \log \det T_xF_t(-,\omega)$$

their existence depends only on the usual additive ergodic theorem, rather than the more sophisticated multiplicative theorem

The covariant equation (77): $Dvt = \nabla Y(v_t) \circ dz_t$ can be interpreted after parallel translation back to x_0 and the addition of an S.D.E. for the horizontal lift of x_t as an equation on $OM \times \mathbb{L}(\mathbb{R}^n;\mathbb{R}^n)$. To describe det TF_t we can therefore use the Ito formula, and with the notation used in §8A of Chapter II we must calculate d/dt log det $\delta S(t,-)$at $t = 0$ and its second derivative. In fact it is a classical result for flows of ordinary differential equations, (the *continuity equation*), which is left as an exercise, that

$$d/dt \det \delta S(t,-) = \operatorname{div} Y_e(S(t,-)e).\det \delta S(t,-). \tag{133}$$

where Y_e is the vector field $Y(-)e$. See Lemma 2B of Chapter V.

Thus, at $t = 0$,

$$d/dt \log \det \delta S(t,-) = \operatorname{div} Y_e$$

and

$$d^2/dt^2 \log \det \delta S(t,-) = <\nabla(\operatorname{div} Y_e), Y_e>.$$

Consequently

$$\log \det T_x F_t = \int_0^t \sum \operatorname{div} X^p(x_s) dB^p_s + \int_0^t \operatorname{div} A(x_s) ds$$
$$+ \tfrac{1}{2} \int_0^t \sum d<\nabla \operatorname{div} X^p(x_s), X^p(x_s)> ds \tag{134}$$

Since M is compact Σ div $X^p(s)$ is bounded, so the Ito integral in (134) is a time changed Brownian motion $B_{\tau(t)}$, say with $\tau(t) \le$ const.t for all t. Therefore $t^{-1}B_{\tau(t)} \to 0$ as $t \to \infty$. Applying the ergodic theorem, with the notation of §1A, for ρ almost all x:

$$\lim 1/t \log \det T_x F_t = \int_{M\times\Omega} (\operatorname{div} A(x) + \tfrac{1}{2} \sum <\nabla \operatorname{div} X^p(x), X^p(x)>) \; \rho(dx)\mathbb{P}(d\omega)$$

i.e.

$$\lambda_\Sigma = \int_M \operatorname{div} A(x)\, \rho(dx) + \tfrac{1}{2} \sum \int_M <\nabla \operatorname{div} X^p(x), X^p(x)> \rho(dx) \tag{135}$$

This is a special case of formulae by Baxendale for sums of the first k exponents in [12].

For a Brownian flow, i.e. when $\mathcal{A} = \frac{1}{2}\Delta$, the invariant measure ρ is just the normalized Riemannian measure. We can use the divergence theorem, equation (1.5), to dispose of the first term of (135). For the second term we can *integrate by parts*: in general if $f: M \to \mathbb{R}$ is C^1 and Z is a C^1 vector field applying the divergence thoerem to fZ together with the formula

$$\operatorname{div} fZ = f \operatorname{div} Z + <\nabla f, Z> \tag{136}$$

yields

$$\int_M f \operatorname{div} Z\, dx = - \int_M <\nabla f, Z> dx \tag{137}$$

Thus

$$\lambda_\Sigma = - \quad (2|M|)^{-1} \quad \int_M \Sigma \ (\text{div}\ X^P(x))^2 dx \tag{138}$$

where $|M|$ denotes the volume of M.

B. From (138) we see that in the Brownian case $\lambda_\Sigma \le 0$ with equality if and only if div $X^P = 0$ for each p. More general results are obtained by Baxendale in [12a] and we consider a simple version of those, assuming now that $\mathcal{A}$ is non-degenerate.

For Borel probability measure λ,μ on a Polish space X define the *relative entropy* $h(\lambda;\mu) \in \mathbb{R}\ (\ge 0) \cup \{+\infty\}$ by $h(\lambda;\mu) = \infty$ unless $\mu \le \lambda$ and

$$\int_X d\mu/d\lambda \ |\log d\mu/d\lambda\ | d\lambda < \infty \tag{140}$$

in which case

$$\begin{aligned} h(\lambda;\mu) &:= \int_X (d\mu/d\lambda \ \log d\mu/d\lambda) d\lambda \\ &= \int_X (\log d\mu/d\lambda)\ d\mu \end{aligned} \tag{141}$$

To see that $h(\lambda;\mu) \ge 0$ observe that $x \to x \log x$ is convex on $(0;\infty)$ so by Jensen's inequality if $h(\lambda,\mu) < \infty$ then

$$h(\lambda;\mu) \ge (\int d\mu/d\lambda\ d\lambda)\ \log (\int d\mu/d\lambda\ d\lambda) = 0$$

with equality if and only if $\mu = \lambda$.

We shall be particularly interested in the case where $\lambda = \rho$, the invariant measure of our S.D.S. on M, and $\mu = \rho_t$ where ρ_t is the random measure on M defined by

$$\rho_t(\omega)(A) = \rho(F_t(-,\omega)^{-1}(A))$$

for A a Borel set in M and $\omega \in \Omega$.

Following Baxendale [12a] and LeJan [69] we will consider $h(\rho;\rho_t)$:

<u>Theorem 2B</u> *For a non-degenerate system*

$$\lambda_\Sigma = - 1/t\ \mathbb{E}h(\rho;\rho_t) \tag{142}$$

Consequently $\lambda_\Sigma \le 0$ *with equality if and only if* ρ *is invariant under the sample flow* $\{t \to F_t(-,\omega): t \ge 0\}$ *for almost all* $\omega \in \Omega$.

<u>Proof</u>

Let λ denote the Riemannian measure of M, and for $t \ge 0$ abuse notation so that $\rho_t(dx)$ is written $\rho_t(x)dx$: by standard results, since $\mathcal{A}$ is elliptic, $\rho_t(-,\omega) : M \to \mathbb{R}$ is smooth and positive. Then $\rho_t(x) = \rho_o(y)(\det T_yF_t)^{-1}$ for $y = F_t^{-1}(x)$, and so

$$\mathbb{E}h(\rho;\rho_t) = \mathbb{E} \int (\log \rho_t(x) - \log \rho_o(x))\rho_t(x)dx$$

$$= \mathbb{E}\int (\log \rho_t(F_t(x)) - \log \rho_0(F_t(x)))\rho_0(x)dx$$
$$= \mathbb{E}\int (\log \rho_0(x) - \log\det T_xF_t - \log \rho_0(F_t(x)))\rho_0(x)dx$$
$$= -\,\mathbb{E}\int (\log\det T_xF_t)\rho_0(x)dx$$
$$= -\,t\lambda_\Sigma \text{ by (134) and (135). //}$$

In [12a], Baxendale gives a formula for $h(\rho;\rho_t)$ analogous to (134) and using analogous computations which can be based on the continuity equation

$$d/dt\,\{\rho_0(S(t,x)e)\det \delta S(t,-)\} = \mathrm{div}(\rho_0 Y(-)e)(S(t,x)e).\det \delta S(t,-) \quad (143)$$

See equation (285) in §2B of Chapter V below.

As he points out some non-degeneracy conditions are needed: in the completely degenerate case of an ordinary dynamical system, when ρ is the point mass at a source $\lambda_\Sigma > 0$, and when it is a point mass at a sink $\lambda_\Sigma < 0$, while in either case $h(\rho;\rho_t) = 0$.

C. For a gradient Brownian system (see §7 Chapter II) it is possible to get a neat upper bound. For $\varphi : M \to \mathbb{R}^m$ the isometric embedding, so that $X^p = \nabla\varphi^p$, equation (138) gives, by the Cauchy-Schwarz inequality,

$$\lambda_\Sigma = -\,1/(2|M|)\int_M \Sigma_p\,(\Delta\varphi^p)^2 dx$$
$$\le -\,1/(2|M|)\,\Sigma\,(\textstyle\int_M \varphi^p\Delta\varphi^p\,dx)^2\,(\int_M (\varphi^p)^2\,dx)^{-1}.$$

By a translation in $\mathbb{R}^m$ we can assume that $\int \varphi^p\,dx = 0$ for each p, i.e. that φ^p is orthogonal in L^2 to the solutions to $\Delta f = 0$ (i.e. the constants, since M is compact). Therefore, with an integration by parts

$$\lambda_\Sigma \le 1/(2|M|)\,\Sigma\int_M |X^p(x)|^2_x\,dx\;\int_M \varphi^p\,\Delta\varphi^p\,dx/\textstyle\int_M(\varphi^p)^2 dx$$
$$\le 1/(2|M|) \times (\text{leading eigenvalue of } \Delta)\int_M \Sigma_p\,|X^p(x)|^2\,dx.$$

Since $\Sigma_p |X^p(x)|^2 = n$ for all x, this yields Chappell's result [25]:

<u>Proposition 2C</u> Let μ be largest non-zero eigenvalue of Δ. Then

$$1/n\,\lambda_\Sigma \le \tfrac{1}{2}\mu. \qquad (144)$$

Moreover there is equality if and only if $\frac{1}{2}\Delta\varphi^p = \mu\varphi^p$ *for each* p, [25], [22].

Such embeddings have been studied by Takahashi, see [65] Note 14. The simplest examples are the spheres S^n with their standard embeddings in $\mathbb{R}^{n+1}$. Another example is the torus in $\mathbb{R}^4$ which is the image of

$$\varphi : S^1(1/\sqrt{2}) \times S^1(1/\sqrt{2}) \to \mathbb{R}^4$$

$$\varphi(u,v) = (1/\sqrt{2}\cos u,\ 1/\sqrt{2}\sin u,\ 1/\sqrt{2}\cos v,\ 1/\sqrt{2}\sin v) \tag{145}$$

Here $\lambda_\Sigma = -2$.

Note that by (74), in the gradient Brownian case, (138) is just

$$\lambda_\Sigma = -\ 1/(2|M|) \int_M |\operatorname{trace} \alpha_x|^2\, dx \tag{146}$$

For the sphere $S^n(r)$ of radius r in $\mathbb{R}^{n+1}$ this shows directly that

$$1/n\ \lambda_\Sigma = -\tfrac{1}{2}\, n/r^2 \tag{147}$$

since

$$\alpha_x(v,v) = -\frac{1}{r}|v|^2\,(x/|x|) \tag{148}$$

for $v \in T_x S^n(r)$.

§3. Exponents for gradient Brownian flows: the difficulties of estimating exponents in general

A. For a gradient Brownian flow, if $v_t = T_{x_0}F_t(v_0)$, equation (95) reduces, when $|v_0| = 1$, to

$$d|v_t|^2 = 2\langle v_t, \nabla X(v_t)dB_t\rangle - \operatorname{Ric}(v_t,v_t)dt + \Sigma_p\langle \nabla X^p(v_t), \nabla X^p(v_t)\rangle dt \tag{149}$$

giving

$$\begin{aligned} 1/t \log|v_t| = {} & 1/t \int_0^t \langle \eta_s, \nabla X(\eta_s)dB_s\rangle - 1/(2t)\int_0^t \operatorname{Ric}(\eta_s,\eta_s)ds \\ & - 1/t \int_0^t \Sigma_p \langle \eta_s, \nabla X^p(\eta_s)\rangle^2\, ds \\ & + 1/(2t)\int_0^t \Sigma_p \langle \nabla X^p(v_s), \nabla X^p(v_s)\rangle ds \end{aligned} \tag{150}$$

where $\eta_s = v_s/|v_s|$ in the sphere bundle SM for $S_xM = \{v \in T_xM: |v|_x = 1\}$. By (70) for $v \in T_xM$

$$\nabla X^p(v) = A_x(v, e_p - X^p(x))$$

so

$$\Sigma_p\langle \eta_s, \nabla X^p(\eta_s)\rangle^2 = |\alpha_x(\eta_s,\eta_s)|^2$$

and

$$\Sigma_p\, |\nabla X^p(v)|^2 = |\alpha_x(\eta_s,-)|^2.$$

Thus

$$\lim \frac{1}{t} \log|v_t| = \lim \frac{1}{t}\int_0^t \{\tfrac{1}{2}|\alpha_x(\eta_s,-)|^2 - |\alpha_x(\eta_s,\eta_s)|^2 - \tfrac{1}{2}\operatorname{Ric}(\eta_s,\eta_s)\}ds \quad (151)$$

(almost surely).

This can be modified by the use of Gauss's theorem that for $v \in T_xM$

$$\operatorname{Ric}(v,v) = -|\alpha_x(v,-)|^2 + \langle \alpha_x(v,v), \operatorname{trace} \alpha_x \rangle \quad (152)$$

so as to get an expression entirely in terms of the second fundamental form and process $\{\eta_t : 0 \le t < \infty\}$.

For $S^n(r)$ in $\mathbb{R}^{n+1}$, if $u, v \in T_xS^n(r)$

$$\operatorname{Ric}(u,v) = \frac{(n-1)}{r^2}\langle u,v\rangle \text{ and } \alpha_x(u,v) = -\frac{1}{r}\langle u,v\rangle\frac{x}{r}$$

so

$$\lim_{t\to\infty} 1/t \log|v_t| = -\tfrac{1}{2}n/r^2.$$

Thus

$$\lambda^1 = -\tfrac{1}{2}n/r^2 \quad (153)$$

and so $\lambda^1 = 1/n\, \lambda_\Sigma$ by (147). This shows that all the exponents for the spheres are the same. Bougerol [19] has shown that among all hypersurfaces it is only the spheres which possess this property.

B. The process $\{\eta_t : 0 \le t < \infty\}$ is given by an S.D.S. on SM. In fact this is just $d\eta_t = P(\eta_t)\delta X(\eta_t) \circ dB_t$ where $P(\eta)$ is the orthogonal projection in T_xM of T_xM onto TS_xM. By compactness of SM it will possess ergodic probability measures which project onto ρ. If ν is one of these we get for ν-almost all v_0

$$\lim 1/t \log|v_t| = \int_{SM} \{\tfrac{1}{2}|\alpha_x(\eta,-)|^2 - |\alpha_x(\eta,\eta)|^2 - \tfrac{1}{2}\operatorname{Ric}(\eta,\eta)\}\nu(d\eta) \quad (154)$$

As the right hand side of (154) varies over the ergodic measures ν which project onto ρ it gives a subset of the set Lyapunov exponents, sometimes called the Markovian, or deterministic, spectrum. They correspond to elements of the filtration which are non-random: see [23], [62] for details. In particular the top exponent λ^1 lies in this set.

Thus if the integrand of (154) is strictly negative for all $\eta \in SM$ the top

LISTE DES AUDITEURS

Mle ALPIUM M.T.	(85)	Lisbonne (Portugal)
Mle AMINE S.	(87)	Paris VI
Mr. ANDERSSON L.	(87)	Stockholm (Suède)
Mr. ANGULO IBANEZ J.	(87)	Grenade (Espagne)
Mle ARENAS C.	(86)	Barcelone (Espagne)
Mr. ARTZNER P.	(85)	Strasbourg I
Mle ATHAYDE E.	(85)	Lisbonne (Portugal)
Mr. AZEMA J.	(87)	Paris VI
Mle BABILLOT M.	(85)	Paris VII
Mr. BADRIKIAN A.	(85-86-87)	Clermont II
Mr. BALDI P.	(86-87)	Pise (Italie)
Mr. BARTOSZEWICZ J.	(86)	Wroclaw (Pologne)
Mr. BENASSI A.	(86-87)	Paris VI
Mr. BERNARD P.	(85-86-87)	Clermont II
Mr. BERTHUET R.	(85)	Clermont II
Mr. BOIVIN D.	(87)	Brest
Mr. BOUAZIZ M.	(87)	Clermont II
Mle BOUTON C.	(86-87)	Palaiseau
Mle BOXLER P.	(86)	Bremen (R.F.A.)
Mr. CANDELPERGHER B.	(87)	Nice
Mr. CANELA M.	(85)	Barcelone (Espagne)
Mme CANTO E CASTRO L.	(86)	Lisbonne (Portugal)
Mme CHALEYAT-MAUREL M.	(85-86)	Paris VI
Mr. CHASSAING P.	(86)	Nancy I
Mle CHEVET S.	(85-87)	Clermont II
Mr. CHOJNACKI W.	(87)	Varsovie (Pologne)
Mle CHOU S.	(87)	Paris
Mr. COMETS F.	(85)	Paris XI
Mr. DARWICH A.	(85)	Paris VI
Mr. DELL'ANTONIO G.	(87)	Rome (Italie)
Mr. DERMOUNE A.	(86-87)	Paris VI
Mme DEWITT-MORETTE	(87)	Austin(U.S.A.)
Mr. DOSS H.	(85)	Paris VI
Mr. DZIUBDZIELLA W.	(87)	Wroclaw (Pologne)
Mme ELIE L.	(85)	Paris VII
Mme EL KAROUI N.	(86)	Paris VI
Mr. EMERY	(87)	Strasbourg I

Mr. FEUERVERGER A.	(86)	Toronto (Canada)
Mr. FOUQUE J.P.	(86)	Paris VI
Mr. FOURT G.	(85-86-87)	Clermont II
Mme GABRIELLA DEL GROSSO	(87)	Rome (Italie)
Mr. GALLARDO L.	(85-86)	Nancy II
Mr. GOLDBERG J.	(86-87)	Lyon
Mr. GRAHAM C.	(86)	Palaiseau
Mr. GRORUD A.	(87)	Marseille
Mr. GUISSE M.	(86)	Abidjan (Cote d'Ivoire)
Mr. GUTIERREZ J.	(87)	Grenade (Espagne)
Mr. HENNEQUIN P.L.	(85-86-87)	Clermont II
Mr. HU Y.	(87)	Strasbourg I
Mr. HOPFNER R.	(86)	Ludwigs (R.F.A.)
Mr. HUARD A.	(86)	Besançon
Mme JOLIS M.	(86)	Barcelone (Espagne)
Mr. KERKYACHARIAN G.	(87)	Nancy I
Mr. KIPNIS C.	(85)	Paris VII
Mr. KOLSRUD T.	(87)	Stockholm (Suède)
Mle KOUKIOU F.	(86)	Lausanne (Suisse)
Mr. KREE P.	(87)	Paris VI
Mr. LAFFERTY J.	(87)	Cambridge (U.S.A.)
Mme LAPEYRE H.	(85)	Paris VI
Mr. LAPEYRE B.	(85)	Paris
Mr. LAPIDUS M.	(87)	Georgia (U.S.A.)
Mle LAWNICZAK A.	(85-86-87)	Toronto (Canada)
Mr. LEANDRE R.	(85)	Besançon
Mr. LE GALL J.F.	(85-86-87)	Paris VI
Mr. LE GLAND F.	(85)	Sophia Antipolis
Mr. LOBRY C.	(87)	Nice
Mr. LOTI-VIAUD	(85)	Paris VI
Mr. MANSMANN U.	(87)	Berlin (R.F.A.)
Mme MASSAM H.	(86)	Ontario (Canada)
Mr. MASSART P.	(85)	Paris XI
Mr. McGILL	(86)	Dublin (Irlande)
Mle MESSACI F.	(85)	Rouen
Mle MIGUENS M.	(85)	Lisbonne (Portugal)
Mle MILLET A.	(85)	Angers
Mr. MOGHA G.	(85-86)	Clermont II
Mle MORA M.	(86)	Toulouse

Mr. NUALART D.	(86)	Barcelone (Espagne)
Mme PARK S.	(87)	Austin (U.S.A.)
Mr. PEREZ	(87)	Grenade (Espagne)
Mme PERRIN Y.	(85-87)	Clermont II
Mr. PETRITIS D.	(86)	Lausanne (Suisse)
Mme PICARD D.	(86-87)	Paris VI
Mr. PICARD J.	(85-86-87)	Sophia Antipolis
Mr. PISTONE G.	(86)	Gênes (Italie)
Mme PONTIER M.	(87)	Orléans
Mr. ROSENBERG S.	(87)	Boston (U.S.A.)
Mr. ROUX D.	(85-86-87)	Clermont II
Mr. ROYER G.	(85)	Clermont II
Mr. RUSSO F.	(85-86)	Lausanne (Suisse)
Mle SAADA E.	(85)	Paris VI
Mr. SADI E. H.	(85-86)	Paris VI
Mme SANZ M.	(87)	Barcelone (Espagne)
Mle SAVONA C.	(86-87)	Lyon
Mr. SONG S.	(87)	Paris VI
Mr. SZCZOTKA W.	(87)	Worclaw (Pologne)
Mr. SZNITMAN A.	(86-87)	Paris VI
Mr. TERRANOVA D.	(87)	Milan (Italie)
Mr. TOUATI A.	(85)	Bizerte (Tunisie)
Mr. UGO	(87)	Italie
Mr. USTUNEL	(87)	Issy les Moulineaux
Mr. VALLOIS P.	(85-86)	Paris VI
Mle WEINRYB S.	(85-86)	Palaiseau
Mr. WICK D.	(87)	Colorado (U.S.A.)
Mr. WU L.	(86-87)	Paris VI
Mr. YCART B.	(87)	Pau

Vol. 1201: Curvature and Topology of Riemannian Manifolds. Proceedings, 1985. Edited by K. Shiohama, T. Sakai and T. Sunada. VII, 336 pages. 1986.

Vol. 1202: A. Dür, Möbius Functions, Incidence Algebras and Power Series Representations. XI, 134 pages. 1986.

Vol. 1203: Stochastic Processes and Their Applications. Proceedings, 1985. Edited by K. Itô and T. Hida. VI, 222 pages. 1986.

Vol. 1204: Séminaire de Probabilités XX, 1984/85. Proceedings. Edité par J. Azéma et M. Yor. V, 639 pages. 1986.

Vol. 1205: B.Z. Moroz, Analytic Arithmetic in Algebraic Number Fields. VII, 177 pages. 1986.

Vol. 1206: Probability and Analysis, Varenna (Como) 1985. Seminar. Edited by G. Letta and M. Pratelli. VIII, 280 pages. 1986.

Vol. 1207: P.H. Bérard, Spectral Geometry: Direct and Inverse Problems. With an Appendix by G. Besson. XIII, 272 pages. 1986.

Vol. 1208: S. Kaijser, J.W. Pelletier, Interpolation Functors and Duality. IV, 167 pages. 1986.

Vol. 1209: Differential Geometry, Peñíscola 1985. Proceedings. Edited by A.M. Naveira, A. Ferrández and F. Mascaró. VIII, 306 pages. 1986.

Vol. 1210: Probability Measures on Groups VIII. Proceedings, 1985. Edited by H. Heyer. X, 386 pages. 1986.

Vol. 1211: M.B. Sevryuk, Reversible Systems. V, 319 pages. 1986.

Vol. 1212: Stochastic Spatial Processes. Proceedings, 1984. Edited by P. Tautu. VIII, 311 pages. 1986.

Vol. 1213: L.G. Lewis, Jr., J.P. May, M. Steinberger, Equivariant Stable Homotopy Theory. IX, 538 pages. 1986.

Vol. 1214: Global Analysis – Studies and Applications II. Edited by Yu.G. Borisovich and Yu.E. Gliklikh. V, 275 pages. 1986.

Vol. 1215: Lectures in Probability and Statistics. Edited by G. del Pino and R. Rebolledo. V, 491 pages. 1986.

Vol. 1216: J. Kogan, Bifurcation of Extremals in Optimal Control. VIII, 106 pages. 1986.

Vol. 1217: Transformation Groups. Proceedings, 1985. Edited by S. Jackowski and K. Pawalowski. X, 396 pages. 1986.

Vol. 1218: Schrödinger Operators, Aarhus 1985. Seminar. Edited by E. Balslev. V, 222 pages. 1986.

Vol. 1219: R. Weissauer, Stabile Modulformen und Eisensteinreihen. III, 147 Seiten. 1986.

Vol. 1220: Séminaire d'Algèbre Paul Dubreil et Marie-Paule Malliavin. Proceedings, 1985. Edité par M.-P. Malliavin. IV, 200 pages. 1986.

Vol. 1221: Probability and Banach Spaces. Proceedings, 1985. Edited by J. Bastero and M. San Miguel. XI, 222 pages. 1986.

Vol. 1222: A. Katok, J.-M. Strelcyn, with the collaboration of F. Ledrappier and F. Przytycki, Invariant Manifolds, Entropy and Billiards; Smooth Maps with Singularities. VIII, 283 pages. 1986.

Vol. 1223: Differential Equations in Banach Spaces. Proceedings, 1985. Edited by A. Favini and E. Obrecht. VIII, 299 pages. 1986.

Vol. 1224: Nonlinear Diffusion Problems, Montecatini Terme 1985. Seminar. Edited by A. Fasano and M. Primicerio. VIII, 188 pages. 1986.

Vol. 1225: Inverse Problems, Montecatini Terme 1986. Seminar. Edited by G. Talenti. VIII, 204 pages. 1986.

Vol. 1226: A. Buium, Differential Function Fields and Moduli of Algebraic Varieties. IX, 146 pages. 1986.

Vol. 1227: H. Helson, The Spectral Theorem. VI, 104 pages. 1986.

Vol. 1228: Multigrid Methods II. Proceedings, 1985. Edited by W. Hackbusch and U. Trottenberg. VI, 336 pages. 1986.

Vol. 1229: O. Bratteli, Derivations, Dissipations and Group Actions on C*-algebras. IV, 277 pages. 1986.

Vol. 1230: Numerical Analysis. Proceedings, 1984. Edited by J.-P. Hennart. X, 234 pages. 1986.

Vol. 1231: E.-U. Gekeler, Drinfeld Modular Curves. XIV, 107 pages. 1986.

Vol. 1232: P.C. Schuur, Asymptotic Analysis of Soliton Problems. VIII, 180 pages. 1986.

Vol. 1233: Stability Problems for Stochastic Models. Proceedings, 1985. Edited by V.V. Kalashnikov, B. Penkov and V.M. Zolotarev. VI, 223 pages. 1986.

Vol. 1234: Combinatoire énumérative. Proceedings, 1985. Edité par G. Labelle et P. Leroux. XIV, 387 pages. 1986.

Vol. 1235: Séminaire de Théorie du Potentiel, Paris, No. 8. Directeurs: M. Brelot, G. Choquet et J. Deny. Rédacteurs: F. Hirsch et G. Mokobodzki. III, 209 pages. 1987.

Vol. 1236: Stochastic Partial Differential Equations and Applications. Proceedings, 1985. Edited by G. Da Prato and L. Tubaro. V, 257 pages. 1987.

Vol. 1237: Rational Approximation and its Applications in Mathematics and Physics. Proceedings, 1985. Edited by J. Gilewicz, M. Pindor and W. Siemaszko. XII, 350 pages. 1987.

Vol. 1238: M. Holz, K.-P. Podewski and K. Steffens, Injective Choice Functions. VI, 183 pages. 1987.

Vol. 1239: P. Vojta, Diophantine Approximations and Value Distribution Theory. X, 132 pages. 1987.

Vol. 1240: Number Theory, New York 1984–85. Seminar. Edited by D.V. Chudnovsky, G.V. Chudnovsky, H. Cohn and M.B. Nathanson. V, 324 pages. 1987.

Vol. 1241: L. Gårding, Singularities in Linear Wave Propagation. III, 125 pages. 1987.

Vol. 1242: Functional Analysis II, with Contributions by J. Hoffmann-Jørgensen et al. Edited by S. Kurepa, H. Kraljević and D. Butković. VII, 432 pages. 1987.

Vol. 1243: Non Commutative Harmonic Analysis and Lie Groups. Proceedings, 1985. Edited by J. Carmona, P. Delorme and M. Vergne. V, 309 pages. 1987.

Vol. 1244: W. Müller, Manifolds with Cusps of Rank One. XI, 158 pages. 1987.

Vol. 1245: S. Rallis, L-Functions and the Oscillator Representation. XVI, 239 pages. 1987.

Vol. 1246: Hodge Theory. Proceedings, 1985. Edited by E. Cattani, F. Guillén, A. Kaplan and F. Puerta. VII, 175 pages. 1987.

Vol. 1247: Séminaire de Probabilités XXI. Proceedings. Edité par J. Azéma, P.A. Meyer et M. Yor. IV, 579 pages. 1987.

Vol. 1248: Nonlinear Semigroups, Partial Differential Equations and Attractors. Proceedings, 1985. Edited by T.L. Gill and W.W. Zachary. IX, 185 pages. 1987.

Vol. 1249: I. van den Berg, Nonstandard Asymptotic Analysis. IX, 187 pages. 1987.

Vol. 1250: Stochastic Processes – Mathematics and Physics II. Proceedings 1985. Edited by S. Albeverio, Ph. Blanchard and L. Streit. VI, 359 pages. 1987.

Vol. 1251: Differential Geometric Methods in Mathematical Physics. Proceedings, 1985. Edited by P.L. García and A. Pérez-Rendón. VII, 300 pages. 1987.

Vol. 1252: T. Kaise, Représentations de Weil et GL_2 Algèbres de division et GL_n. VII, 203 pages. 1987.

Vol. 1253: J. Fischer, An Approach to the Selberg Trace Formula via the Selberg Zeta-Function. III, 184 pages. 1987.

Vol. 1254: S. Gelbart, I. Piatetski-Shapiro, S. Rallis. Explicit Constructions of Automorphic L-Functions. VI, 152 pages. 1987.

Vol. 1255: Differential Geometry and Differential Equations. Proceedings, 1985. Edited by C. Gu, M. Berger and R.L. Bryant. XII, 243 pages. 1987.

Vol. 1256: Pseudo-Differential Operators. Proceedings, 1986. Edited by H.O. Cordes, B. Gramsch and H. Widom. X, 479 pages. 1987.

Vol. 1257: X. Wang, On the C*-Algebras of Foliations in the Plane. V, 165 pages. 1987.

Vol. 1258: J. Weidmann, Spectral Theory of Ordinary Differential Operators. VI, 303 pages. 1987.

Vol. 1259: F. Cano Torres, Desingularization Strategies for Three-Dimensional Vector Fields. IX, 189 pages. 1987.

Vol. 1260: N.H. Pavel, Nonlinear Evolution Operators and Semi-groups. VI, 285 pages. 1987.

Vol. 1261: H. Abels, Finite Presentability of S-Arithmetic Groups. Compact Presentability of Solvable Groups. VI, 178 pages. 1987.

Vol. 1262: E. Hlawka (Hrsg.), Zahlentheoretische Analysis II. Seminar, 1984–86. V, 158 Seiten. 1987.

Vol. 1263: V.L. Hansen (Ed.), Differential Geometry. Proceedings, 1985. XI, 288 pages. 1987.

Vol. 1264: Wu Wen-tsün, Rational Homotopy Type. VIII, 219 pages. 1987.

Vol. 1265: W. Van Assche, Asymptotics for Orthogonal Polynomials. VI, 201 pages. 1987.

Vol. 1266: F. Ghione, C. Peskine, E. Sernesi (Eds.), Space Curves. Proceedings, 1985. VI, 272 pages. 1987.

Vol. 1267: J. Lindenstrauss, V.D. Milman (Eds.), Geometrical Aspects of Functional Analysis. Seminar. VII, 212 pages. 1987.

Vol. 1268: S.G. Krantz (Ed.), Complex Analysis. Seminar, 1986. VII, 195 pages. 1987.

Vol. 1269: M. Shiota, Nash Manifolds. VI, 223 pages. 1987.

Vol. 1270: C. Carasso, P.-A. Raviart, D. Serre (Eds.), Nonlinear Hyperbolic Problems. Proceedings, 1986. XV, 341 pages. 1987.

Vol. 1271: A.M. Cohen, W.H. Hesselink, W.L.J. van der Kallen, J.R. Strooker (Eds.), Algebraic Groups Utrecht 1986. Proceedings. XII, 284 pages. 1987.

Vol. 1272: M.S. Livšic, L.L. Waksman, Commuting Nonselfadjoint Operators in Hilbert Space. III, 115 pages. 1987.

Vol. 1273: G.-M. Greuel, G. Trautmann (Eds.), Singularities, Representation of Algebras, and Vector Bundles. Proceedings, 1985. XIV, 383 pages. 1987.

Vol. 1274: N. C. Phillips, Equivariant K-Theory and Freeness of Group Actions on C*-Algebras. VIII, 371 pages. 1987.

Vol. 1275: C.A. Berenstein (Ed.), Complex Analysis I. Proceedings, 1985–86. XV, 331 pages. 1987.

Vol. 1276: C.A. Berenstein (Ed.), Complex Analysis II. Proceedings, 1985–86. IX, 320 pages. 1987.

Vol. 1277: C.A. Berenstein (Ed.), Complex Analysis III. Proceedings, 1985–86. X, 350 pages. 1987.

Vol. 1278: S.S. Koh (Ed.), Invariant Theory. Proceedings, 1985. V, 102 pages. 1987.

Vol. 1279: D. Ieşan, Saint-Venant's Problem. VIII, 162 Seiten. 1987.

Vol. 1280: E. Neher, Jordan Triple Systems by the Grid Approach. XII, 193 pages. 1987.

Vol. 1281: O.H. Kegel, F. Menegazzo, G. Zacher (Eds.), Group Theory. Proceedings, 1986. VII, 179 pages. 1987.

Vol. 1282: D.E. Handelman, Positive Polynomials, Convex Integral Polytopes, and a Random Walk Problem. XI, 136 pages. 1987.

Vol. 1283: S. Mardešić, J. Segal (Eds.), Geometric Topology and Shape Theory. Proceedings, 1986. V, 261 pages. 1987.

Vol. 1284: B.H. Matzat, Konstruktive Galoistheorie. X, 286 pages. 1987.

Vol. 1285: I.W. Knowles, Y. Saitō (Eds.), Differential Equations and Mathematical Physics. Proceedings, 1986. XVI, 499 pages. 1987.

Vol. 1286: H.R. Miller, D.C. Ravenel (Eds.), Algebraic Topology. Proceedings, 1986. VII, 341 pages. 1987.

Vol. 1287: E.B. Saff (Ed.), Approximation Theory, Tampa. Proceedings, 1985–1986. V, 228 pages. 1987.

Vol. 1288: Yu. L. Rodin, Generalized Analytic Functions on Riemann Surfaces. V, 128 pages, 1987.

Vol. 1289: Yu. I. Manin (Ed.), K-Theory, Arithmetic and Geometry. Seminar, 1984–1986. V, 399 pages. 1987.

Vol. 1290: G. Wüstholz (Ed.), Diophantine Approximation and Transcendence Theory. Seminar, 1985. V, 243 pages. 1987.

Vol. 1291: C. Mœglin, M.-F. Vignéras, J.-L. Waldspurger, Correspondances de Howe sur un Corps p-adique. VII, 163 pages. 1987

Vol. 1292: J.T. Baldwin (Ed.), Classification Theory. Proceedings, 1985. VI, 500 pages. 1987.

Vol. 1293: W. Ebeling, The Monodromy Groups of Isolated Singularities of Complete Intersections. XIV, 153 pages. 1987.

Vol. 1294: M. Queffélec, Substitution Dynamical Systems – Spectral Analysis. XIII, 240 pages. 1987.

Vol. 1295: P. Lelong, P. Dolbeault, H. Skoda (Réd.), Séminaire d'Analyse P. Lelong – P. Dolbeault – H. Skoda. Seminar, 1985/1986. VII, 283 pages. 1987.

Vol. 1296: M.-P. Malliavin (Ed.), Séminaire d'Algèbre Paul Dubreil et Marie-Paule Malliavin. Proceedings, 1986. IV, 324 pages. 1987.

Vol. 1297: Zhu Y.-l., Guo B.-y. (Eds.), Numerical Methods for Partial Differential Equations. Proceedings. XI, 244 pages. 1987.

Vol. 1298: J. Aguadé, R. Kane (Eds.), Algebraic Topology, Barcelona 1986. Proceedings. X, 255 pages. 1987.

Vol. 1299: S. Watanabe, Yu.V. Prokhorov (Eds.), Probability Theory and Mathematical Statistics. Proceedings, 1986. VIII, 589 pages. 1988.

Vol. 1300: G.B. Seligman, Constructions of Lie Algebras and their Modules. VI, 190 pages. 1988.

Vol. 1301: N. Schappacher, Periods of Hecke Characters. XV, 160 pages. 1988.

Vol. 1302: M. Cwikel, J. Peetre, Y. Sagher, H. Wallin (Eds.), Function Spaces and Applications. Proceedings, 1986. VI, 445 pages. 1988.

Vol. 1303: L. Accardi, W. von Waldenfels (Eds.), Quantum Probability and Applications III. Proceedings, 1987. VI, 373 pages. 1988.

Vol. 1304: F.Q. Gouvêa, Arithmetic of p-adic Modular Forms. VIII, 121 pages. 1988.

Vol. 1305: D.S. Lubinsky, E.B. Saff, Strong Asymptotics for Extremal Polynomials Associated with Weights on ℝ. VII, 153 pages. 1988.

Vol. 1306: S.S. Chern (Ed.), Partial Differential Equations. Proceedings, 1986. VI, 294 pages. 1988.

Vol. 1307: T. Murai, A Real Variable Method for the Cauchy Transform, and Analytic Capacity. VIII, 133 pages. 1988.

Vol. 1308: P. Imkeller, Two-Parameter Martingales and Their Quadratic Variation. IV, 177 pages. 1988.

Vol. 1309: B. Fiedler, Global Bifurcation of Periodic Solutions with Symmetry. VIII, 144 pages. 1988.

Vol. 1310: O.A. Laudal, G. Pfister, Local Moduli and Singularities. V, 117 pages. 1988.

Vol. 1311: A. Holme, R. Speiser (Eds.), Algebraic Geometry, Sundance 1986. Proceedings. VI, 320 pages. 1988.

Vol. 1312: N.A. Shirokov, Analytic Functions Smooth up to the Boundary. III, 213 pages. 1988.

Vol. 1313: F. Colonius, Optimal Periodic Control. VI, 177 pages. 1988.

Vol. 1314: A. Futaki, Kähler-Einstein Metrics and Integral Invariants. IV, 140 pages. 1988.

Vol. 1315: R.A. McCoy, I. Ntantu, Topological Properties of Spaces of Continuous Functions. IV, 124 pages. 1988.

Vol. 1316: H. Korezlioglu, A.S. Ustunel (Eds.), Stochastic Analysis and Related Topics. Proceedings, 1986. V, 371 pages. 1988.

Vol. 1317: J. Lindenstrauss, V.D. Milman (Eds.), Geometric Aspects of Functional Analysis. Seminar, 1986–87. VII, 289 pages. 1988.

Vol. 1318: Y. Felix (Ed.), Algebraic Topology – Rational Homotopy. Proceedings, 1986. VIII, 245 pages. 1988

Vol. 1319: M. Vuorinen, Conformal Geometry and Quasiregular Mappings. XIX, 209 pages. 1988.

exponent will be negative. Using (152) it is straightforward to see that this holds for hypersurfaces if the principal curvatures $\ell_1(x),\dots,\ell_n(x)$ at each point x satisfy

$$l_j(x) > \frac{1}{n}(\tfrac{1}{2}+\varepsilon)(l_1(x)+\dots+l_n(x)) \qquad x \in M,\ j = 1,\dots,n$$

for some $\varepsilon > 0$, see [26]. This is a convexity condition. It seems reasonable to guess that $\lambda^1 < 0$ when M is the boundary of a convex domain.

B. Formula (150) is a version of Carverhill's version of Khasminski's formula (see his article in [B]). From it we get more general versions of (151) and (154). A major difficulty in extracting information from (154) is the lack of knowledge of the behaviour of the invariant measures ν, in particular lack of knowledge about their supports (one cannot expect the infinitesimal generator of $\{\eta_t : t \geq 0\}$ to be elliptic or even hypoelliptic in general). Control theory gets involved here: see the article by Arnold et al in [B], and also more recent work by L. Arnold and San Martin.

C. Rather than considering the process $\{\eta_t : t \geq 0\}$ on SM it is often more convenient to take its projection onto the projective bundle PM which is simply the quotient of SM obtained by identifying antipodal points in each fibre S_xM. It is shown in [12a] that given ellipticity of $\mathcal{A}$ (for example) there is an invariant measure ν for this process such that with ν_t its shift by the flow of the process on PM

$$\mathbb{E}\{h(\nu;\nu_1) - h(\rho;\rho_1)\} \leq n\,\lambda_1 - \lambda_\Sigma \tag{155}$$

where h is the relative entropy as in §2B. Using this Baxendale showed that all the exponents are equal given some non-degeneracy of $\mathcal{A}$ (e.g. ellipticity), if and only if there is a Riemannian metric such that the sample flows $F_t(-,\omega)$ are conformal diffeomorphisms. See also [19].

D. For gradient Brownian flows the exponents and their multiplicities are geometric invariants of the embedding of M into $\mathbb{R}^m$. We have seen that in general there are non trivial filtrations of tangent spaces T_xM. These are dependent on the embedding and the particular sample path: it is rather difficult to imagine what, necessarily long time, property of the sample path will determine the position of say $V^{(2)}(x,\omega)$ in T_xM.

§4. Exponents for canonical flows

A. Consider the canonical flow on the orthonormal frame bundle OM of the Riemannian manifold M, (or on SOM if M is orientable, for some orientation). There is a natural metric on OM defined by requiring that the trivialization ψ of TOM, given by equation (107) in §9 Chapter II gives isometries $\psi_u : T_uOM \to \mathbb{R}^n \times \underline{o}(n)$ for each $u \in OM$. Here the inner product on $\underline{o}(n)$ is taken to be

$$\langle A,B\rangle = -\tfrac{1}{2}\,\mathrm{trace}\,AB \tag{156}$$

for $A,B \in \underline{o}(n)$ identified as skew-symmetric matrices. The factor of $\frac{1}{2}$ has some advantages e.g. if $e \in \mathbb{R}^n$

$$|Ae| \le |A|\,|e| \tag{157}$$

with this definition. (A disadvantage is that it was not used in [24]). The corresponding measure on OM is sometimes called the Liouville measure. Since $T_u\pi : T_uOM \to T_{\pi(u)}M$ is an isometry on the horizontal subspace H_uOM and vanishes on its orthogonal complement, we see $T\pi$ maps the Liouville measure onto the Riemannian measure of M. Also by the invariance $\tilde{\omega}\circ TR_g = ad(g^{-1})\circ\tilde{\omega}$ of connection forms and the invariance under $ad(g^{-1})$ of the given inner product on $\underline{o}(n)$ it follows that the Liouville measure is invariant under the right action of $O(n)$ on OM.

It is a standard result, observed by Malliavin, that the canonical flow has sample flows $F_t(-,\omega)$ which preserve the Liouville measure. Rather than check that $\operatorname{div} X^p = 0$ for each p we can see this from the Stratonovich equations (121a) and (121b)

$$d\xi_t = A_t \circ dB_t$$

$$dA_t = u_t^{-1} R(u_t \circ dB_t, u_t\, \xi_t)u_t$$

of §9C Chapter II for $\xi_t = \theta(\delta F_t(V))$, $A_t = \tilde{\omega}(\delta F_t(V))$. Indeed the equation for ξ_t involves only A_t and conversely, so the trace of the right hand side considered as a linear transformation of (ξ_t, A_t) vanishes identically. Therefore the Stratonovich equation for $\det \delta F_t(-)$ shows that the determinant is identically 1, and so the Liouville measure is preserved.

Our Lyapunov spectrum will be taken with this as basic measure. However in general it will not be ergodic: for example it will not be if M is the product $M_1 \times M_2$ of two Riemannian manifolds, or when M is flat (i.e. has vanishing curvature). In the latter case we noted in §9B of Chapter II that OM is foliated by horizontal submanifolds: each of these will be invariant under the flow. More generally the *holonomy bundle*, see [64], is invariant.

C. Since the system can degenerate we must first show that the exponents $\lambda^{(r)}(u) < \ldots < \lambda^{(1)}(u)$ can be taken to be independent of $u \in OM$. Here, and for the rest of this discussion of canonical flows we are following [24]. To do this observe (as in the proof of Theorem 4C, Chapter II) that for $g \in O(n)$

$$F_t(u,\omega)\cdot g = F_t(u\cdot g,\ g^{-1}\omega) \tag{158}$$

and so for $V \in TOM$

$$TR_g(TF_t(V,\omega)) = TF_t(TR_g(V),\ g^{-1}\omega) \tag{159}$$

The measure one subset Γ of $OM \times \Omega$ in Theorem 1B, consisting of points for which convergence to the exponents occurs can therefore be taken to be invariant under $(u,\omega) \to (u\cdot g,\ g^{-1}\omega)$ for $g \in O(n)$, with corresponding invariance for the filtrations i.e. $V^{(j)}(u,\omega) = V^{(j)}(u\cdot g,\ g^{-1}\omega)$, and so for the exponents: $\lambda^{(j)}(u\cdot g) = \lambda^{(j)}(u)$, since they are non-random. Thus we obtain maps $\lambda^{(j)}{}_0 : M \to \mathbb{R}$ with $\lambda^{(j)}{}_0(\pi(u)) = \lambda^{(j)}(u)$ for u in OM, defined almost surely. These are measurable. Also since each $\lambda^{(j)}$ is invariant under $\Phi_t: OM \times \Omega \to OM \times \Omega$, we have

$$\lambda^{(j)}(u) = \mathbb{E}\,\lambda^{(j)}(F_t(u,\omega)) = \mathbb{E}\,\lambda^{(j)}{}_0(\pi\, F_t(u,\omega))$$
$$= P_t\, \lambda^{(j)}{}_0(x)$$

for $x = \pi(u)$, where $\{P_t : t \geq 0\}$ is the heat semigroup for M (solving $\partial/\partial t = \frac{1}{2}\Delta$). Thus $P_t\, \lambda^{(j)}{}_0$ is independent of t, and so $\lambda^{(j)}{}_0$ is constant (for example by the ergodicity of the Riemannian measure: but this itself is usually proved by observing that $P_t f$ independent of t implies $\Delta P_t f = 0$ for $t > 0$, since $P_t f$ is C^2 for $t > 0$, which implies $P_t f$ is constant for each positive t, which implies by strong continuity of P_t in t that f is constant).

D. From our equations (122a,b) for $\xi_t = \theta(\delta F_t(V_0)) \in \mathbb{R}^n$ and $A_t = \tilde{\omega}(\delta F_t(V_0)) \in o(\underline{n})$ we could write down an expression for $\log|\delta F_t(V_0)| = \frac{1}{2}\log(|\xi_t|^2 + |A_t|^2)$. However that does not seem very illuminating, and we shall resist doing so (but see equation (172) below when dim $M = 2$). To start with we shall just consider the horizontal component ξ_t. For this set $v_t = T\pi(\delta F_t(V_0))$, so $v_t = u_t(\xi_t)$ and $v_t \in T_{x_t}M$ for $x_t = \pi(u_t)$ the Brownian motion induced on M. In particular $|v_t|_{x_t} = |\xi_t|$. By (122a)

$$|\xi_t|^2 = |\xi_0|^2 + 2\int_0^t \langle \xi_s, A_s dB_s\rangle - \int_0^t Ric(v_s,v_s)ds + 2\int_0^t |A_s|^2\, ds \quad (160)$$

$$\log|\xi_t| = \log|\xi_0| + \int_0^t \langle \xi_s/|\xi_s|, A_s/|\xi_s|\, dB_s\rangle - \tfrac{1}{2}\int_0^t Ric(v_s/|v_s|, v_s/|v_s|)ds$$
$$+ \int_0^t |A_s|^2/|\xi_s|^2\, ds - \int_0^t |A_s\xi_s|^2/|\xi_s|^4\, ds \quad (161)$$

(at least until the first hitting time τ of 0 by ξ_t).

Therefore by (157) and the observation that $|Ae| = |A|\,|e|$ when $n = 2$

$$\log|\xi_t| \geq \log|\xi_0| + M_t - \tfrac{1}{2}\int_0^t Ric\,(v_s/|v_s|, v_s/|v_s|)ds \quad (162)$$

with equality when $n = 2$, where $\{M_t : t \geq 0\}$ is the local martingale

$$M_t = \int_0^t \langle \xi_s/|\xi_s|, A_s/|\xi_s|\, dB_s\rangle \quad (163)$$

Now $\{M_t : T \geq 0\}$ is a time changed Brownian motion and for ξ_t to vanish in finite time τ (assuming $\xi_0 \neq 0$), we would have to have $\underline{\lim}_{t\to\tau-} M_t = -\infty$. Then $\overline{\lim}_{t\to\tau-} M_t = \infty$ and so

$$\overline{\lim}_{t\to\tau-} \log(|\delta F_t(V_0)|) \geq \overline{\lim}_{t\to\tau-} \log|\xi_t| = \infty$$

which cannot be true for finite τ. Thus $|\xi_t|$ never vanishes and (161) holds for all time.

Theorem 4D [24]. *Let* $\overline{Ric}(x) = \sup\{Ric(v,v) : v \in T_xM$ *and* $|v| = 1\}$ *for each* $x \in M$. *Then the top exponent* λ^1 *of the canonical flow satisfies*

$$\lambda^1 \geq \overline{\lim}_{t\to\infty} 1/t \log|v_t| \geq -1/(2|M|)\int_M \overline{Ric}(x)dx \quad (164)$$

Proof: Since M_t is a time changed 1-dimensional Brownian motion

$$\underline{\lim}_{t\to\infty} 1/t\, M_t \leq 0 \leq \overline{\lim}_{t\to\infty} 1/t\, M_t$$

Therefore by (162)

$$\overline{\lim}_{t\to\infty} 1/t \log|v_t| \geq \overline{\lim}(-1/(2t)\int_0^t Ric(v_s/|v_s|, v_s/|v_s|))\,ds)$$
$$\geq -\underline{\lim}\; 1/(2t)\int_0^t \overline{Ric}(x_s)ds$$
$$= -1/(2|M|)\int_M \overline{Ric}(x)dx$$

almost surely, by the ergodic theorem, since ($1/|M| \times$ the Riemannian measure) is ergodic for Brownian motion. //

<u>Remark 4D</u> For dim $M = 2$ the Ricci curvature is essentially the Gaussian curvature $K_p(x)$ for each x. The Gauss-Bonnet theorem states that

$$1/(2\pi) \int_M K_p(x)dx = \chi(M) \tag{165}$$

where $\chi(M)$ is the Euler characteristic of M, a topological invariant (e.g. $\chi(S^2) = 2$, $\chi(S^1 \times S^1) = 0$). It is proved in Chapter VI below. From this, (162), and the argument above: *if* dim $M = 2$ *then for* $\rho \otimes \mathbb{P}$ *almost all* (x,ω)

$$\lambda^1 \geq \overline{\lim}_{t\to\infty} 1/t \log |v_t| = \overline{\lim}_{t\to\infty} 1/t\, M_t - (\pi/|M|)\, \chi(M)$$
$$\geq -(\pi/|M|)\, \chi(M) \tag{166}$$

Also

$$\underline{\lim}_{t\to\infty} 1/t \log |v_t| = \underline{\lim}\, 1/t\, M_t - (\pi/|M|)\, \chi(M) \leq -(\pi/|M|)\, \chi(M) \tag{167}$$

(The $\frac{1}{2}$ in the corresponding formula in [24] should not be there).

From (167) we get

$$\lambda^1 = \underline{\lim}\, 1/(2t) \log(|\xi_t|^2) + |A_t|^2 = \underline{\lim}\, 1/t\, (\log|\xi_t| + \tfrac{1}{2} \log(1 + |A_t|^2/|\xi_t|^2))$$
$$\leq -(\pi/|M|)\chi(M) + \overline{\lim}\, 1/(2t) \log(1 + |A_t|^2/|\xi_t|^2).$$

Since $\lambda^1 \geq 0$ this shows: *for dim* $M = 2$

$$\overline{\lim}\, 1/t \log(1 + |A_t|^2/|\xi_t|^2) \geq (2\pi/|M|)\, \chi(M). \tag{168}$$

E. Next we consider the case dim $M = 2$ in more detail.
Write $k(x)$ for the Gauss curvature $K_p(x)$ (with $P = T_xM$) so that

$$\mathrm{Ric}\,(u,u) = |u|^2\, k(x) \tag{169a}$$

and

$$R(u,v)w = k(x)(\langle w,v\rangle u - \langle w,u\rangle v) \tag{169b}$$

for u,v,w in T_xM. The following formulae are given for completeness. They come from (160) and (122b): the rather straightforward proof is left an exercise; there are details in [24] (using the scalar curvature $S(x) = 2k(x)$)

$$|\xi_t|^2 = |\xi_0|^2 + 2\int_0^t \langle \xi_s, A_s dB_s\rangle - \int_0^t k(x_s)\, |\xi_s|^2\, ds + 2\int_0^t |A_s|^2\, ds \tag{170a}$$

$$|A_t|^2 = |A_0|^2 - 2\int_0^t k(x_s)\langle \xi_s, A_s dB_s\rangle - 2\int_0^t k(x_s)\, |A_s|^2\, ds$$
$$+ \int_0^t dk(u_s A_s \xi_s) ds + \int_0^t k(x_s)^2\, |\xi_s|^2 ds \tag{170b}$$

The following formulae from [24] are useful:

<u>Proposition 4E</u>

For dim $M = 2$ *there is the Stratonovich equation*

$$d|A_t|^2 + k(x_t) \circ d|\xi_t|^2 = 0 \tag{171}$$

and the Ito equation

$$|A_t|^2 + k(x_t)|\xi_t|^2 = |A_0|^2 + k(x_0)|\xi_0|^2 + \int_0^t |\xi_s|^2\, dk(u_s dB_s)$$
$$+ \tfrac{1}{2}\int_0^t |\xi_s|^2\, \Delta k(x_s)ds - \int_0^t dk(u_s A_s \xi_s)ds \tag{172}$$

Proof The Stratonovich equations (121a,b):

$$d\xi_t = A_t \circ dB_t \text{ and } dA_t = u_t^{-1} R(u_t \circ dB_t, u_t\xi_t)u_t$$

give

$$d|\xi_t|^2 = 2\langle \xi_t, A_t \circ dB_t\rangle \tag{173}$$

$$d|A_t|^2 = -\operatorname{trace} A_t\, u_t^{-1} R(u_t \circ dB, u_t\xi_t)u_t$$

$$= \sum_{p=1}^{n} \langle u_t^{-1}R(u_t \circ dB_t, u_t\xi_t)u_t e_p, A_t e_p\rangle$$

$$= \sum_{p=1}^{n} k(x_t)\,\{\langle u_t e_p u_t, \xi_t\rangle\langle u_t \circ dB_t, u_t A_t e_p\rangle - \langle u_t\, e_p, u_t \circ dB_t\rangle\langle u_t\, \xi_t, u_t\, A_t\, e_p\rangle\}$$

$$= -\,2k(x_t)\langle \xi_t, A_t \circ dB_t\rangle \tag{174}$$

Equation (171) follows immediately. On integrating it by parts and then using the Ito formula for $k(x_t)$ (and hoping the use of d for stochastic differentials as in $d(k(x_t))$ and for ordinary differentials as in dk will not cause confusion):

$$|A_t|^2 + k(x_t)|\xi_t|^2 - |A_0|^2 + k(x_0)|\xi_0|^2$$
$$= \int_0^t \{d|A_s|^2 + k(x_s) \circ d|\xi_s|^2 + |\xi_s|^2 \circ d(k(x_s))\}$$
$$= \int_0^t |\xi_s|^2 \circ d(k(x_s))$$
$$= \int_0^t |\xi_s|^2\, dk(u_s \circ dB_s) + \tfrac{1}{2}\int_0^t |\xi_s|^2\, \Delta k(x_s)ds + \tfrac{1}{2}\int_0^t d|\xi_s|^2 dk(u_s dB_s)$$

giving (172) by (173). //

Theorem 4E

When dim M = 2 *and* $k(x) > 0$ *for all* x

$$\lambda^1 \le 1/(4|M|) \int_M \{|\nabla k(x)|/\sqrt{k(x)} + |\Delta k(x)|/k(x)\}dx \tag{175}$$

Proof

Since $k(x) > 0$ we can take $\sqrt{\{|A|^2 + k(x)\, |\xi\,|^2\}}$ as the norm of (ξ,A) when computing the exponents. Write it as $\|(\xi,A)\|$. By (172)

$$\log \|(\xi_t,A_t)\| = \log \|(\xi_0,A_0)\| + \tfrac{1}{2}\int_0^t |\xi_s|^2/\|(\xi_s,A_s)\|^2\, dk(u_s dB_s)$$
$$+ 1/4 \int_0^t |\xi_s|^2/\|(\xi_s,A_s)\|^2\, \Delta k(x_s)ds - \tfrac{1}{2}\int_0^t dk(u_s A_s \xi_s)ds/\|(\xi_s,A_s)\|^2 ds$$
$$- 1/4 \int_0^t |\xi_s|^4\, |(dk)_{x_s}|^2/\|(\xi_s,A_s)\|^4 ds \qquad (176)$$

Since

$$|dk(u_s A_s \xi_s)| = \langle \nabla k(x_s), u_s A_s \xi_s\rangle \le |\nabla k(x_s)|\,|A_s|\,|\xi_s|$$
$$\le \tfrac{1}{2}|\nabla k(x_s)|\sqrt{k(x_s)}^{-1}\|(\xi_s,A_s)\|^2 \qquad (177)$$

and the coefficient of the Ito integral in (176) is bounded, the ergodic theorem gives the result. //

This corrects the upper bounds in [24]. It must be possible to do better.

F. Next we consider the case of constant curvature. An important point here, and later, is the idea of a *covering* $p : \mathbb{M} \to M$. This is a C^∞ map of manifolds which is surjective and such that each $x \in M$ has a connected open neighbourhood U with p mapping each component of $p^{-1}(U)$ diffeomorphically onto U. The typical examples are $p : S^1 \to S^1$ given by $p(e^{i\theta}) = e^{2i\theta}$ and $p:\mathbb{R} \to S^1$ given by $p(\theta) = e^{i\theta}$. The covering is *Riemannian* if $\mathbb{M}$ and M are Riemannian and $T_z p : T_z\mathbb{M} \to T_{p(z)}M$ preserves the inner product. Clearly if M is Riemannian and p is a covering map then we can define a Riemannian metric on $\mathbb{M}$ so that p becomes Riemannian. In general coverings have the *path lifting property*: if $\sigma : [a,b) \to M$ is a continuous path and $z \in p^{-1}(\sigma(a))$ then there is a unique continuous $\sigma^\sim : [a,b) \to \mathbb{M}$ with $\sigma^\sim(a) = z$ and $p \circ \sigma^\sim = \sigma$. The lifting gives a continuous map from the space of continuous paths in M starting from $\sigma(a)$ to the corresponding space of paths from z in $\mathbb{M}$. Thus stochastic processes can be lifted from M to $\mathbb{M}$. Also it is easy to construct an S.D.S. $(\tilde{Y},z)$ p-related to a given one (Y,z) on M.

For a Riemannian covering a Brownian motion on $\mathbb{M}$ maps by p to a Brownian motion on M: to see this choose an S.D.S. (Y,z) on M which has Brownian motions as its solutions. The lift $(\tilde{Y},z)$ will then have Brownian motions on $\mathbb{M}$ as its solutions since the condtions of §4A Chapter II for this to happen are purely local, and locally $(\tilde{Y},z)$ and (Y,z) are the same, as are $\mathbb{M}$ and M. Since $(\tilde{Y},z)$ and (Y,z) are p-related the result follows by §3D Chapter 1. (For a generalization of this see [43], p. 256.) Since curvature is a local property p will also map the curvature tensors of $\tilde{M}$ to that of M.

Given the equation $dx_t = X(x_t) \circ dB_t + A(x_t)dt$ on a compact Riemannian M and a Riemannian covering $p : \tilde{M} \to M$, it is immediate that the Lyapunov filtrations and any stable manifolds lift to corresponding objects for the lift of the SDS : the filtration of $T_z\tilde{M}$ will map by T_zp to the filtration for $T_{p(z)}M$ and will exist when the latter exists, with the same exponents, and p will map stable manifolds to stable manifolds as a covering. In particular $\tilde{M}$ need not be compact.

The other two main ingredients we need are:

(i) there exists a covering $p : \tilde{M} \to M$ with $\tilde{M}$ simply connected, and this is essentially unique

(ii) if M is simply connected and of constant curvature k then: M is isometric to $\mathbb{R}^n$ if k = 0, to S^n if k = 1, and to hyperbolic space H^n if k = 1. (e.g. see [64]).

If we now note that for a Riemannian covering $p : \tilde{M} \to M$ the map $u \to Tp \circ u$: $O\tilde{M} \to OM$ is also a covering, we see that to investigate the Lyapunov exponents and stable manifolds when M has constant curvature k = +1 or k = -1 it suffices to take $M = S^n$ or $M = H^n$.

G. Suppose now that M has constant curvature k. Equations (173) (174) are valid with $k(x_p)$ replaced by k, because of formula (9) for the curvature. Thus, [24],

$$d|A_t|^2 + k\, d|\xi_t|^2 = 0$$

whence

$$|A_t|^2 + k|\xi_t|^2 = |A_0|^2 + k|\xi_0|^2 \tag{178}$$

When k > 0 we see immediately that $\lambda^1 = 0$, whence $\lambda^1 = \lambda_\Sigma$, and so *all the exponents vanish given constant positive curvature.*

For k < 0 we see that except perhaps for some exceptional V_0

$$\lambda^1 = \lim_{t\to\infty} 1/t \log |\xi_t| = \lim_{t\to\infty} 1/t \log |A_t|$$

and in particular these limits exist.

From (164) this yields

$$\lim_{t\to\infty} 1/t \log |\xi_t| \geq -(n-1)k/2 > 0$$

so that almost surely

$$\lim_{t\to\infty} |A_t|^2/|\xi_t|^2 = -k + \lim_{t\to\infty} (|A_0| + k|\xi_0|^2)/|\xi_t|^2 = -k$$

From (161) (162) we see immediately that *if* <u>dim M = 2</u> *then for* $k < 0$

$$\lambda^1 = -\tfrac{1}{2}k \tag{179}$$

G. For constant negative curvature since $\lambda^1 > 0$ and $\lambda_\Sigma = 0$ there must be some negative exponent and correspondingly some stable manifolds. We will investigate these for $k = -1$ and dim $M = 2$ following [24]. By the discussion in the previous paragraph we need only take M to be the hyperbolic plane H^2, even though it is not compact.

In §5C of Chapter II we described H^2 as the hyperboloid $\{(t,x,y) \in \mathbb{R}^3 : t \geq 1$ and $x^2 + y^2 - t^2 = -1\}$ with Riemannian metric induced from the Lorentz metric of $\mathbb{R}^3$.

Writing $N = (1,0,0)$ the tangent space T_NH^2 can be identified with $\{0\} \times \mathbb{R}^2$ in $\mathbb{R}^3$. For $v = (v^1, v^2)$ in $\mathbb{R}^2$ the path $\gamma_v : \mathbb{R} \to H^2$ given by

$$\gamma_v(\alpha) = (\cosh(|v|\alpha),\ v^1/|v| \sinh(|v|\alpha),\ v^2/|v| \sinh(|v|\alpha))$$

has

$$\frac{d}{d\alpha}\gamma_v(\alpha) = (|v| \sinh(|v|\alpha),\ v^1\cosh(|v|\alpha),\ v^2 \cosh(|v|\alpha))$$

so that

$$\left|\frac{d}{d\alpha}\gamma_v(\alpha)\right|_{\gamma_v(\alpha)} = |v|.$$

Differentiating this we see $D/\partial\alpha\ \ d/d\alpha\ \gamma_v(\alpha)$ is orthogonal to $d/d\alpha\ \gamma_v(\alpha)$ and so vanishes by symmetry. Thus γ_v is a geodesic through N and using our identification of T_NH^2 with $\mathbb{R}^2$

$$\exp_N v = \gamma_v(1).$$

For $V \in T_v\mathbb{R}^2 \approx \mathbb{R}^2$, with $v \neq 0$

$$D\exp_N(v)(V) = \Big(\langle v,V\rangle/|v| \sinh|v|,\ (\sinh|v|)DP(v)(V) + (\langle v,V\rangle/|v|)(\cosh|v|)(v/|v|)\Big) \in \mathbb{R} \times \mathbb{R}^2$$

where $P : \mathbb{R}^2 - \{0\} \to S^1$ is $P(v) = v/|v|$. Thus

$$|D\exp_N(v)(V)|^2 = \langle V,P(v)\rangle^2 + |DP(v)(V)|^2(\sinh|v|)^2$$

since $DP(v)V$ is orthogonal to v in $\mathbb{R}^2$. This gives the induced metric in the chart given by $\exp_N$ i.e. in normal co-ordinates at N

$$\langle V,W\rangle_v = \langle V,P(v)\rangle\langle W,P(v)\rangle + \sinh(|v|)^2\langle DP(v)V,DP(v)W\rangle_{\mathbb{R}^2} \qquad (180)$$

which is most easily considered in polar co-ordinates: in classical notation

$$ds^2 = dr^2 + (\sinh r)^2 d\theta^2 \ . \qquad (181)$$

(One way to see this is to interpret (18)) as meaning that if a curve σ in M is given in normal co-ordinates by $\sigma(t) = (r_t \cos\theta_t, r_t \sin\theta_t)$ then $|\dot\sigma(t)|^2_{\sigma(t)} = \dot r_t^2 + (\sinh r_t)^2 \dot\theta_t^2$).)

Let D^0 be the open unit disc in $\mathbb{R}^2$. Define $f: D \to \mathbb{R}^2$ by

$$f(r,\theta) = (2\tanh^{-1} r, \theta)$$

in polar co-ordinates. The metric (181) on $\mathbb{R}^2$ induces the metric

$$ds^2 = 4(1-r^2)^{-2}(dr^2 + r^2 d\theta^2) \qquad (182)$$

on D^0, or in Cartesian co-ordinates

$$ds^2 = 4(1-r^2)^{-2}(dx^2 + dy^2). \qquad (183)$$

The disc with this metric is the Poincaré disc model of H^2. It represents N as $(0,0)$, but since the subgroup G of the Lorentz group acted transitively on the hyperboloid as isometries we can compose $\exp_N \circ f : D^0 \to H^2$ with an isometry to get an isometry which maps $(0,0)$ to any given point of H^2.

There is also the representation of H^2 by the upper $\frac{1}{2}$-plane $U = \{(x,y) : y > 0\}$ in $\mathbb{R}^2$. For this choose some point c of S^1. Then there is an analytic diffeomorphism $d^c : U \to D^0$

$$d^c(z) = c(z-i)/(z+i) \qquad (184)$$

which maps the closure $\bar U$ to the closed disc D with the point at infinity in U mapped to c. The metric induced on U is

$$ds^2 = y^{-2}(dx^2 + dy^2) \qquad (185)$$

The disc model shows how to talk about "points at infinity" on H^2: they can be taken to be the points of S^1. For $c \in S^1$ write U^c for U when d^c has been used to give it its metric.

H. Since any $p \in H^2$ can be identified with $(0,0)$ in D we can identify the points of 'the circle at ∞', N_∞, in H^2 with $\frac{1}{2}$ rays γ emanating from p and parametrized by arc length. The Buseman function of such γ corresponding to c is

$$\beta_p(c,-) : H^2 \to \mathbb{R}$$

for

$$\beta_p(c,z) = \lim_{t\to\infty} [t - d(z,\gamma(t))] \tag{186}$$

(Since $t \to t - d(z,\gamma(t))$ is increasing and bounded above by $d(z,p)$ this limit exists). This is sometimes given the opposite sign.

In the model U^c with $p = (0,1)$ we have $\gamma(t) = (0,e^t)$ and if $z = (x,y) \in U^c$ then

$$t - \log y \le d(z,\gamma(t)) \le t - \log y + e^{-2t}|x|$$

since log y is the distance of z from the line $\{(\alpha,1) : \alpha \in \mathbb{R}\}$. Thus in this case

$$\beta_p(c,z) = \log y \tag{187}$$

<u>Lemma 4H</u> [24]

Let $\{z_t : t \ge 0\}$ *be a Brownian motion starting from* p. *Then with probability* 1

$$\text{(i)}\quad z_\infty(\omega) = \lim_{t\to\infty} z_t(\omega) \in N_\infty \text{ exists}$$

and

$$\text{(ii)}\quad \lim_{t\to\infty} 1/t\, \beta_p(z_\infty(\omega), z_t(\omega)) = \tfrac{1}{2} \tag{188}$$

Proof Part (i) is a very special case of Prat's result for not necessarily constant curvature. In our case it follows because in D^0 our Brownian motion is just the time change of an ordinary Brownian motion in $\mathbb{R}^2$, and the latter almost surely leaves D^0 in finite time.

For (ii) it is enough to show that

$$\mathbb{P}\{1/t\, \beta_p(c,z_t(\omega)) \to \tfrac{1}{2} \mid z_\infty(\omega) = c\} = 1.$$

To condition z to tend to c we can use the Doob h-transform. Now, as described in [81] 2X9 the standard Brownian motion $\mathbb{R}^2$ conditioned to exit from D^0 at a point c of S^1 is the h-transform of that Brownian motion, h-transformed by the Poisson kernel

$$h(z) = (1 - |z|^2)/|c-z|^2 \qquad |z| < 1.$$

This means it has the law of the diffusion process with generator $\frac{1}{2}\Delta + \nabla \log h$ for Δ, ∇ the Euclidean operators. Since time changing commutes with our conditioning the hyperbolic Brownian motion of M conditioned to tend to c as $t \to \infty$ is a diffusion process with generator $\frac{1}{2}\Delta + \nabla \log h$ where now Δ and ∇ refer to the hyperbolic metric.

In the model U_c the Laplacian is given, for $x^1 = x$, $x^2 = y$, by

$$\Delta f(z) = \sum_{j=1}^{2} g(z)^{-1/2} \, \partial/\partial x^i \, \{g(z)^{\frac{1}{2}} g^{ij}(z) \, \partial f/\partial x^j\}$$

$$= y^2 \, (\partial^2 f/\partial x^2 + \partial^2 f/\partial y^2) \tag{189}$$

while h is represented by $\tilde{h}$ for $\tilde{h} = \tilde{h} \circ d^c$ i.e.

$$\tilde{h}(z) = 1/4(|z+i|)^2 - |z-i|^2) = y \quad \text{for } z = x + iy.$$

Thus

$$\nabla \log \tilde{h}(z) = (0, y^2 . \, 1/y) = (0,y)$$

and the conditioned diffusion can be represented by $z_t = (x_t, y_t)$ for

$$dx_t = y_t \, dB^1_t, \; dy_t = y_t \, dB^2_t + y_t \, dt \tag{190}$$

where $\{(B^1_t, B^2_t) : t \geq 0\}$ is a Brownian motion on $\mathbb{R}^2$. Then $y_t = y_0 \exp(B^2_t + \frac{1}{2} t)$ and so (ii) follows by (187). //

We can now give the basic result from [24] on the stable manifolds of the canonical flow on OM for hyperbolic space:

<u>Theorem 4H</u> [24] *For* $M = H^2$ *take* $u \in OM$. *Let* $F_t(-,\omega):OM \to OM$ *be the canonical flow. Then for almost all* $\omega \in \Omega$ *the following holds:*

The limit $c(\omega) = \lim_{t\to\infty} \pi \, F_t(u,\omega)$ *exists in* N_∞ *and if* $\mathcal{V}(u,\omega)$ *is the submanifold of OM given by* $\{Tg \circ u \text{ s.t. } g: U^c \to U^c$ *is a horizontal translation*$\}$ *then for* $u' \in \mathcal{V}(u,\omega)$

$$\lim_{t\to\infty} 1/t \log d(F_t(u,\omega), F_t(u',\omega)) = -\tfrac{1}{2} \tag{191}$$

and for all other frames u'

$$\underline{\lim}_{t\to\infty} 1/t \log d(F_t(u,\omega), F_t(u',\omega)) \geq 0 \tag{192}$$

<u>Proof:</u> Choose $\omega \in \Omega$ so that the conclusions of Lemma 4H are true, and so that

the flow $F_t(-,\omega)$ exists and satisfies $F_t((Tg) \circ u,\omega) = Tg \circ F_t(u,\omega)$ for all isometries g of M. The latter is possible either by general principles, because the canonical S.D.S. is invariant under the action of such Tg, or by noting the special properties of the flow $F_t(-,\omega)$ on OM when OM, or rather SO(M), is identified with our subgroup G of the Lorentz group: see Remark 4H(i) below. Then $c = z_\infty(\omega)$ exists. We will work in U^c.

It is necessary only to consider oriented frames i.e. restrict ourselves to the component SOM of OM. Such a frame at $(x,y) \in U^c$ can be identified with a tangent vector to U^c of unit Euclidean length. Using this we shall write frames as $(x,y,\lambda) \in U^c \times S^1$. Let $\tilde{d}$ be the metric on SOM which is the product of the Euclidean metric on U^c with the standard one for S^1. Over the compact subset W of U^c

$$W = \{(x,y) \in U^c : |x| + |1-y| \le \tfrac{1}{2}\}$$

this will be equivalent to the standard metric of OM described previously (or to any other metric).

Set $(x_t,y_t,\lambda_t) = F_t(u,\omega)$.

If $u' \in \mathcal{V}(u,\omega)$ there exists $a \in \mathbb{R}$ with

$$F_t(u',\omega) = (x_t + a, y_t,\lambda_t) \qquad t \ge 0$$

Horizontal translation in U^c is an isometry and so is the dilation $(x,y) \to (\alpha x, \alpha y)$ for $\alpha > 0$. Therefore

$$\begin{aligned} d(F_t(u,\omega), F_t(u',\omega)) &= d((0,y_t,\lambda_t), (a,y_t,\lambda_t)) \\ &= d((0,1,\lambda_t), (a/y_t, 1, \lambda_t)) \end{aligned} \qquad (193)$$

since isometries on M induce isometries on OM. For sufficiently large t both (0,1) and $(ay_t^{-1},1)$ lie in W, and so d may be replaced by $\tilde{d}$ in estimating (193) for such t. However

$$\tilde{d}((0,1,\lambda_t), (a/y_t, 1, \lambda_t)) = |a|/|y_t| \qquad (194)$$

and by Lemma 4H(ii) and equation (187)

$$\lim_{t\to\infty} 1/t \log (|a|/|y_t|) = -\tfrac{1}{2}$$

This proves (191).

For (192) first suppose $u' = (\alpha x_0, \alpha y_0, \lambda_0)$ for some $\alpha > 0$, $\alpha \neq 1$ where u =

(x_0,y_0,λ_0). Then $F_t(u',\omega) = (\alpha x_t, \alpha y_t, \lambda_t)$, giving (by a horizontal translation)

$$d(F_t(u,\omega), F_t(u',\omega)) \geq d((0,,\alpha y_t), (0,y_t))$$
$$= d((0,\alpha), (0,1))$$

from which (192) follows.

Combining this with (191) we see the same holds for any u' which is obtained from u by the action of the isometries of M generated by the horizontal translations and the dilations $(x,y) \to (\alpha x, \alpha y)$ of H^c. These isometries correspond to a subgroup G_c, say, of G when we use the identification of SOM with our subgroup G of the Lorentz group. The group G itself was identified with isometries of M in the hyperboloid model and in this model it is easy to see that G_c is precisely the subgroup of G which leaves the point at infinity c fixed (the latter subgroup is just the natural embedding in G of the identity component of the Lorentz group of the 1 + 1-dimensional space time acting in the plane orthogonal to c in $\mathbb{R}^3$: this is two dimensional as is G_c and the former is known to be connected).

For other u' in SOM = G there is the isometry corresponding to $g = u'u^{-1}$ which sends u to u'. Since $F_t(u',\omega) = g\, F_t(u,\omega)$ and g is not in G_c

$$\lim_{t\to\infty} F_t(u',\omega) \neq c.$$

Consequently

$$d(F_t(u,\omega), F_t(u',\omega)) \geq d(\pi\, F_t(u,\omega), \pi F_t(u',\omega))$$
$$\geq d(\pi\, F_t(u,\omega), \{(x,1):x \in \mathbb{R}\})$$

or sufficiently large t. This is just $\beta_p(c,\pi,F_t(u,\omega))$ for $p = (0,1)$ by (187). Thus in this case, by (188)

$$\underline{\lim}\ 1/t \log d(F_t(u,\omega), F_t(u',\omega)) \geq \tfrac{1}{2}. \quad //$$

This theorem, together with the fact that we know there must be at least one negative exponent with corresponding stable manifolds, shows that there is precisely one, namely $-\frac{1}{2}$, and that the stable manifold through u is $\mathcal{V}(u,\omega)$. Consequently the multiplicity of the exponent $-\frac{1}{2}$ is $\dim \mathcal{V}(u,\omega)$ i.e. 1. Since dim OM = 3, and $\lambda_\Sigma = 0$ and $\lambda^1 = \frac{1}{2}$ by (179), because $2(\frac{1}{2}) + (-\frac{1}{2}) = \frac{1}{2} \neq 0$ there must be another exponent. It can only have multiplicity 1 and it must be 0. Thus *the exponents for the canonical flow on* H^2 *are* $-\frac{1}{2}, 0, \frac{1}{2}$.

As for the filtration of T_uM we know that $V^{(3)}(u,\omega) = T_u\mathcal{V}(u,\omega)$. It will now be no surprise that $V^{(2)}(u,\omega)$ is the tangent to the orbit of G_c (i.e. the tangent to the coset $G_c u$ in G): for a proof see [24]. More detailed information 'stability' properties of the flow can be found in [24] and [10], especially the latter.

Remark 4H(i) Identifying SOM with G the canonical S.D.E. becomes a left invariant stochastic differential equation

$$du_t = X(u_t) \circ dB_t$$

with $X(1)(\xi) = \xi$ for ξ in $\mathbb{R}^2$ as in equation (56) of Chapter II §5B. The flow is then $F_t(u.\omega) = u.g_t(\omega)$ where $\{g_t : t \geq 0\}$ is the solution starting from 1. Equation (178) showing that $|A_t|^2 - |\xi_t|^2$ is constant follows from the invariance of the Cartan-Killing form: see [64] p. 155. The metric we have taken on SOM corresponds to a left invariant metric on G so our exponents are measuring how right multiplication by $g_t(\omega)$ spreads out or contracts the space (at least infinitesimally). Use of the Lie group structure of G gives a good way to obtain the result about the exponents given above and especially for their higher dimensional analogues. This is carried out in [10]. See [71], [73] for earlier work for symmetric spaces.

The vanishing of the exponents for S^n comes out particularly simply by the corresponding representation of SOS^n as $SO(n+1)$. This time the metric on $SO(n+1)$ is bi-invariant and so $F_t(-,\omega)$ consists of isometries: this is the reason for the constancy of $|\xi_t|^2 + |A_t|^2$ in equation (178) for $k = 1$.

Remark 4H(ii) The projections onto M of the stable manifolds $\mathcal{V}(u,\omega)$ are *horocycles.* In the disc model the horocycle $H_p(c)$ for $p \in M$ and c on the circle at infinity is the circle tangent to S^1 at c which goes through p. The horocycles are precisely the level surfaces of the Buseman functions defined by (186). Equivalently they can be defined as the boundary of the *horoballs* defined as the union $\bigcup_{t>0} B_t(\gamma(t))$ of balls radius t about $\gamma(t)$ for γ a unit speed geodesic. These definitions make sense in greater generality: in particular for simply connected manifolds of non-positive curvature. For more details see [2], [8]. However there is no reason to believe that stable manifolds for the canonical flows of these more general manifolds project onto these horocycles.

Remark 4H(iii) For results about the non-triviality of the spectrum for the canonical flow when $M \neq S^n$ see [24a].

Remark 4H(iv) The characteristic exponents for the geodesic flow on the unit sphere bundle in TM have been studied a lot [2], [106]. The results are analogous for constant negative curvature: especially for dim M = 2 when the bundle SOM can be identified with the sphere bundle. See also [105].

§5. Moment exponents

In this section we no longer require M to be compact.

A. Consider a process $\{x_t : t \geq 0\}$ on M and a process $\{v_t : t \geq 0\}$ on some space B with projection $p : B \to M$ such that $p(v_t(\omega)) = x_t(\omega)$ for $t \geq 0$. If the fibres $p^{-1}(x)$ of B are normed vector spaces for each $x \in M$ we can consider

$$\nu_q := \overline{\lim_{t\to\infty}} 1/t \log \mathbb{E}|v_t|^q \tag{195}$$

for $q \in \mathbb{R}$. Typical cases of interest are:

(i) $B = M \times \mathbb{R}$ and p the projection with v_t defined by

$$dv_t/dt = V(x_t)v_t \tag{196}$$

for given v_0, for $V : M \to \mathbb{R}$. We should then write ν_p as $\nu_p(v_0)$ etc. This is the situation of the "Kac-functionals" studied extensively in [15], [54], [55] especially in the non-compact case, i.e. the behaviour as $t \to \infty$ of

$$1/t \log \mathbb{E}\, e^{\int_0^t V(x_s)ds} \qquad (\text{for } v_0 \neq 0).$$

(ii) The analogue of (i) for $B = M \times \mathbb{R}^n$ and $V : M \to \mathbb{L}(\mathbb{R}^n; \mathbb{R}^n)$

(iii) $p : B \to M$ the tangent bundle or a tensor bundle like $\Lambda^p T^*M$ with v_t defined by a covariant equation

$$Dv_t/\partial t = V(x_t)v_t \tag{197}$$

where $V(x) \in \mathbb{L}(p^{-1}(x), p^{-1}(x))$ for each x in M. One could equally well take other vector bundles over M with a linear connection: this would then include (ii) as the special case of the trivial bundle.

(iv) $v_t = T_{x_0} F_t(v_0)$ where $\{F_t(-,\omega) : t \geq 0, \omega \in \Omega\}$ is the flow of an S.D.E. on M.

The last example is somewhat more complicated than the previous ones since the equation for v_t is a stochastic differential equation in general. We shall look in more detail at situations related to cases (i) and (iii) in the next chapter. Case (ii) was investigated in [3]: there is the following general result essentially taken from there as in [8]. In case (iv) it relates these moment exponents to the Lyapunov exponents of the flow.

<u>Proposition 5A</u>

Let ν^- *and* ν_- *be the random variables*

$$\nu^- = \overline{\lim}_{t\to\infty} \; 1/t \log |v_t|$$

and

$$\nu_- = \underline{\lim}_{t\to\infty} \; 1/t \log |v_t|$$

Then

(i) $q \to \nu_q$ *is convex*

(ii) $q \to 1/q \; \nu_q$ *is increasing*

Also if $\mathbb{E}|\nu^-| < \infty$ and $\mathbb{E}|\nu_-| < \infty$

(iii) $\nu_q \geq \underline{\lim} \; 1/t \, \mathbb{E} \, |v_t|^q \geq q \, \mathbb{E}\nu_- \qquad q \geq 0$

$\nu_q \geq q \, \mathbb{E} \, \nu_- \qquad q \geq 0$

(iv) $$\frac{d}{dq^-}\nu_q\Big|_{q=0} \leq \mathbb{E} \, \nu_- \leq \frac{d}{dq^+}\nu_q\Big|_{q=0} .$$

<u>Proof</u> [3] Part (i) comes from the convexity of $q \to \log \mathbb{E} \, |Z|^q$ for any random variable Z and (ii) comes from the monotonicity of $q \to (\mathbb{E}|Z|^q)^{1/q}$ for $q > 0$ and of $q \to (\mathbb{E}(1/|Z|)^{-q})^{-1/q}$ for $q < 0$. Also by Jensen's inequality

$$1/t \log \mathbb{E} \, |v_t|^q \geq \mathbb{E} \, 1/t \log |v_t|^q = q \, \mathbb{E} \, 1/t \log |v_t|$$

if $\log |v_t|$ is integrable, so that (iii) and hence (iv) follows by Fatou's lemma. //

B. To show one reason for studying the moment exponents let us go back to the canonical flow on OM of a Riemannian manifold. Assume it is stochastically complete so the solutions of the canonical S.D.E. exist for all time. There is then the formal derivative flow which can be represented by $(\xi_t, A_t) \in \mathbb{R}^n \times \underline{o}(n)$, as before, satisfying (121a,b) and (122a,b). Using the notation of §4 set $v_t = u_t \xi_t$. We can consider a 1-form φ on M as a section of T^*M or as $\varphi : TM \to \mathbb{R}$ with the restrictions $\varphi_x : T_xM \to \mathbb{R}$ linear. The element in T_xM dual to φ_x will be written $\varphi_x^{\#}$.

<u>Lemma 5B</u>

For a C^2 *1-form* φ *on* M

$$\varphi(v_t) = \varphi(v_0) + \int_0^t \nabla\varphi(u_s dB_s)v_s + \int_0^t \varphi(u_s A_s dB_s)$$

$$+ \tfrac{1}{2}\int_0^t (\text{trace}\,\nabla^2\varphi(v_s)ds - \text{Ric}(\varphi^*_{x_s}, v_s))ds$$

$$+ \int_0^t \sum_{i=1}^n d\varphi(u_s e_i, u_s A_s e_i)ds \tag{198}$$

<u>Proof</u> Coming back to the S(t,u)e and $\delta S(t,V)e$ notation of §9B Chapter II recall that by Lemma 9B of Chapter II if $J(t,V) = T\pi(\delta S(t,V))$ then $J(-,V)$ is a Jacobi field with $D/\partial t\, J(t,V) = u\tilde{\omega}(V)e$ at t = 0. Therefore, with $\gamma(t) = \pi S(t,u)e$,

$$d/dt\; \varphi(T\pi\delta S(t,V)) = \nabla\varphi(\dot\gamma(t))(J(t,V)) + \varphi(D/\partial t\, J(t,V))$$
$$= \nabla\varphi(ue)(T\pi V) + \varphi(u\tilde{\omega}(V)e)$$

at t = 0; and, at t = 0,

$$d^2/dt^2\; \varphi(T\pi\delta S(t,V)) = \nabla^2\varphi(ue,ue)(T\pi V) + 2\nabla\varphi(ue)(u\tilde{\omega}(V)e)$$
$$- \varphi(R(T\pi V,ue)ue).$$

Now for an orthonormal basis $e_1,\dots,e_n$ of $\mathbb{R}^n$ if $f_i = ue_i$ and S is the skew adjoint operator $u\tilde{\omega}(V)u^{-1}$

$$\Sigma_i\; \nabla\varphi(ue_i)(u\tilde{\omega}(V)e_i) = \Sigma_i\; \nabla\varphi(f_i)(Sf_i) = \Sigma_i\; \langle\nabla\varphi^*(f_i),Sf_i\rangle$$
$$= -\,\text{trace}\; S\nabla^*\varphi = -\,\text{trace}(\nabla^*\varphi)S$$
$$= -\;\Sigma_i\; \langle(\nabla\varphi^*)S\,f_i, f_i\rangle = -\,\Sigma\,\nabla\varphi(Sf_i)f_i$$
$$= -\,\Sigma\, d\varphi(Sf_i,f_i) - \Sigma\,\nabla\varphi(f_i)Sf_i\,.$$

Thus (198) holds by Ito's formula. //

The (*de Rham-Hodge*) *Laplacian* $\Delta\varphi$ of a 1-form φ satisfies the *Weitzenbock formula*

$$\Delta\varphi = \text{trace}\;\nabla^2\varphi - \text{Ric}(-,\varphi^*) \tag{199}$$

(with non-standard sign conventions), see Proposition 3D of Chapter V, below, for the proof. The following result is discussed in [75], [43], [77].

<u>Theorem 5B</u> *Suppose the family* $(\varphi_t : t \geq 0)$ *of 1-forms on* M *satisfies:*

(i) φ_t *is* C^2 *on* M *and* C^1 *in* t, *with the partial derivatives jointly continuous,*

(ii) $\partial\varphi_t/\partial t = \tfrac{1}{2}\Delta\varphi_t \qquad t > 0$

(iii) $d\varphi_t = 0$ $\qquad t > 0$

(iv) φ_t *is bounded uniformly in* $t \in [0,T]$ *each* $T > 0$.

Assume M *is stochastically complete and* $|v_t|$ *lies in* L^1 *for each* t *where* $v_t = u_t \xi_t$ *for* $\xi_0 = u_0^{-1} v_0$ *any frame* u_0 *at* x_0, *and* $A_0 = 0$.

Then

$$\varphi_t(v_0) = \mathbb{E}\varphi_0(v_t) \qquad v_0 \in T_{x_0}M, \ t \geq 0 .$$

Proof Set $\psi_t = \varphi_{T-t}$ for $0 \leq t \leq T$ and apply the time dependent version of Lemma 5B to ψ_t. //

Note that (iv) holds automatically if M is compact as does the integrability of $|v_t|$. It is also true that $d\varphi_0 = 0$ implies $d\varphi_t = 0$ when (ii) is satisfied, at least for M compact. Furthermore, as we will see below, $\Delta\varphi = 0$ implies $d\varphi = 0$ for M compact. Thus

Corollary 5B(1). *If* M *is compact and* $\lim_{t\to\infty} \mathbb{E}\, |v_t| = 0$ *there are no harmonic 1-forms except* 0. //

Note that from the analogue of (170a) for constant curvature k if we substitute $k|\xi_s|^2 + |A_s|^2 = 0$ we see

$$\mathbb{E}|v_t|^2 = \mathbb{E}|\xi_t|^2 = e^{-3kt}\, |v_0|$$

and so the conditions of 5B(i) hold if $k > 0$. By (160) we have

$$\mathbb{E}\, |v_t|^2 = |\xi_0|^2 - \int_0^t \mathbb{E}\, \mathrm{Ric}\, (v_s, v_s) ds + \int_0^t \mathbb{E}\, |A_s|^2\, ds$$

so its hypotheses cannot hold if $\mathrm{Ric}\,(v,v) < \alpha|v|^2$, for all v, for some $\alpha < 0$. On the other hand if the Ricci curvature is strictly positive everywhere Bochner's theorem implies that there are no non-zero harmonic 1-forms. This will be discussed in detail below.

C. For more about moment exponents and also their relationships with large deviation theory see [3], [4], [11], [13], [25], [45].

CHAPTER IV. THE HEAT FLOW FOR DIFFERENTIAL FORMS AND THE TOPOLOGY OF M.

§1. A Class of semigroups and their solutions.

A. Let $p : B \to M$ be some tensor bundle over a Riemannian manifold M e.g. $B = TM, T^*M, \Lambda^p TM$, or a trivial bundle $M \times \mathbb{R}^n$, with induced inner product on each $B_x := p^{-1}(x)$ (in fact any Riemannian vector bundle with a Riemannian connection would do). For $x \in M$ suppose we have a linear map $J_x : B_x \to B_x$ depending measurably on x. Let $\{x_t : t \geq 0\}$ be Brownian motion on M from the point x_0: we will assume M is stochastically complete.

For $v_0 \in B_{x_0}$ define the process $\{v_t ; t \geq 0\}$ over $\{x_t : t \geq 0\}$ by

$$Dv_t/\partial t = J_{x_t}(v_t) \tag{200}$$

as in equation (47). Assuming J is bounded above (i.e. the map j defined below is bounded above) the solution of (200) will exist for all time and

$$d/dt\, |v_t|^2 = 2 \langle J_{x_t}(v_t), v_t \rangle_{x_t} \tag{201}$$

$$\leq 2\, j(x_t)\, |v_t|^2 \tag{202}$$

if $j(x) = \sup \{\langle J_x v, v\rangle : v \in B_x \text{ and } |v| = 1\}$.

Thus $$|v_t| \leq e^{\int_0^t j(x_s) ds} . \tag{203}$$

By a $C^{2,1}$ section $\{\varphi_t : t \geq 0\}$ of B^* we mean a time dependent section of the dual bundle to B: so $\varphi_{t,x} \in \mathbb{L}(B_x;\mathbb{R})$ for $x \in M$ and the map $\varphi_. : B \times [0,\infty) \to \mathbb{R}$ given by $(v,t) \to \varphi_t(v)$ has two partial derivatives in the first variable and one in t, all of them continuous. The following can be considered as a uniqueness result:

<u>**Proposition 1A.**</u> *Suppose $\{\varphi_t : t \geq 0\}$ is a $C^{2,1}$ section of B^* such that*

$$\partial\varphi_t/\partial t = \tfrac{1}{2}\, \text{Trace}\, \nabla^2 \varphi_t + J^*(\varphi_t) \tag{204}$$

with φ_t bounded (i.e. $\{|\varphi_{t,x}| : x \in M\}$ *bounded) uniformly on each* $0 \leq t \leq T$, *for* $T > 0$. *Then if J is bounded above, and $v_0 \in B_{x_0}$ some $x_0 \in M$*

$$\varphi_t(v_0) = \mathbb{E}\,\varphi_0(v_t) \tag{205}$$

where $\{v_t : t \geq 0\}$ *is the solution to* (200).

<u>**Proof:**</u> To interpret (200) we can suppose $x_t = \pi(u_t)$ for $\{u_t : t > 0\}$ a solution to the canonical S.D.E. on OM. For $S(t,u)e$ as before (e.g. §9A Chapter II) and $\gamma_t = \pi S(t,u)e$ suppose

$$D/\partial t\; v_t(e) = J_{\gamma_t}(v_t(e))$$

with $v_0(e) = w_0$ some $w_0 \in B_{\gamma_0}$. Then for $\psi : B \to \mathbb{R}$ of class C^2 and linear on the fibres, if $x = \pi(u) = \gamma_0$

$$\begin{aligned} d/dt\;\; \psi(v_t(e)) &= \nabla\psi(\dot\gamma_t)(v_t(e)) + \psi(J_{\gamma_t}(v_t(e))) \\ &= \nabla\psi(u(e))(w_0) + \psi(J_x(w_0)) \text{ at } t = 0 \end{aligned} \tag{206}$$

and, at $t = 0$,

$$d^2/dt^2\; \psi(v_t(e))$$

$$= \nabla^2\psi(ue,ue)(w_0) + 2\nabla\psi(ue)(J_x(w_0)) + \psi(\nabla J(ue)(w_0)) + J_x(J_x(w_0)) \tag{207}$$

if J is differentiable. At first sight it is not obvious how to interpret this to obtain the Ito formula for $\psi(v_t)$ using Proposition 3A of Chapter I. In fact our system does fit into that result but with $z_t = (B_t,t)$, and as a system on OM $\times$ B_{x_0}, namely

$$du_t = X(u_t) \circ dB_t$$

$$dw_t = //_t^{-1}\, J_{\pi(u_t)}\, (//_t w_t)dt$$

where $//_t$ is parallel translation of the tensors along $\{\pi(u_s) : 0 \leq s \leq t\}$: (this is $\rho(u_0\, u_t^{-1})$ for a suitable representation ρ of $O(n)$ on B_{x_0}). Thus the terms above without adequate e's will have a 'dt' in Ito's formula and the terms involving ∇J will have a '$dt\; dB^i_t$' and so not appear. The assumption of differentiability of J is therefore not needed (remember the global Ito formula depends on local formulae, the way we are working out its coefficients is just formalism: a method of obtaining a formula whose coefficients have geometric content). Thus

$$\psi(v_t) = \psi(v_0) + \int_0^t \nabla\psi(u_s dB_s)(v_s) + \int_0^t \left(\tfrac{1}{2}\text{ trace } \nabla^2\psi(v_s) + \psi(J_{x_s}(v_s))\right)ds\,. \tag{208}$$

Alternatively this can be derived from the Stratonovich equation

$$\psi(v_t) = \psi(v_0) + \int_0^t \nabla\psi(u_s \circ dB_s)(v_s) + \int_0^t \psi(J_{x_s}(v_s))ds \tag{209}$$

The result follows by applying the time dependent form of (208) with $\psi_t = \varphi_{T-t}$ where $T > 0$. //

Versions of (208) for more general systems than (200) are given in [14].

B. After a 'Feynman-Kac type' formula here is a 'Girsanov-Cameron-Martin' formula. Let A and Z be C^1 vector fields on M. Let $\{x_t : t \geq 0\}$ denote Brownian motion on M from x_0 with drift A, assumed non-explosive, and $\{u_t : t \geq 0\}$ its horizontal lift to OM: so we can take it that

$$du_t = X(u_t) \circ dB_t + \tilde{A}(u_t)dt \tag{210}$$

for (X,B) the canonical S.D.S. on OM and $\tilde{A}$ the horizontal lift of A, with $\pi(u_t) = x_t$. Let M_t be the process on $\mathbb{R}$ given by $M_0 = 1$ and

$$dM_t = M_t\langle Z(x_t), u_t \circ dB_t\rangle - \tfrac{1}{2} M_t \{\operatorname{div} Z(x_t) + |Z(x_t)|^2\}dt \tag{211}$$

so

$$M_t = \exp\{\int_0^t \langle Z(x_s), u_s \circ dB_s\rangle - \tfrac{1}{2}\int_0^t \{\operatorname{div} Z(x_s) + |Z(x_s)|^2\}ds \tag{212}$$

In the more familiar Ito formalism

$$dM_t = M_t\langle Z(x_t), u_t\, dB_t\rangle_{x_t} \tag{213}$$

and

$$M_t = \exp\{\int_0^t \langle Z(x_s), u_s\, dB_s\rangle - \tfrac{1}{2}\int_0^t |Z(x_s)|^2 ds\} \tag{214}$$

Proposition 1B *Suppose $\{\varphi_t : t \geq 0\}$ is a $C^{2,1}$ section of B^* such that*

$$\partial\varphi_t/\partial t = \tfrac{1}{2}\operatorname{trace}\nabla^2\varphi_t + \nabla\varphi_t(A) + \nabla\varphi_t(Z) + J^*(\varphi_t) \tag{215}$$

and φ_t is bounded uniformly on $0 \leq t \leq T$ for each $T > 0$. Then with the assumptions and notation above, if also the process with generator $\frac{1}{2}\Delta + A + Z$ is complete and if J is bounded above

$$\varphi_t(v_0) = \mathbb{E}M_t\varphi_0(v_t)$$

for each x_0 in M and $v_0 \in Bx_0$ where $\{v_t : t \geq 0\}$ satisfies the covariant equation along the paths of $\{x_t : t \geq 0\}$

$$Dv_t/\partial t = J(v_t)\,. \tag{216}$$

<u>**Proof.**</u> First consider the case $B = M \times \mathbb{R}$ with p the projection and $J \equiv 0$. This is the classical theorem: for $\varphi : M \to \mathbb{R}$ which is C^2 and bounded the Ito formula for $M_t\varphi(x_t)$ shows $Q_t\varphi$ defined by $Q_t\varphi(x_o) = \mathbb{E}M_t\varphi(x_t)$ is a minimal semigroup on L^∞ with differential generator $\frac{1}{2}\Delta + A + Z$; there is a unique such semi-group so the change of probability to $\tilde{\mathbb{P}}$ with $\tilde{\mathbb{P}} = M_T\mathbb{P}$ (on paths ω restricted to $0 \le t \le T$) *is* a change to a probability measure and under $\mathbb{P}^\sim$ the process $\{x_t : 0 \le t \le T\}$ has generator $\frac{1}{2}\Delta + A + Z$.

Now Proposition 1A extends with essentially the same proof to the case where the process $\{x_t : t \ge 0\}$ has a drift. Applying this to $\{x_t : t \ge 0\}$ under the probability $\mathbb{P}^\sim$ gives (215). //

<u>**Note:**</u>

(i) Under the completeness conditions $\{M_t : t \ge 0\}$ is a martingale.

(ii) we can allow A and Z to be time dependent provided their sum A + Z is not.

C. A case which we will be particularly interested in is B = TM and $J(v) = \mathrm{Ric}(v,-)^*$. From the Weitzenbock formula (199) proved later (Chapter V, §3), and elliptic regularity which shows that solutions to the heat equation for forms are $C^{2,1}$ (in fact C^∞) we have from Proposition 1A and Yau's result on the stochastic completeness of M when M is complete with Ricci curvature bounded below:

<u>**Theorem 1C.**</u> *If* M *is complete with Ricci curvature bounded below then any solution* $\{\varphi_t : t \ge 0\}$ *to the heat equation for 1-forms*

$$\partial\varphi_t/\partial t = \tfrac{1}{2}\,\Delta\varphi_t$$

with φ_t *uniformly bounded on compact intervals* $[0,T]$ *of* $\mathbb{R}$ *is given by*

$$\varphi_t(v_o) = \mathbb{E}\varphi_o(v_t)$$

where $\{v_t : t \ge 0\}$ *satisfies the covariant equation along Brownian paths*

$$Dv_t/\partial t = -\tfrac{1}{2}\,\mathrm{Ric}(v_t,-)^*. \ //$$

We will discuss the analogous situation for p-forms $p > 1$ later.

§2. The top of the spectrum of Δ

A. For complete M it is a standard result that Δ is essentially self-adjoint on the space of C^∞ functions with compact support (as is trace ∇^2 acting on sections of B^* as in §1 and the de Rham-Hodge Laplacian on forms, [91]). Since, for f of compact support,

$$\int_M f\Delta f = -\int_M \langle\nabla f, \nabla f\rangle \le 0 ,$$

Δ is non-positive and so there is a semi-group induced on $L^2(M)$ by $\frac{1}{2}\Delta$ which we will write $\{e^{\frac{1}{2}t\Delta} : t \ge 0\}$. This semigroup restricted to $L^\infty \cap L^2$ extends to a contraction semigroup on $L^\infty(M)$, e.g. see [86] p.209. By elliptic regularity and the simplest case of Proposition 1A, this implies that

$$e^{\frac{1}{2}t\Delta} f(x_0) = \mathbb{E}f(x_t) \tag{217}$$

for $f \in L^\infty \cap L^2$. We will not distinguish between Δ and its (self adjoint) closure.

There is the heat kernel $p_t(x,y)$ for $t > 0$ and $x,y \in M$. It satisfies

$$p_t(x,y) = \lim_{i\to\infty} p_t^{D_i}(x,y) \tag{218}$$

where $\{D_i\}_{i=1}^{\infty}$ is an increasing sequence of bounded domains in M with smooth boundaries whose union is M and where $p_t^{D_i}(x,y)$ denotes the heat kernel in D_i with Dirichlet boundary conditions. Equation (218) holds because the corresponding result holds for the transition probabilities of Brownian motion on M and the Brownian motions in D_i killed on the boundary.

For an incomplete manifold (218) can be taken as the *definition* of $p_t(x,y)$, each D_i having compact closure.

B. Since Δ is a negative operator $\lambda_0 := \sup\{\lambda \in \operatorname{Spec}\Delta\} \le 0$. When M is compact or has finite volume $\lambda_0 = 0$ since the constants lie in L^2. There are various characterizations of $\lambda_0(M)$ e.g. see [93]: in particular

$$\lambda_0(M) = -\inf\{\int_M |\nabla\varphi|^2 / \int_M |\varphi|^2 : \varphi \text{ is } C^\infty \text{ with compact support}\}.$$

Let $\{D_i\}_{i=1}^{\infty}$ be an exhaustion of M by pre-compact domains with smooth boundaries as before. The spectrum of the Laplacian with Dirichlet boundary conditions for functions on D_i is discrete. Let $\lambda_0(D_i)$ be the first eigenvalue, so $\lambda_0(D_i) < 0$. Then

$$\lambda_0(D_i) = -\inf\{\textstyle\int_{D_i} |\nabla\varphi|^2 / \int_{D_i} |\varphi|^2 : \varphi \text{ is } C^\infty \text{ with compact support in } D_i\}$$

e.g. see [28]. Thus $\{\lambda_0(D_i)\}_{i=1}^\infty$ is increasing and

$$\lambda_0(M) = \lim_{i\to\infty} \lambda_0(D_i).$$

It is shown in [31], see also [93], that if $p \in D_1$ and $h^i : D_i \to \mathbb{R}$ satisfies $h^i(p) = 1$ and $\Delta h^i = \lambda_0(D_i)h^i$ then on any compact set in M the sequence $\{h^i\}_{i=1}^n$ has a uniformly convergent subsequence giving a limit $h : M \to \mathbb{R}$ which is positive and satisfies $\Delta h = \lambda_0(M)h$. A smooth function h is a *λ-harmonic function* if $\Delta h = \lambda h$. A basic result [93] is (for non-compact M) :

There are positive λ-harmonic functions if and only if $\lambda \geq \lambda_0(M)$.

Note that for $\lambda \neq \lambda_0$ such functions cannot be in L^2 .

C. The *Green's region* consists of those λ with

$$g^\lambda(x,y) = \tfrac{1}{2}\int_0^\infty e^{-\frac{1}{2}\lambda t}\, p_t(x,y)dt < \infty$$

for all x,y with $x \neq y$. From functional analysis if $\lambda > \lambda_0$ then λ lies in the Green's region. See [6], [93]. On the other hand if $\lambda < \lambda_0(M)$ then $\lambda < \lambda_0(D_i)$ for some D_i. Writing λ^i for $\lambda_0(D_i)$ and Δ^i for the Dirichlet Laplacian for D^i, if λ were in the Green's region this would imply from

$$h^i(x) = e^{-\frac{1}{2}\lambda^i t}\, e^{\frac{1}{2}t\Delta^i}\, h^i(x)$$

that

$$h^i(x) = 1/t\int_0^t h^i(x)ds = 1/t\int_0^t e^{-\frac{1}{2}\lambda^i s}\left(\int_M p^{D_i}{}_s(x,y)h^i(y)dy\right)ds$$

$$\leq 1/t\int_0^t e^{-\frac{1}{2}\lambda s}\left(\int_{D_i} p_s(x,y)h^i(y)dy\right)ds$$

$$= \int_{D_i} (1/t \int_0^t e^{-\frac{1}{2}\lambda s} p_s(x,y)h^i(y)ds)dy$$

$$\to 0 \text{ as } t \to \infty.$$

Thus [93], *the Green's region consists of* $[\lambda_0(M),\infty)$ *or* $(\lambda_0(M),\infty)$.

Given a positive λ-harmonic function $h:M \to \mathbb{R}\ (>0)$ we can *h-transform* Brownian motion, and the heat semigroup P_t. For this define, for measurable $f : M \to \mathbb{R}$, the function $P^h{}_t f:M \to \mathbb{R}$ by

$$P^h{}_t\, f(x) = 1/h(x)\, e^{-\lambda t/2}\, P_t(hf)(x) \tag{219}$$

when it exists. This gives a semigroup with differential generator $\mathcal{A}^h$ where

$$\mathcal{A}^h f(x) = \tfrac{1}{2}\Delta f(x) + \langle \nabla \log h(x), \nabla f(x)\rangle_x \tag{220}$$

so the corresponding Markov process is Brownian motion with drift $\nabla \log h$; this is *the* h-*transformed Brownian motion on* M. The fundamental solution is given by

$$p_t{}^h(x,y) := 1/h(x)\, e^{-\lambda t/2}\, p_t(x,y)h(y) \tag{221}$$

From general results about transience and the existence of Green's operators in [6], or from [93], we see λ is *in the Green's region if and only if the h-transformed Brownian motion is transient.*

D. The following result from [93] will be very useful, as will its method of proof which comes from [46]. See also [85]

__Proposition 2D__ *If* λ *belongs to the Green's region then for every* $x_0 \in M$ *and compact set* K *of* M, *if* $\{x_t : t \geq 0\}$ *is Brownian motion from* x_0

$$\lim_{t\to\infty} e^{-(\lambda/2)t}\, \mathbb{P}\{x_t \in K\} = 0\,. \tag{222}$$

__Proof__ Choose a λ-harmonic function $h : M \to \mathbb{R}(>0)$. Let $\{y_t : t \geq 0\}$ be the h-transformed Brownian motion from x_0. It is transient and so $\lim_{t\to\infty} \mathbb{E}\, \chi_K(y_t) = 0$

i.e. $\lim_{t\to\infty} P^h{}_t(\chi_K)(x_0) = 0$

By definition of $P^h{}_t$ this means $\lim_{t\to\infty} e^{-\frac{1}{2}\lambda t}\, \mathbb{E}h(x_t)\chi_K(x_t) = 0$

which gives (222) since h is bounded away from 0 on K.

<u>Corollary 2D</u> *For* K *compact and* $\{x_t : t \geq 0\}$ *Brownian motion on* M

$$\overline{\lim_{t\to\infty}}\ 1/t \log \mathbb{P}\{x_t \in K\} \leq \tfrac{1}{2}\lambda_0(M) \tag{223}$$

<u>Proof</u> By (222) if $\lambda > \lambda_0(M)$ then $\overline{\lim\limits_{t\to\infty}}\ 1/t \log \mathbb{P}\{x_t \in K\} \leq \frac{1}{2}\lambda$. //

§3. Bochner theorems for L^2 harmonic forms

A. There is a wide literature in differential geometry relating curvature conditions on M to the existence of functions, forms or tensors, of particular kinds e.g. see [14], of these Bochner's theorem has been particularly important: the simplest case of it says that if the Ricci curvature of M is positive definite and M is compact then there are no harmonic 1-forms. The importance is because of Hodge's theorem which states, in particular, that the dimension of the space of harmonic 1-forms is the 1st Betti number of M when M is compact i.e. it is the dimension of the first cohomology group $H^1(M;\mathbb{R})$ of M with real coefficients. This case of Bochner's theorem is very simple from what we have done in Theorem 1C with equation (202). It is also almost immediate functional analytically from Weitzenbock's formula and the following. Let $C_0^\infty \mathbb{L}(TM; T^*M)$ be the space of C^∞ sections of the tensor bundle over M whose fibres consists of $\mathbb{L}(T_xM; T_x{}^*M)$ for x in M, with corresponding notation for other spaces of sections.

<u>Lemma 3A</u> *Let* $\nabla^* : C_0^\infty \mathbb{L}(TM; T^*M) \to C_0^\infty(T^*M)$ *be the formal adjoint of* $\nabla : C_0^\infty T^*M \to C_0^\infty \mathbb{L}(TM; T^*M)$. *Then for* $\varphi \in C_0^\infty \mathbb{L}(TM;T^*M)$,

$$(\nabla^*\varphi)_x = -\operatorname{trace}(\nabla\varphi)_x = -\textstyle\sum_{i=1}^n \nabla\varphi(e_i)(e_i) \tag{224}$$

where $e_1,\dots,e_n$ *is an orthonormal base for* T_xM.

<u>Proof</u> Given $\varphi \in C_0^\infty \mathbb{L}(TM; T^*M)$ and $\psi \in C_0^\infty(T^*M)$ there is the one-form given by $v \to \langle\varphi_x(v),\psi_x\rangle$ for $v \in T_xM$ with corresponding vector field $x \to \langle\varphi_x(-), \psi_x\rangle^\#$. By the divergence theorem

$$0 = \int_M \operatorname{div} \langle \varphi_-(-), \psi_- \rangle^* = \int_M \operatorname{trace} \nabla \langle \varphi_-(-), \psi_- \rangle^*$$

$$= \int_M \sum_i \{\langle \nabla \varphi(e_i)(e_i), \psi \rangle + \langle \varphi(e_i), \nabla \psi(e_i) \rangle\}.$$

But $\sum_i \langle \varphi(e_i), \nabla \psi(e_i) \rangle = \langle \varphi, \nabla \psi \rangle$ and so (224) follows. //

From Lemma 3A the Weitzenbock formula (199) can be written

$$\Delta \varphi = -\nabla^* \nabla \varphi - \operatorname{Ric}(\varphi^*, -) \tag{225}$$

for a 1-form φ. Thus if φ has compact support and is smooth

$$\langle \Delta \varphi, \varphi \rangle_{L^2} = -\langle \nabla \varphi, \nabla \varphi \rangle_{L^2} - \int_M \operatorname{Ric}(\varphi^*, \varphi^*) \tag{226}$$

from which Bochner's theorem for compact M with $\operatorname{Ric}(v,v) > 0$ for all $v \neq 0$ follows. In fact it clearly extends to the case of non-compact M if we consider only L^2 forms φ (and are careful about the existence of $\nabla \varphi$ in L^2 if $\Delta \varphi = 0$). See [39], for example, for generalities about L^2 harmonic forms etc.

B. The following is an improved version of the L^2 Bochner theorem, taken from [46]. The case of strict equality can also be fairly easily obtained analytically, using the method of domination of semi-groups (a direct parallel of the proof given here) as in [40] §4. For $x \in M$ set $\underline{\operatorname{Ric}}(x) = \inf\{\operatorname{Ric}(v,v) : |v| = 1, v \in T_xM\}$.

<u>Theorem 3B</u> [46]. *Assume* M *is complete with* $\underline{\operatorname{Ric}}(x) \geq \lambda_0(M)$ *for all* $x \in M$ *and also either:*

(i) $\underline{\operatorname{Ric}}(x) > \lambda_0(M)$ *for some* $x \in M$ or

(ii) $\lambda_0(M)$ *is in the Green's region.*

Then there are no L^2 *harmonic 1-forms except* 0.

Before giving the proof we need a few more facts about the Laplacian on forms. It is usually defined (although with the opposite sign) on smooth forms by

$$\Delta = -(d\delta + \delta d) \tag{227}$$

where d is exterior differentiation and δ is the L^2 adjoint of d. These are discussed in more detail later, and the Weitzenbock formula proved. For the moment simply observe that if φ is a sufficiently regular one-form then

$$\langle\Delta\varphi,\varphi\rangle_{L^2} = -\langle d\varphi,d\varphi\rangle_{L^2} - \langle\delta\varphi,\delta\varphi\rangle_{L^2} \tag{228}$$

since dd = 0. Thus Δ is a negative operator. It is essentially self adjoint on C_0^∞ and so there is a naturally defined semigroup $\{e^{\frac{1}{2}t\Delta} : t > 0\}$. By regularity theory, if the Ricci curvature is bounded below, Theorem 1C identifies $e^{\frac{1}{2}t\Delta}\varphi$ with $P_t\varphi$, when φ is in $L^2 \cap L^\infty$, for $P_t\varphi(v_0) = \mathbb{E}\varphi(v_t)$ with the notation of Theorem 1C. See [91a] and the discussion in [46].

When we wish to distinguish between the Laplacian on forms and on functions we will use Δ^1 and Δ^0, with $P^1{}_t$ and $P^0{}_t$ for the corresponding probabilistically defined semigroup.

<u>Proof of Theorem 3B.</u> Suppose there is a non-zero L^2 harmonic 1-form φ_0. Choose a smooth $\mu : M \to \mathbb{R}(\geq 0)$ with support in some compact set K such that $\varphi := \mu\varphi_0$ is not identically zero. The space of L^2 harmonic 1-forms is closed in L^2 (it is $\{\varphi \in L^2 : e^{t\Delta}\varphi = \varphi$ for all $t > 0\}$). Let H be the projection in L^2 onto it. Then $H\varphi \neq 0$ since

$$\langle H\varphi, \varphi_0\rangle_{L^2} = \langle\varphi,\varphi_0\rangle_{L^2} = \int_M \mu\langle\varphi_0,\varphi_0\rangle > 0.$$

By abstract operator theory $e^{\frac{1}{2}t\Delta}\varphi \to H\varphi$ in L^2 as $t \to \infty$. A subsequence therefore converges almost surely on M, say on some subset M_0 of M. Choose $x_0 \in M_0$. Set $v_0 = (H\varphi)^{\#}{}_{x_0} \in T_{x_0}M$. Then

$$\overline{\lim_{t\to\infty}}\, (e^{\frac{1}{2}t\Delta}\varphi)(v_0) > 0 \tag{229}$$

Set $C = \inf \underline{\mathrm{Ric}}(x)$. For $Dv_t/\partial t = -\frac{1}{2}\,\mathrm{Ric}(v_t,-)^{\#}$ along Brownian paths, by Theorem 1C and estimate (202), equation (229) gives

$$0 < \overline{\lim_{t\to\infty}}\, \mathbb{E}\varphi(v_t) \leq \overline{\lim_{t\to\infty}}\, |\varphi|_{L^\infty}\, \mathbb{E}(\chi_K(x_t)|v_t|)$$

$$\leq |\varphi|_{L^\infty}\, \overline{\lim}\, e^{-\frac{1}{2}Ct}\, \mathbb{P}\{x_t \in K\} \tag{230}$$

which implies C is not in the Green's region by Proposition 2D. By assumption

$C \geq \lambda_0(M)$. Therefore $C = \lambda_0(M)$ and $\lambda_0(M)$ is not in the Green's region i.e. (ii) does not hold.

To contradict (i) assume it holds and take a λ_0-harmonic function $h : M \to \mathbb{R}(>0)$ for $\lambda_0 = \lambda_0(M) = C$. Let $z_t = x^h{}_t$, the h-transformed Brownian motion starting at x_0. Its generator is $\frac{1}{2}\Delta + \nabla \log h$. Since λ_0 is not in the Green's region $\{z_t : t \geq 0\}$ is recurrent and hence complete and we can apply the Girsanov theorem, Proposition 1B, to $P^1{}_t(\varphi)$ to get

$$e^{-\frac{1}{2}t\Delta}\varphi(v_0) = \mathbb{E}\, M_t\, \varphi(v_t)$$

where $\{v_t : t \geq 0\}$ satisfies $Dv_t/\partial t = -\frac{1}{2}\operatorname{Ric}(v_t,-)^*$ along the paths of $\{z_t : t \geq 0\}$ and

$$M_t = \exp\{-\int_0^t \langle \nabla \log h(z_s), u_s \circ dB_s\rangle - \frac{1}{2}\int_0^t (-\Delta \log h(z_s) ds + |\nabla \log h(z_s)|^2) ds$$

$$= \exp\{-\log h(z_t) + \log h(x_0) + \frac{1}{2}\int_0^t |\nabla \log h(z_s)|^2 ds + \frac{1}{2}\int_0^t \Delta \log h(z_s) ds\}$$

$$= h(x_0) h(z_t)^{-1} e^{\frac{1}{2}\lambda_0 t} \tag{240}$$

since $\Delta \log h = h^{-1}\Delta h - |\nabla \log h|^2 = \lambda_0 - |\nabla \log h|^2$. Here u_t refers to the horizontal lift of $\{z_t : 0 \leq t < \infty\}$.

Take a bounded open set V of M with $\underline{\operatorname{Ric}}(x) > \delta + C$ for x in V, some $\delta > 0$. Let $A_t = \{s \in [0,t] : z_s \notin V\}$ and $B_t = \{s \in [0,t] : z_s \in V\}$. Then

$$\exp\{-\frac{1}{2}\int_0^t \underline{\operatorname{Ric}}(z_s) ds\} \leq \exp\{-\frac{1}{2} C|A_t| - \frac{1}{2}(C+\delta)|B_t|\}$$

$$\leq \exp\{-\frac{1}{2} Ct - \frac{1}{2}\delta|B_t|\}$$

where $|A_t|$, $|B_t|$ denote the Lebesgue measures of the random sets A_t, B_t.

Consequently

$$0 < \overline{\lim_{t\to\infty}}\ \mathbb{E}\, M_t \varphi(v_t)$$

$$\leq |\varphi|_{L^\infty}\, h(x_0)(\inf\{h(x) : x \in K\})^{-1} \lim\ \mathbb{E} \exp(-\frac{1}{2}\delta|B_t|)$$

since $\lambda_0 = C$. However $|B_t| \to \infty$ as $t \to \infty$ almost surely by the recurrence of $\{z_t : 0 \leq t < \infty\}$ (e.g. see [5] proof of Lemma 1), so this is impossible. Thus (i) cannot hold either. //

Corollary 3B *If* M *is complete and has a non-trivial* L^2 *harmonic one-form then*

$$\lambda_0(M) \geq \inf_x \underline{Ric}(x) \tag{241}$$

with strict inequality if either Ric is non-constant or $\lambda_0(M)$ *is in the Green's reason.* //

This compares with Cheng's estimate for λ_0 when $\inf \underline{Ric}(x) < 0$, [30]:

$$\lambda_0(M) \geq 1/4\,(n-1) \inf_x \underline{Ric}(x) \tag{242}$$

and improves it for $n > 5$ given some non-trivial L^2 harmonic 1-form

These results are discussed in relation to quotients of hyperbolic spaces in [46].

Remark 3B

(i) Corresponding results for p-forms can be proved in the same way given the Weitzenbock formula for the Laplacian on p-forms (see below), and similarly for the Dirac operator, [46]. The discussion in §1 shows how to formulate a general theorem.

(ii) For compact manifolds Theorem 3B reduces to the classical Bochner theorem. Note that the flat torus $S^1 \times S^1$ has $\lambda_0(M) = \underline{Ric}(x) = 0$ for all x but has harmonic 1-forms, e.g. $d\theta_1$ and $d\theta_2$ where (θ_1, θ_2) parametrize $S^1 \times S^1$ by angle. Thus some additional conditions like (i) or (ii) are needed.

§4. de Rham cohomology, Hodge theory, and cohomology with compact support.

A. Let A^p be the space of C^∞ p-forms on M. (See Chapter V, §3.) Exterior differentiation d gives a map

$$d : A^p \to A^{p+1}$$

and the p-th de Rham cohomology group $H^p(M;\mathbb{R})$ is defined by

$$H^p(M;\mathbb{R}) = \frac{\ker(d : A^p \to A^{p+1})}{\mathrm{Im}(d : A^{p-1} \to A^p)} \tag{243}$$

It is a classical result that it is isomorphic to any of the standard cohomology groups with real coefficients (e.g. simplicial or singular). The de Rham-Hodge

Laplacian on p-forms, Δ, or Δ^p to be precise, is given by

$$\Delta^p = -(d\delta + \delta d) \tag{244}$$

where δ is the formal adjoint of d in the L^2 sense. On the space A_0^p of p-forms with compact support it is known to be essentially self-adjoint, e.g. see [91], and so we can take its closure which will be self-adjoint. This will still be written as Δ^p. There is then the corresponding heat semigroup $e^{\frac{1}{2}t\Delta^p}$ acting on the space $L^2 A^p$ of L^2 p-forms since Δ^p is non-negative by the same argument as for Δ^1; see equation (228).

Let $H = H^p : L^2 A^p \to L^2 A^p$ be the projection onto the space of harmonic p-forms. Then, as before, $e^{\frac{1}{2}t\Delta} \to H$ strongly on $L^2 A^p$. For $\varphi \in L^2 A^p$ set

$$G\varphi = \int_0^\infty (e^{\frac{1}{2}t\Delta} - H)\varphi \, dt.$$

Then, leaving aside rigour for the moment,

$$\Delta G\varphi = \int_0^\infty \Delta e^{\frac{1}{2}t\Delta}\varphi \, dt = \int_0^\infty \partial/\partial t(e^{\frac{1}{2}t\Delta}\varphi)dt = H\varphi - \varphi.$$

Thus we have the decomposition for $\varphi \in L^2 A^p$

$$\varphi = -\Delta G\varphi + H\varphi \tag{245}$$

From (244) we may believe the *Hodge decomposition theorem*, at least for compact manifolds (when Δ^p has discrete spectrum): any $\varphi \in A^p$ has a decomposition into three orthogonal summands

$$\varphi = H\varphi + d\alpha + \delta\beta \tag{246}$$

for $\alpha \in L^2 A^{p-1}$ and $\beta \in L^2 A^{p+1}$. In particular if $d\varphi = 0$ then $\varphi = H\varphi + d\alpha$ since $\langle\varphi, \delta\beta\rangle_{L^2} = \langle d\varphi, \beta\rangle_{L^2} = 0$. Thus we have Hodge's theorem: *every cohomology class has a unique harmonic representative.* In particular the p-th Betti-number β_p

$$\beta_p := \dim H^p(M;\mathbb{R}) = \dim \text{(space of harmonic p-forms)}$$

when M is compact.

For non-compact manifolds the heat equation method outlined above was used by Gaffney to get a version of Hodge's theorem [53]; see also [37], [38].

B. The operators d and δ can be restricted to the space A_0^p to give the *cohomology groups* $H^p{}_K(M;\mathbb{R})$ of M *with compact support* by the analogue of (243). There is a natural inclusion

$$i_p : H^p{}_K(M;\mathbb{R}) \to H^p(M;\mathbb{R})$$

whose kernel has elements $[\varphi]_K$ represented by forms $\varphi \in A_0^p$ with $\varphi = d\alpha$ some α in A_0^p. If $\varphi' = d\alpha'$ is also in A_0^p then $[\varphi']_K = [\varphi]_K$ if and only if $\alpha = \alpha'$ outside of some compact set.

In the special case p = 1 this gives a linear surjection

$$\mathcal{E}(M) \overset{d_*}{\to} \ker i_1$$

where $\mathcal{E}(M)$ is the quotient of the vector space of bounded C^∞ functions $f : M \to \mathbb{R}$ with $df \in A^1{}_0$ by the space $A^0{}_0$ of C^∞ functions with compact support. The kernel of this map consists of {[f] : f is constant} and so we have an exact sequence

$$0 \to \mathbb{R} \to \mathcal{E}(M) \overset{d_*}{\to} H^1{}_K(M;\mathbb{R}) \overset{i_1}{\to} H^1(M;\mathbb{R})$$

The *set of ends* EndM of M is the projective limit of the inverse system whose terms are the sets of connected components of M−K as K ranges over all compact subsets of M, directed by inclusion. Thus an end is an indexed set $\{\mathcal{E}_K : K \text{ compact}\}$ such that $\mathcal{E}_K$ is a component of M−K and if $K \subset K'$ then $\mathcal{E}_{K'} \subset \mathcal{E}_K$. They can be thought of as the components of M at infinity and we will say that a continuous path $\sigma : [0,\infty) \to M$ *goes out to infinity through the end* $\mathcal{E}$ or '$\lim_{t\to\infty} \sigma(t) = \mathcal{E}$' if for each compact K there exists t_K with $\sigma(t) \in \mathcal{E}_K$ for $t > t_K$.

There is a natural map $j : \mathcal{E}(M) \to \mathbb{R}^{\mathrm{End}(M)}$ given by $j([f])(\mathcal{E}) = \lim_K \{f(x) : x \in \mathcal{E}_K\}$ which is clearly injective. Thus M is connected at infinity (i.e. has a unique end) iff $\mathcal{E}(M) = \mathbb{R}$.

C. A simple example is that of the cylinder $M = S^1 \times \mathbb{R}$. This has two ends, so $\mathcal{E}(M) \approx \mathbb{R} \oplus \mathbb{R}$. If it is parametrized by (θ,x) where θ represents the angle, there is the one form $d\theta$ which determines the generator of $H^1(M;\mathbb{R})$, which is isomorphic to $\mathbb{R}$ since M is homotopy equivalent to S^1. Now $[d\theta]$ clearly does not lie in the image of i_1 (otherwise it would have to be exact outside a compact set i.e. equal to df, for some function f, on the complement of some compact set). Therefore $H^1{}_0(M;\mathbb{R}) \approx \mathbb{R}$ generated by $[df]$ where $f(x) = +1$ for $x > 1$ and $f(x) = -1$ for $x < -1$.

D. The relevance of these concepts to L^2 harmonic form theory and Brownian motion comes from the Hodge decomposition for $\varphi \in A_0^p$, [37], with $d\varphi = 0$, which gives

$$\varphi = H\varphi + d\alpha$$

for $\alpha \in A^{p-1}$. From this we see that if *there are no non-trivial* L^2 *harmonic* p-*forms then the map*

$$i_p : H^p{}_0(M;\mathbb{R}) \to H^p(M;\mathbb{R})$$

is identically zero. The point is that the latter is a topological condition independent of the Riemannian metric (and in fact even of the differentiable structure). From Corollary 3B we can now say that if i_1 is not identically zero then $\lambda_0(M) \geq \inf_x \underline{Ric}(x)$; see [46] for some examples. This relationship between L^2 harmonic forms and i_1 was exploited by Yau [10] for complete manifolds with non-negative Ricci curvature. Using properties of such manifolds (in particular the Gromoll-Cheeger splitting theorem) he was able to give conditions for the vanishing of ker i_1 and hence for $H^1{}_0(M;\mathbb{R})$ itself rather than just its image in $H^1(M;\mathbb{R})$. Analogous results for $\underline{Ric}(x) > \lambda_0(M)$ are given in [46], but different methods are needed, and additional conditions of 'bounded geometry' appear to be needed for these methods to work. One such result is described next.

§5. Brownian motion and the components of M at infinity

A. It is shown in [46] that if M has bounded sectional curvatures and a positive injectivity radius (i.e. there exists $r > 0$ such that $\exp_x : T_xM \to M$ is a diffeomorphism of the ball of radius r about 0 onto some open set in M for each $x \in M$) e.g. if M covers a compact manifold, then $Ric(x) > \lambda_0(M)$ for all x implies that M is connected at infinity, and so by the previous discussion $H^1{}_0(M;\mathbb{R}) = 0$. The proof uses some Green's function estimates by Ancona. Here we will discuss another result from [46] which is very similar but has a slightly different emphasis.

Theorem 5B [46]. *Let* M_0 *be a compact Riemannian manifold. Assume there exist* $\varepsilon > 0$ *with*

$$\int_{M_0} \underline{\mathrm{Ric}}(x)f(x)^2dx > -\int_{M_0} |\nabla f(x)|^2dx + \varepsilon \qquad (247)$$

for all C^∞ *functions on* M_0 *with* $|f|_{L^2} = 1$. *Then every covering manifold* M *of* M_0 *is connected at infinity* (i.e. *has at most one end*). *Moreover* $H^1{}_0(M;\mathbb{R}) = 0$ *and* M, *with covering Riemannian structure, has no non-trivial harmonic 1-forms* in L^2.

Note: The condition on M_0 is precisely the condition that the top of the spectrum of Δ-$\underline{\mathrm{Ric}}$ is negative.

Proof: To show connectness at infinity take $f : M \to \mathbb{R}$ smooth and bounded with df having compact support. It suffices to show that such a function f is constant outside of any sufficiently large compact set.

Let U_1, U_2 be among the unbounded components of M-supp(df). For $p : M \to M_0$ the covering map, take $x_0 \in M$ and choose $\{x_i\}_{i=1}^\infty$ in U_1 and $\{y_i\}_{i=1}^\infty$ in U_2 with $x_i \to \infty$ and $y_i \to \infty$ and $p(x_i) = x_0$, $p(y_i) = x_0$ for each i. (We can assume M non-compact of course.) Take a finite set of generators for the fundamental group $\pi_1(M_0,x_0)$ and let $g_1,\dots,g_r$ denote them together with their inverses. Choose smooth loops $\gamma_1,\dots,\gamma_r$ at x_0, with each γ_j in the class g_j and with $|\dot\gamma_j(s)| = 1$ for all s and j e.g. the γ_j could be geodesics. Take a shortest path α from x_i to x_{i+1}. Then $p \circ \alpha$ is a loop at x_0 and we can write its homotopy class $[p \circ \alpha]$ as a product $g_{j_1} \dots g_{j_s}$ for some s. Lift γ_{j_s} to a path in M starting from x_i and lift the other corresponding paths in turn to start where the previous lift ended and so give a continuous piecewise C^1 path from x_i to x_{i+1}. Do the same for x_1 to y_1 and y_i to y_{i+1} for each i. Let $\sigma : (-\infty,\infty) \to M$ be the curve obtained from the union of these lifts: it is piecewise C^1 and satisfies $|\dot\sigma_s| = 1$ for all s where $\dot\sigma(s)$ is defined and there exists $T \in \mathbb{R}$ such that $\sigma(s) \in U_1$ if $s < -T$ and $\sigma(s) \in U_2$ if $s > T$.

Let $\mathfrak{D}$ be the set of all open subsets of M on which p is injective. For compact K in M let c(K) be the minimum number of elements in $\mathfrak{D}$ needed to cover K, and suppose $U_1,\dots,U_c$ are in $\mathfrak{D}$ and cover K. Then

$$\int_{-\infty}^{\infty} \chi_K(\sigma(s))ds \le \sum_{j=1}^{c} \int_{-\infty}^{\infty} \chi_{U_j}(\sigma(s))ds$$

Now the intersection of the curve σ with U_j decomposes into portions $P_1,\dots,P_r$ where each P_i consists of pieces which come from lifts of γ_i. Since p is injective on U_j it maps each P_i injectively into γ_i. Therefore

$$\int_{-\infty}^{\infty} \chi_{U_j}(\sigma(s))ds \le \sum_{i=1}^{r} \int_{-\infty}^{\infty} (\chi_{P_i}(\sigma(s)))ds$$

$$\le \sum_{i=1}^{r} \ell(\gamma_i)$$

where $\ell(\gamma_i)$ is the length of γ_i. Thus for any compact K in M

$$\int_{-\infty}^{\infty} \chi_K(\sigma(s))ds \le c(K) \sum_{i=1}^{r} \ell(\gamma_i). \tag{248}$$

We can now show that $\int_\sigma P_t^1(df)$ exists for $t \ge 0$ and converges to zero as $t \to \infty$. To do this take a smooth flow of diffeomorphisms on M_0 of Brownian motions, $F^0{}_t(-,\omega) : M_0 \to M_0$ for $t > 0$, $\omega \in \Omega$, e.g. a gradient Brownian flow as in Chapter III. This is possible because M_0 is compact. It lifts to a smooth Brownian flow of diffeomorphisms $F_t(-,\omega) : M \to M$, $t \ge 0$, $\omega \in \Omega$. Set K = supp (df). Then by (202) and Theorem 1C

$$\left|\int_\sigma P_t^1(df)\right| \le \int_{-\infty}^{\infty} |P^1{}_t(df)|(\sigma_s)ds$$

$$\leq \int_{-\infty}^{\infty} \mathbb{E}\Big(|df|(F_t(\sigma(s)))\exp\{-\int_0^t \underline{\mathrm{Ric}}(F_r(\sigma(s))dr\}\Big)ds$$

$$\leq |df|_{L^\infty} \int_{-\infty}^{\infty} \mathbb{E}\Big(\chi_K(F_t(\sigma(s)))\exp\{-\int_0^t \underline{\mathrm{Ric}}(F_r(\sigma(s))dr\}\Big)ds \qquad (249)$$

where $K = \operatorname{supp}(df)$.

To avoid worrying about regularity properties stemming from the possible lack of smoothness of $\underline{\mathrm{Ric}}$ choose a smooth map $\rho_0 : M_0 \to \mathbb{R}$ with $\underline{\mathrm{Ric}}(x) \geq \rho_0(x)$ for all x and such that condition (247) holds for $\underline{\mathrm{Ric}}$ replaced by ρ_0. Let ν denote the top of the spectrum of $\Delta - \rho_0$ on M_0. The revised condition (247) implies that $\nu < 0$. From Perron-Frobenius theory (e.g. see [86]) there is a strictly positive $h_0 : M_0 \to \mathbb{R}(> 0)$ with

$$\Delta h_0(x) - \rho_0(x)h_0(x) = \nu h_0(x)$$

for $x \in M_0$.

Let $h = h_0 \circ p : M \to \mathbb{R}(> 0)$ and $\rho = \rho_0 \circ p$. There is a flow for the h_0-transformed Brownian motion on M_0 and a lift of it to a flow $F^h{}_t$, $t \geq 0$, say, on M of h-transformed Brownian motions. By the Girsanov theorem using the analogous computation as that which led to (240), from (249) we get

$$\Big|\int_\sigma P_t^1(df)\Big| \leq |df|_{L^\infty} \int_{-\infty}^{\infty} \Big(\mathbb{E}\, \chi_K(F^h{}_t(\sigma_s))e^{\frac{1}{2}\nu t}\, h(\sigma_s)/(h(F^h{}_t(\sigma_s)))$$

$$\exp\{-\int_0^t (\underline{\mathrm{Ric}} - \rho)(F^h{}_r(\sigma_s))dr\}\Big)ds$$

$$\leq \text{const.}\, e^{\frac{1}{2}\nu t} \int_{-\infty}^{\infty} \mathbb{E}\, \chi_{K_t}(\sigma_s)ds$$

where K_t is the random compact set $(F^h{}_t)^{-1}(K)$. However $c(K_t) = c(K)$ since $\mathfrak{D}$ is invariant under those diffeomorphisms of M which cover diffeomorphisms of M_0. Therefore

$$\left|\int_\sigma P^1_t\,(df)\right| \le \text{const.}\ e^{\frac{1}{2}\nu t}\ c(K)\ \sum_{i=1}^{r} \ell(\gamma_i)$$

$$\to 0 \ \text{as}\ t \to \infty.$$

Thus

$$\lim_{t\to\infty} \int_\sigma P^1_t\,(df) = 0\,.$$

On the other hand, by the compactness of M_0, all the curvature tensors and their covariant derivatives are bounded on M_0, so by Theorem 5B of Chapter III, $P^1{}_s\,\Delta(df) = P^1{}_s\,d(\Delta f) = dP^0{}_s\,\Delta f$ (or alternatively by [52], $P^1{}_s\,\Delta\,df = \Delta P^1{}_s\,df = -(d\delta + \delta d)P^1{}_s\,df = dP^0{}_s\,\Delta f$). Therefore

$$\int_\sigma (P_t\,df - df) = \lim_{R\to\infty} \int_{-R}^{R} \{d \int_0^t P^0{}_s(\Delta f)ds\}\ \dot\sigma(\tau)d\tau$$

$$= \lim_{R\to\infty} \{\int_0^t P^0{}_s(\Delta f)(\sigma(R))ds - \int_0^t P^0{}_s(\Delta f)(\sigma(-R))ds\}$$

$$= 0$$

by dominated convergence and the 'C^0-property' of the semigroup $\{P^0{}_t : t \ge 0\}$. This last property says that $P_t(g)(x) \to 0$ as $x \to \infty$ for each t whenever g is continuous with $g(x) \to 0$ as $x \to \infty$. It was shown by Yau to hold for complete manifolds with Ricci curvature bounded below, e.g. see [100]. Alternatively it follows rather easily from the existence of a Brownian flow of diffeomorphisms [46].

Thus

$$0 = \lim_{t\to\infty} \int_\sigma P^1_t\,df = \int_\sigma df = \lim_{R\to\infty} f(\sigma(R)) - \lim_{R\to\infty} f(\sigma(-R)),$$

and so $f|U_1 = f|U_2$, proving the first part of the theorem.

Next we observe that M has no non-trivial harmonic forms in L^2 by arguing by contradiction as in the proof of Theorem 3B but using the h-transform this time for h as above. The triviality of $H^1{}_0(M;\mathbb{R})$ follows from the discussion in §4. //

CHAPTER V HEAT KERNELS: ELEMENTARY FORMULAE, INEQUALITIES, AND SHORT TIME BEHAVIOUR

§1. The elementary formula for the heat kernel for functions.

A. We will be following [47] and [81] fairly closely in this section. For a Riemannian manifold M and continuous $V : M \to \mathbb{R}$, bounded above there is a continuous map

$$(t,x,y) \to p_t(x,y)$$

$$(\mathbb{R} > 0) \times M \times M \to \mathbb{R}$$

such that the minimal semigroup $\{P_t : t \geq 0\}$ for $\frac{1}{2}\Delta + V$ has

$$P_t f(x) = \int_M p_t(x,y) f(y) dy \qquad t > 0 \tag{250}$$

for bounded measurable f. This is the fundamental solution. If $\{D_i : i = 1 \text{ to } \infty\}$ is an increasing sequence of domains exhausting M, with smooth boundaries, and if $p_t^{D_i}(x,y)$ denotes the fundamental solution to the equation

$$\partial f_t / \partial t = \tfrac{1}{2}\Delta f_t + V f_t \tag{251}$$

on D_i with Dirichlet boundary conditions then

$$p_t(x,y) = \lim_{i\to\infty} p_t^{D_i}(x,y) \tag{252}$$

and the right hand side is an increasing limit. This is clear from the Feynman-Kac formula, or alternatively we can define $p_t(x,y)$ by (252), with compact D_i, and then P_t by (250). In either case in order to obtain an expression for $p_t(x,y)$ it will be enough to find one for the fundamental solutions on each D_i, with D_i compact, and then take the limit.

B. To obtain exact formulae for these fundamental solutions we will need some rather strong conditions on the domains, and on M. However these conditions will turn out to be irrelevant when the asymptotic behaviour of $p_t(x,y)$ as $t \downarrow 0$ is being considered, at least for generic x and y and complete M. To describe these conditions we need to look in slightly more detail at the exponential map.

First suppose M is complete. For $p \in M$ let

$U(p) = \{v \in T_pM : d(\exp_p v, p) = |v|\}$

and let $\partial U(p)$ be its boundary and $U^\circ(p)$ its interior. The following facts can be found in [16], [29], [63]: The image Cut(p) of $\partial U(p)$ is a closed subset of M known as the *cut locus* of p, moreover:

(a) $U^\circ(p)$ is star shaped from the origin in T_pM

(b) $\exp_p$ maps $U^\circ(p)$ diffeomorphically onto the open subset M − Cut(p) of M.

<u>Example 1:</u> $M = S^1$. Here the exponential map wraps $T_pS^1 \approx \mathbb{R}$ around S^1 as a covering map (it is locally a diffeomorphism), and Cut(p) is the point antipodal to p.

<u>Example 2.</u> $M = S^n$ for $n > 1$. This is quite different from $M = S^1$ since the exponential map is no longer a local diffeomorphism: it maps the whole sphere radius π to the antipodal point of p. Again Cut(p) is this antipodal point.

<u>Example 3</u>. Real projective space: $M = \mathbb{RP}(n)$. This is the quotient space of S^n under the equivalence relation $x \sim y$ if x is antipodal to y. It is given the differentiable structure and Riemannian metric which makes the projection $p : S^n \to \mathbb{RP}(n)$ a Riemannian covering. If $x \in \mathbb{RP}(n)$ corresponds to the North (and therefore the South) pole of S^n then Cut(x) is the image under p of the equator, a copy of S^{n-1}. Thus Cut(x) is a submanifold, isometric to $\mathbb{RP}(n-1)$, in $\mathbb{RP}(n)$. It has co-dimension one and so will almost surely be hit by Brownian paths from x in $\mathbb{RP}(n)$.

<u>Example 4.</u> $M = H^n$, hyperbolic space. In §4G of Chapter III we saw that there are global exponential co-ordinates about a general point p. Thus Cut(p) = ∅.

<u>Example 5.</u> Complete manifolds with non-positive sectional curvatures ("Cartan-Hadamard manifolds"). The Cartan-Hadamard theorem e.g. [65], [79] states that for such manifolds (e.g. $M = S^1$) each exponential map $\exp_p : T_pM \to M$ is a covering map. In particular it is a local diffeomorphism. (To prove this see Exercise 1A below.) It follows that if M is simply connected then $\exp_p$ is a diffeomorphism and Cut(p) = ∅ for each p in M.

When Cut(p) = ∅, so that there exists a global exponential chart about p, the point p is said to be a *pole* of M. If so, M is diffeomorphic to $\mathbb{R}^n$ and so is essentially $\mathbb{R}^n$ with a different metric. The images under $\exp_p : T_pM \to M$ of a point v such that the derivative $T_v \exp_p$ of $\exp_p$ at v is singular is called a *conjugate point* of p along the geodesic $\{\exp_p tv : 0 \le t < \infty\}$, and v itself is said to be conjugate to p in T_pM.

Exercise 1A Show that the derivative of $\exp_p$ at v in the direction w is given by

$$T_v \exp_p(w) = J_1$$

where $\{J_t : 0 \le t \le 1\}$ is a vector field along $\{\exp tv : 0 \le t \le 1\}$ with $J_0(0) = 0$ and $DJ_t/\partial t|_{t=0} = w$. (Here T_vT_pM is identified with T_pM using the vector space structure of T_pM). Hint: look at the proof of Lemma 9B of Chapter III. Thus v is conjugate to p in T_pM if and only if there is a non-trivial Jacobi field along $\{\exp tv : 0 \le t \le 1\}$ which vanishes at $t = 0$ and at $t = 1$. See [65], [79] for example.

A basic result is that $x \in \mathrm{Cut}(p)$ if and only if either x is the first conjugate point to p along some geodesic from p, or there exist at least two minimizing geodesics from p to x. For example when $M = S^n$ and p and x are antipodal then both possibilities hold.

If $r : M \to \mathbb{R}$ is given by $r(x) = d(x,p)$ then r is C^∞ on $M - (\mathrm{Cut}(p) \cup \{p\})$ since there

$$r(x) = |\exp_p{}^{-1}(x)|_p \tag{253}$$

B. Suppose now that D is a domain in M - Cut(p) with $D \subset W$ for W open with $\bar{W}$ compact and in M-Cut(p). We can use $\exp_p{}^{-1}$ to identify M-Cut(p) with the star-shaped open set U° of T_pM, and give U° the induced Riemannian metric. Then D and W are considered as sets in T_pM. Using spherical polar coordinates in T_pM the Riemannian metric at a point v has the form

$$ds^2 = dr^2 + \sum_{i,j=1}^{n-1} \tilde{g}_{ij}(v)d\sigma^i\, d\sigma^j \tag{254}$$

where $\sigma^1, \dots, \sigma^{n-1}$ refer to coordinates on the sphere S^{n-1}. Since the space of Riemannian metrics on any manifold (and on S^{n-1} in particular) is a convex set in a linear space, it is easy to first modify $\tilde{g}_{ij}$ outside of W, if necessary so that it extends to a metric on the whole of S^{n-1} for each sphere in T_pM about p which intersects W, and then modify this family of metrics (one for each relevant radius |v|) outside of D and extend so that we obtain a Riemannian metric on the whole of T_pM of the form

$$ds^2 = dr^2 + \sum_{i,j=1}^{n-1} h_{ij}(v)d\sigma^i\, d\sigma^j \tag{255}$$

which agrees with the original one on D and agrees with the standard Euclidean one coming from $\langle\, ,\, \rangle_p$ on T_pM outside of some compact set.

This gives T_pM a Riemannian structure for which it is complete since the geodesics from p are easily seen to be the straight lines from p, by the distance minimizing characterization of geodesics, and the existence of all geodesics from some point for all time is known to be equivalent to metric completeness. The point p is now a pole and the curvature tensors are all C^∞ with compact support. Moreover the heat kernel for the Dirichlet problem in D is unchanged since all these modifications took place outside of D. We can therefore assume that M was T_pM with this metric.

C. Assuming the metric and manifold M has been changed in this way, and M identified with T_pM,

$$p_t^D(x,p) = \lim_{\lambda\downarrow 0} \int_M (2\pi\lambda)^{-n/2}\, p_t^D(x,y)\exp\{-\, d(y,p)^2/(2\lambda)\}\theta_p(y)dy \tag{256}$$

where dy refers to the Lebesgue measure of T_pM, identified with M, using $\langle\, ,\, \rangle_p$, and θ_p is the volume element from the Riemannian metric (255): in terms of our original metric it is given on T_pM by

$$\theta_p(v) = |\det_M T_v \exp_p|$$

and is known sometimes as *Ruse's invariant.* See [16] for more details about it.

Thus

$$p_t^D(x,p) = \lim_{\lambda\downarrow 0} P_t^D f_\lambda(x) \tag{257}$$

where $\{P_t^D : t \geq 0\}$ is the Dirichlet semigroup for $\frac{1}{2}\Delta + V$ and

$$f_\lambda(x) = (2\pi\lambda)^{-n/2} \exp\{-r(x)^2/(2\lambda)\}.$$

To evaluate $P_t^D f_\lambda$ we will use the Girsanov theorem. Fix $T > 0$ and for $\lambda \geq 0$ let $\{z^\lambda{}_t : 0 \leq t < T + \lambda\}$ be a Brownian motion on M from a point x_0 of D with time dependent drift $Z^\lambda{}_s$ for $Z^\lambda{}_s = \nabla Y^\lambda{}_s$ with

$$Y^\lambda{}_s(x) = -\, r(x)^2/(2(\lambda+T-s)) - \tfrac{1}{2}\log\theta_p(x) \qquad 0 \leq s < T + \lambda$$

Now if $f: M \to \mathbb{R}$ is smooth with $f(x) = F(r(x))$ for some smooth $F: (0,\infty) \to \mathbb{R}$, by equation (4) for Δ in local coordinates, we have

$$\Delta f(x) = \partial^2 F/\partial r^2 + \big((n-1)/(r(x)) + \partial/\partial r\,(\log \theta_p)(x)\big)\partial F/\partial r \tag{258}$$

In particular

$$\Delta r = (n-1)/r + \partial/\partial r \log \theta_p. \tag{259}$$

By Ito's formula for $r^\lambda_t = r(z^\lambda_t)$ we have

$$r^\lambda_t = r(x_0) + \int_0^t dr(u^\lambda_s\, dB_s) + \int_0^t \partial/\partial r\ \ \gamma^\lambda_s(z^\lambda_s)ds + \tfrac{1}{2}\int_0^t \Delta r(z^\lambda_s)ds$$

$$= r(x_0) + \int_0^t dr(u^\lambda_s\, dB_s) - \int_0^t r^\lambda_s/(\lambda+T-s)ds + \tfrac{1}{2}\int_0^t (n-1)/(r^\lambda_s)ds \tag{260}$$

where $\{u^\lambda_s ; 0 \le s < T + \lambda\}$ is the horizontal lift of $\{z^\lambda_s : 0 \le s < t\}$ to the frame bundle OM and $\{B_s : 0 \le s < \infty\}$ is a Brownian motion (which can be taken to be independent of λ by taking the canonical construction of z^λ_s from an S.D.E. on OM).

Since $|dr| = 1$ and $u^\lambda_s(\omega)$ is an orthonormal frame the martingale term in (260) is just a 1-dimensional Brownian motion, so $\{r^\lambda_t : 0 \le t < T + \lambda\}$ satisfies a stochastic differential equation which is essentially independent of the manifold M. In fact from Ito's formula it satisfies essentially the same equation (to be precise it is a weak solution of the same equation) as for the radial distance $\{|x^\lambda_s| : 0 \le s < T + \lambda\}$ where $\{x^\lambda_s : 0 \le s \le T + \lambda\}$ is the Euclidean Brownian bridge from x_0 to 0 in $\mathbb{R}^n \approx T_pM$ in time $T + \lambda$, given by

$$x^\lambda_s = x_0 - sx_0/(T+\lambda) + (s-T-\lambda)\int_0^s d\beta_t/(t-T-\lambda) \tag{261}$$

where $\{\beta_s : 0 \le s < \infty\}$ is Brownian motion on $\mathbb{R}$, e.g. see [59]. This Brownian bridge is itself equal in law to

$$s \to x_0 + B_s - s\,(B_{T+\lambda} + x_0)\,(T+\lambda)^{-1} \qquad 0 \le s \le T + \lambda \tag{262}$$

Thus $\{r^\lambda_s : 0 \le s < T + \lambda\}$ is equal in law to $\{|x^\lambda_s| : 0 \le s \le T + \lambda\}$. In particular it is non-explosive: as $s \uparrow T + \lambda$ so it converges to p (now identified with the origin).

Lemma 1C *As* $\lambda \downarrow 0$ *so* $\{z^\lambda_s : 0 \le s \le T\}$ *converges in law to the process* $\{z_s : 0 \le s \le t\}$ *which is sample continuous, agrees with* z^0_s *for* $0 \le s < t$ *and has* $z_T = p$. *Furthermore* $\{z_s : 0 \le s \le T\}$ *has radial component* $\{r(z_s) : 0 \le s \le t\}$ *which has the same distributions as the radial component of the Euclidean Brownian bridge in* $\mathbb{R}^n$ *starting from a point distance* $r(x_0)$ *from* 0 *and ending at* 0 *in time* T.

Proof Let $dx_t = X(x_t) \circ dB_t + A(x_t)dt$ be a smooth stochastic differential equation on M whose solutions are Brownian motions on M and, identifying M with T_pM and so with $\mathbb{R}^n$, such that A has compact support and $X(x) = 0$ outside some compact set. Then we can represent $\{z^\lambda_t : 0 \le t < T + \lambda\}$ as the solution to

$$dz^\lambda_t = X(z^\lambda_t) \circ dB_t + A(z^\lambda_t)dt + Z^\lambda_t(z^\lambda_t)dt. \tag{263}$$

Fix $t_0 \in (0,T)$. Then $\{z^\lambda_t : 0 \le t \le t_0\}$ converges uniformly in probability to $\{z^0_t : 0 \le t \le t_0\}$. (Indeed we can choose versions so that it converges almost surely since the coefficients of (260) have derivatives bounded uniformly on $[0,t_0]$.) Now suppose $0 < \varepsilon < 1$ and $\delta > 0$. Set $\varepsilon_1 = \min\{\varepsilon/3, \varepsilon\delta/18\}$ and choose $\delta_1 > 0$ such that

$$\mathbb{P}\{r^0_s < \varepsilon_1 \text{ for } t-\delta_1 \le s \le T\} > 1-\varepsilon_1 > 1 - \varepsilon/3,$$

(which is possible by the continuity of r^0_s at $s = T$).

Take $\delta_2 > 0$ such that for $0 < \lambda < \delta_2$

$$\mathbb{P}\{|z^\lambda_s - z_s| \le \delta \text{ for } 0 \le s \le T-\delta_1\} \ge 1-\varepsilon/3$$

and such that

$$\mathbb{E}[\sup\{r^\lambda_s \wedge 1 : T-\delta_1 \le s \le T\}] \le \varepsilon_1 + \mathbb{E}[\sup\{r^0_s \wedge 1 : T-\delta_1 \le s \le T\}]$$

(which is possible because $r^\lambda_s \to r^0_s$ in probability uniformly on $[0,T]$).

Then, for $0 < \lambda < \delta/2$

$$\mathbb{E}[\sup\{r^\lambda_s \wedge 1 : T-\delta_1 \le s \le T] \le 2\varepsilon_1 + (1-\varepsilon_1)\varepsilon_1 \le 3\varepsilon_1$$

whence

$$\mathbb{P}[\sup\{r^\lambda_s \wedge 1 : T-\delta_1 \le s \le T\} > \delta/2] \le 6\varepsilon_1/\delta \le \varepsilon/3$$

and so

$$\mathbb{P}\{|z^\lambda_s - z_s| \le \delta \text{ for } 0 \le s \le T\} \ge 1-\varepsilon/3 - \varepsilon/3 - \varepsilon/3 = 1-\varepsilon.$$

This prove uniform convergence of $\{z^\lambda_s : 0 \le s \le T\}$ to $\{z_s : 0 \le s \le T\}$ in probability.

If $x_0 = p$, we should have been a bit more careful because of the singularity of the distance function at p, but this is no problem since it is only the convergence for t near T that causes any difficulties. //

The process $\{z_t : 0 \le t \le T\}$ will be called the *semi-classical bridge from* x_0 *to* p *in time* T. In [42] it was called the *Brownian-Riemannian bridge* but this suggestion by K.D. Watling seems preferable. In particular it emphasizes the fact that it will not in general coincide with Brownian motion from x_0 conditioned to arrive at p at time T .

D. The following comes from [81], following earlier results with more restrictive conditions in [42], [43].

<u>**Theorem 1D**</u> *Suppose* $\mathbb{M}$ *is a Riemannian manifold which has a pole* p *in the sense that its exponential map* $\exp_p$ *maps an open star-shaped region of* T_pM *diffeomorphically onto* $\mathbb{M}$. *Let* M *be some open subset of* $\mathbb{M}$ *(possible* $M = \mathbb{M}$*) with* $p \in M$. *Then for* $V : M \to \mathbb{R}$ *bounded above and continuous, the fundamental solution to the minimal semigroup for* $\frac{1}{2}\Delta + V$ *on* M *is given by*

$$p_t(x_0,p) = (2\pi t)^{-n/2}\, \theta_p(x_0)^{-\frac{1}{2}}\, e^{-d(x_0,p)^2/(2t)}$$

$$\mathbb{E}[\chi_{\{t<\tau\}}\exp\{\int_0^t (\tfrac{1}{2}\, \theta_p^{\frac{1}{2}}(z_s)\Delta\theta_p^{-\frac{1}{2}}(z_0) + V(z_s))ds\}] \qquad (264)$$

where $\{z_s : 0 \le s \le t\}$ *is the semi-classical bridge in* M *from* x_0 *to* p *in time* t, *defined up to its explosion time* τ. *In particular the expectation on the right hand side of* (264) *is finite.*

Remark: By the 'semi-classical bridge' here we mean a process which is a Brownian motion with drift $\{Z^0{}_s : 0 \le s \le t\}$, Z^0 as before, in the interval $\{0 \le s < t \wedge \tau\}$ where τ is its explosion time in M (so that if $\tau < t$ it either goes out to infinity or leaves M as $s \uparrow \tau$), and which is sample continuous with value p at time t if $t < \tau$.

<u>**Proof**</u>: Choose a nested sequence of domains $\{D^i\}_{i=1}^{\infty}$ with smooth boundaries, such that D_i is compact, both p and x_0 lie in D_1, and M is the union of the D_i. Let $p^i{}_t(x_0,p)$ be the Dirichlet fundamental solution for $\partial f/\partial t = \frac{1}{2}\Delta f_t + Vf_t$ in D_i.

Let $\tau^\lambda{}_i$ be the explosion time from D_i of the Brownian motion with drift

$\{z^\lambda_s : 0 \le s < T + \lambda\}$ starting from x. While we obtain an expression for $p^i_t(x_0,p)$ we can assume $\mathbb{M}$ modified outside of D_i as in §1B. Thus τ^λ_i is the first exit time from D_i of $\{z^\lambda_s : 0 \le s \le T + \lambda\}$.

By (257) and the Girsanov theorem

$$p^i_T(x_0,p) = \lim_{\lambda\downarrow 0} \mathbb{E}[\chi\{T\le\tau^\lambda_i\} M^\lambda_T f_\lambda(z^\lambda_T) \exp\{\int_0^T V(z^\lambda_s)ds\}] \quad (265)$$

where

$$M^\lambda_T = \exp\{\int_0^T -\langle Z^\lambda_s(z^\lambda_s), u^\lambda_s \circ dB_s\rangle$$

$$- \tfrac{1}{2}\int_0^T (-\mathrm{div}\, Z^\lambda_s(z^\lambda_s) + |Z^\lambda_s(z^\lambda_s|^2)ds\}$$

for $\{u^\lambda_s : 0 \le s < T + \lambda\}$ the horizontal lift of $\{z^\lambda_s : 0 \le s < T + \lambda\}$ in OM from some frame u_0 at x_0; see equation (212).

Since $Z^\lambda_s = \nabla Y^\lambda_s$, writing θ for θ_p:

$$M^\lambda_T = \exp\Big(-Y^\lambda_T(z^\lambda_T) + Y^\lambda_0(x_0)$$

$$+ \int_0^T \{(\partial/\partial s\, (Z^\lambda_s))(z^\lambda_s) + \tfrac{1}{2}|Z^\lambda_s(z^\lambda_s)|^2 + \tfrac{1}{2}\Delta Y^\lambda_s(z^\lambda_s)\}ds\Big)$$

$$= \theta(z^\lambda_T)^{\frac{1}{2}}\, \theta(x_0)^{-\frac{1}{2}} \exp\Big(\tfrac{1}{2}\int_0^T \theta^{\frac{1}{2}}(z^\lambda_s)\ \Delta\theta^{-\frac{1}{2}}(z^\lambda_s)\, ds$$

$$- \tfrac{1}{2} r(x_0)^2/(\lambda+T) + r(y^\lambda_T)^2/(2\lambda) - n/2 \int_0^T ds/(\lambda+T-s))ds\Big)$$

since $\qquad -\Delta\, \tfrac{1}{2} \log\theta(x) = \theta^{\frac{1}{2}}(x)\Delta\theta^{-\frac{1}{2}}(x) - |\nabla \log \theta^{\frac{1}{2}}(x)|^2$

and $\qquad -\Delta r^2 = -2 - 2(n-1) - 2r\, \partial/\partial r \log\theta$ by (258).

Thus $p^i_T(x_0,p) = \lim_{\lambda\downarrow 0} \theta(z^\lambda_T)^{\frac{1}{2}}\, \theta(x_0)^{-\frac{1}{2}}\, \lambda^{n/2}/(\lambda+T)^{n/2}\ 1/(2\pi\lambda)^{n/2}$

$$\exp\{-\tfrac{1}{2}r(x_0)^2/(\lambda+T)\}\ \mathbb{E}\chi_{\{T<\tau^{\lambda}_i\}}\exp\int_0^T V_{eff}(z^{\lambda}_s)ds \qquad (266)$$

where

$$V_{eff}(x) = V(x) + \tfrac{1}{2}\,\theta^{\frac{1}{2}}(x)\Delta\,\theta^{-\frac{1}{2}}(x) \qquad (267)$$

To treat this limit carefully let C be the space of continuous paths $\sigma : [0,T] \to \mathbb{M}^+$, where $\mathbb{M}^+$ is the one point compactification of $\mathbb{M}$, with $\sigma(0) = x_0$. Let $\xi : C \to \mathbb{R} \cup \{\infty\}$ be the first exit time from D_i. Let $\mathbb{P}^{\lambda}$, $\mathbb{P}$ be probabilities induced on C by $\{z^{\lambda}_s : 0 \le s \le T\}$ and $\{z_s : 0 \le s \le T\}$. By Lemma 1C, $\mathbb{P}^{\lambda} \to \mathbb{P}$ narrowly (= 'weakly'). Since D_i is compact and has smooth boundary

$$\sigma \to \chi_{\{T<\xi\}}(\sigma)\exp\int_0^T V_{eff}(\sigma(s))ds$$

is bounded and Riemann integrable for $\mathbb{P}$ in the sense that it is continuous except on a set of $\mathbb{P}$-measure zero (this is because there is probability zero of the path of a non-degenerate diffusion hitting ∂D_i without leaving D_i). It follows, e.g. see [88] p. 375, that

$$p^i_T(x_0,p) = \theta(x_0)^{-\frac{1}{2}}(2\pi T)^{-n/2}\exp\{-r(x_0)^2/(2T)\}\ \mathbb{E}\chi_{\{T<\tau_i\}}\exp\int_0^T V_{eff}(z_s)ds \qquad (268)$$

If we now let $i \to \infty$ the left hand side of (268) converges and the term under the expectation on the right hand side is positive and non-decreasing in i. The theorem follows. //

<u>Remark 1D.</u> The above proof shows that the upper bound on V was not essential provided we know that $p_t(x_0,p)$ exists and is given as the limit of the Dirichlet heat kernels.

<u>Corollary 1D</u>. (i) [47] *For a complete manifold* M, *if* $x_0 \notin$ Cut(p), *then*

$$p_t(x_0,p) \ge (2\pi t)^{-n/2}\,\theta_p(x_0)^{-\frac{1}{2}}\,e^{-d(x_0,p)^2/(2t)}\,\mathbb{E}\chi_{\{t<\tau\}}\exp\int_0^t V_{eff}(z_s)ds \qquad (269)$$

where V_{eff} *is in* (267) *and* τ *is now the first exit time of the semi-classical bridge from* M - Cut(p).

Proof: Replace both $\mathbb{M}$ and M in the theorem by M-Cut(p) respectively and observe that $p_t(x_0,p)$ is greater than the corresponding value of the kernel for M-Cut(p). //

Corollary 1D was used to get results about the limiting behaviour as $t \downarrow 0$ of the trace, $\int_M p_t(x,x)dx$, in [47]. It is especially useful for small t since in the limit it becomes an equality as in Corollary 1D(iii) below. When Cut(p) is codimension 2 it has capacity zero (the Brownian motion never hits it, e.g. see [51]) and so fundamental solutions at (x_0,p) for M and M-Cut(p) are the same if $x_0 \notin$ Cut(p). Thus:

Corollary 1D(ii) c.f. [81]. *If* M *is complete and* Cut(p) *has codimension* 2 *(or capacity zero more generally) then if* $x_0 \notin$ Cut(p) *there is equality in* (269). //

The following is a well known result with both analytical and probabilistic proofs e.g. see [7], [17], [59a], [80], [84].

Corollary 1D (iii) *Suppose* M *is complete and* $x_0 \notin$ Cut(p). *Then as* $t \downarrow 0$

$$p_t(x_0,p) = (2\pi t)^{-n/2} e^{-d(x_0,p)^2/(2t)} \theta_p(x_0)^{-\frac{1}{2}}(1 + o(t)) \tag{270}$$

Proof: First choose a compact domain D with smooth boundary in M which contains the geodesic of shortest length from x_0 to p. We need now quote the result that as $t \downarrow 0$

$$p_t(x_0,p) = p^D{}_t(x_0,p)(1 + O(t^k)) \tag{271}$$

for k = 1,2,.... For this see [7], [80], or [33] when $x_0 = p$, (the $O(t^k)$ can be replaced by $O(\exp(-\delta/t))$ for some $\delta > 0$). Thus we need only examine the behaviour of $p^D{}_t(x_0,p)$ as $t \downarrow 0$. Choosing D with D inside M-Cut(p) we can therefore modify M outside of D as in §1B, so that p is a pole and it is flat outside of a compact set, and also we modify V outside D to give it compact support. Following this we can use (271) in the reverse direction and consider $p_t(x_0,p)$ for the modified M. For this we have

$$p_t(x_0,p) = (2\pi t)^{-n/2} \theta_p(x_0)^{-\frac{1}{2}} e^{-d(x_0,p)^2/(2t)} \mathbb{E}\, e^{\int_0^t V_{eff}(z_s)ds}$$

and so the result follows since V_{eff} is now bounded on M. //

The following corollary was noted in [42], [43]. It can be compared with the trace formula and asymptotics in [28], [32], [94] for example. When M is complete and p has no conjugate points then $\exp_p : T_pM \to M$ is a covering map and if T_pM is given the induced metric to make it a Riemannian manifold, M_0, say, the origin O is a pole. If $x_0 \in M$ there are at most countably many points x^γ_0 in M_0 with $\exp_p(x^\gamma_0) = x_0$, one for each geodesic γ from x_0 to p: we take γ to be the geodesic $t \to \exp_p tx^\gamma_0$ in the reverse direction. For fixed t there is a semi-classical bridge in M_0 from each x^γ_0 to 0 in time t. Let $\{z^\gamma_s : 0 \le s \le t\}$ be its image in M under $\exp_p$. This will be called the "*semi-classical bridge from* x_0 *to* p *along the geodesic* γ, *in time* t". There is also a corresponding $\theta^\gamma_p(x_0)$ which is just $\theta_0(x^\gamma_0)$ evaluated in M_0.

<u>Corollary 1D(iv)</u>. *Suppose* M *is complete and the point* p *has no conjugate points. Then*

$$p_t(x_0,p) = (2\pi t)^{-n/2} \sum_\gamma \theta^\gamma_p(x_0)^{-\frac{1}{2}} e^{-\ell(\gamma)^2/2t}\, \mathbb{E} e^{\int_0^t (V(z^\gamma_s) + \alpha^\gamma_s)ds} \tag{272}$$

where the sum is over all geodesics γ *from* x_0 *to* p, *with* $\ell(\gamma)$ *the length of* γ, *and* $\{z^\gamma_s : 0 \le s \le t\}$ *the corresponding semi-classical bridge; also* $\{\alpha^\gamma_s : 0 \le s \le t\}$ *is* $\frac{1}{2}\theta_0(x^\gamma_s)^{\frac{1}{2}} \Delta\theta_0^{-\frac{1}{2}}(x^\gamma_s)$ *where* θ_0 *is Ruse's invariant in* T_pM *from* 0 *computed using its induced metric and* x^γ_s *is the semi-classical bridge in* T_pM *from* x^γ_0 *to* p *in time* t.

<u>**Proof.**</u> Let U be a sufficiently small open neighbourhood of x_0 so that its inverse image under $\exp_p$ consists of open neighbourhoods U^γ of x^γ_0 in T_pM. Let f be the characteristic function of U and f^γ that of U^γ. Then for the semigroups $\{P_t : t \ge 0\}$ and $\{\tilde{P}_t : t \ge 0\}$ for $\frac{1}{2}\Delta + V$ and $\frac{1}{2}\Delta + V \circ \exp_p$ on M and T_pM respectively we see

$$P_tf(p) = \sum_\gamma \tilde{P}_t f^\gamma(p) \tag{273}$$

by the Feynman-Kac formula since Brownian motion on T_pM from 0 covers Brownian motion from p in M. Because U^γ is mapped isometrically to U it has the same volume as U and so we can let U be a ball radius ε about x_0 and let $\varepsilon \downarrow 0$ to obtain

$$p_t(p,x_0) = \sum_\gamma \tilde{p}_t(p,x^\gamma_0)$$

in the obvious notation. However $p_t(p,x_0) = p_t(x_0,p)$ and similarly for $\tilde{p}_t(p,x\gamma_0)$, and so the corollary follows from the theorem. //

Note: The Cartan-Hadamard theorem assures us that the hypotheses on p and M in Corollary 1D(iv) are always true when M is a complete manifold with all sectional curvatures non-positive.

Example 1D (i) [92], [93]. The simplest non-trivial example of a manifold with a pole is n-dimensional hyperbolic space H^n. From §4G of Chapter III, equation (181), we see that

$$\theta_p(x_0) = (\sinh r\,(x_0)/(r(x_0)))^{n-1} \tag{274}$$

from which, using (258), we have

$$\tfrac{1}{2}\theta^{\frac{1}{2}}(x_0)\Delta\theta^{-\frac{1}{2}}(x_0) = -\tfrac{1}{8}(n-1)^2 + \tfrac{1}{8}(n-1)(n-3)(r(x_0)^{-2} - (\sinh r(x_0))^{-2}) \tag{275}$$

When n = 3 and V ≡ 0 we can deduce the well known formula for the heat kernel of H^3:

$$p_t(x,y) = (2\pi t)^{-3/2}\, e^{-t/2}\, e^{-d(x,y)^2/(2t)}\; d(x,y)/(\sinh d(x,y)) \tag{276}$$

with corresponding exact formulae for non-simply connected 3-manifolds of constant negative curvature obained by using Corollary 1D(iv). The heat kernel for the hyperbolic plane H^2 is computed analytically in [28]. For a recurrence relation between the kernels for hyperbolic spaces of different dimensions see [28], with [35] for more details.

Example 1D(iii) [81], [83], [49]. For $M = S^{n-1}$ note that if p is the North pole, say, in polar coordinates (r,σ) in $\mathbb{R}^{n-1}$ (so $\sigma \in S^{n-2}$) the exponential map is essentially the map $(r,\sigma) \to (\cos r, (\sin r,\sigma)) \in \mathbb{R} \times \mathbb{R}^{n-1}$. In particular it maps the sphere about 0 radius r to an embedding in $\mathbb{R}^n$ onto an isometric copy of the sphere in $\mathbb{R}^{n-2}$ radius sin r. Thus the metric in normal polar coordinates is

$$ds^2 = dr^2 + (\sin r)\ \ (\text{standard metric of } S^{n-2}).$$

Thus

$$\theta_p(x_0) = (\sin r(x_0)/(r(x_0))^{n-1} \tag{277}$$

and

$$\tfrac{1}{2}\theta^{\frac{1}{2}}(x_0)\,\Delta\theta^{-\frac{1}{2}}(x_0) = \tfrac{1}{8}\,(n-1)^2 + \tfrac{1}{8}\,(n-1)(n-3)(1/r^2 - 1/\sin^2 r) \tag{278}$$

for $r = r(x_0)$.

Again n = 3 is an especially nice case. Since Cut(p) has co-dimension 2 we can use Corollary 1D(ii) to get, for $V \equiv 0$, if x is not antipodal to y

$$p_t(x,y) = (2\pi t)^{-n/2}\,(r/\sin r)\,e^{\frac{1}{2}t}\ e^{-r^2/(2t)}\ \mathbb{P}\{t < \tau\}$$

where $r = d(x,y)$ and where τ is the first hitting time of Cut(y) by the semi-classical bridge from x to y in time t. However as we saw in Lemma 1C the radial distributions of this bridge are the same as those of a Brownian bridge $\{\beta_s : 0 \le s \le t\}$, say, in $\mathbb{R}^3$ from a point distance r from 0, to 0, in time t. Thus, [81],

$$p_t(x,y) = (2\pi t)^{-n/2}\,(r/\sin r)\,e^{\frac{1}{2}t}\,e^{-r^2/2t}\,\mathbb{P}\{\sup_{0\le s\le t}|\beta_s| \le \pi\} \quad (279)$$

This formula is discussed in [83]. In [43] it is used to obtain the exact formula, for x,y not antipodal,

$$p_t(x,y) = (2\pi t)^{-3/2}\,e^{\frac{1}{2}t}\,\Sigma_\gamma\ \ell(\gamma)/(\sin\ell(\gamma))\ e^{-\ell(\gamma)^2/(2t)} \qquad (280)$$

where the sum is over all geodesics γ from x to y and $\ell(\gamma)$ is the length of γ. Note the similarity here with the case of S^1, or the situation in Corollary 1D(iv). However in this case we no longer have a sum of positive terms. This formula is a special case of a general formula for compact Lie groups, [50], proved using harmonic analysis on such groups.

§2. General remarks about the elementary formula method and its extensions.

A. The way we were able to get a tractable formula for the heat kernel in the last section depended on a suitable choice of drift $\{Z^\lambda{}_s : 0 \le s < T + \lambda\}$ for which there were convenient cancellations after the use of the Girsanov theorem, and which gave processes with a very nice radial behaviour. In fact the choice of $Z^\lambda{}_s$ came from a general philosophy outlined in [42], which is explained below. However first it should be noted that there are various ways of getting 'bridges' from x_0 to p e.g. see [17]. The standard one is Brownian motion from x_0 conditioned to be at p at time T. This can be described as the h-transform of (space time) Brownian motion where $h_s(x) = p^o{}_{T-s}(p,x)$ for $p^o{}_t(x,y)$ the fundamental solution for M when $V \equiv 0$. This is used in [36], [80] and [104]; it is Brownian motion with drift $\nabla \log p_{T-s}(p,-)$. Writing it as $\{x_s :$

$0 \le s \le T\}$ it is immediate from the Feynman-Kac formula that the kernel with a potential V is given by

$$p_T(x_0,p) = p^o{}_T(x_0,p)\, \mathbb{E} \exp \int_0^T V(x_s)ds.$$

It has the advantage over the semi-classical bridge of symmetry in x_0, p i.e. the reversed time bridge is the bridge from p to x_0. However the radial behaviour will not be so pleasant in general.

B. Let $\{P_t : t \ge 0\}$ be the heat semigroup associated to $\frac{1}{2}\Delta + V$ on M. Suppose $g_0(x) = \exp(-S_0(x)) \cdot f_0(x)$ where S_0 also are smooth functions on M with S_0 bounded below and with f_0 of compact support. The drift terms for the semi-classical bridge arose, [42], from seeking a nice expression for $P_t\, g_0$ which would exhibit its behaviour as $\lambda \downarrow 0$ when $S_0 = \lambda^{-1} R_0$ some R_0. Here is a brief description. Assume for simplicity that M is complete and V, ∇S_0 and the curvature tensor are all bounded on M.

First we associate to g_0 the classical mechanical system with trajectories $\{\Phi_t(a) : t \ge 0\}$ for each a in M, satisfying

$$D/\partial t\, \dot{\Phi}_t(a) = 0 \qquad (281)$$

with $\Phi_0(a) = a$ and $\dot{\Phi}_0(a) = \nabla S_0(a)$. Under our assumptions it is shown in [42a] that there exists $T > 0$ such that $\{\Phi_t(a) ; t \ge 0\}$ is defined for all $0 \le t \le T$ and determines a diffeomorphism

$$\Phi_t : M \to M.$$

This is a '*no caustics*' assumption.

For this T we can define the *Hamiltonian-Jacobi principle function.*

$$S : [0,T] \times M \to \mathbb{R}$$

given by

$$S(t,a) = S_0(\Phi_t^{-1}(a)) + \frac{1}{2}\int_0^t |\dot{\Phi}_s \circ \Phi_t^{-1}(a)|^2\, ds \qquad (282)$$

There is then the following standard lemma, as in [42a]:

<u>Lemma 2B</u>

(i) $\dot{\Phi}_t(a) = \nabla S(\Phi_t(a),t) \qquad 0 \le t \le T,\ a \in M \qquad (283)$

(ii) S *satisfies the Hamilton-Jacobi equation*

$$\tfrac{1}{2}|\nabla S(x,t)|^2 + \partial S/\partial t\,(x,t) = 0 \qquad 0 \le t \le T \tag{284}$$

with $S(x,0) = S_0(x)$

(iii) *Define* $\varphi : N \times [0,T] \to \mathbb{R}$ *by*

$$\varphi(x,t) = |\det T_x \Phi^{-1}{}_t|$$

(*using the Riemannian metric of* M). *Then* φ *satisfies the continuity equation*

$$\partial\varphi/\partial t\,(x,t) + \mathrm{div}(\varphi(x,t)\,\nabla S(x,t)) = 0 \tag{285}$$

Proof:

$$\nabla S(x,t) = (T\Phi^{-1}{}_t)^* \,\nabla S_0(\Phi^{-1}{}_t(x)) + \int_0^t D/\partial s\,\{T(\Phi_s \circ \Phi_t{}^{-1})^* \,(\dot\Phi_s \circ \Phi^{-1}{}_t(x)\}ds.$$

Integrate by parts to obtain $\nabla S(x,t) = \dot\Phi_t \circ \Phi_t{}^{-1}(x)$ yielding (i). Also

$$\partial/\partial t\, S(x,t) = dS_0(\Phi^{-1}{}_t(x)) + \tfrac{1}{2}\,|\dot\Phi_t \circ \Phi_t{}^{-1}(x)|^2 + \int_0^t \langle D/\partial s\, T\Phi_s(\dot\Phi_t{}^{-1}(x)),\, \dot\Phi_s \circ \Phi_t{}^{-1}(x)\rangle ds.$$

Integrate by parts again and use (i) together with the identity

$$T\Phi_t \circ \dot\Phi^{-1}{}_t + \dot\Phi_t \circ \Phi_t{}^{-1} = 0 \tag{286}$$

(which comes from differentiating $\Phi_t \circ \Phi_t{}^{-1} = 1d$) to obtain (ii).

For (iii) take any C^∞ function $f\colon M \to \mathbb{R}$ with compact support. Integrating by parts

$$\int_M \mathrm{div}(\varphi(-,t)\nabla S(-,t))(x) f(x)dx$$
$$= -\int_M \varphi(x,t)\langle \nabla S(x,t), \nabla f(x)\rangle dx$$
$$= -\int_M \langle \nabla S(\Phi_t(x),t), \nabla f(\Phi_t(x)\rangle dx$$
$$= -\int_M df(\dot\Phi_t(x))dx \qquad \text{(by (i))}$$
$$= -d/dt \int_M f(\Phi_t(x))dx = -d/dt \int_M \varphi(x,t) f(x)dx = -\int_M \partial/\partial t\, \varphi(x,t) f(x)dx$$

giving (iii). //

Now run the classical mechanical flow backwards. Take t in (0,T] and set

$$\Theta_s(a) = \Phi_{t-s}(\Phi^{-1}{}_t(a)) \qquad 0 \le s \le t,\ a \in M.$$

Then

$$\partial/\partial s\, \Theta_s(a) = -\nabla S_{t-s}(\Theta_s(a)).$$

Let $\{y_t : t \ge 0\}$ be a Brownian motion on M from x_0 with time dependent drift $\{\nabla Y_s(x) : 0 \le s \le t,\ x \in M\}$ for $Y_s(x) = -S(a,t-s)$. We can think of it as $\{\Theta_s(x_0) :$

$0 \le s \le t\}$ perturbed by white noise.

<u>Formula A</u> c.f. [42] For $0 \le t \le T$

$$P_t\, g_0(x_0) = \exp\{-S(x_0,t)\}\, \mathbb{E}\left[\exp\{\int_0^t V(y_s) - \tfrac{1}{2}\Delta S_{t-s}(y_s)ds\}\, f_0(y_t)\right] \quad (287)$$

<u>Proof:</u>

$$P_t g_0(x_0) = \mathbb{E}\exp\{\int_0^t V(x_s)ds - S_0(x_t)\} f_0(x_t).$$

Apply Girsanov's formula to obtain an expectation with respect to $\{y_s : 0 \le s \le t\}$. The exponential martingale which comes in is

$$\exp(Y_0(x_0) - Y_t(y_t) + \int_0^t (\partial/\partial s\, Y_s(y_s) + \tfrac{1}{2}|\nabla Y_s(y_s)|^2 + \tfrac{1}{2}\Delta Y_s(y_s))ds$$

i.e. $$\exp\{S_0(y_t) - S(x_0,t) - \tfrac{1}{2}\int_0^t \Delta S(y_s,t-s)ds\}$$

using the Hamilton-Jacobi equation. //

This method can be modified in various ways. To obtain information about the limiting behaviour of ${P^\lambda}_t\, {g^\lambda}_0$ as $\lambda \downarrow 0$ where ${g^\lambda}_0$ is as g_0 but with S_0 replaced by $\lambda^{-1} S_0$ and where ${P^\lambda}_t$ refers to the semigroup generated by $\frac{1}{2}\lambda\Delta + \lambda V$ one proceeds in essentially the same way and Formula A gives the 'W.K.B' approximation. However in this case, and for us, a slight modification gives a more useful formula, [42a]:

<u>Formula B</u>

$$P_t g_0(x_0) = \sqrt{\varphi_t(x_0)}\exp\{-S(x_0,t)\}\mathbb{E}[\exp\{\tfrac{1}{2}\int_0^t \varphi(z_s,t-s)^{-\frac{1}{2}}\,\Delta\varphi^{\frac{1}{2}}(z_s,t-s)ds\} f_0(z_t)] \quad (288)$$

where $\{z_s : 0 \le s \le t\}$ *is Brownian motion on* M *from* x_0 *with drift* $\{\nabla Y_s(x)\,;\, 0 \le s \le t,\, x \in M\}$ *for*

$$Y_s(a) = -S(a,t-s) + \tfrac{1}{2}\log\varphi(a,t-s),$$

assuming this process is complete.

<u>Proof.</u> From the continuity equation

$$\Delta S(x,t) = -\,\partial/\partial t \log \varphi\,(x,t) - \langle \nabla \log \varphi(x,t), \nabla S(x,t)\rangle$$

so that

$$\Delta S(\Theta_s(a),t-s) = \partial/\partial s \log \varphi\,(\Theta_s(a), t-s).$$

If we use this as well as the Hamilton-Jacobi equation after the Girsanov transformation the formula follows. //

The reason for introducing these formulae (which have many variations) here is that to obtain the elementary formula for the heat kernel we needed the case $S_0(x) = d(x,p)^2/(2\lambda)$. For this, given that p is a pole, we have $\Phi_t(x) = (\lambda + t)\lambda^{-1}x$ in normal coordinates about p and $S_t(x) = \frac{1}{2}\, d(x,p)^2/(\lambda + t)$. Then

$$\varphi_t(x) = (\lambda/(\lambda+t))^n\, \theta_p(x)^{-1}\, \theta_p(\lambda/(\lambda+t)x)$$

in normal coordinates. To obtain the 'elementary formula' we could have used the process $z_s = z^\lambda{}_s$ of Formula B as in [42]. However the actual process $z^\lambda{}_s$ we used is easier to handle and gives the same limiting process, the semi-classical bridge, as $\lambda \downarrow 0$.

C. This very simple approach to the study of asymptotic behaviour seems to have wide applicability, applying to both the Schrodinger and the heat equations. In the former there is no Girsanov theorem, but this is made up for by unitarity of the semigroup, and the use of a transformation of semigroups: essentially an h-transform. This semigroup approach was worked out by Watling [98a] to deal with both types of equation almost simultaneously. He showed how it could be used to obtain full asymptotic expansions with exact remainders. This was extended by Ndumu [82], and here we give a brief description of how to get the asymptotics of the heat kernel $p_t(x_0,p)$ for $\frac{1}{2}\Delta + V$.

Assume that p is a pole for M, with M complete and Euclidean outside some compact region for simplicity, and that V is bounded and smooth. Consider

$$q_t(x,p) = (2\pi t)^{-n/2}\, \theta_p(x)^{-\frac{1}{2}}\, \exp\{-\,d(x,p)^2/(2t)\} \qquad (289)$$

The first observation is that as a function of x, writing θ for θ_p it satisfies

$$(\partial/\partial t) f_t(x) = \tfrac{1}{2}\,\Delta\, f_t\,(x) - \tfrac{1}{2}\theta^{\frac{1}{2}}(x)\, \Delta\theta^{-\frac{1}{2}}(x)\, f_t(x) \qquad (291)$$

and moreover as $t \downarrow 0$ it converges to the Dirac delta function at p. Next define the 'semi-classical' evolution $\{Q_p(t,s) : t \geq s > 0\}$ on bounded measurable functions by

$$Q_p(t,s)(f)(x) = q_t(x,p)^{-1}\, P_{t-s}(q_s(-,p)f)(x) \qquad (292)$$

where $\{P_t : t \geq 0\}$ is the semigroup for $\frac{1}{2}\Delta + V$. Another (this time standard) computation yields for f smooth and with compact support

$$(\partial/\partial t)\, Q_p(t,s)(f)(x) = \{\tfrac{1}{2}\Delta + \nabla \log q_t(-,p) + V_{eff}\} Q_y(t,s)(f)(x)$$

where $V_{eff}(x) = V(x) + \frac{1}{2}\theta^{\frac{1}{2}}(x)\,\Delta\theta^{-\frac{1}{2}}(x)$ as usual. Consequently, now by a Feynman-Kac formula rather than a Girsanov theorem, for $t > s \geq 0$

$$Q_p(t,t-s)\,(f)(x_0) = \mathbb{E}[\exp\{\int_0^t V_{eff}(z_r)dr\} f(z_s)] \qquad (293)$$

where $\{z_s : 0 \leq s \leq t\}$ is the semi-classical bridge from x_0 to p in time t. Letting $s \uparrow t$ we obtain another proof of the 'elementary formula' (264).

To get the asymptotic expansion assume now that *each* pair of points x and y in M can be joined by a unique geodesic. Let $\gamma(x,y)$ denote this path parametrized proportionally to arc length so that $\gamma(x,y)(0) = x$ and $\gamma(x,y)(1) = y$. For $f : M \to \mathbb{R}$ and $r \geq s \geq 0$ define

$$F(r,s)(f) : M \to \mathbb{R}$$

by

$$F(r,s)(f)(x) = f(\gamma(p,x)(s/r)) = f(s/r\; x)$$

in normal coordinates at p. Then, as for (292), for smooth f of compact support

$$(\partial/\partial s)\,[Q_p(t,t-s)F(t-s,t-r)(f)](x) = Q_p(t,t-s)\mathfrak{G}\,[F(t-s,t-r)(f)](x) \qquad (294)$$

where

$$\mathfrak{G} = \tfrac{1}{2}\Delta - \tfrac{1}{2}\nabla \log\theta.\nabla + V_{eff}.$$

From (294) we have on integrating

$$Q_p(t,t-s)\,[F(t-s,t-r)(f)](x) - Q_p(t,t)[F(t,t-r)(f)](x)$$

$$= \int_0^s Q_p(t,t-s_1)\,\mathfrak{G}[F(t-s_1,\, t-r)(f)](x)ds_1.$$

Setting r = s we get an expression for $Q_p(t,t-s)(f)$ since $F(t-s,t-s) = 1d$. This can be iterated arbitrarily many times by replacing f by $\mathfrak{G}[F(t-s,t-r)(f)]$ and substituting in the integrand, to yield a rather complicated expansion for $Q_p(t,t-s)(f)$ to arbitrarily many terms with a remainder consisting of a time integral of various iterations of the operators. Knowledge of $Q_p(t,t-s)(f)(x_0)$ gives knowledge of $p_t(x_0,p)$, c.f. (292). This way Watling's expansion [98a],

[98] is obtained, see also [82], for N = 1,2,... .

$$p_t(x_0,p) = (2\pi t)^{-n/2}\theta_p(x_0)^{-\frac{1}{2}}\exp(-d(x_0,p)^2/(2t))[1 + a_1(x_0,p)t + \dots + a_N(x_0,p)t^N] + R_{N+1}(x_0,p,t)t^{N+1} \qquad (295)$$

where

$$a_1(x_0,p) = \int_0^1 F(1, 1-r_1)(V_{eff})(x)dr_1$$

and

$$a_j(x_0,p) = \int_0^1 \int_0^{r_1} \dots \int_0^{r_{j-1}} F(1,1-r_j)\mathfrak{G}F(1-r_j,1-r_{j-1}) \dots \mathfrak{G}F(1-r_2,1-r_1)(V_{eff})(x)dr_j \dots dr_1$$

for $2 \le j \le N$ and

$$R_{N+1}(x_0,p,t) = \mathbb{E}[\int_0^1 \int_0^{r_1} \dots \int_0^{r_N} \{\mathfrak{G}F(1-r_{N+1},1-r_N) \dots \mathfrak{G}F(1-r_2, 1-r_1)(V_{eff})\}(z_{tr_{N+1}}) \exp\{\int_0^{tr_{N+1}} V_{eff}(z_s)ds\}dr_{N+1} \dots dr_1].$$

As before this gives an exact expression when M has a pole and given some additional bounds on its geometry [82], and furnishes an asymptotic expansion when $x_0 \notin \mathrm{Cut}(p)$ for general complete M. It is easily modified to deal with the fundamental solution to $\frac{1}{2}\Delta + A + V$ where A is a first order operator (i.e. a vector field). Essentially the only differences are: (i) that $\theta_p^{-\frac{1}{2}}(x)$ is replaced by

$$\theta_p^{-\frac{1}{2}}(x)\exp\{\int_0^1 \langle\dot\gamma(s), A(\gamma(s))\rangle ds\}$$

where γ is the geodesic from x to p parametrized to take unit time, and (ii) $\frac{1}{2}\Delta$ is replaced by $\frac{1}{2}\Delta + A$ throughout; see [98].

§3. The fermionic calculus for differential forms, and the Weitzenbock formula

A. The use of creation and annihilation operations for differential forms was exploited by Witten for his approach to Morse theory [99]. The notation is very useful in stating, and proving, the Weitzenbock formula for the Laplacian on p-forms, as described in [33]. For our purposes it will enable us to give an 'elementary formula' for the heat kernel for forms especially suited to the 'supersymmetric' approach to the Gauss-Bonnet-Chern theorem which is discussed later. It would be difficult to improve on the exposition in [33] and it will be followed closely, as in [48] on which these sections are based.

B. First let us fix some notation and recall some basic facts. If V is a real finite dimensional vector space the space $A^p(V)$ of antisymmetric linear maps $\varphi : V \times \ldots \times V \to \mathbb{R}$ can be identified with the space of linear maps $\mathbb{L}(\wedge^p V; \mathbb{R})$ i.e. $(\wedge^p V)^*$. If $\varphi \in A^p(V)$ and $\alpha \in A^1(V)$ there is $\alpha \wedge \varphi \in A^{p+1}V$ given by

$$\alpha \wedge \varphi(v_1,\ldots,v_{p+1}) = \sum_{j=1}^{p+1} (-1)^{j+1} \alpha(v_j)\, \varphi(v_1,\ldots,\hat{v}_j,\ldots,v_{p+1}) \qquad (296)$$

where $\wedge$ indicates that the indicated term is omitted. This determines an isomorphism of $(\wedge^p V)^*$ with $\wedge^p V^*$, every element of the latter being representible as a linear combination of terms of the form $\alpha^1 \wedge \ldots \wedge \alpha^p$ for $\alpha^j \in V^*$. If V has an inner product there is an induced inner product on $\wedge^p V$ and $\wedge^p V^*$ determined by

$$\langle \alpha^1 \wedge \ldots \wedge \alpha^p, \beta^1 \wedge \ldots \wedge \beta^p \rangle = \det [\langle \alpha^i, \beta^j \rangle]_{i,j=1}^p$$

for α^i, β^j in V or V^* respectively.

For such V, given $\varphi \in \wedge^p V^*$ and $e \in V$ define the "creation operator" $a(e)^* : \wedge^p V^* \to \wedge^{p+1} V^*$ by

$$a(e)^* \varphi = e^{\#} \wedge \varphi$$

where $e^{\#} \in V^*$ is dual to e (it is most convenient to formulate it this way to avoid a plethora of $^{\#}$'s later on). Let its adjoint be

$$a(e) : \wedge^{p+1} V^* \to \wedge^p V^*.$$

Then

$$a(e)(\alpha^1 \wedge \ldots \wedge \alpha^{p+1}) = \sum_{j=1}^{p+1} (-1)^{j+1}\alpha^j(e)\alpha^1 \wedge \ldots \wedge \hat{\alpha}^j \wedge \ldots \wedge \alpha^{p+1} \quad (297)$$

or as an antisymmetric linear map, for $\varphi \in A^{p+1}(V)$

$$a(e)\varphi(v_1,\ldots,v_p) = \varphi(e,v_1,\ldots,v_p) \quad (297a)$$

There is the anti-commutation relation for

$$\{a(e),a(f)^*\} := a(e)a(f)^* + a(f)^*a(e) : \Lambda^p V^* \to \Lambda^p V^*$$

with e, f $\in$ V:if $\varphi \in \Lambda^p V^*$ then, from (297)

$$\{a(e),a(f)^*\}\varphi = \langle e,f\rangle\varphi \quad (298)$$

C. A p-form φ on M gives an anti-symmetric p-linear map

$$\varphi_x : T_xM \times \ldots \times T_xM \to \mathbb{R}$$

for each $x \in M$. Thus we can consider $\varphi_x \in \Lambda^p T_x{}^*M$ and consider φ as a section of the tensor bundle $\Lambda^p T^*M$. Let A^p denote the space of such sections which are C^∞ and let $A = \bigoplus_p A^p$ be the space of sections of $\Lambda T^*M := \bigoplus^p \Lambda^p T^*M$. Supposing M is given a Riemannian structure (as we will from now on) we can use the Riemannian measure and the inner products in each $\Lambda^p T^*{}_xM$ defined as above to obtain a space L^2A^p of L^2 p-forms with inner product

$$\langle\varphi,\psi\rangle_{L^2} = \int_M \langle\varphi_x, \psi_x\rangle dx.$$

Exterior differentiation $d : A \to A$ restricts to

$$d : A^p \to A^{p+1}$$

for each p with

$$(d\varphi)_x(v_1,\ldots,v_{p+1}) = \sum_{j=1}^{p+1}(-1)^{j+1}D\varphi(x)(v_j)(v_1,\ldots,\hat{v}_j,\ldots,v_{p+1})$$

in local coordinates, where D is the Fréchet derivative. Since the covariant derivative agrees with the ordinary derivative at the centre of normal coordinates for the Levi-Civita connection

$$(d\varphi)_x(v_1,\ldots,v_{p+1}) = \sum_{j=1}^{p+1}(-1)^{j+1}\nabla\varphi(v_j)(v_1,\ldots,\hat{v}_j,\ldots,v_{p+1}) \quad (299)$$

From the definition we gave of covariant differentiation by lifting to OM it is almost immediate that if $\varphi, \psi \in A$ then for $v \in TM$

$$\nabla_v(\varphi \wedge \psi) = \nabla_v\varphi \wedge \psi + \varphi \wedge \nabla_v\psi \quad (300)$$

Let $e_1,\ldots,e_n$ be an orthonormal basis for T_xM then if $\varphi \in A^p$ and $v_1,\ldots,v_p \in T_xM$

$$\sum_{j=1}^{n}(a(e_j)^* \nabla_{e_j}\varphi)_x(v_1,\ldots,v_p) = \sum_{j=1}^{n} e_j^* \wedge \nabla_{e_j} \varphi(v_1,\ldots,v_p)$$

$$= \Sigma_j \Sigma_k(-1)^{k+1}\langle e_j,v_k\rangle \nabla_{e_j}\varphi(v_1,\ldots,\hat{v}_k,\ldots,v_p)$$

$$= \Sigma_k(-1)^{k+1} \nabla_{v_k} \varphi (v_1,\ldots,\hat{v}_k,\ldots,v_p).$$

Thus from (299) the two types of differentiation are related by

$$(d\varphi)_x = \Sigma^n{}_{j=1} (a(e_j)^* \nabla_{e_j} \varphi)_x \tag{301}$$

Since $a(e_j)^*(\varphi \wedge \psi) = a(e_j)^*(\varphi) \wedge \psi = (-1)^p \varphi \wedge (a(e_j)^*\psi)$ when $\varphi \in A^p$ equations (300) and (301) immediately yield

$$d(\varphi \wedge \psi) = d\varphi \wedge \psi + (-1)^p \varphi \wedge d\psi \tag{302}$$

Consequently if $\alpha^1,\ldots,\alpha^p$ are 1-forms

$$d(\alpha^1 \wedge \ldots \wedge \alpha^p) = \Sigma(-1)^{j+1} \alpha_1 \wedge \ldots \wedge d\alpha^j \wedge \alpha^{j+1} \wedge \ldots \wedge \alpha^p \tag{303}$$

Let d^* be the formal L^2 adjoint of d so $d^* : A \to A$ restricting to $d^* : A^{p+1} \to A^p$ for each p. Thus $d^* = \delta$ in the notation of Chapter IV.

We already know, by (39) and (40), that the formal adjoint of ∇ acting on functions is minus the divergence: for a vector field A

$$\nabla^*A(x) = -\text{div } A(x) = - \Sigma_j \langle\nabla_{e_j}A(x),e_j\rangle.$$

Thus on 1-forms α

$$(d^*\alpha)_x = - \Sigma_j\, a(e_j) \nabla_{e_j}\alpha.$$

From (303) it follows that for $\varphi \in A$ the same formula holds:

$$(d^*\varphi)_x = - \Sigma_j\, a(e_j) \nabla_{e_j} \varphi \tag{304}$$

D. Let A be a section of $\mathbb{L}(TM;TM)$, so for each x we have $A_x{:}T_xM \to T_xM$. It has adjoint

$$A^*{}_x : T_x{}^*M \to T_x{}^*M$$

and can operate on A by

$$(A^\wedge (\alpha^1 \wedge \ldots \wedge \alpha^p))_x = -\sum_{j=1}^{p} \alpha^1{}_x \wedge \ldots \wedge A^*{}_x(\alpha^j{}_x) \wedge \ldots \wedge \alpha^p{}_x \tag{305}$$

Observe that

$$A^\wedge{}_x = -\Sigma_{l,k} A^*{}_{kl}\, a(e_k)^*\, a(e_l) \tag{306}$$

where $A^*_{kl} = \langle e_l, Ae_k\rangle$.

For vector fields A, B the curvature tensor R determines $R(A(x), B(x)) : T_xM \to T_xM$ for each $x \in M$, which is skew symmetric. Recall from equation (83) that if V is another vector field

$$R(A(x), B(x))(V(x)) = ([\nabla_A, \nabla_B] - \nabla_{[A,B]})(V)(x).$$

It is therefore immediate from (300) that as operators on A

$$[\nabla_A, \nabla_B] - \nabla_{[A,B]} = R(A(\cdot), B(\cdot))^{\wedge} \tag{307}$$

Recall that the (de-Rham-Hodge) Laplace operator $\Delta : A \to A$ is defined by

$$\Delta = -(dd^* + d^*d)$$

using the sign which makes it negative definite since $d^2 = 0$: see equation (228).

Set $R_{ijkl} = R(e_i,e_j,e_k,e_l) = \langle R(e_i,e_j)e_l,e_k\rangle$, see (86). Note the sign difference from [33], [88].

<u>**Proposition 3D**</u> (Weitzenbock formula)

$$\Delta = \text{trace } \nabla^2 - W \tag{310}$$

where W, the Weitzenbock term, is the zero order operator given at x by

$$W_x = -\Sigma_{i,j,k,l} R_{ijkl}\, a(e_i)^* a(e_j) a(e_k)^* a(e_l) \tag{311}$$

<u>**Proof**</u> Take normal coordinates at x, and using $e_1,\dots,e_n$ as a basis for T_xM, take $E_1,\dots,E_n$ to be vector fields on M which are C^∞ with compact support and agree with the Gram-Schmidt orthonormalization of the fields $\partial/\partial x^1,\dots,\partial/\partial x^n$ at each point near x. Then for y near x, $E_1(y),\dots,E_n(y)$ forms an orthonormal base for T_yM. Moreover the covariant derivative of each e_j vanishes at x. Write a(j) for $a(E_j(\cdot))$ acting on forms by $(a(j)\varphi)_x = a(E_j(x))\varphi_x$, etc. By (301) and (304) if $\varphi \in A^p$, summing over repeated suffices and working near x

$$dd^*\varphi = -a(j)^* \nabla_j a(k) \nabla_k \varphi$$

so

$$(dd^*\varphi)_x = -a(e_j)^* a(e_k)(\nabla_j \nabla_k \varphi)_x.$$

Also

$$d^*d\varphi = -a(k) \nabla_k a(j)^* \nabla_j \varphi$$

so

$$(d^*d\varphi)_x = -a(e_k)a(e_j)^* (\nabla_k \nabla_j \varphi)_x.$$

Thus

$$(\Delta\varphi)_x = \{a(e_j)^*, a(e_k)\} (\nabla_j\nabla_k\varphi)_x + a(e_k)a(e_j)^* ([\nabla_k,\nabla_j]\varphi)_x$$
$$= \text{trace} (\nabla^2\varphi)_x + a(e_k)a(e_j)^* \widehat{R(e_k,e_j)}$$

by (307) since $[E_k,E_j] = 0$ near x. Thus (310) holds with

$$W = - a(e_i)a(e_j)^* \widehat{R(e_i,e_j)} \tag{312}$$
$$= -a(e_i)a(e_j)^* \langle R(e_i,e_j)e_k,e_l\rangle a(e_k)^*a(e_l)$$

by (306), which agrees with (311) since $\{a(e_i), a(e_j)^*\} = \delta_{ij}$ and $R(e_i,e_i) = 0$. //

Let $\Delta^p : A^p \to A^p$ denote the Laplacian acting on p-forms (i.e. the restriction of Δ) and W^p the corresponding Weitzenbock term.

The following special case of the Weitzenbock formula was used in Chapter III:

<u>Corollary 3D</u> *For a smooth* 1-*form* φ

$$\Delta^1\varphi = \tfrac{1}{2} \text{ trace } \nabla^2\varphi - \text{Ric}(\varphi^*,-),$$

<u>Proof</u> By (312) for $v \in T_xM$

$$(W^2\varphi)(v) = a(e_i)a(e_j)^* R(e_i,e_j)^*(\varphi)v$$
$$= a(e_i) (e_j{}^* \wedge (\varphi \circ R(e_i,e_j)(-))) (v)$$
$$= -\langle e_j,v\rangle\, \varphi(R(e_i,e_j)(e_i))$$
$$= -\langle R(e_i,v)(e_i),\varphi^*\rangle = \text{Ric}(\varphi^*,v). \quad //$$

§4. An elementary formula for the heat kernel on forms

A. Assume M is complete. Then the de Rham-Hodge Laplacian on C^∞ forms with compact supports is known to be essentially self-adjoint, [52], [91], [33], and so determines a semigroup $\{e^{\frac{1}{2}t\Delta} : t \geq 0\}$ on L^2A which by elliptic regularity has a kernel $k_t(x,y)$ which is C^∞ in $t > 0$, and x,y in M such that for φ in L^2A

$$(e^{\frac{1}{2}t\Delta}\varphi)_x = \int_M k_t(x,y)\varphi_y\, dy \tag{313}$$

with

$$k_t(x,y) : \Lambda T_y{}^*M \to \Lambda T_x{}^*M.$$

By the same argument described for 1-forms before the proof of Theorem 3B of Chapter IV the Weitzenbock formula implies that if W^p is bounded below then $e^{\frac{1}{2}t\Delta}$ determines a bounded map from $L^\infty A^p$ to $L^\infty A^p$ and if $\varphi \in L^2A^p$

is also bounded then

$$e^{\frac{1}{2}t\Delta}\,\varphi = P_t\varphi$$

where

$$P_t\varphi(v_0) = \mathbb{E}\varphi(v_t) \tag{314}$$

for $v_0 \in \Lambda^p\, T_{x_0}M$ with $v_t(\omega) \in \Lambda^p T_{x_t(\omega)}M$ given by

$$(D/\partial t)(v_t) = -\tfrac{1}{2}\,(W^P_{x_t})^*(v_t) \tag{315}$$

along the paths of the Brownian motion $\{x_t : t \geq 0\}$ from x_0. Here we must also assume that M is stochastically complete (e.g. that Ric is bounded below on M). Here $(W^P_x)^*$ is the dual of $W^P_x : (\Lambda^p T_x M)^* \to (\Lambda^p T_x M)^*$.

B. To obtain a formula for the kernel like that of §1, assume that p is a pole for M, work with q-forms (to avoid confusion!), take $\alpha \in \Lambda^q\, T_p{}^*M$ and choose $\varphi \in A^q$ with compact support and such that $\varphi_p = \alpha$.

Fix $\tau > 0$ and for $\lambda > 0$ define $\varphi_\lambda \in A^q$ by

$$\varphi_{\lambda,x} = (2\pi\lambda\tau)^{-n/2}\exp\{-d(x,p)^2/(2\lambda\tau)\}\varphi_x \tag{316}$$

Observe that the kernel $k^q{}_t(x,y)$ for q-forms satisfies

$$k^q{}_t(x_0,p)(\alpha)(v_0) = \lim_{\lambda\downarrow 0} P_t\varphi_\lambda(v_0) \tag{317}$$

In fact we will obtain a formula in a slightly different form to that in §1 and more adapted to describing the asymptotics as $t \downarrow 0$. For this let $\{H_t : t \geq 0\}$ be the semigroup $\{e^{t\tau\Delta/2} : t \geq 0\}$. Thus $H_t = P_{t\tau}$. Let $\{x_t : t \geq 0\}$ now have generator $\frac{1}{2}\tau\Delta^0$ (where Δ^0 is the Laplacian on functions) and let $\{v_t : t \geq 0\}$ be defined by

$$(D/\partial t)\,(v_t) = -\tfrac{1}{2}\,\tau(W_{x_t})^*(v_t) \tag{318}$$

Then

$$P_\tau\,\varphi_\lambda\,(v_0) = \mathbb{E}\varphi_\lambda(v_1) \tag{319}$$

As in §1 apply the Girsanov theorem to obtain

$$P_\tau\,\varphi_\lambda(v_0) = \theta(z^\lambda{}_1)^{\frac{1}{2}}\,\theta(x_0)^{-\frac{1}{2}}\,(2\pi\tau(1+\lambda))^{-n/2}\exp\{-d(x_0,p)^2/(2(1+\lambda)\tau)\}$$

$$\mathbb{E}[\exp\{\tau\int_0^1 \tfrac{1}{2}\,\theta^{\frac{1}{2}}(z^\lambda{}_s)\,\Delta^0\theta^{-\frac{1}{2}}(z^\lambda{}_s)ds\}\varphi(v^\lambda{}_1)] \tag{320}$$

where the processes $\{z^\lambda{}_s : 0 \leq s < 1+\lambda\}$ now have generators $\frac{1}{2}\tau\Delta^0 + \nabla Y^\lambda{}_s$ for

$$Y^\lambda_s(x) = -\tfrac{1}{2}\, d(x,p)^2/(\lambda+1-s) - \tfrac{1}{2}\tau \log \theta(x) \tag{321}$$

and $\{v^\lambda_s : 0 \le s < 1+\lambda\}$ satisfies the analogue of (318) but along the paths of $\{z^\lambda_s : 0 \le s < 1+\lambda\}$.

To take the limit as $\lambda \downarrow 0$ we need to know that $\lim v^\lambda_1$ exists, and to get a sensible answer we would like this limit to be $\lim_{s\uparrow 1} v^0_s$. In particular the latter should exist. For simplicity assume now that in normal coordinates about p the manifold M is Euclidean outside some compact set.

__Proposition 4A__ *For each* $v_0 \in \Lambda^p T_{x_0} M$ *the limit* $\lim_{s\uparrow 1+\lambda} v^\lambda_s$ *exists almost surely. Moreover in normal coordinates about* p *it exists in* L^2 *uniformly in* $0 \le \lambda \le 1$ *and* $0 < \tau \le 1$.

__Proof__ From §4E of Chapter II and the stochastic version of equation (12), in normal coordinates about p for $0 \le s \le t < 1+\lambda$

$$v^\lambda_t - v^\lambda_s = -\int_s^t \Gamma(x^\lambda_r)(\circ dx^\lambda_r) v^\lambda_r - \tfrac{1}{2}\tau \int_s^t (W^q{}_{x^\lambda_r})^*(v^\lambda_r)\,dr$$

Since W^p is bounded $|v^\lambda_r|$ is bounded independently of chance, and of λ, τ, r. Also $|\Gamma(x)| <$ const. $|x|$ since it vanishes at the origin and has compact support. Thus there exist constants $c_1, c_2, \ldots$ independent of s,t, λ, τ and of chance such that using the Euclidean inner product of our coordinates:

$$\begin{aligned} |v^\lambda_t - v^\lambda_s| \le\; & c_1 \left|\int_s^t \Gamma(x^\lambda_r)(\mu\, u^\lambda_r \circ dB_r)\right| \\ & + c_2 \int_s^t |x^\lambda_r|/(1+\lambda-r)\,dr + c_3 \int_s^t |\nabla \log \theta(x^\lambda_r)|\,dr \\ & + c_4(t-s) \end{aligned} \tag{322}$$

where $\mu^2 = \tau$, $\{u^\lambda_r : 0 \le r < \lambda+1\}$ is the horizontal lift of $\{x^\lambda_r : 0 \le r < \lambda+1\}$, and $\{B_r : 0 \le r < \infty\}$ is a Brownian motion on $\mathbb{R}^n$. The radial component $|x^\lambda_r|$ is now such that if $\rho^\lambda_r = \mu^{-1} |x^\lambda_r|$ then $\{\rho^\lambda_r : 0 \le r \le \lambda+1\}$ has the same law as the Euclidean Brownian bridge from $\mu^{-1} x_0$ to 0 in time $1+\lambda$, by the same argument as for the case $\mu = 1$. Thus it itself is equal in law to

$$r \to |x_0 + \mu B_r - \mu r(B_{1+\lambda} + \mu^{-1}x_0)/(1+\lambda)|$$

i.e. $$r \to |(1+\lambda-r)(1+\lambda)^{-1}x_0 + \mu(1+\lambda)^{-1}(1+\lambda-r)B_{1+\lambda} + \mu(B_r - B_{1+\lambda})|$$

Thus the second term on the right hand side of (322) can be estimated by

$$c_2 \int_s^t |(1+\lambda)^{-1}x_0 + \mu(1+\lambda)^{-1}B_{1+\lambda} + \mu(1+\lambda-r)^{-1}(B_r - B_{1+\lambda})|dr$$

$$\leq c_2(t-s) + c_3|B_{1+\lambda}||t-s| + c_4 \int_s^t |B_{1+\lambda} - B_r|(1+\lambda-r)^{-1}dr$$

$$= O(|1+\lambda-s|^{\alpha})$$

as $s \uparrow 1+\lambda$ both almost surely, and in L^2 independently of μ, λ, for any $\alpha \in (0,\frac{1}{2})$, (using the pathwise Holder continuity of Brownian paths for the almost sure case). The martingale term in (322) is equally tractable: since Γ is bounded and so is $|u^{\lambda}{}_r|$, it is a time changed Brownian motion with a bounded time change. Thus, as $s \to 1+\lambda$,

$$|v^{\lambda}{}_t - v^{\lambda}{}_s| = O(|1+\lambda-s|^{\alpha})$$

for any $\alpha \in (0,\frac{1}{2})$ both almost surely, and in L^2 uniformly in $\mu \in (0,1]$ and $\lambda \in [0,1]$. This gives the required result. //

<u>Theorem 4B</u> *Let* M *be a complete manifold with pole* p *such that the Weitzenbock term for* q*-forms,* W^q, *is bounded below. Let* $\{z_s : 0 \leq s \leq 1\}$ *be the semi-classical bridge from* x_0 *to* p *in time 1 with diffusion constant* τ: *so it has time dependent generator* $\frac{1}{2}\tau\Delta + \nabla Y^0{}_s$, $0 \leq s < 1$, *for* Y^0 *given by* (321). *For* $v_0 \in \Lambda^q T_{x_0}M$ *let* $\{v_t : 0 \leq t < 1\}$ *be the solution to* (318) *along the paths of* $\{z_t : 0 \leq t < 1\}$. *Then*

(i) $v_1 = \lim_{t\to\infty} v_t$ *exists almost surely as an element of* $\Lambda^q T_pM$;

(ii) *the fundamental solution to the heat equation for* q*-forms is given by*

$$k^q{}_\tau(x_0,p)(\alpha)(v_0) = (2\pi\tau)^{-n/2}\,\theta_p(x_0)^{-\frac{1}{2}}\exp\{-d(x_0,p)^2/(2\tau)\}$$

$$\mathbb{E}[\exp\{\tfrac{1}{2}\tau \int_0^1 \theta_p(z_s)^{\frac{1}{2}}\Delta\theta_p{}^{-\frac{1}{2}}(z_s)ds\}\,\alpha(v_1)]. \qquad (323)$$

for $\alpha \in \Lambda^q T^*{}_pM$.

Proof Part (i) follows from the previous proposition by progressively modifying the metric of M to be Euclidean outside of larger and larger domains as described in §1B.

For (ii) take domains $\{D_i\}_{i=1}^{\infty}$ as in the proof of Theorem 1D and observe that $k^q{}_\tau(x_0,p) = \lim_{i\to\infty} k^{q,i}{}_\tau(x_0,p)$ where $k^{q,i}{}_\tau$ refers to the fundamental solution on D_i with Dirichlet boundary conditions. To compute $k^{q,i}{}_\tau$ we can assume M is Euclidean outside of a compact set, in normal coordinates about p. Use the canonical S.D.E. on OM with added drift to define $\{x^\lambda{}_s : 0 \le s < 1 + \lambda\}$. Then by standard results about S.D.E. with parameters as in [57] or [90] there are versions of $\{x^\lambda{}_s : 0 \le s < t\}$ and $\{v^\lambda{}_s : 0 \le s \le t\}$ for each $t < 1$ such that $v^\lambda{}_s$ converges in L^2 (in our coordinates) to v_s uniformly in $s \in [0,t]$. By the previous proposition it follows that $v^\lambda{}_1 \to v_1$ in L^2 and so arguing as in the proof of Theorem 1D

$$k^{q,i}{}_\tau(x_0,p) = (2\pi\tau)^{-n/2}\,\theta(x_0)^{-\frac{1}{2}} \exp\{-d(x_0,p)^2/(2\tau)\}$$

$$\mathbb{E}[\chi_i \exp\{\tfrac{1}{2}\tau\int_0^1 \theta^{\frac{1}{2}}(z_s)\,\Delta\theta^{-\frac{1}{2}}(z_s)ds\}\,\alpha(v_1)]$$

where χ_i is the characteristic function of $\{\omega \in \Omega : z_s \in D_i \text{ for } 0 \le s \le 1\}$. Now take the limit as $i \to \infty$: the result follows by dominated convergence since $v_1 \in L^\infty$ and the exponential term is in L^1 by Theorem 1D. //

Example 4B (Hyperbolic space). For $M = H^n$ the Weitzenbock term $W^q{}_x$ is just multiplication by the constant $-q(n-q)$. It follows that in (323)

$$v_1 = \exp\{\tfrac{1}{2}q(n-q)\tau\}\ //v_0$$

where $//v_0$ refers to the parallel translate of v_0 along the paths of $\{z_t : 0 \le t \le 1\}$. Thus for $\alpha \in \Lambda^q T^*{}_pM$ and $v_0 \in \Lambda^q T_{x_0}M$

$$k^q{}_\tau(x_0,p)\alpha(v_0) = (2\pi\tau)^{-n/2}\,(r/\sinh r)^{\frac{1}{2}(n-1)}\,e^{-r^2/(2\tau)}e^{\frac{1}{2}q(n-q)\tau}$$

$$\mathbb{E}[\alpha(//v_0)\exp\{\tfrac{1}{2}\tau\int_0^1 \theta^{\frac{1}{2}}(z_s)\Delta\theta^{-\frac{1}{2}}(z_s)ds\}] \quad (324)$$

with $\theta^{\frac{1}{2}}(z_s)\Delta\theta^{-\frac{1}{2}}(z_s)$ given by (275), and in particular equal to -1 when $n=3$.

CHAPTER VI. THE GAUSS-BONNET-CHERN THEOREM

§1 Supertraces and the heat flow for forms

A. Let V be a real inner product space. If $B \in \mathbb{L}(\Lambda V^*; \Lambda V^*)$, with B^q its restriction to $\Lambda^q V^*$, define its *supertrace*, str B, by

$$\text{str } B = \Sigma_q (-1)^q \text{ trace } B^q.$$

For a fixed orthonormal basis $e_1,\dots,e_n$ of V we have the annihilation and creation operators $a(e_i)$ and $a(e_i)^*$ defined in §3 of the previous chapter. For $I = \{i_1,\dots,i_k\}$ a naturally ordered subset of $\{1,\dots,n\}$ define

$$a^I = a(e_{i_1}) \dots a(e_{i_k}) \in \mathbb{L}(\Lambda V^*; \Lambda V^*).$$

Followers of quantum probability will recognise these and recall, e.g. from [78a] p. 221, that the collection $\{(a^I)^* a^J\}_{I,J}$ forms a basis for $\mathbb{L}(\Lambda V^*;\Lambda V^*)$. Indeed there are the correct number of them, 2^{2n}, and they act transitively on ΛV^* since given $\alpha \in \Lambda V^*$, with $\alpha \neq 0$, it can be annihilated down to a non-zero element of $\Lambda^0 V^* \approx \mathbb{R}$ by a suitable a^J and any non-zero element of ΛV^* can be created from a non-zero element of $\Lambda^0 V^*$ by a suitable $(a^I)^*$. Alternatively see [33] page 248.

A basic result, emphasized, and called the *Berezin-Patodi formula* in [33] is:

Proposition 1A. For $B = \Sigma_{I,J} \beta_{I,J}(a^I)^* a^J$,

$$\text{str } B = (-1)^n \beta_{\{1,\dots,n\},\ \{1,\dots,n\}} \tag{325}$$

B. The de Rham cohomology groups, or vector spaces, $H^q(M,\mathbb{R})$ were defined by (243) in §4A of Chapter IV. As described there, when M is compact, the Hodge theorem shows that

$$\dim H^q(M;\mathbb{R}) = \dim \ker \Delta^q.$$

(The latter is finite by ellipticity of Δ^q). The *Euler characteristic* for compact M can be defined by

$$\chi(M) = \Sigma_q (-1)^q \dim H^q(M;\mathbb{R}) = \Sigma_q (-1)^q \dim \ker \Delta^q.$$

Compactness of M implies that if $P_t^q = \exp\{\frac{1}{2} t\Delta^q\}$ acting on $L^2 A^q$ then P_t^q

is trace class for $t > 0$ and

$$\text{trace } P_t^q = \int_M \text{trace } k_t^q(x,x)\,dx$$

e.g. see [33]. Define the *supertrace* of $P_t = \bigoplus_q P_t^q$ acting on $L^2 A$ by

$$\text{str } P_t = \Sigma_q (-1)^q P_t^q \qquad t > 0.$$

There is the following remarkable fact due to McKean and Singer:

Proposition 1B *For* M *compact and all* $t > 0$

$$\text{Str} P_t = \chi(M). \tag{326}$$

Proof Divide A into $A^+ \oplus A^-$ where $A^+ = \bigoplus_{p \text{ even}} A^p$ and $A^- = \bigoplus_{p \text{ odd}} A^p$ respectively and for $\lambda \in \mathbb{R}$ let $n_+(\lambda)$ and $n_-(\lambda)$ be the multiplicities of λ as an eigenvalue of Δ^+ and Δ^-, (the spectrum of Δ consists of a discrete set of eigenvalues increasing to ∞, see [28] for example).

Note that $\Delta = -(d + d^*)^2$ since $d^2 = 0$. Therefore if $\Delta\varphi = \lambda\varphi$ then $\Delta(d+d^*)\varphi = \lambda(d+d^*)\varphi$. Also $(d+d^*)\varphi \neq 0$ unless $d\varphi = 0$ and $d^*\varphi = 0$ i.e. unless $\Delta\varphi = 0$. Thus $n_+(\lambda) = n_-(\lambda)$ for $\lambda \neq 0$. However

$$\text{Str } P_t = \Sigma_\lambda (n_+(\lambda) - n_-(\lambda))e^{-\frac{1}{2}t\lambda}$$

Therefore

$$\text{Str } P_t = n_+(0) - n_-(0) = \chi(M)$$

as required. //

From the proposition there is the following corollary

Corollary 1B(1) *For all* $t > 0$, *when* M *is compact*

$$\chi(M) = \int_M \text{str } k_t(x,x)\,dx \tag{327}$$ //

§2. Proof of the Gauss-Bonnet-Chern Theorem

A. Suppose M is compact with even dimension, $\dim M = 2\ell$, say. The G-B-C. theorem expresses $\chi(M)$ as an integral over M of a certain function of the curvature of M. We can do this by looking at (327) as $t \downarrow 0$ using the results of §1 and the 'elementary formula' (323) for the heat kernel. This is essentially Patodi's proof as described in [33]. The difference from the treatment in [33] is simply the use of the elementary formula. The Malliavin calculus approach as in [59a] has the same structure but the cancellations take place at the level of distributions on Wiener space i.e. *before* taking expectations. Related proofs of other classical index theorems are in [18], [56], [59a], [68], [103], [104], and discussed in [33]. The approach of [104] fits in particularly well here, especially if it is simplified somewhat by using semi-classical bridges: an important technique used in [104] is a rescaling of Brownian bridges, and this seems to work equally well with semi-classical bridges. The Atiyah-Singer index theorem describes the index of an elliptic differential operator $\mathcal{E}$ (i.e. dim ker $\mathcal{E}$ - dim ker $\mathcal{E}^*$) in terms of the coefficients of $\mathcal{E}$. For the G-B-C. theorem $\mathcal{E} = d + d^* : A^+ \to A^-$. (Remember $\Delta\varphi = 0$ iff $d\varphi = d^*\varphi = 0$). The other 'classical index theorems' are for other geometrically defined operators $\mathcal{E}$. It turns out that the general result, for $\mathcal{E}$ of arbitrary order, follows from these special cases.

B. To examine the behaviour of str $k_\tau(x,x)$ as $\tau \downarrow 0$, or more generally str $k_\tau(x_0,p)$ when $x_0 \notin \mathrm{Cut}(p)$ we can argue as Corollary 1D(iii) of Chapter IV and assume that p is a pole for M and M is Euclidean outside some compact set in normal coordinates about p. (Of course we have lost the compactness of M after this modification.) Having done this, rewrite (323) as

$$k^q_\tau(x_0,p) = (2\pi\tau)^{-n/2}\,\theta(x_0)^{-\frac{1}{2}} \exp\{-d(x_0,p)^2/(2\tau)\}$$

$$\mathbb{E}[\exp\{\tfrac{1}{2}\tau \int_0^1 V_{eff}(z(\tau)_s)ds\}\Phi^q_\tau] \qquad (328)$$

where

$$\Phi^q_\tau : \Lambda^q T^*_p M \to \Lambda^q T^*_{x_0} M$$

is given by

$$\Phi^q_\tau(\alpha)(v_0) = \alpha(v_1).$$

using the notation of (323), but now we write z_s as $z(\tau)_s$ to make clear its τ-dependence. Writing Φ_τ for $\bigoplus_q \Phi^q_\tau$ note that $\Phi_\tau = \Phi_{\tau,1}$ where $\{\Phi_{\tau,s} : 0 \le s \le 1\}$ is the solution to

$$(D/\partial s)\,\Phi_{\tau,s} = -\tfrac{1}{2}\tau\,\Phi_{\tau,s}\,W_{z(\tau)_s} \qquad (329)$$

along the paths of $\{z(\tau)_s : 0 \le s \le 1\}$ with $\Phi_{\tau,0} = \mathrm{Id}$. This exists by Proposition 4A of the last chapter.

Now take an orthonormal base $e_1(0),\dots,e_n(0)$ for $T_{x_0}M$ and let $e_j^\tau(s) = //_s^\tau$

$e_j(0)$ be the parallel translate of $e_j(0)$ along $z(\tau)$ from x_0 to $z(\tau)_s$. Omitting the superscript τ unless it is needed, write

$R_{ijkl}(s) = \langle R(e_i(s), e_j(s))\, e_l(s), e_k(s)\rangle$

and set $a^j_s = a(e_j(s))$, a random annihilation operator. If we parallel translate $\Phi_{\tau,s}$ back to x_0 and define $H_{\tau,s} \in \mathbb{L}(\Lambda T^*_{x_0}M; \Lambda T^*_{x_0}M)$ by

$$H_{\tau,s}\,\alpha = \Phi_{\tau,s}(\alpha \circ (//_s)^{-1}),$$

then

$$\frac{d}{ds} H_{\tau,s} = -\tfrac{1}{2}\,\tau\, H_{\tau,s} \circ \mathcal{R}(s)$$

for

$$\mathcal{R}(s) = -\Sigma\, R_{ijkl}(s)(a^i)^*\, a^j (a^k)^* a^l$$

where $a^i = a^i{}_0 = a(e_i(0))$.

C. If we iterate the formula

$$H_{\tau,t} = \mathrm{Id} - \tfrac{1}{2}\,\tau \int_0^t H_{\tau,s} \circ \mathcal{R}(s)ds \tag{330}$$

by substituting the corresponding expression for $H_{\tau,s}$ back into the integrand we obtain

$$H_{\tau,1} = \mathrm{Id} + Z_1 + \ldots Z_\ell + O(\tau^{\ell+1})$$

where

$$Z_\ell = (-\tfrac{1}{2}\,\tau)^\ell \int_0^1 \int_0^{s_\ell} \ldots \int_0^{s_2} \mathcal{R}(s_\ell) \circ \ldots \circ \mathcal{R}(s_1)ds_1 \ldots ds_\ell$$

for $\ell = 2,3,\ldots$, and analogously for $\ell = 1$.

Thus by the Berezin-Patodi formula (325)

$$\mathrm{str}\, H_{\tau,1} = \mathrm{str}\, Z_\ell + O(\tau^{\ell+1}) \tag{331}$$

and

$$\mathrm{str}\, Z_\ell = (-\tfrac{1}{2}\,\tau)^\ell \int_0^1 \int_0^{s_\ell} \ldots \int_0^{s_2} Z^\tau(s_1,\ldots,s_\ell)ds_1 \ldots ds_\ell \tag{332}$$

where

$$Z^{\tau}(s_1,\ldots,s_l) = (-1)^{\ell}\, \Sigma\, \mathrm{sgn}(\pi)\mathrm{sgn}(\sigma)\, R_{\pi(1)\sigma(1)\pi(2)\sigma(2)}(s_\ell)\ldots$$
$$\ldots R_{\pi(n-1)\sigma(n-1)\pi(n)\sigma(n)}(s_1)$$

where the sum is over all permutations π and σ of $\{1,\ldots,n\}$. (This is still a random variable.) To get Z_ℓ into the form in which we could read off its supertrace by the Berezin–Patodi formula we have used the anti-commutation relations $a^i(a^j)^* + (a^j)^* a^i = \delta^{ij}$ and $a^i a^j = -a^j a^i$.

Now $\Phi_t = (//^1{}_0)^*\, H_{\tau,1} = H_{\tau,1} + [(//^1{}_0)^* - \mathrm{Id}]H_{\tau,1}$. We claim that $\mathrm{str}[(//^1{}_0)^*-\mathrm{Id}]H_{\tau,1} = O(\tau)$, see §E below. Then from (328), (331) and (332)

$$\mathrm{str}\, k_\tau(x_0,p) = (-4\pi)^{-\ell}\, \theta_p(x_0)^{-\frac{1}{2}} \exp\{-d(x_0,p)^2/(2\tau)\}$$

$$\mathbb{E}[\exp\{\tfrac{1}{2}\tau\int_0^1 V_{eff}(z(\tau)_s)ds\}\int_0^1\int_0^{s_\ell}\cdots\int_0^{s_2} Z^{\tau}(s_1,\ldots,s_\ell)ds_\ell\ldots ds_\ell] + O(\tau) \quad (333)$$

Now let $\tau \downarrow 0$. Still with our assumptions on M we can choose versions of $\{z(\tau)_s : 0 \le s \le 1\}$ so that, almost surely, they converge uniformly on [0,1] to the geodesic from x_0 to p parametrized to take unit time. (Recall from (321) that $z(\tau)$ had generator $\frac{1}{2}\tau\Delta^0 + \nabla Y_s{}^0$ with

$$Y_s{}^0(x) = -\tfrac{1}{2}\, d(x,p)^2/(1-s) - \tfrac{1}{2}\tau \log \theta(x).)$$

Correspondingly by Proposition 4A of Chapter V the horizontal lifts $\{u(\tau)_s: 0\le s\le 1\}$ will converge to that of the geodesic, and Z^τ will converge to the corresponding non-random term Z^0. In particular as $\tau \downarrow 0$ so $\mathrm{str}\, k_\tau(p,p) \to E(p)$ where

$$E(p) = (4\pi)^{-\ell} 1/(\ell!) \sum \mathrm{sgn}\,\pi\, \mathrm{sgn}\,\sigma\, R_{\pi(1)\sigma(1)\pi(2)\sigma(2)}(p)\ldots$$
$$R_{\pi(n-1)\sigma(n-1)\pi(n)\sigma(n)}(p) \qquad (334)$$

for $R_{ijkl}(p) = \langle R(e_i(0), e_j(0))e_l(0), e_k(0)\rangle$.

In particular the right hand side of (334) does not depend on the basis $e_1(0),\ldots,e_n(0)$ for T_pM. Note that, by equations (85) and (84),

$$R_{\pi(1)\sigma(1)\pi(2)\sigma(2)} = -R_{\sigma(1)\sigma(2)\pi(2)\pi(1)} - R_{\sigma(2)\pi(1)\pi(2)\sigma(1)}$$
$$= R_{\pi(1)\pi(2)\sigma(1)\sigma(2)} + R_{\pi(1)\sigma(2)\pi(2)\sigma(1)}\,.$$

Thus $$E(p) = \frac{2^{-1}}{(4\pi)^{\ell}\ell!}\sum \mathrm{sgn}\pi\,\mathrm{sgn}\sigma\, R_{\pi(1)\pi(2)\sigma(1)\sigma(2)}(p)\, R_{\pi(3)\sigma(3)\pi(4)\sigma(4)}(p)$$
$$\ldots R_{\pi(n-1)\sigma(n-1)\pi(n)\sigma(n)}(p)$$

$$= \dots = \frac{2^{-\ell}}{(4\pi)^{\ell}\ell!}\sum \operatorname{sgn}\pi \operatorname{sgn}\sigma\, R_{\pi(1)\pi(2)\sigma(1)\sigma(2)} \cdots R_{\pi(n-1)\pi(n)\sigma(n-1)\sigma(n)}$$

which is the more standard expression for it.

D. From (327), Corollary 1B(i), we obtain:

Gauss-Bonnet-Chern Theorem. For a compact even dimensional manifold M

$\chi(M) = \int_M E(x)dx$ *where* E *is defined by* (330).

The only thing we need to be careful about is the uniformity in x of the analogue $k_\tau(x,x) = k_\tau^D(x,x)\,(1 + O(\tau^k))$ of (271) for suitable domains D which enabled us to replace M by a manifold with x as a pole.

<u>Special case: dim M = 2</u> Here

$$E(x) = (4\pi)^{-1}\{-R_{1221}(x) - R_{2112}\} = (2\pi)^{-1} K(x)$$

where K is the Gauss curvature, see (87). Thus we have the classical Gauss-bonnet theorem $\chi(M) = (2\pi)^{-1}\int_M K(x)dx$.

E. To see that $\operatorname{str}((//^1{}_0)^*-\mathrm{Id})H_{\tau,1} = O(\tau)$ consider the expansion of $(//^1{}_0)^*-\mathrm{Id}$ in terms of a_i, a_j^* analogous to that for $H_{\tau,1}$, using normal co-ordinates as in the proof of Proposition 4A. As an operator on $\Lambda T^*_{x_0}M$, if $u_s \equiv u(\tau)_s$ is the horizontal lift of $z(\tau)_s$, $0 \le s \le 1$, then

$$(//^1{}_0)^*-\mathrm{Id} = \int_0^1 \Gamma^k_{ji}(z(\tau)_s)(\circ dz^j(\tau)_s)\, a_i^*\, a_k +$$

$$\int_0^1\!\!\int_0^1 \Gamma^{k_1}_{j_1 i_1}(z(\tau)_{s_1})(\circ dz^{j_1}(\tau)_{s_1})\,\Gamma^{k_2}_{j_2 i_2}(z)(\tau)_{s_2})(\circ dz^{j_2}(\tau)_{s_2})\, a^*_{i_1} a_{k_1} a^*_{i_2} a_{k_2} + \dots .$$

If we substitute $dz(\tau)_s = \mu u(\tau)_s \circ dB_s - (z(\tau)_s/(1-s) + \tfrac{1}{2}\tau\nabla \log \theta(z(\tau)_s))ds$, and use the facts that $\Gamma(x_0) = 0$ and $\mu^{-1}|z(\tau)_s|$, $0 \le s \le 1$, is a Bessel Bridge in $\mathbb{R}^n$ from 0 to 0 it is easy to see that the pth term in this expansion is $O(\tau^p)$, $p = 1,2,\dots$. Now multiply by the expansion of $H_{\tau,1}$ and use the Berezin-Patodi formula (325). See also [59a]. //

The expansion of $(//^1{}_0)^*-\mathrm{Id}$ plays a more important role in other index theorems: [18], [59a], [104].

REFERENCES

A. Séminaire de Probabilités XVI, 1980/81, Supplément: Géométrie Différentielle Stochastique. Lecture Notes in Maths., 921, (1981).

B. Lyapunov Exponents. Proceedings, Bremen 1984. Eds. L. Arnold & V. Wihstutz. Lecture Notes in Mathematics, 1186, (1986).

C. From Local Times to Global Geometry, Control and Physics. Ed. K.D. Elworthy, Pitman Research Notes in Mathematics Series, 150, Longman and Wiley, 1986.

[1] Airault, H. (1976). Subordination de processus dans le fibré tangent et formes harmoniques. C.R. Acad. Sc. Paris, Sér. A, 282 (14 juin 1976), 1311-1314.

[2] Arnold, V.I. & Avez, A. (1968). Ergodic problems of classical mechanics. New York: Benjamin.

[3] Arnold, L. (1984). A formula connecting sample and moment stability of linear stochastic systems. SIAM J. Appl. Math., 44, 793-802.

[4] Arnold, L. & Kliemann, W. Large deviations of linear stochastic differential equations, *In* "Proceedings of the Fifth IFIP Working Conference on Stochastic Differential Systems, Eisenach 1986" ed. Engelbert, Lecture Notes in Control and Information Sciences. Springer-Verlag.

[5] Azéma, J, Kaplan-Duflo, M. & Revuz, D. (1966). Récurrence fine des processus de Markov. Ann. Inst. H. Poincaré, Sect. B, II, no. 3, 185-220.

[6] Azencott, R. (1974). Behaviour of diffusion semigroups at infinity. Bull. Soc. Math. France, 102, 193-240.

[7] Azencott, R. et al. (1981). Géodésiques et diffusions en temps petit. Séminaire de probabilités, Université de Paris VII. Astérique 84-85. Société Mathématique de France.

[8] Ballman, W., Gromov, & M. Schroeder, V. (1985). Manifolds of non positive curvature. Boston-Basel-Stuttgart: Birkhauser.

[9] Baxendale, P. (1980). Wiener processes on manifolds of maps. Proc. Royal Soc. Edinburgh, 87A, 127-152.

[10] Baxendale, P.H. (1986). Asymptotic behaviour of stochastic flows of diffeomorphisms: two case studies. Prob. Th. Rel. Fields, 73, 51-85.

[11] Baxendale, P. (1985). Moment stability and large deviations for linear stochastic differential equations. *In* Proceedings of the Taniguchi Symposium

on Probabilistic Methods in Mathematical Physics, Kyoto 1985, ed. N. Ikeda. To appear.

[12] Baxendale, P.H. (1986). The Lyapunov spectrum of a stochastic flow of diffeomorphisms. *In* [B] pp. 322-337.

[12a] Baxendale, P.H. (1986). Lyapunov exponents and relative entropy for a stochastic flow of diffeomorphisms. Preprint: University of Aberdeen.

[13] Baxendale, P.H. & Stroock, D.W. (1987). Large deviations and stochastic flows of diffeomorphisms. Preprint.

[14] Bérard, P. & Besson, G. (1987). Number of bound states and estimates on some geometric invariants. Preprint: Institut Fourier, B.P. 74, 38402, St. Martin d'Heres Cedex, France.

[15] Berthier, A.M. & Gaveau, B. (1978). Critère de convergence des fonctionnelles de Kac et applications en mécanique quantique et en géométrie. J. Funct. Anal. 29, 416-424.

[16] Besse, A.-L. (1978). Manifolds all of whose geodesics are closed. Ergebruisse der Mathematik 93, Berlin, Heidelberg, New York: Springer-Verlag.

[17] Bismut, J.-M. (1984). Large deviations and the Malliavin calculus. Progress in Mathematics, 45, Boston, Basel, Stuttgard: Birkhauser.

[18] Bismut, J.-M. (1984). The Atiyah-Singer theorems: a probabilistic approach. I & II. J. Funct. Anal. 57, 56-99 & 329-348.

[19] Bougerol, P. (1986). Comparaison des exposants de Lyapunov des processus Markoviens multiplicatifs.

[20] Carverhill, A.P. (1985). Flows of stochastic dynamical systems: Ergodic Theory. Stochastics, 14, 273-317.

[21] Carverhill, A.P. (1985). A formula for the Lyapunov numbers of a stochastic flow. Application to a perturbation theorem. Stochastics, 14, 209-226.

[22] Carverhill, A.P., Chappell, M. & Elworthy, K.D. (1986). Characteristic exponents for stochastic flows. *In* Stochastics Processes - Mathematics and Physics. Proceedings, Bielefeld 1984. Ed. S. Albeverio et al. pp. 52-72. Lecture Notes in Mathematics 1158. Springer-Verlag.

[23] Carverhill, A.P. (1986) A non-random Lyapunov spectrum for non-linear stochastic systems. Stochastics 17, 253-287.

[24] Carverhill, A.P. & Elworthy, K.D. (1986). Lyapunov exponents for a stochastic analogue of the geodesic flow. Trans. A.M.S., 295, no. 1, 85-105.

[24a] Carverhill, A.P. & Elworthy, K.D. (1983). Flows of stochastic dynamical systems: the functional analytic approach. Z. fur Wahrscheinlichkeitstheorie 65, 245-267.

[24b] Carverhill, A.P. (1987). The stochastic geodesic flow: nontriviality of the Lyapunov spectrum. Preprint: Department of Mathematics, University of North Carolina at Chapel Hill, Chapel Hill NC 27514, U.S.A.

[25] Chappell, M.J. (1986). Bounds for average Lyapunov exponents of gradient stochastic systems. *In* [B] pp. 308-321.

[26] Chappell, M.J. (1987). Lyapunov exponents for certain stochastic flows Ph.D. Thesis. Mathematics Institute, University of Warwick, Coventry CV4 7AL, England.

[27] Chappell, M.J. & Elworthy, K.D. (1987). Flows of Newtonian Diffusions. *In* Stochastic Mechanics and Stochastic Processes, ed. A. Truman. Lecture Notes in Maths. To appear.

[28] Chavel, I. (1984). Eigenvalues in Riemannian Geometry. Academic Press.

[29] Cheeger, J. and Ebin, D. (1975). Comparison Theorems in Riemannian Geometry. Amsterdam: North Holland.

[30] Cheng, S.-Y. (1975). Eigenvalue comparison theorems and its geometric applications. Math. Z., 143, 289-297.

[31] Cheng, S.Y. & Yau, S.T. (1975). Differential equations on Riemannian Manifolds and their Geometric Applications. Comm. Pure Appl. Maths., XXVIII, 333-354.

[32] Colin de Verdiere, Y., (1973) Spectre du Laplacien et longueurs des géodésiques périodiques I. Compositio Math., 27, 83-106.

[33] Cycon, H., Froese, R., Kirsch, W. & Simon, B. (1987). Schrodinger Operators with applications to quantum mechanics and global geometry. Texts and Monographs in Physics. Springer-Verlag.

[34] Darling, R.W.

[35] Davies, E.B. & Mandouvalos, N. (1987). Heat kernel bounds on hyperbolic space and Kleinian groups. Preprint: Maths Department, Kings college, The Strand, London WC2R 2LS.

[36] Debiard, A., Gaveau, B. & Mazet, E. (1976). Théorèmes de comparaison en

géométrie riemannienne. Publ. RIMS. Kyoto Univ., 12, 391-425.

[37] De Rham, G. (1955). Varietes Differentiables, Paris: Herman et Cie.

[38] Dodziuk, J. (1983). Maximum principle for parabolic inequalities and the heat flow on open manifolds. Indiana U. Math. J., 32, 703-716.

[39] Dodziuk, J. (1982). L^2 harmonic forms on complete manifolds. *In*: Seminar on Differential Geometry pp. 291-302. Princeton University Press.

[40] Donnelly, H. and Li, P (1982). Lowr bounds for the eigenvalues of Riemannian manifolds. Michigan Math. J. 29, 149-161.

[41] Doob, J.L. (1984). Classical Potential Theory and its Probabilistic Counterpart. Grund. der math. Wiss. 262. New York, Berlin, Heidelberg, Tokyo: Springer-Verlag.

[42] Elworthy, K.D. & Truman, A. (1982). The diffusion equation and classical mechanics: an elementary formula. *In* 'Stochastic Processes in Quantum Physics' ed. S. Albeverio et al. pp. 136-146. Lecture Notes in Physics 173. Springer-Verlag.

[42a] Elworthy, K.D. & Truman, A. (1981). Classical mechanics, the diffusion (heat) equation and the Schrodinger equation on a Riemannian manifold. J. Math. Phys. 22, no. 10, 2144-2166.

[43] Elworthy, K.D. (1982). Stochastic Differential Equations on Manifolds. London Math. Soc. Lecture Notes in Mathematics 70, Cambridge University Press.

[44] Elworthy, K.D. (1982). Stochastic flows and the C_0 diffusion property. Stochastics 6, no. 3-4, 233-238.

[45] Elworthy, K.D. & Stroock, D. (1984). Large deviation theory for mean exponents of stochastic flows. Appendix to [22] above.

[46] Elworthy, K.D. & Rosenberg, S. (1986). Generalized Bochner theorems and the spectrum of complete manifolds. Preprint: Boston University, M.A., U.S.A.

[47] Elworthy, K.D., Ndumu, M. & Truman, A. (1986). An elementary inequality for the heat kernel on a Riemannian manifold and the classical limit of the quantum partition function. *In* [C], pp. 84-99.

[48] Elworthy, K.D. (1987). Brownian motion and harmonic forms. To appear in proceedings of the workshop on stochastic analysis at Silivri, June 1986,

eds. H.K. Korezlioglu and A.S. Ustunel, Lecture Notes in Maths.

[49] Elworthy, K.D. (1987). The method of images for the heat kernel of S^3. Preprint, University of Warwick.

[50] Eskin, L.D. (1968). The heat equation and the Weierstrass transform on certain symmetric spaces. Amer. Math. Soc. Transl., 75, 239-254.

[51] Friedman, A. (1974). Non-attainability of a set by a diffusion process. Trans. Amer. Math. Soc., 197, 245-271.

[52] Gaffney, M.P. (1954). A special Stoke's theorem for complete Riemannian manifolds. Ann. of Math., 60, 140-145.

[53] Gaffney, M.P. (1954). The heat equation method of Milgram and Rosenbloom for open Riemannian manifolds. Annals of Mathematics 60, no. 3. 458-466.

[54] Gaveau, B. (1979). Fonctions propres et non-existence absolee d'etats liés dans certains systèmes quantiques. Comm. Math. Phys. 69, 131-169.

[55] Gaveau, B. (1984). Estimation des fonctionelles de Kac sur une variété compacte et premièr valeur propre de Δ + f. Proc. Japan Acad., 60, Ser.A, 361-364.

[56] Getzler, E. (1986). A short proof of the local Atiyah-Singer Index Theorem. Topology, 25, no. 1., 111-117.

[57] Gikhman, I.I. & Skorohod, A.V. (1972). Stochastic Differential Equations. Berlin, Heidelberg, New York: Springer-Verlag.

[58] Greene, R.E. and Wu, H. (1979). Function Theory on Manifolds which Possess a Pole. Lecture Notes in Maths., 699. Berlin, Heidelberg, New York: Springer-Verlag.

[59] Ikeda, N. & Watanabe, S. (1981). Stochastic Differential Equations and Diffusion Processes. Tokyo: Kodansha. Amsterdam, New York, Oxford: North-Holland.

[59a] Ikeda, N. & Watanabe, S. (1986). Malliavin calculus of Wiener functionals and its applications. *In* [C], pp. 132-178.

[60] Ito, K. (1963). The Brownian motion and tensor fields on a Riemannian manifold. Proc. Internat. Congr. Math. (Stockholm, 1962), pp. 536-539. Djursholm: Inst. Mittag-Leffler.

[61] Kendall, W. (1987). The radial part of Brownian motion on a manifold: a semi-martingale property. Annals of Probability, 15, no.4, 1491-1500.

[62] Kifer, Yu. (1986). Ergodic Theory of Random transformations. Basel: Birkhauser.

[63a] Kifer, Y. (1987). A note on integrability of C^r-norms of stochastic flows and applications. Preprint: Institute of Mathematics, Hebrew University, Jerusalem.

[63b] Kifer, Y. & Yomdin, Y. (1987). Volume growth and topological entropy for random transformations. Preprint: Institute of Mathematics, Hebrew University, Jerusalem.

[64] Kobayashi, S. & Nomizu, K. (1963). Foundations of Differential Geometry. Volume 1. New York, London: John Wiley, Interscience.

[65] Kobayashi, S. & Nomizu, K. (1969). Foundations of differential geometry, Vol. II. As vol. I, above.

[66] Kunita, H. (1982). On backward stochastic differential equations. Stochastics, 6, 293-313.

[67] Kunita, H. (1984). Stochastic differential equations and stochastic flows of diffeomorphisms. *In* Ecole d'Eté de Probabilités de Saint-Flour XII - 1982, ed. P.L. Hennequin, pp. 143-303. Lecture Notes in Maths. 1097. Springer.

[68] Leandre, R. (1986). Sur le theoreme d'Atiyah-Singer. Preprint: Dept. de Mathematiques, Faculté des Sciences 25030 Besacon, France.

[69] Le Jan, Y. Equilibre statistique pour les produits de difféomorphismes aléatoires independants. Preprint: Laboratoire de Probabilités, Université, Paris 6.

[70] Lott, J. (1987). Supersymmetric path integrals. Commun. Math. Phys., 108, 605-629.

[71] Malliavin, P. (1977b). Champ de Jacobi stochastiques. C.R. Acad. Sc. Paris, 285, série A, 789-792.

[71] Malliavin, M.-P. & Malliavin, P. (1974). Factorisations et lois limites de la diffusion horizontale an-dessus d'un espace Riemannien symmetrique. *In* Theory du Potential et Analyse Harmonique, ed. J. Faraut, Lecture Notes in Maths. 404, Springer-Verlag.

[72] Malliavin, P. (1974). Formule de la moyenne pour les formes harmoniques. J. Funct. Anal., 17, 274-291.

[73] Malliavin, M.-P, & Malliavin, P. (1975). Holonomie stochastique au-dessus d'un espace riemannien symmétrique. C.R. Acad. Sc. Paris, 280, Série

A, 793-795.

[75] Malliavin, P. (1978). Géométrie différentielle stochastique. Séminaire de Mathématiques Supérieures. Université de Montréal.

[76] Markus, L. (1986). Global Lorentz geometry and relativistic Brownian motion. *In* [C], pp. 273-287.

[77] Meyer, P.A. (1981). Géométrie différentielle stochastique (bis). *In* [A], pp. 165-207.

[78] Meyer, P.A. (1981). Flot d'une equation différentielle stochastique. *In* Séminaire de Probabilités XV, 1979/80, eds. J. Azema and M. Yor, 103-117. Lecture Notes in Maths 860. Berlin, Heidelberg, New York: Springer-Verlag.

[78a] Meyer, P.A. (1986). Elements de probabilites quantiques *In* Séminaire de Probabilités XX, 1984/85, eds. J. Azema and M. Yor. Lecture Notes in Maths. 1204. Springer.

[79] Milnor, J. (1963). Morse Theory. Annals of Math. Studies, 51, Princeton: Princeton University Press.

[80] Molchanov, S.A., (1975). Diffusion processes and Riemannian geometry. Usp. Math. Nauk, 30, 3-59. English translation: Russian Math. Surveys, 30, 1-63.

[81] Ndumu, M.N. (1986). An elementary formula for the Dirichlet heat kernel on Riemannian manifolds. *In* [C], pp.320-328.

[82] Ndumu, M.N. Ph.D. Thesis, Maths Dept. University of Warwick. In preparation.

[83] Ndumu, M.N. (1987). The heat kernel of the standard 3-sphere and some eigenvalue problems. Submitted to Proc. Edinburgh Math. Soc.

[84] Pinsky, M. (1978). Stochastic Riemannian geometry. *In* Probabilistic Analysis and Related Topics, 1, ed. A.T. Bharucha Reid. London, New York: Academic Press.

[85] Pinsky, M.A. (1978). Large deviations for diffusion processes. *In* Stochastic Analysis, eds. Friedman and Pinsky, pp. 271-283. New York, San Francisco, London: Academic Press.

[86] Reed, M. and Simon, B. (1978). Methods of Modern Mathematical Physics IV: Analysis of Operators. New York, San Francisco, London: Academic Press.

[87] Ruelle, D. (1979). Ergodic Theory of Differentiable Dynamical Systems, Publications I.H.E.S., Bures-sur-Yvette, France.

[88] Schwartz, L. (1973). Radon Measures on Arbitrary Topological Spaces and Cylindrical Measures. Tata Institute Studies in Mathematics 6. Bombay: Oxford University Press.

[89] Schwartz, L. (1980). Semi-martingales sur des variétés et martingales conformes sur des variétés analytiques complexes. Lecture Notes in Maths., 780, Springer-Verlag.

[90] Schwartz, L. (1982). Géométrie différentielle du 2-eme ordre, semi-martingales et equations différentielles stochastique sur une variété différentielle. In [A] pp. 1-149.

[91] Strichartz, R.S. (1983). Analysis of the Laplacian on the complete Riemannian manifold, J. of Functional Anal. 52, 48-79.

[91a] Strichartz, R.S. (1986). L^p Contractive projections and the heat semigroup for differential forms. Jour. of Functional Anal., 65, 348-357.

[92] Stroock, D.W. & Varadhan, S.R.S. (1979). Multidimensional Diffusion Processes. Berlin, Heidelberg, New York: Springer-Verlag.

[93] Sullivan, D. (1987). Related aspects of positivity in Riemannian geometry. J. Differential Geometry, 25, 327-351.

[94] Sunada, T. (1982). Trace formula and heat equation asymptotics for a non-positively curved manifold. American J. Math. 104, 795-812.

[95] Van den Berg, M. & Lewis, J.T. (1985). Brownian motion on a hypersurface. Bull. London Math. Soc., 17, 144-150.

[96] Vauthier, J. (1979). Théoremes d'annulation et de finitude d'espaces de 1-formes harmoniques sur une variété de Riemann ouverte. Bull Sc. Math., 103, 129-177.

[97] Vilms, J. (1970). Totally geodesic maps. J. Differential Geometry, 4, 73-99.

[98a] Watling, K.D. (1986). Formulae for solutions to (possibly degenerate) diffusion equations exhibiting semi-classical and small time asymptotics. Ph.D. Thesis, University of Warwick.

[98] Watling, K.D. (1987). Formulae for the heat kernel of an elliptic operator exhibiting small time asymptotics. *In* Stochastic Mechanics and Stochastic Processes, ed. A. Truman. Lecture Notes in Maths. To appear.

[99] Witten, E. Supersymmetry and Morse theory, J. Diff. Geom. 17 (1982), 661-692.

[100] Yau, S.-T. (1975). Harmonic functions on complete Riemannian manifolds. Comm. Pure Appl. Math., 28, 201-228.

[101] Yau, S.-T. (1976). Some function-theoretic properties of complete Riemannian manifolds and their applications to geometry. Indiana Univ. Math. J., 25, No. 7, 659-670.

[102] Yosida, K. (1968). Functional Analysis. (Second Edition). Grundlehren der math. Wissenschaften, 123, Berlin, Heidelberg, New York: Springer-Verlag.

[103] Azencott, R. (1986) Une Approche Probabiliste du Théoréme d'Atiyah-Singer, d'après J.M. Bismut. *In* Séminaire Bourbaki, 1984-85. Astérisque, 133-134, 7-8. Société Mathématique de France.

[104] Hsu, P. (1987). Brownian motion and the Atiyah-Singer index theorem. Preprint (present address: University of Illinois at Chicago).

[105] Mañé, R. & Freire, A. (1982). On the entropy of the geodesic flow in manifolds without conjugate points. Invent. Math., 69, 375-392.

[106] Pesin Ya. B. (1981). Geodesic flows with hyperbolic behaviour of the trajectories and objects connected with them. Russian Math. Surveys, 36, no.4, 1-59.

[107] Rogers, L.C.G. & Williams, D. (1987). Diffusions, Markov processes and martingales, Vol.2: Itô calculus. Wiley series in probability and mathematical statistics. Chichester: Wiley.

NOTATION

INDEX

INDEX

INDEX

STOCHASTIC MECHANICS AND RANDOM FIELDS

Edward NELSON

E. NELSON : "STOCHASTIC MECHANICS AND RANDOM FIELDS"

This is intended as a guide to stochastic mechanics, and not as a systematic account. It is addressed to mathematicians with a background in probability theory; in particular, to those with some knowledge of Markov processes, martingales, and the Wiener process. It is by no means self-contained, for one of its purposes is to provide a guide to the literature.

Let us begin with some mathematical questions. There will be occasion later for historical remarks and a discussion of physical and philosophical issues.

I. Kinematics of Diffusion

1. The Wiener process and diffusions on R^n

To fix some notation, recall that a *stochastic process indexed by* I is a function ξ from I to random variables on some probability space. Two stochastic processes ξ and η indexed by the same set I are *equivalent* in case they have the same finite joint distributions; that is, in case for all n and all $t_1, \ldots, t_n$ in I, the $\xi(t_i)$ and $\eta(t_i)$ are equidistributed. We use E to denote the expectation, and if $\mathcal{B}$ is a σ-algebra of measurable sets, we use $\mathrm{E}\{\ \mid \mathcal{B}\}$ to denote the conditional expectation with respect to $\mathcal{B}$.

A *difference process* is a stochastic process δ indexed by $\mathsf{R} \times \mathsf{R}$ such that for all t_1, t_2, and t_3 in R we have $\delta(t_1,t_2) + \delta(t_2,t_3) = \delta(t_1,t_3)$. Let δ be a difference process, choose t_0 in R, choose a random variable $w(t_0)$ independent of the process δ, and define $w(t) = w(t_0)+\delta(t,t_0)$. Then w is a stochastic process indexed by R such that $\delta(t_1,t_2) = w(t_1)-w(t_2)$ for all t_1 and t_2 in R. It is determined up to equivalence by specifying the *initial time* t_0 and the *initial distribution*, i.e., the probability distribution of the initial value $w(t_0)$. We shall write difference processes $\delta(t_1,t_2)$ simply as $w(t_1) - w(t_2)$, although this notation is abusive unless the initial time and initial distribution have been specified.

The one dimensional *Wiener difference process* w has independent increments that are Gaussian of mean 0 and variance the length of the interval; this description determines the process uniquely up to equivalence. The n dimensional Wiener difference process is the R^n-valued process whose components w^i are independent one dimensional Wiener difference processes.

Let ξ be a stochastic process indexed by R. For t in R, let $\mathcal{P}_t$ be the σ-algebra generated by all $\xi(s)$ with $s \leq t$ and let $\mathcal{F}_t$ be the σ-algebra generated by all $\xi(s)$ with $s \geq t$. Then $\mathcal{P}$ is a *forward filtration* (an increasing family of σ-algebras) and $\mathcal{F}$ is a *backward filtration* (a decreasing family of σ-algebras). Let $\mathcal{N}_t$ be the σ-algebra generated by $\xi(t)$. We use E_t as an abbreviation for $\mathrm{E}\{\ \mid \mathcal{N}_t\}$. We call $\mathcal{P}_t$ the *past*, $\mathcal{F}_t$ the *future*, and $\mathcal{N}_t$ the *present* at time t. Then ξ is a *Markov process* in case for all t the past and the future are conditionally independent given the present.

The theory of Markov processes is completely independent of a choice of the direction of time. This fact is of central importance in stochastic mechanics. The kinematics of diffusion is utterly symmetric with respect to the two directions of time (though it is not usually developed with this fact in mind), and the conservative dynamics that we shall develop likewise gives no preference to either direction of time.

We use dt as a strictly positive variable. If F is any function defined on R, we define its *forward increment* by

$$d_+F(t) = F(t+dt) - F(t) \tag{1.1}$$

and its *backward increment* by

$$d_-F(t) = F(t) - F(t-dt). \tag{1.2}$$

An L^1 stochastic process ξ indexed by R is a *forward martingale* with respect to a forward filtration $\mathcal{P}$ in case

$$\mathrm{E}\{\xi(t) \mid \mathcal{P}_s\} = \xi(s), \qquad s \le t, \tag{1.3}$$

and a *backward martingale* with respect to a backward filtration $\mathcal{F}$ in case

$$\mathrm{E}\{\xi(t) \mid \mathcal{F}_s\} = \xi(s), \qquad s \ge t. \tag{1.4}$$

An L^1 difference process w is a *forward difference martingale* with respect to a forward filtration $\mathcal{P}$ in case

$$\mathrm{E}\{w(t) - w(s) \mid \mathcal{P}_s\} = 0, \qquad s \le t, \tag{1.5}$$

and a *backward difference martingale* with respect to a backward filtration $\mathcal{F}$ in case

$$\mathrm{E}\{w(t) - w(s) \mid \mathcal{F}_s\} = 0, \qquad s \ge t. \tag{1.6}$$

The Wiener process on R^n with an initial distribution at the initial time t_0 is a Markov process. Restricted to $t \ge t_0$ it is a forward martingale, and restricted to $t \le t_0$ it is a backward martingale.

Heuristically, if a particle moving according to the Wiener process is at x at the time $t \ge t_0$, then to find its position at an infinitesimal time dt later, choose a direction in R^n at random and move a distance $\sqrt{n\,dt}$ from x in that direction, so that each component $dw^i(t)$ of the Wiener increment is of mean 0 and variance dt; then repeat, choosing the directions independently each time. The paths will be very rough (non-differentiable) because of the large increments, but continuous due to cancellations among the different random directions. For times later than t_0, this process has no tendency to head in any particular direction; its forward drift is 0 at times $t \ge t_0$. A similar description holds for times $t \le t_0$ and its position an infinitesimal time dt earlier; the backward drift is 0 at times $t \le t_0$. The heuristic description of a general diffusion process is similar, except that there are time-dependent vector fields $b_\pm(x,t)$, the forward and backward drifts, and the particle is displaced by them in addition to the Wiener increments.

With this as motivation, we say that a (Markovian) *diffusion* on R^n is an R^n-valued Markov process ξ indexed by R for which there exist Borel-measurable functions b_+ (the *forward drift*) and b_- (the *backward drift*) from $\mathrm{R}^n \times \mathrm{R}$ to R^n and Wiener difference processes w_+ and w_- such that for all $t_0 \le t_1$ the increment $w_+(t_1) - w_+(t_0)$ is independent of $\mathcal{P}_{t_0}$ and $w_-(t_1) - w_-(t_0)$ is independent of $\mathcal{F}_{t_1}$, and

$$\xi(t_1) - \xi(t_0) = \int_{t_0}^{t_1} b_+(\xi(t),t)\,dt + w_+(t_1) - w_+(t_0) \tag{1.7}$$

$$= \int_{t_0}^{t_1} b_-(\xi(t),t)\,dt + w_-(t_1) - w_-(t_0). \tag{1.8}$$

The processes w_+ and w_-, being Wiener difference processes, are equivalent, but they are not equal. We can express one process in terms of the other:

$$w_-(t_1) - w_-(t_0) = \int_{t_0}^{t_1} (b_+ - b_-)(\xi(t), t)\,dt + w_+(t_1) - w_+(t_0). \tag{1.9}$$

A diffusion is said to be a *finite energy diffusion* in case for all $t_0 \le t_1$

$$A = \int_{t_0}^{t_1} \mathrm{E}\tfrac{1}{4}(b_+^2 + b_-^2)(\xi(t), t)\,dt < \infty. \tag{1.10}$$

We call A the *kinetic energy integral*. It depends on the process and on the interval $[t_0, t_1]$.

By a *smooth diffusion* I shall mean a finite energy diffusion such that b_+ and b_- are smooth (i.e., C^∞). Let $\rho(x,t)$ be the probability density of the smooth diffusion $\xi(t)$, so that for any measurable set B in $\mathbb{R}^n$ we have

$$\Pr\{\,\xi(t) \in B\,\} = \int_B \rho(x,t)\,dx \tag{1.11}$$

where dx is Lebesgue measure on $\mathbb{R}^n$. The function ρ exists and is strictly positive and smooth on $\mathbb{R}^n \times \mathbb{R}$; this follows from the theory of partial differential equations because ρ is a solution of a parabolic equation with smooth coefficients, the *forward Fokker-Planck equation*

$$\frac{\partial \rho}{\partial t} = -\nabla\cdot(\rho b_+) + \tfrac{1}{2}\Delta\rho. \tag{1.12}$$

It also satisfies the anti-parabolic *backward Fokker-Planck equation*

$$\frac{\partial \rho}{\partial t} = -\nabla\cdot(\rho b_-) - \tfrac{1}{2}\Delta\rho. \tag{1.13}$$

There is a mechanical procedure for deriving backward equations from forward equations. Define the *time-reversed process* $\check{\xi}$ by $\check{\xi}(t) = \xi(-t)$, write the forward equation for $\check{\xi}$, and express the result in terms of ξ to obtain the backward equation for ξ.)

Although the paths of a smooth diffusion are non-differentiable, the forward drift and the backward drift express the *mean forward velocity* and the *mean backward velocity:*

$$\lim_{dt\downarrow 0} \mathrm{E}_t \frac{d_\pm \xi(t)}{dt} = b_\pm(\xi(t), t). \tag{1.14}$$

The *current velocity* v is the average of b_+ and b_- and the *osmotic velocity* u is one half their difference:

$$v = \tfrac{1}{2}(b_+ + b_-), \qquad u = \tfrac{1}{2}(b_+ - b_-). \tag{1.15}$$

The probability density ρ is related to the current velocity by the *current equation* (or equation of continuity)

$$\frac{\partial \rho}{\partial t} = -\nabla\cdot(\rho v), \tag{1.16}$$

and to the osmotic velocity by the *osmotic equation*

$$u = \tfrac{1}{2}\nabla \log \rho. \tag{1.17}$$

The current equation is obtained simply by averaging the forward and backward Fokker-Planck equations. If we take the difference of these two equations, we find that $-\rho u + \frac{1}{2}\nabla\rho$ is divergence-free, but to see that it is actually 0 (i.e., that (1.17) holds) takes a longer argument; see [N85]. We write

$$R = \tfrac{1}{2}\log \rho, \tag{1.18}$$

so that $\rho = e^{2R}$ and $\nabla R = u$.

2. An estimate for finite energy diffusions

Let ξ be a smooth diffusion on $\mathbf{R}^n$, and let f be a smooth function with compact support on $\mathbf{R}^n \times \mathbf{R}$. Then I claim that

$$\begin{aligned} f(\xi(t_1),t_1) - f(\xi(t_0),t_0) &= \int_{t_0}^{t_1} \left(\partial_t + b_+(\xi(t),t)\cdot\nabla + \tfrac{1}{2}\Delta\right) f(\xi(t),t)\,dt + \int_{t_0}^{t_1} \nabla f(\xi(t),t)\cdot d_+w_+(t) \qquad (2.1)\\ &= \int_{t_0}^{t_1} \left(\partial_t + b_-(\xi(t),t)\cdot\nabla - \tfrac{1}{2}\Delta\right) f(\xi(t),t)\,dt + \int_{t_0}^{t_1} \nabla f(\xi(t),t)\cdot d_-w_-(t). \qquad (2.2)\end{aligned}$$

These are the forward and backward *Itô equations.* The second integral in (2.1) is a *forward Itô stochastic integral*, and that in (2.2) is a *backward Itô stochastic integral.* They are defined as limits in L^2 of Riemann sums using forward and backward increments respectively, and will be discussed more in detail in §3. To establish (2.1), expand f in a Taylor polynomial of degree 2 and substitute into $d_+f(\xi(t),t)$, obtaining

$$d_+f = \partial_t f\,dt + \nabla f\cdot d_+\xi + \tfrac{1}{2}\partial_{x^i}\partial_{x^j} f\,d_+\xi^i d_+\xi^j + o(dt), \qquad (2.3)$$

where f is evaluated at $(\xi(t),t)$, the summation convention is used, and $o(dt)$ refers to the L^2 norm. Now $d_+\xi^i d_+\xi^j = d_+w_+^i d_+w_+^j + o(dt)$, and $d_+w_+^i d_+w_+^j$ can be replaced by $\delta^{ij}dt$, since for $i \neq j$ it is a random variable of mean 0 and variance $o(dt)$ (so that although its L^2 norm is not $o(dt)$, since these random variables for different values of t in the Riemann sum are orthogonal their sum is negligible), while for $i = j$ it differs from dt by such a random variable; this gives rise to the term $\frac{1}{2}\Delta$ in (2.1). These equations continue to hold for smooth f without compact support, since the trajectories of ξ are continuous with probability one. The Fokker-Planck equations are consequences of the Itô equations: the expectations of the stochastic integrals are 0; the other expectations can be expressed in terms of ρ; since the resulting equation holds for all test functions f, the ρ is by definition a weak solution of the Fokker-Planck equation.

In terms of v and u, we can write (1.10) as

$$\begin{aligned} A &= \int_{t_0}^{t_1} \mathrm{E}\tfrac{1}{2}(v^2+u^2)(\xi(t),t)\,dt \\ &= \int_{t_0}^{t_1}\int_{\mathbf{R}^n} \tfrac{1}{2}(v^2+u^2)(x,t)\rho(x,t)\,dx\,dt < \infty. \qquad (2.4)\end{aligned}$$

The current velocity v satisfies the current equation (1.16). This equation is still satisfied if we add to v any time-dependent vector field z such that for each t we have $\int z^2\rho\,dx < \infty$ and $\nabla\cdot(\rho z) = 0$ in the sense of distributions. Among all these vector fields $v + z$ there will be a unique one v_0 such that $\int_{\mathbf{R}^n} v_0^2\rho\,dx$ is a minimum. (A closed convex set in a Hilbert space has a unique element of smallest norm.) The condition on this v_0 is that

$$\frac{d}{d\lambda}\int_{\mathbf{R}^n} (v_0+\lambda z)^2\rho\,dx\Big|_{\lambda=0} = 0; \qquad (2.5)$$

i.e., $\int_{\mathbf{R}^n} v_0\cdot z\rho\,dx = 0$. Notice that for a smooth diffusion ρ is strictly positive, since the osmotic velocity $u = \frac{1}{2}\nabla\log\rho$ is smooth. Since v_0 is orthogonal to the general divergence-free vector field $z\rho$, the field v_0 is a gradient. We set

$$\nabla S = v_0; \qquad (2.6)$$

then S is uniquely determined up to a function of t. By definition, S satisfies

$$-\nabla\cdot(\rho\nabla S) = \frac{\partial\rho}{\partial t}. \tag{2.7}$$

If we divide by 2ρ we find

$$-(\tfrac{1}{2}\Delta + u\cdot\nabla)S = \frac{\partial R}{\partial t}, \tag{2.8}$$

so that S is smooth for each t. By construction,

$$\int_{\mathsf{R}^n} (\nabla S)^2 \rho\, dx \leq \int_{\mathsf{R}^n} v^2 \rho\, dx. \tag{2.9}$$

The osmotic velocity is automatically a gradient for each t. We shall be interested primarily, but not exclusively, in *gradient diffusions*, for which v is also a gradient. (This is the same as saying that b_+, or equivalently b_-, is a gradient.) If v is a gradient, then $\nabla S = v$.

Let ξ be a smooth diffusion. Our goal is to estimate

$$\Pr\{\sup_{t_0\leq t\leq t_1} |R(\xi(t),t) - R(\xi(t_0),t_0)| \geq \lambda\} \tag{2.10}$$

purely in terms of λ and the energy integral A. (In [N85a] there is a similar estimate, but it involves more than just A.) I hope that this will be useful in constructing general finite energy diffusions whose drifts are not assumed to be smooth. The difficulty in doing this is that the drifts become exceedingly singular on the *nodes*, i.e., the region in $\mathsf{R}^n \times \mathsf{R}$ where $\rho = 0$ (and $R = -\infty$). But our estimate will show that it is impossible ever to reach the nodes; R cannot become $-\infty$.

For a smooth diffusion ξ, take the average of (2.1) and (2.2) where f is the smooth function R. This gives

$$\begin{aligned} R(\xi(t_1),t_1) - R(\xi(t_0),t_0) &= \int_{t_0}^{t_1} \Big(\partial_t + v(\xi(t),t)\cdot\nabla\Big) R(\xi(t),t)\,dt \\ &+ \tfrac{1}{2}\int_{t_0}^{t_1} \nabla R(\xi(t),t)\cdot d_+w_+(t) + \tfrac{1}{2}\int_{t_0}^{t_1} \nabla R(\xi(t),t)\cdot d_-w_-(t). \end{aligned} \tag{2.11}$$

Take one half the difference of (2.1) and (2.2) where now f is the smooth function S to find

$$\begin{aligned} &-\int_{t_0}^{t_1} \Big(\tfrac{1}{2}\Delta + u(\xi(t),t)\cdot\nabla\Big) S(\xi(t),t)\,dt \\ &= \int_{t_0}^{t_1} \nabla S(\xi(t),t)\cdot d_+w_+(t) - \int_{t_0}^{t_1} \nabla S(\xi(t),t)\cdot d_-w_-(t). \end{aligned} \tag{2.12}$$

But by (2.8), the integrand of the left hand side of this is $\partial_t R$, so we can substitute the right hand side for the integral of $\partial_t R$ in (2.11). Recalling that $\nabla R = u$, we find

$$\begin{aligned} R(\xi(t_1),t_1) - R(\xi(t_0),t_0) &= \int_{t_0}^{t_1} v(\xi(t),t)\cdot u(\xi(t),t)\,dt \\ &+ \tfrac{1}{2}\int_{t_0}^{t_1} (u+\nabla S)(\xi(t),t)\cdot d_+w_+(t) + \tfrac{1}{2}\int_{t_0}^{t_1} (u-\nabla S)(\xi(t),t)\cdot d_-w_-(t). \end{aligned} \tag{2.13}$$

This expresses the difference in R as the sum of three processes: an ordinary integral, a forward martingale, and a backward martingale. Replace t_1 by t, constrained to lie in the interval $[t_0, t_1]$, and let $X_1(t)$ be the ordinary integral, $X_2(t)$ the forward martingale, and $X_3(t)$ the backward martingale. Clearly

$$\sup_t |X_1(t)| \leq \int_{t_0}^{t_1} \left| v(\xi(t),t) \cdot u(\xi(t),t) \right| dt, \tag{2.14}$$

which has expectation $\leq A$. Hence by Chebyshev's inequality,

$$\Pr\{ \sup_t |X_1(t)| \geq \lambda \} \leq \frac{A}{\lambda}. \tag{2.15}$$

By a martingale inequality,

$$\Pr\{ \sup_t |X_2(t)| \geq \lambda \} \leq \frac{1}{\lambda} ||X_2(t_1)||_1 \leq \frac{1}{\lambda} ||X_2(t_1)||_2. \tag{2.16}$$

The estimate for X_3 is similar, but the roles of t_0 and t_1 are interchanged. By the triangle inequality, we have

$$\Pr\{\sup_t |X_3(t)| \geq \lambda\} \leq \frac{2}{\lambda} ||X_3(t_1)||_2. \tag{2.17}$$

But

$$||X_2(t_1)||_2^2 + ||X_3(t_1)||_2^2 = \tfrac{1}{4} \mathrm{E} \int_{t_0}^{t_1} \left((u + \nabla S)^2 + (u - \nabla S)^2 \right) (\xi(t), t) dt \leq A. \tag{2.18}$$

If the event in (2.10) holds, then at least one of $\sup_t |X_\alpha(t)| \geq \lambda/3$, for $\alpha = 1$, 2, or 3, must hold. Consequently,

$$\Pr\{ \sup_t \left| R(\xi(t),t) - R(\xi(t_0),t_0) \right| \geq \lambda \} \leq 6 \frac{A + \sqrt{A}}{\lambda}. \tag{2.19}$$

We have proved the following slight extension of a theorem of Wallstrom [W88].

Theorem 1. *Let ξ be a smooth diffusion with kinetic energy integral A. Then for all $\lambda > 0$ and all $t_0 \leq t_1$ we have* (2.19).

3. Existence of finite energy diffusions

We must say more about how Itô stochastic integrals are defined.

If $t \mapsto \mathcal{B}_t$ is a family of σ-algebras, we say that the stochastic process ξ is *adapted* to the family in case each $\xi(t)$ is $\mathcal{B}_t$ measurable. Although we are interested only in stochastic integrals in which the integrand is adapted to the present, it is necessary to discuss integrands that are adapted to the past to define forward stochastic integrals. Let ξ be a diffusion. The set of all strongly measurable $\mathbf{R}^n$-valued functions η from $\mathbf{R}$ to L^2 of the underlying probability space such that

$$\int \mathrm{E}\eta^2(t)dt < \infty \tag{3.1}$$

and that are adapted to the past is a real Hilbert space $\mathcal{H}$. Let $\mathcal{H}_0$ be the subspace consisting of step functions. It is not difficult to see that $\mathcal{H}_0$ is a dense linear subspace of $\mathcal{H}$. (Partition a large compact interval with a small mesh and let η_0 be the step function that on each subinterval is the average of η over the previous subinterval.) For η in $\mathcal{H}_0$, define the forward stochastic integral $\int \eta(t) \cdot d_+ w_+(t)$ to be the obvious finite sum. This is a linear mapping from $\mathcal{H}_0$ to L^2 and a simple computation shows that it is isometric. Therefore it extends uniquely to an isometric linear map from $\mathcal{H}$ into L^2; this defines the forward stochastic integral when η is adapted to the past and satisfies (3.1). For η in $\mathcal{H}_0$ and $t_0 \leq t_1$ we have

$$\mathrm{E}_{t_0} \left(\int_{t_0}^{t_1} \eta(t) \cdot d_+ w_+(t) \right)^2 = \mathrm{E}_{t_0} \int_{t_0}^{t_1} \eta^2(t) dt, \tag{3.2}$$

so this continues to hold for all η in $\mathcal{H}$.

Suppose that η is adapted to the past and satisfies the weaker condition that

$$\int \eta^2(t) dt < \infty \tag{3.3}$$

with probability one. Let $\chi_m(t)$ be the indicator function of the event that $\int_{-\infty}^{t} \eta^2(s) ds \leq m$. Then $\chi_m \eta$ is in $\mathcal{H}$, so its forward stochastic integral is well defined. But with probability one, $\chi_m(t)\eta(t)$ is equal to $\eta(t)$ for all t for m sufficiently large, so we define its forward stochastic integral to be the common value for large m. In fact, the same argument defines the forward stochastic integral for any η adapted to the past on the set where (3.3) holds.

For η in $\mathcal{H}_0$ define ς by

$$\varsigma = \exp \left[\int \eta(t) \cdot d_+ w_+(t) - \tfrac{1}{2} \int \eta^2(t) dt \right]. \tag{3.4}$$

Then ς is a positive random variable, and $\mathrm{E}\varsigma = 1$. To see this, use the elementary fact that for a Gaussian random variable W of mean 0 we have

$$\mathrm{E} \exp \left[W - \tfrac{1}{2} \mathrm{E} W^2 \right] = 1, \tag{3.5}$$

write ς as a product, and take conditional expectations E_{t_i} where t_i is the left endpoint of an interval for the step function, working from right to left, to conclude the desired result. Any η in $\mathcal{H}$ is a limit of a sequence of elements of $\mathcal{H}_0$ and the corresponding stochastic integrals converge in L^2; by picking a subsequence, we can ensure that they and hence also the corresponding ς's converge with probability one. By Fatou's lemma, therefore, $\mathrm{E}\varsigma \leq 1$ for any η in $\mathcal{H}$. For ς adapted to the past and satisfying (3.3) with probability one, an entirely similar argument shows that $\mathrm{E}\varsigma \leq 1$. For a general η adapted to the past, we define ς to be 0 on the set where $\int \eta^2(t) dt = \infty$, so that we always have $\mathrm{E}\varsigma \leq 1$.

Let $\mathcal{M}_0$ be the set of all triples $D = \langle \rho, v, u \rangle$ of smooth functions mapping $\mathbf{R}^n \times \mathbf{R}$ into $\mathbf{R}$, $\mathbf{R}^n$, and $\mathbf{R}^n$, respectively, such that for each t the function $\rho(t) = \rho(\cdot, t)$ is a probability density, the current and osmotic equations are satisfied, and the kinetic energy integral is finite on every interval $[t_0, t_1]$.

There is a very simple construction of the diffusion associated to a triple D in $\mathcal{M}_0$, by means of a *Girsanov transformation* using the density ς. Let Ω be the set of all continuous functions from $\mathbf{R}$ to $\mathbf{R}^n$ and let $\mathcal{B}$ be the σ-algebra generated by the $\xi(t)$ given by $\xi(t)(\omega) = \omega(t)$ for t in $\mathbf{R}$, and similarly for $\mathcal{B}_{t_0,t_1}$ with $t \in [t_0, t_1]$. Let Pr_0 be the probability measure on $\langle \Omega, \mathcal{B} \rangle$ for the Wiener process with initial probability density $\rho(t_0)$ at the initial time t_0, and let $t_0 \leq t_1$. Let

$$\varsigma = \exp \left[\int_{t_0}^{t_1} b_+(\xi(t), t) \cdot d_+ \xi(t) - \tfrac{1}{2} \int_{t_0}^{t_1} b_+^2(\xi(t), t) dt \right], \tag{3.6}$$

where $b_+ = v + u$. Notice that this is a particular case of (3.4); here the diffusion ξ is the Wiener process, so for times after t_0 we have $d_+\xi = d_+w_+$. Let E_0 denote the expectation with respect to Pr_0. Then it can be shown that $\mathrm{E}_0\varsigma = 1$. Let Pr_{t_0,t_1} be the probability measure $\varsigma\,\mathrm{Pr}_0$. Then there is a unique probability measure Pr on $\langle \Omega, \mathcal{B}\rangle$ that agrees with each Pr_{t_0,t_1} on $\mathcal{B}_{t_0,t_1}$, and the process $\xi(t)$ is the diffusion associated to D. This is *Girsanov's formula.*

The heuristic reason for this is that

$$\mathrm{E}_t d_+\xi(t) = \mathrm{E}_t^0 \exp\left[b_+\big(\xi(t),t\big)\cdot d_+\xi(t) - \tfrac{1}{2}b_+^2\big(\xi(t),t\big)dt\right] d_+\xi(t) \tag{3.7}$$

where E_t^0 is the conditional expectation with respect to the present for the measure Pr_0. When the exponential is expanded to order dt we obtain

$$\begin{aligned} &1 + b_+\big(\xi(t),t\big)\cdot d_+\xi(t) + \tfrac{1}{2}\left[b_+\big(\xi(t),t\big)\cdot d_+\xi(t)\right]^2 - \tfrac{1}{2}b_+^2\big(\xi(t),t\big)dt = \\ &1 + b_+\big(\xi(t),t\big)\cdot d_+\xi(t) + o(dt), \end{aligned} \tag{3.8}$$

so that $\mathrm{E}_t d_+\xi(t) = b_+\big(\xi(t),t\big)dt + o(dt)$ as desired.

Let $\mathcal{M}$ be the set of all triples $D = \langle \rho, v, u\rangle$ of Borel measurable functions mapping $\mathbb{R}^n \times \mathbb{R}$ into $\mathbb{R}$, $\mathbb{R}^n$, and $\mathbb{R}^n$, respectively, such that for all t the function $\rho(t) = \rho(\cdot,t)$ is a probability density and the current and osmotic equations are satisfied weakly, and the kinetic energy integral is finite on every interval $[t_0,t_1]$. Carlen [C84] constructed the diffusion associated to the general element of $\mathcal{M}$. His proof was by means of a partial differential equations approach, exploiting through intricate estimates the maximum principle and energy integral estimates. I conjecture that the Girsanov formula continues to be valid for all D in $\mathcal{M}$, in the following sense: There is a unique probability measure Pr on Ω such that for all $t_0 \le t_1$ it agrees on $\mathcal{B}_{t_0,t_1}$ with $\varsigma\,\mathrm{Pr}_0$ (where ς is put equal to 0 on the set where $\int_{t_0}^{t_1} b_+^2\big(\xi(t),t\big)dt = \infty$) such that with respect to it ξ is a finite energy diffusion with probability density ρ, current velocity v, and osmotic velocity u. This would strengthen Carlen's result. Notice that according to the conjecture, for each interval $[t_0,t_1]$ the diffusion is absolutely continuous with respect to the Wiener process, so that every property known to hold almost surely for the Wiener process on a finite time interval would hold almost surely for the general finite energy diffusion. The converse is false, since we can have $\varsigma = 0$ on a set of large Wiener measure.

Theorem 1 should play a role in the proof of this conjecture. But the situation is perhaps not as simple as it appears at first sight. Theorem 1 was established for diffusions corresponding to D in $\mathcal{M}_0$, but is $\mathcal{M}_0$ dense in any suitable sense in $\mathcal{M}$? Guerra [G85] introduced a metric on $\mathcal{M}$, which makes this question precise.

4. Action

For a particle of unit mass traveling along a smooth curve $t \mapsto \xi(t)$ during the time interval $[t_0,t_1]$, its kinetic action is

$$\int_{t_0}^{t_1} \frac{1}{2}\left(\frac{d\xi}{dt}\right)^2 dt. \tag{4.1}$$

For the trajectories of a diffusion, this makes no sense. Nevertheless, we can calculate $\big(d_+\xi(t)\big)^2$ to order dt^2 and see what it looks like.

Recall that $\big(d_+\xi(t)\big)^2$ is $\big(\xi(t_1) - \xi(t_0)\big)^2$ where $t_0 = t$ and $t_1 = t + dt$, and recall (1.7):

$$\xi(t_1) - \xi(t_0) = \int_{t_0}^{t_1} b_+\big(\xi(t),t\big)dt + w_+(t_1) - w_+(t_1).$$

We can estimate the integral by $(t_1 - t_0)b_+(\xi(t_0), t_0)$, but this does not give us the desired accuracy. To improve the accuracy, apply (1.7) itself to $\xi(t)$ in the integrand to obtain

$$\begin{aligned}\xi(t_1) &- \xi(t_0)\\ &= \int_{t_0}^{t_1} b_+\Big(\xi(t_0) + \int_{t_0}^{t} b_+(\xi(s), s)\,ds + w_+(t) - w_+(t_0), t\Big)dt + w_+(t_1) - w_+(t_0). \quad (4.2)\end{aligned}$$

Now take a Taylor expansion to first order for b_+ at $(\xi(t_0), t_0)$ and let

$$W^k = \int_{t_0}^{t_1} (w_+^k(t) - w_+^k(t_0))\,dt. \tag{4.3}$$

Then

$$\begin{aligned}\xi(t_1) &- \xi(t_0)\\ &= b_+(\xi(t_0), t_0)(t_1 - t_0) + \frac{\partial b_+}{\partial x^k}(\xi(t_0), t_0)W^k + w_+(t_1) - w_+(t_0) + o((t_1 - t_0)^{3/2}). \quad (4.4)\end{aligned}$$

Take the inner product of this with itself and revert to the t, $t + dt$ notation. We find

$$\tfrac{1}{2}\left(\frac{d_+\xi}{dt}\right)^2 = \tfrac{1}{2}b_+^2 + \frac{b_+ \cdot d_+w_+}{dt} + \frac{\partial b_+}{\partial x^k}\frac{W^k \cdot d_+w_+}{dt} + \tfrac{1}{2}\left(\frac{d_+w_+}{dt}\right)^2 + o(1), \tag{4.5}$$

where everything is evaluated at $(\xi(t), t)$. Now take the conditional expectation with respect to the present. Notice that $\mathrm{E}_t b_+ \cdot d_+w_+ = 0$ and that $\mathrm{E}_t(d_+w_+)^2 = n\,dt$. Also,

$$\mathrm{E}_t \frac{\partial b_+}{\partial x^k} W^k \cdot d_+w_+ = \tfrac{1}{2}\nabla \cdot b_+ dt. \tag{4.6}$$

Therefore

$$\mathrm{E}_t \tfrac{1}{2}\left(\frac{d_+\xi}{dt}\right)^2 = \tfrac{1}{2}b_+^2 + \tfrac{1}{2}\nabla \cdot b_+ + \tfrac{1}{2}\frac{n}{dt} + o(1). \tag{4.7}$$

This contains the singular term $n/2dt$, but this term is a constant that is the same for all diffusions. When we study action, we are leaving kinematics and entering into dynamics, where action has played a fundamental role that has survived the revolutions of twentieth century physics. But action enters into *variational principles* that are not affected by an additive constant. We call

$$\mathrm{E}\int_{t_0}^{t_1} \Big(\tfrac{1}{2}b_+^2(\xi(t), t) + \tfrac{1}{2}\nabla \cdot b_+(\xi(t), t)\Big)dt \tag{4.8}$$

the *expected kinetic action*. If we express b_+ in terms of u and v, the expectation in terms of ρ integration, and integrate by parts, we find that the expected kinetic action is

$$\int_{t_0}^{t_1}\int_{\mathbf{R}^n} \tfrac{1}{2}(v^2 - u^2)(x, t)\rho(x, t)\,dx\,dt. \tag{4.9}$$

It differs from the kinetic energy integral (2.4) by having a minus sign instead of a plus sign. The derivation was for a smooth diffusion, but the result is meaningful for any finite energy diffusion.

A *potential* on $\mathbf{R}^n$ is a pair of Borel measurable functions $\phi: \mathbf{R}^n \to \mathbf{R}$ (the *scalar potential*) and $A: \mathbf{R}^n \to \mathbf{R}^n$ (the *vector potential*). Then the *expected potential action* is

$$\mathrm{E}\left[\int_{t_0}^{t_1} \phi(\xi(t),t)\,dt - \int_{t_0}^{t_1} A(\xi(t),t)\cdot\tfrac{1}{2}\big(d_+\xi(t) + d_-\xi(t)\big)\right]. \tag{4.10}$$

The minus sign is a matter of convention, but the use of

$$\tfrac{1}{2}\big(d_+\xi(t) + d_-\xi(t)\big) = v(\xi(t),t)\,dt + \tfrac{1}{2}\big(d_+w_+(t) + d_-w_-(t)\big) \tag{4.11}$$

is necessary to avoid asymmetry in the two directions of time. Of course, the expected potential action only exists if the expectation exists, in which case it can be written as

$$\int_{t_0}^{t_1}\int_{\mathbf{R}^n} [\phi(x,t) - A(x,t)\cdot v(x,t)]\,\rho(x,t)\,dx\,dt. \tag{4.12}$$

The *expected action* is the difference of the expected kinetic action and the expected potential action. Let $\chi: \mathbf{R}^n \times \mathbf{R} \to \mathbf{R}$ be smooth and of compact support (for simplicity of exposition). Then the transformation

$$\begin{cases} \phi \to \phi + \partial_t \chi \\ A \to A - \nabla\chi \end{cases} \tag{4.13}$$

is called a *gauge transformation*. The expected action is the same after a gauge transformation; this follows from the current equation after integration by parts.

I shall close this chapter with some brief comments on the extension of the theory of the kinematics of diffusion to Riemannian manifolds. The first observation to make is that this is not done simply for the sake of complicating the theory; the Riemannian metric is an intrinsic part of the probabilistic structure. Consider a diffusion ξ on a differentiable manifold M, and in local coordinates at a point x at which the diffusion starts, define σ^{ij} by

$$\sigma^{ij}\,dt = \mathrm{E}\,d\xi^i(t)\,d\xi^j(t) + o(dt).$$

Then the inverse matrix σ_{ij} is a Riemannian metric.

The notion of the Wiener difference process does not generalize to the context of a Riemannian manifold except in terms of its differential, but this is enough to enable us to define intrinsically the notions of mean forward and backward velocities. The current and osmotic equations hold in the more general context.

Dankel [D70] was the first to develop stochastic mechanics on a Riemannian manifold, with applications to spin. One problem was that to differentiate tensor fields along diffusion trajectories, one needs a notion of stochastic parallel transport. Itô's notion developed in [I62] was unsuitable, and a notion more adapted to the needs of stochastic mechanics was developed by Dohrn and Guerra [DG78, 79]. The Ricci curvature, but not the full uncontracted Riemannian curvature, plays a role in the Dohrn-Guerra notion of parallel transport.

When the expected kinetic action is computed, there is a term involving the scalar curvature, the Pauli-DeWitt term familiar to the physicists; see [D57]. All of these questions are discussed in [N85].

II. Conservative Dynamics of Diffusion

5. The variational principle

We seek a dynamics for diffusions that shall be analogous to conservative deterministic dynamics. Therefore we shall base it on a variational principle. Since we are seeking a dynamical law, in the beginning we make as many simplifying assumptions of smoothness as possible. Let the potential ϕ, A be smooth with compact support, and let $D = \langle \rho, v, u \rangle$ be in $\mathcal{M}_0$. What shall it mean for the diffusion D to be critical for the potential?

As we have seen, the expected action is

$$I = \int_{t_0}^{t_1} \int_{\mathbf{R}^n} (\tfrac{1}{2}v^2 - \tfrac{1}{2}u^2 - \phi + A \cdot v)\rho \, dx \, dt. \tag{5.1}$$

We shall say that D is critical in case this expected action is stationary, for every interval $[t_0, t_1]$, for variations in which $\rho(t_0)$ and $\rho(t_1)$ are held fixed. This is a direct analogue of Hamilton's principle of least action in deterministic mechanics.

More precisely, suppose that for all α in some neighborhood of 0 in $\mathbf{R}$ we have a $D(\alpha)$ in $\mathcal{M}_0$ such that $\rho(\alpha, x, t)$, $v(\alpha, x, t)$, and $u(\alpha, x, t)$ are smooth, that $D(0) = D$, and that $\rho(\alpha, t_0) = \rho(t_0)$ and $\rho(\alpha, t_1) = \rho(t_1)$ for all α. This is called a *variation* of D (for the interval $[t_0, t_1]$). We say that D is *critical* for the potential in case for all intervals $[t_0, t_1]$ and all variations of D,

$$\frac{d}{d\alpha} I(\alpha) \bigg|_{\alpha=0} = 0. \tag{5.2}$$

To distinguish this notion from other related notions, we also say that D is *critical in the sense of Lafferty*; see [L87] for this formulation and the result about to be derived.

Let D be critical and let $[t_0, t_1]$ be fixed. One possible type of variation is one in which ρ, and consequently also u, is fixed for all t in $[t_0, t_1]$, and only v varies with α, with $v(\alpha) = v + \alpha z$ where $\nabla \cdot (z\rho) = 0$ for each t (so that the current equation is satisfied for each α). Then

$$I'(0) = \int_{t_0}^{t_1} \int_{\mathbf{R}^n} (v + A) \cdot z\rho \, dx \, dt. \tag{5.3}$$

This must vanish for all choices of z with $\nabla \cdot (z\rho) = 0$, so a necessary condition for D to be critical is that for each t the vector field $v + A$ be a gradient vector field; i.e., that there exist a function S, uniquely defined up to a function of t, such that

$$v + A = \nabla S. \tag{5.4}$$

This is called the *stochastic Hamilton-Jacobi condition* because of its role in the generalization by Guerra and Morato of Hamilton-Jacobi theory to conservative diffusions; see [GM83].

Now we compute $I'(\alpha)$. In the following computation ρ, v, and u all depend on α, and for each α the current and osmotic equations hold. The idea of the computation is to bring all α-derivatives to ρ. Now $I'(\alpha)$ is the sum of three integrals:

$$\int_{t_0}^{t_1} \int_{\mathbf{R}^n} (v + A) \cdot \partial_\alpha (v\rho) \, dx \, dt, \tag{5.5}$$

$$\int_{t_0}^{t_1} \int_{\mathbf{R}^n} (-u) \cdot \partial_\alpha (u\rho) \, dx \, dt, \tag{5.6}$$

$$\int_{t_0}^{t_1} \int_{\mathbf{R}^n} (-\tfrac{1}{2}v^2 + \tfrac{1}{2}u^2 - \phi) \partial_\alpha \rho \, dx \, dt, \tag{5.7}$$

where we chose to include ρ with v and u in differentiating them and then to compensate by subtracting in the third integral.

The first integral at $\alpha = 0$ is

$$\int_{t_0}^{t_1}\int_{\mathbf{R}^n} \nabla S\cdot\partial_\alpha(v\rho)dx\,dt = -\int_{t_0}^{t_1}\int_{\mathbf{R}^n} S\partial_\alpha\nabla\cdot(v\rho)dx\,dt$$

$$= \int_{t_0}^{t_1}\int_{\mathbf{R}^n} S\partial_\alpha\partial_t\rho\,dx\,dt = -\int_{t_0}^{t_1}\int_{\mathbf{R}^n} \partial_t S\partial_\alpha\rho\,dx\,dt \tag{5.8}$$

by the stochastic Hamilton-Jacobi condition, a spatial integration by parts, the current equation, and a temporal integration by parts. This last is justified by the fact that $\partial_\alpha\rho = 0$ at t_0 and t_1. Using the osmotic equation in the form $u = \nabla\rho/2\rho$, we see that the second integral is

$$\int_{t_0}^{t_1}\int_{\mathbf{R}^n} \tfrac{1}{2}\nabla\cdot u\partial_\alpha\rho\,dx\,dt \tag{5.9}$$

by a spatial integration by parts. Therefore at $\alpha = 0$ we have

$$I'(\alpha) = \int_{t_0}^{t_1}\int_{\mathbf{R}^n} \left[-\partial_t S + \tfrac{1}{2}\nabla\cdot u - \tfrac{1}{2}v^2 + \tfrac{1}{2}u^2 - \phi\right]\partial_\alpha\rho\,dx\,dt. \tag{5.10}$$

The only constraints on $\partial_\alpha\rho$ at $\alpha = 0$ are that its spatial integral be 0 (since each $\rho(\alpha)$ is a probability density for all t) and that it vanish at t_0 and t_1 (by the definition of a variation). Consequently, $I'(0) = 0$ if and only if the term in brackets is a function of t alone. We choose the additive function of t in S so that the term in brackets is 0; now S is uniquely determined up to a constant. Expressing u as ∇R and rearranging the terms in brackets, we find that D is critical if and only if

$$\partial_t S + \tfrac{1}{2}(\nabla S - A)^2 + \phi - \tfrac{1}{2}(\nabla R)^2 - \tfrac{1}{2}\Delta R = 0. \tag{5.11}$$

This is known as the *stochastic Hamilton-Jacobi equation.*

The triple $D = \langle\rho, v, u\rangle$ is determined by R and S (when the vector potential A is known and $v + A = \nabla S$). The current equation is

$$\partial_t R + \nabla R\cdot(\nabla S - A) + \tfrac{1}{2}\Delta S - \tfrac{1}{2}\nabla\cdot A = 0. \tag{5.12}$$

Together, (5.11) and (5.12) constitute necessary and sufficient conditions for a diffusion to be critical for a potential. They are a coupled system of nonlinear partial differential equations. But make the change of dependent variables

$$\psi = e^{R+iS}. \tag{5.13}$$

Then they are equivalent to the linear equation

$$\frac{\partial\psi}{\partial t} = -i\left[\tfrac{1}{2}(-i\nabla - A)^2 + \phi\right]\psi. \tag{5.14}$$

This is the *Schrödinger equation.* When the vector potential A is zero, it is customary to write V instead of ϕ for the scalar potential, so that the Schrödinger equation takes the form

$$\frac{\partial\psi}{\partial t} = -i[-\tfrac{1}{2}\Delta + V]\psi. \tag{5.15}$$

6. Stochastic mechanics

What is conserved by conservative diffusions? First it is necessary to say that the adjective "conservative" does not apply to the diffusion process itself—a diffusion is just a diffusion—but to the dynamics, i.e., the rules for associating a class of diffusions to a potential. The same comment applies to the assertion that stochastic mechanics is time-symmetric: it is not the diffusions themselves that are time-symmetric (whatever that might mean), but the dynamics.

The *current energy* is $\frac{1}{2}v^2$, the *osmotic energy* is $\frac{1}{2}u^2$, the *potential energy* is ϕ, and the *stochastic energy* E is their sum. Thus $t \mapsto E(\xi(t),t)$ is a stochastic process, and it certainly is not a constant. But if the potential ϕ, A is constant in time, a straightforward computation shows that the expected value $\mathrm{E}\,E(\xi(t),t)$ is constant in time.

From now on let us suppose that the vector potential A is zero, so that $\nabla S = v$ and we have a gradient diffusion. Then if one takes the gradient of the stochastic Hamilton-Jacobi equation (5.11) one finds

$$-\nabla V = (\partial_t + v\cdot\nabla)v - (\tfrac{1}{2}\Delta + u\cdot\nabla)u. \tag{6.1}$$

The left hand side is the familiar expression for the force in conservative dynamics. For any stochastic process η we define

$$D_{\pm}\eta(t) = \lim_{dt\downarrow 0} \mathrm{E}_t \frac{d_{\pm}\eta(t)}{dt} \tag{6.2}$$

when the limits exist. Thus $D_{\pm}\xi(t) = b_{\pm}(\xi(t),t)$. A simple computation shows that the right hand side of (6.1), evaluated at $(\xi(t),t)$, is

$$\tfrac{1}{2}(D_+D_- + D_-D_+)\,\xi(t), \tag{6.3}$$

which we call the *stochastic acceleration*. Hence (6.1) is called the *stochastic Newton equation*, because it is the analogue of the Newton equation $F = ma$. (What happened to the m? It is hidden away in the inner product $\cdot$ to simplify the formulas, and we can do this even when $\mathbf{R}^n$ is the configuration space of a system of particles with different masses. One other brief remark on physical dimensions: the inner product is also used to set the scale of the local fluctuations of our diffusions, via $\mathrm{E}(d_+w_+)^2 = dt$. But to make this dimensionally correct one must introduce a factor $\hbar$ with the dimensions of action. Throughout this account we have chosen units so that $\hbar = 1$.)

Here is an example of a critical diffusion, which I call the *one-slit process*. In this example $V = 0$, so the motion is said to be *free*. Let

$$v = \frac{t}{1+t^2}x, \tag{6.4}$$

$$u = \frac{-1}{1+t^2}x. \tag{6.5}$$

Then one can verify the stochastic Newton equation and the current equation, or equivalently one can verify that $\psi_0 = e^{R+iS}$ satisfies the free Schrödinger equation. (As always, R is determined by $\nabla R = u$ and the requirement that $\rho = e^{2R} = |\psi|^2$ be a probability density. The function S is determined up to an additive constant, which in this example is the constant value of S at $t = 0$.) The random variable $\xi(0)$ is Gaussian with mean 0 and variance $\frac{1}{2}$ times the identity matrix. The current velocity v is 0 at time 0; the particle is just resting near the origin. The random variable $\xi(t)$ is also Gaussian with mean 0, but its variance has grown to $\frac{1}{2}\sqrt{1+t^2}$ (whether t is positive or negative). The expected current and osmotic energies are

$$\mathbf{E}\tfrac{1}{2}v^2(\xi(t),t) = \frac{t^2}{4(1+t^2)}, \tag{6.6}$$

$$\mathbf{E}\tfrac{1}{2}u^2(\xi(t),t) = \frac{1}{4(1+t^2)}. \tag{6.7}$$

Their sum, the expected stochastic energy, is constantly $\frac{1}{4}$. Initially this is all osmotic energy, but as $t \to \pm\infty$ it changes into current energy.

So far in this example we have discussed only random variables at a single time. What about the stochastic process ξ itself? Let

$$f(t) = \frac{t-1}{1+t^2}, \tag{6.8}$$

so that $b_+(\xi(t),t) = (v+u)(\xi(t),t) = f(t)\xi(t)$. Then ξ satisfies the stochastic differential equation

$$d_+\xi(t) = f(t)\xi(t)dt + d_+w_+(t), \tag{6.9}$$

which is simply the differential form of (1.7). This is an inhomogeneous linear equation, whose solution is

$$\begin{aligned} \xi(t) &= e^{\int_0^t f(s)ds}\xi(0) + \int_0^t e^{\int_s^t f(r)dr} d_+w_+(s) \\ &= \sqrt{1+t^2}e^{-\arctan t}\left[\xi(0) + \int_0^t \frac{e^{\arctan s}}{\sqrt{1+s^2}} d_+w_+(s)\right]. \end{aligned} \tag{6.10}$$

Notice that $f(t) \sim 1/t$ as $t \to \infty$. Consequently the forward drift $b_+(\xi(t),t) = f(t)\xi(t)$, evaluated at the position of the particle, converges in L^2 and with probability one to

$$\pi_+ = e^{-\pi/2}\left[\xi(0) + \int_0^\infty \frac{e^{\arctan s}}{\sqrt{1+s^2}} d_+w_+(s)\right]. \tag{6.11}$$

Asymptotically as $t \to \infty$, the particle travels with a constant drift (constant in time, but random). The asymptotic motion is that of a particle traveling in a straight line with constant velocity plus a Wiener process.

How does the process, which is Markovian and knows only its present position, remember to maintain the constant asymptotic average velocity? At time 0 the particle does not know what asymptotic regime it will eventually enter, but once it is there the drift, which is a function of position and time, happens to be just right to maintain the particle in the same asymptotic regime.

Now let us look at a more interesting free process, the *two-slit process*. Initially the particle is to be at rest ($v = 0$ for $t = 0$) and localized with equal probabilities near two points $\pm a$. We could take as the initial probability density

$$\tfrac{1}{2}\left(|\psi_0(x-a,0)|^2 + |\psi_0(x+a,0)|^2\right) \tag{6.12}$$

where ψ_0 is the wave function for the one-slit process. We assume that $|a|$ is several multiples of the standard deviation $1/\sqrt{2}$ of the Gaussian at time 0 for the one-slit process; then there is practically no overlap between the two probability densities at a and $-a$. But let us be a bit more clever: let $\psi_1(x,t) = \psi_0(x-a,t)$ and $\psi_2(x,t) = \psi_0(x+a,t)$. Then these describe the one-slit process but shifted by a and $-a$ respectively. Let $\psi = \gamma(\psi_1 + \psi_2)$, where γ is a normalization constant, very close to $1/\sqrt{2}$, to make ψ a unit vector in L^2. Then $|\psi(x,0)|^2$ is, as desired, a probability density concentrated equally near a and $-a$, and it is practically the same as (6.12).

But now we can exploit the fact that the Schrödinger equation is linear, so that if ψ_1 and ψ_2 are solutions, describing critical diffusions for a given potential, so is their sum $\psi_1 + \psi_2$ (multiplied by a normalization constant γ to make it a unit vector). We can ask for the current and osmotic velocities of the new process in terms of the old, and we find

$$v = \tfrac{1}{2}(v_1 + v_2) + \tfrac{1}{2}\frac{(v_1 - v_2)\sinh(R_1 - R_2) + (u_1 - u_2)\sin(S_1 - S_2)}{\cosh(R_1 - R_2) + \cos(S_1 - S_2)}, \tag{6.13}$$

$$u = \tfrac{1}{2}(u_1 + u_2) + \tfrac{1}{2}\frac{(u_1 - u_2)\sinh(R_1 - R_2) - (v_1 - v_2)\sin(S_1 - S_2)}{\cosh(R_1 - R_2) + \cos(S_1 - S_2)}. \tag{6.14}$$

Let us see how the particle in the two-slit process moves according to stochastic mechanics. For its forward drift we find

$$b_+(x,t) = \frac{1}{1+t^2}\left[(t-1)x + \frac{(1-t)\sinh\frac{a\cdot x}{1+t^2} - (1+t)\sin\frac{2ta\cdot x}{1+t^2}}{\cosh\frac{a\cdot x}{1+t^2} + \cos\frac{2ta\cdot x}{1+t^2}}\,a\right]. \tag{6.15}$$

Only the direction of a is interesting, so we take the dimension n to be 1. The denominator never vanishes; there are no nodes. But when t is reasonably large compared to a, it comes close to vanishing on the hyperbolas

$$\frac{2tax}{1+t^2} = (2m+1)\pi, \qquad m \in \mathbf{Z}, \tag{6.16}$$

where the cos is -1 and the cosh is close to 1. But in this region the hyperbolas are practically equal to their asymptotes, the straight lines $2ax = (2m+1)\pi t$. For very small times t, the drift is practically $(t-1)x/(1+t^2)$, because of the large term $\cosh(ax/(1+t^2))$. (Remember that the particle starts near $x = \pm a$.) But for larger times t, there is an enormous drift repelling the particles from the asymptotes; the particle finds itself trapped in one of the channels between these straight lines and its probability density has alternate peaks and troughs resembling those produced by interference phenomena in wave motion.

Let us discuss one other example of a critical diffusion. In this example $n = 1$ and the potential is $V(x) = \omega^2 x^2$, where ω is a constant. This is the *harmonic oscillator*. If we take $v = 0$ and $u = -\omega x$ we obtain a stationary Gaussian process of mean 0 and covariance

$$\mathrm{E}\xi(t)\xi(s) = \frac{1}{2\omega}e^{-\omega|t-s|}, \tag{6.17}$$

called the *ground state process* for the harmonic oscillator. For the harmonic oscillator, the stochastic Newton equation

$$\tfrac{1}{2}(D_+D_- + D_-D_+)\xi(t) = -\omega^2\xi(t) \tag{6.18}$$

is linear, so if we add to the ground state process a solution $\mu(t) = c_1\cos\omega t + c_2\sin\omega t$ of the deterministic harmonic oscillator equation

$$\frac{d^2}{dt^2}\mu(t) = -\omega^2\mu(t) \tag{6.19}$$

we again get a solution, called a *coherent state process*; see [GL81].

Since its beginning 35 years ago in the paper [F52] of Fényes, stochastic mechanics has been applied to a number of topics in quantum physics. There are references in [N85] and the recent book [BCZ87]. Here I shall just mention a few topics. Shucker [S80] showed that the asymptotic motion of a general free conservative diffusion is motion with a constant random velocity (whose probability density is given by the square modulus of the Fourier transform of the wave function) with the Wiener process superimposed, and this was extended to potential scattering by Carlen [C85, 86]. Stochastic mechanics allows one to investigate certain questions that cannot even be formulated in quantum mechanics; for example, one can ask for the probability law of the first time that a particle doing a critical diffusion enters a certain region. For results on this, and a discussion of possible physical tests, see [CT86] and the references cited there. Stochastic mechanics has also been applied to some macroscopic problems; see [DT87] and references in some of the articles appearing there.

7. Stochastic mechanics and nature

Recently I had lunch with a young mathematician who is familiar with stochastic mechanics but not working in the field. I told him about some of the exciting recent developments and he asked me what is the next step in stochastic mechanics. Without hesitation I said that the next step is to throw it away and start over. Let me try to explain why I feel that.

The predictions of stochastic mechanics are thoroughly confirmed by experiment. The outcome of any experiment can be described in terms of the positions of various objects (meters, pointers, marks on photographic plates, etc.) at a fixed time. According to stochastic mechanics, if it is assumed that all of the objects *including the measuring devices* perform a critical diffusion, then the probability density for finding the objects at certain positions x at times t is given by $|\psi(x,t)|^2$. Thus the only aspect of stochastic mechanics that is tested by experiment is the Schrödinger equation itself, and not the fascinating mathematical results on the behavior of the random trajectories.

Now the Schrödinger equation was discovered long before the advent of stochastic mechanics. There have been many attempts to understand its predictions, bringing in such concepts as complementarity, reduction of the wave packet by the consciousness of the observer, many worlds, etc. To these I much prefer the viewpoint so brilliantly expounded in [F85], in which Feynman says, "... we are not going to deal with *why* Nature behaves in the peculiar way that She does; there are no good theories to explain that."

The last phrase is a statement of fact. Nevertheless, one wants to understand why. Stochastic mechanics, as I see it, was an attempt to construct a naïvely realistic model of nature. To decide whether it was successful, I propose the following test. Suppose that we have measuring devices that are not themselves subject to the random fluctuations of stochastic mechanics, and that permit us to observe the random trajectories described by the theory. Is this consistent with what we know about nature?

This is a very stringent test. It goes far beyond showing that the predictions of stochastic mechanics cannot be falsified by experiment; as already remarked, that is not in doubt so long as we have only measuring devices that are themselves subject to the Schrödinger equation. There are two reasons for proposing this test. The first is one of intellectual coherence. It appears pointless to posit certain features of a physical theory and when they appear paradoxical to say that there is no problem because they are unobservable. The second reason is one of physics. So long as we do not understand why Nature behaves in the peculiar way that She does, we have no idea of the limitations of our theory. Quantum theorists are prone to insist that their rules are of universal validity, but that is challenged by many physicists who work on gravitation. I will not repeat here the analogy suggested by Kappler's experiment;

see pp. 117–118 of [N85] and also [BS34]. The point is that perhaps the randomness observed in phenomena on the microscopic scale has a physical cause, and there may be systems that are not subject to it. Alternatively, one may remark that stochastic mechanics is a dynamical theory of diffusion in a frictionless medium. When it is compared with quantum theory, the frictionless medium in question is the vacuum. But liquid helium and superconductors are also frictionless media; are there diffusion phenomena in these domains that can be observed without significantly disturbing the system under observation, and if so could we expect stochastic mechanics to describe them?

Stochastic mechanics fails the proposed test. I refer to the section entitled "The case against stochastic mechanics" of [N86].

III. Random Fields

8. The free Euclidean field

A field is a function defined on R^d, so we would expect a random field to be a stochastic process indexed by R^d. But it turns out that many interesting examples are too singular to be well-defined at points, so we define a *random field* to be a stochastic process ϕ indexed by the *test functions* on R^d, i.e., the space $C_0^\infty(\mathsf{R}^d)$ of smooth functions with compact support, such that ϕ is linear. We shall only consider scalar (real-valued) fields.

Our first example is the *free Euclidean field* of Pitt [P71]. (It is also called the free Markov field.) It is the Gaussian stochastic process of mean 0 and covariance

$$\mathsf{E}\phi(f)\phi(g) = \langle f, (-\tfrac{1}{2}\Delta + m^2)^{-1} g\rangle. \tag{8.1}$$

Here $\langle,\rangle$ is the inner product in $L^2(\mathsf{R}^d)$ and m is a strictly positive constant, the *mass*. To verify that this process is linear, and hence a random field, it suffices to observe that the variance of $\phi(c_1 f_1 + c_2 f_2) - c_1\phi(f_1) - c_2\phi(f_2)$ is 0.

The *Sobolev space* $\mathcal{H}^{-1}$ is the real Hilbert space obtained by completing the test functions in the inner product

$$\langle f, g\rangle_{-1} = \langle f, (-\tfrac{1}{2}\Delta + m^2)^{-1} g\rangle = \int_{\mathsf{R}^d}\int_{\mathsf{R}^d} f(x)G(x-y)g(y)\,dx\,dy, \tag{8.2}$$

where G is the fundamental solution of $(-\frac{1}{2}\Delta + m^2)G = \delta$. This is a space of distributions containing many that are supported by hypersurfaces, but it does not contain delta functions at points if $d > 1$. The random field ϕ clearly extends to be a linear stochastic process indexed by $\mathcal{H}^{-1}$.

We have already seen the free Euclidean field for $d = 1$; it is the ground state process (6.17) for the harmonic oscillator. In fact, for $d = 1$ and $\omega = m$, the fundamental solution G is $e^{-\omega|t|}/2\omega$. This is a Markov process. What is the analogous property in higher dimensions?

Let ϕ be a random field indexed by $\mathcal{H}^{-1}$. For any subset B of R^d, let $\mathcal{O}(B)$ be the σ-algebra generated by all $\phi(f)$ with $\operatorname{supp} f \subseteq B$. We denote the complement of B by B^c and the boundary of B by ∂B. Then we say that ϕ is a *Markov field* in case for all open sets U, the σ-algebras $\mathcal{O}(U)$ and $\mathcal{O}(U^c)$ are conditionally independent given $\mathcal{O}(\partial U)$.

I claim that the free Euclidean field is a Markov field. For any subset B of R^d, let $\mathcal{M}(B)$ be the closed linear subspace of $\mathcal{H}^{-1}$ spanned by the f in $\mathcal{H}^{-1}$ with $\operatorname{supp} f \subseteq B$. Now for any two sets of Gaussian random variables of mean 0, the σ-algebras they generate are independent if and only if the sets are orthogonal. Also, the mapping $f \mapsto \phi(f)$ is an isometry of $\mathcal{H}^{-1}$ into L^2. Consequently, we need only show that for all open sets U, the spaces $\mathcal{M}(U)$ and $\mathcal{M}(U^c)$ are conditionally orthogonal in $\mathcal{H}^{-1}$ given $\mathcal{M}(\partial U)$; that is, we need only show that if $f \in \mathcal{M}(U)$ and g is its orthogonal projection onto $\mathcal{M}(U^c)$, so that

$$\langle h, (-\tfrac{1}{2}\Delta + m^2)^{-1} f\rangle = \langle h, (-\tfrac{1}{2}\Delta + m^2)^{-1} g\rangle, \qquad h \in \mathcal{M}(U^c), \tag{8.3}$$

then $g \in \mathcal{M}(\partial U)$. But let k be smooth with compact support in U^c, and let $h = (-\frac{1}{2}\Delta + m^2)k$. Then $h \in \mathcal{M}(U^c)$, so

$$\langle k, f\rangle = \langle k, g\rangle. \tag{8.4}$$

But the left hand side is 0 since $\operatorname{supp} f \subseteq U$. Hence the right hand side is 0, but since this holds for all such k, we have by definition that $\operatorname{supp} g \subseteq \partial U$, and the claim is established.

Notice that the Euclidean group (the group of all isometries of $\mathbf{R}^d$ as d-dimensional Euclidean space) acts by isometries on $\mathcal{H}^{-1}$ and hence as measure-preserving transformations on the underlying probability space of the free Euclidean field. The free Euclidean field has proved useful in constructive quantum field theory, in which one first replaces relativistic space-time (d-dimensional Minkowski space) by d-dimensional Euclidean space by means of an analytic continuation in time that replaces t by it; see [S74] and [GJ81]. But the free Euclidean field also arises in the extension of stochastic mechanics to the theory of random fields. This was first accomplished by Guerra and Ruggiero [GR73].

The study of a field as a mechanical system goes back to Jeans. But it is an abstract mechanical system; instead of material particles moving in time, Fourier coefficients change in time. Consider the deterministic scalar field satisfying the relativistic field equation

$$(\Box + m^2)\phi = 0. \tag{8.5}$$

Here ϕ is just a function on $\mathbf{R}^d$, and we have represented Minkowski space $\mathbf{R}^d$ as the Cartesian product of a space-like hyperplane $\mathbf{R}^{d-1}$ coordinatized by $\mathbf{x}$ and a time-like axis $\mathbf{R}$ coordinatized by t. The wave operator $\Box$ is $\partial_t^2 - \nabla_{\mathbf{x}}^2$. To simplify the discussion, put the system in a spatial box. That is, choose a number L and consider only functions ϕ that are periodic of period L in each of the $\mathbf{x}$-variables. Now for each t expand $\phi(\mathbf{x}, t)$ in a Fourier series

$$\phi(\mathbf{x}, t) = \sum q_{\mathbf{k}}(t)\cos \mathbf{k}\cdot\mathbf{x} + \sum q'_{\mathbf{k}}(t)\sin \mathbf{k}\cdot\mathbf{x}. \tag{8.6}$$

Then the field equation is satisfied if and only if each *oscillator* $q_{\mathbf{k}}$ and $q'_{\mathbf{k}}$ satisfies the deterministic oscillator equation (6.19) with $\omega^2 = \mathbf{k}^2 + m^2$. Guerra and Ruggiero replaced these deterministic oscillators by independent ground state processes for the harmonic oscillator, computed the covariance, and took the limit as $L \to \infty$ (in such a way that the sum in the Fourier series became a Lebesgue integral). The result was the covariance of the free Euclidean field. Thus stochastic mechanics applied to the free scalar field yields the free Euclidean field.

Their procedure breaks the relativistic symmetry of the problem. Starting with a problem that has relativistic (Poincaré) symmetry, the choice of hyperplane $\mathbf{R}^{d-1}$ leads to a solution that has Euclidean symmetry. Nevertheless, as discussed in §§2,4 of [N86], when one passes from the ground state of the free field to the field coupled to an external current or to the coherent state processes of the free field (which were discussed in the case $d = 1$ in §6), one finds that the expectation values satisfy relativistic equations. Roughly speaking, the means are relativistic and the fluctuations are Euclidean!

9. Bell's inequalities

The inequalities of Bell and their violation in the experiments of Aspect and collaborators are an important part of modern culture, essential to any attempt to understand the world in which we live. I refer to the discussions in [N84, 86]. Here I just want to take the occasion to correct a blunder in the second of these references, which was repeated in [N86a].

In the statement of the theorem on page 445 of [N86], and on pages 533–534 of [N86a], replace $\frac{1}{2}$ by $\frac{1}{3}$. (And delete the remark in the first reference to the effect that in view of Mermin's argument, it is surprising that the result holds with $\frac{1}{2}$ instead of $\frac{1}{3}$!) I am grateful to Lee Newberg for pointing out to me that the $\frac{1}{2}$ had to be wrong on conceptual grounds, precisely because the hypotheses of the theorem can be used to model the situation discussed by Mermin in [M81]. The mistake in the proof is that the center of the $p_1p_2p_3$-cube is a local minimum, where the value is $\frac{1}{2}$, but the absolute minimum occurs at a corner of the cube, where the value is $\frac{1}{3}$. The rest of the proof and the subsequent discussion are unaffected.

10. Speculations on a new starting point

Some of my friends who work on stochastic mechanics are unhappy when I express dissatisfaction with aspects of it. I think it quite likely that it is in some sense an approximation to a physically satisfactory theory, and as mathematics it is a lot of fun. But it is important to find a naïvely realistic model of nature that does not violate locality, or to show that this is impossible. I shall conclude by sketching some ideas that have not yet been explored. They come with no guarantee.

Consider first deterministic mechanics and deterministic field theory. What is the analogue in field theory of a time t? It is a space-like hypersurface Σ (a maximal one, that separates Minkowski space $\mathbb{R}^d$ into a past and a future). The analogue of a configuration at time t is a field configuration on Σ.

A stochastic process is a random configuration for each time t. Therefore a random field should be a random field configuration for each Σ, that is, a linear stochastic process indexed by test functions f each of which is supported on some Σ. Let us call this a *random space field* to distinguish it from a random field (a *random space-time field*) indexed by test functions F on space-time. Given such an F, it can be represented as a limit of linear combinations of f_i supported by Σ_i, but this can be done in many different ways depending on how the support of F is sliced up into space-like pieces. It is a kind of continuity assumption on a space field to require that for any such decomposition of F, the limit exists and is independent of the slicing. That is, a space field may not come from, or give rise to, a space-time field. If the space field ϕ is constructed as a limit in distribution of space-time fields ϕ_κ as a cut-off $\kappa \to \infty$, there is no reason to assume that it comes from a space-time field, especially as the notion of a momentum cut-off is relative to a space-like hypersurface.

Given an inertial frame, one can construct the free Euclidean field in the manner of Guerra and Ruggiero. It can then be restricted to test functions f supported on $t = constant$ hyperplanes. The same can be done for a different inertial frame, but the random variables are defined on a different probability space. Is there a way to knit all of these together as part of a single relativistically invariant random space field? I believe that the answer is yes.

A space-like hypersurface Σ has a space-like tangent plane at each of its points, which can be characterized by the forward unit time-like vector orthogonal to it (in the sense of Minkowski space). Let $\mathbf{S}_d$ be the set of all pairs $\langle x, u \rangle$ where $x \in \mathbb{R}^d$ and u is a forward unit time-like vector. This is a $2d-1$ dimensional manifold. The tangent space at each point of $\mathbf{S}_d$ splits naturally as the direct sum of three spaces: u can vary on the $d-1$ dimensional unit hyperboloid, keeping x fixed; x can move in the direction u with u carried along by parallel transport; or x can move in the $d-1$ directions orthogonal to u, again with u carried along by parallel transport. (This description is valid for any pseudo-Riemannian manifold with signature $(d-1,1)$.) But each of these three spaces has a natural Riemannian metric; taking their direct sum, we endow $\mathbf{S}_d$ in a natural way with a Riemannian (not pseudo-Riemannian) metric that is invariant under the action of the Poincaré group. Let $\widetilde{\Delta}$ be the Laplace-Beltrami operator on $\mathbf{S}_d$ with respect to this metric, and let $\widetilde{G}$ be the kernel of $(-\frac{1}{2}\widetilde{\Delta}+m^2)^{-1}$, so that $\widetilde{G}$ is a function of positive type on $\mathbf{S}_d \times \mathbf{S}_d$. Now let $\widetilde{\phi}$ be the Gaussian stochastic process indexed by test functions on $\mathbf{S}_d$ of mean 0 and covariance given by $\widetilde{G}$.

We want to construct a random space field from $\widetilde{\phi}$. Let f be a test function on Σ, let $\widetilde{\Sigma}$ be the hypersurface in $\mathbf{S}_d$ consisting of all points $\langle x,u\rangle$ in $\mathbf{S}_d$ with $x \in \Sigma$, and let $\widetilde{f}$ be the function supported on $\widetilde{\Sigma}$ such that $\widetilde{f}(\langle x,u\rangle) = f(x)$. Now $\langle \widetilde{f}, \widetilde{G}\widetilde{f}\rangle = \infty$. But let $B_\kappa(u)$ be the ball of radius κ and center u in the unit hyperboloid and let C_κ be its volume (which does not depend on u); then it should be true that the following exists:

$$[f,g]_{-1} = \lim_{\kappa\to\infty} C_\kappa^{-2} \int_\Sigma \int_{B_\kappa(n_x)} \int_{\Sigma'} \int_{B_\kappa(n'_y)} \widetilde{f}(\langle x,u\rangle)\widetilde{G}(\langle x,u\rangle,\langle y,v\rangle)\widetilde{g}(\langle y,v\rangle), \qquad (10.1)$$

where f and g are test functions supported by Σ and Σ', and n_x is the forward unit normal at x in Σ and similarly for n'_y.

Notice that for fixed u (which amounts to fixing an inertial frame up to a spatial rotation), the set of all $\langle x,u\rangle$ is a submanifold of $\mathbf{S}_d$ whose induced metric is the *Euclidean* metric on $\mathbf{R}^d$, and $\mathbf{S}_d$ is isometric to the Cartesian product of Euclidean space and the unit hyperboloid. Therefore, for f and g supported by two $t =$ *constant* hyperplanes in this frame, I would expect that $[f,g]_{-1} = K_d\langle f\otimes\delta_1, g\otimes\delta_2\rangle_{-1}$ where δ_1 and δ_2 are delta functions in the variable orthogonal to the hyperplanes and K_d is a constant depending only on the dimension.

In this picture, the Guerra-Ruggiero field would be a random space field, equally valid in every inertial frame. It would serve to specify in a relativistically invariant way the kinematics of random space fields, and possibly open the way to construct a dynamics of critical random space fields that might avoid problems of non-locality and lack of relativistic covariance that have hitherto troubled the extension of the ideas of stochastic mechanics to field theory.

Some corrections to "Quantum Fluctuations" [N85]

p. 8, line 4: *For* mult *read* multi.
p. 8, line 6: *For* There *read* There-.
p. 77, line 5: The assertion that $\frac{1}{2}(D_+ + D_-)E = 0$, or equivalently that

$$(\partial/\partial t + v\cdot\nabla)E = 0,$$

where E is the stochastic energy $\frac{1}{2}(u^2+v^2)+\phi$, is in general false even for free diffusions. For otherwise we would have

$$(\partial/\partial t + \nabla S\cdot\nabla)\big((\nabla R)^2 + (\nabla S)^2\big) = 0.$$

Eliminate the time derivatives using the stochastic Hamilton-Jacobi equation for free motion and the current equation. What remains is a third order partial differential equation that R and S must satisfy at each time. But this is absurd, since the initial values can be chosen arbitrarily subject only to the normalization condition $\int e^{2R} = 1$. I am grateful to R. Marra for help with this correction.

p. 87, (16.18): Replace $-$ by $+$.

p. 91, (17.5): The coefficient of the sinh term should have a minus sign (or replace the factor $t - 2\lambda^2$ next to it by $2\lambda^2 - t$). I am grateful to J. Fronteau for this correction.

p. 102, end of §20: There should be an acknowledgment of the priority of the work of Schulman [S68] and of Laidlaw and C. M. DeWitt [LD70] on the restrictions (expressed by these authors in terms of Feynman path integrals rather than stochastic mechanics) on the wave function when the configuration space is not simply connected.

References

[BCZ87] Ph. Blanchard, Ph. Combe, and W. Zheng, "Mathematical and Physical Aspects of Stochastic Mechanics," *Springer Lecture Notes in Phys.* **281**, 1987.

[BS34] R. Bowling Barnes and S. Silverman, "Brownian motion as a natural limit to all measuring processes," *Rev. Mod. Phys.* **6** (1934), 162–192.

[C84] Eric A. Carlen, "Conservative diffusions," *Commun. Math. Phys.* **94** (1984), 293–315.

[C85] —, "Potential scattering in stochastic mechanics," *Ann. Inst. H. Poincaré* **42** (1985), 407–428.

[C86] —, "The pathwise description of quantum scattering in stochastic mechanics," in *Stochastic Processes in Classical and Quantum Systems, Proc., Ascona, Switzerland 1985*, ed. S. Albeverio, G. Casati, and D. Merlini, *Springer Lecture Notes in Phys.* **262** (1986), 139–147.

[CT86] E. A. Carlen and A. Truman, "Sojourn times and first hitting times in stochastic mechanics," in *Fundamental Aspects of Quantum Theory*, ed. Vittorio Gorini and Alberto Frigerio, Plenum, New York (1986), 151–161.

[D57] Bryce S. DeWitt, "Dynamical theory in curved spaces I. A review of the classical and quantum action principles," *Rev. Mod. Phys.* **29** (1957), 377–397.

[D70] Thaddeus George Dankel, Jr., "Mechanics on manifolds and the incorporation of spin into Nelson's stochastic mechanics," *Arch. Rational Mech. Anal.* **37** (1970), 192–222.

[DG78] D. Dohrn and F. Guerra, "Nelson's stochastic mechanics on Riemannian manifolds," *Lettere al Nuovo Cimento* **22** (1978), 121–127.

[DG79] —, "Geodesic correction to stochastic parallel displacement of tensors," in *Stochastic Behavior in Classical and Quantum Physics*, ed. G. Casati and J. Ford, *Springer Lecture Notes in Phys.* **93** (1979), 241–249.

[DT87] Ian Davies and Aubrey Truman, eds., Proceedings of a conference held in Swansea 1986, to appear in *Springer Lecture Notes in Math.*

[F52] Imre Fényes, "Eine wahrscheinlichkeitstheoretische Begründung und Interpretation der Quantenmechanik," *Zeitschrift für Phys.* **132** (1952), 81–106.

[F85] Richard P. Feynman, "QED—The Strange Theory of Light and Matter," Princeton University Press, Princeton, NJ, 1985.

[G85] Francesco Guerra, "Carlen processes: a new class of diffusions with singular drifts," in *Quantum Probability and Applications II, Proc. Workshop Heidelberg 1984*, ed. L. Accardi and W. von Waldenfels, *Springer Lecture Notes in Math.* **1136** (1985), 259–267.

[GJ81] James Glimm and Arthur Jaffe, "Quantum Physics—A Functional Integral Point of View," Springer-Verlag, New York, 1981.

[GL81] Francesco Guerra and Maria I. Loffredo, "Thermal mixtures in stochastic mechanics," *Lettere al Nuovo Cimento* **30** (1981), 81–87.

[GM83] Francesco Guerra and Laura M. Morato, "Quantization of dynamical systems and stochastic control theory," *Phys. Rev. D* **27** (1983), 1774–1786.

[GR73] Francesco Guerra and Patrizia Ruggiero, "A new interpretation of the Euclidean-Markov field in the framework of physical Minkowski space-time," *Phys. Rev. Letters* **31** (1973), 1022–1025.

[I62] Kiyosi Itô, "The Brownian motion and tensor fields on a Riemannian manifold," *Proc. Int. Congress Math. (Stockholm)* (1962), 536–539.

[L87] John D. Lafferty, "The density manifold and configuration space quantization," to appear in *Trans. Am. Math. Soc.*

[LD70] Michael G. G. Laidlaw and Cécile Morette DeWitt, "Feynman functional integrals for systems of indistinguishable particles," *Phys. Rev. D* **3** (1971), 1375–1378.

[M81] N. D. Mermin, "Bringing home the quantum world: quantum mysteries for anybody," *Am. J. Phys.* **49** (1981), 940–943.

[N84] Edward Nelson, "Quantum fluctuations—an introduction," in *Mathematical Physics VII, Proc. VIIth Int. Congress on Math. Phys., Boulder, CO, 1983*, ed. W. E. Brittin, K. E. Gustafson, and W. Wyss, North-Holland, Amsterdam, (1984), 509–519.

[N85] —, "Quantum Fluctuations," Princeton University Press, Princeton, NJ, 1985.

[N85a] —, "Critical diffusions," in *Séminaire de Probabilités XIX 1983/84*, ed. J. Azéma and M. Yor, *Springer Lecture Notes in Math.* **1123** (1985), 1–11.

[N86] —, "Field theory and the future of stochastic mechanics," in *Stochastic Processes in Classical and Quantum Systems, Proc., Ascona, Switzerland 1985*, ed. S. Albeverio, G. Casati, and D. Merlini, *Springer Lecture Notes in Phys.* **262** (1986), 438–469.

[N86a] —, "The locality problem in stochastic mechanics," in *New Techniques and Ideas in Quantum Measurement Theory*, ed. Daniel M. Greenberger, *Ann. N. Y. Acad. Sci.* **480** (1986), 533–538.

[P71] L. Pitt, "A Markov property for Gaussian processes with a multidimensional parameter," *Arch. Rational Mech. Anal.* **43** (1971), 367–391.

[S68] Lawrence Schulman, "A path integral for spin," *Phys. Rev.* **176** (1968), 1558–1569.

[S74] Barry Simon, "The $P(\Phi)_2$ Euclidean (Quantum) Field Theory," Princeton University Press, Princeton, NJ, 1974.

[S80] D. S. Shucker, "Stochastic mechanics of systems with zero potential," *J. Functional Anal.* **38** (1980), 146–155.

[W88] Timothy C. Wallstrom, Thesis, Princeton University, to appear.

EXPOSES 1985

O. ADELMAN — Une curiosité géométrique concernant les trajectoires browniennes dans $\mathbb{R}^d$

P. ARTZNER — Les rôles de la loi des grands nombres en théorie des assurances

M. BABILLOT — Potentiel des chaînes semi-markoviennes : que se passe-t-il lorsque le point de départ de la chaîne s'en va à l'infini ?

P. BALDI — Optimisation globale et déstabilisation d'équilibres

L. BIRGE — Estimating decreasing densities. Asymptotics versus non-asymptotics

C. BOUTON — Approximation gaussienne pour des algorithmes stochastiques à dynamique markovienne

P. CHASSAING — Une loi forte des grands nombres pour un produit demi markovien de matrices

F. COMETS — Tunnelling and nucleation for a local mean-field magnetic model

A.R. DARWICH — Sur l'orthogonalité de deux probabilités correspondant à deux processus de Markov

L. GALLARDO — Critère de transience des marches aléatoires sur les hypergroupes commutatifs

P.L. HENNEQUIN — Les probabilités et les statistiques dans l'enseignement secondaire

C. KIPNIS — Un théorème central limite pour un système infini de particules

H. LAPEYRE — Grandes déviations pour certains systèmes differentiels aléatoires

R. LEANDRE	Application du calcul de Malliavin à certaines estimations de densité de processus
J.F. LE GALL	Un théorème central limite pour le nombre de points visités par une marche aléatoire plane récurrente
F. LE GLAND	Méthodes de Monte-Carlo en filtrage non-linéaire
G. LETAC	La loi de Cauchy-conforme et les fonctions qui préservent son type
P. MASSART	Majorations exponentielles pour les processus empiriques et principes d'invariance
F. MESSACI	Estimation de la densité spectrale d'un processus à temps continu par échantillonnage poissonnien
M. MIGUENS	Categorical data and small samples
J. PICARD	Discrétisation en temps du problème de Dobrushin
G. ROYER	Modèles de rotateurs et théorème de Dobrushin
F. RUSSO	Expression intégrale de certaines espérances conditionnelles relatives à des tribus engendrées par le drap brownien
E.H. SADI	Caractéristiques locales et processus à accroissements indépendants tangents
A. TOUATI	m-convergence en loi pour des suites de processus. Application aux semi-martingales
P. VALLOIS	Une extension d'un théorème de Ray-Knight concernant le temps local du mouvement brownien
S. WEINRYB	Etude asymptotique de l'image de la saucisse de Wiener par une mesure de $\mathbb{R}$. Applications à un problème d'homogénéisation et aux temps locaux d'intersection

EXPOSES 1986

P. BALDI	Sur le module de continuité des diffusions
J. BARTOSZEWICZ	Ordre de dispersion et transformation TTT
A. BENASSI	Ponts markoviens Problème de Dirichlet stochastique quasi-linéaire aux limites non homogènes
L. CANTO E CASTRO	Vitesse de convergence en théorie des valeurs extrêmes (modèles Normal et Gamma)
A. DERMOUNE	Minoration de l'état fondamental de l'équation de Schrödinger d'une particule de spin - $\frac{1}{2}$
N. EL KAROUI	Une méthode probabiliste de construction des réduites
J.P. FOUQUE	Hydrodynamique des systèmes de particules monotones et non symétriques : l'exclusion simple et le zero range
G. FOURT	Problèmes statistiques posés par l'estimation des paramètres dans des processus de cristallisation
C. GRAHAM	Système de particules réfléchies avec intéraction sur une frontière collante
R. HOPFNER	Asymptotic inference for continuous-time Markov chains
A. HUARD	Résolution par mixage d'opérateurs de l'équation de la cinétique des neutrons avec approximation probabiliste du terme de diffusion
A.T. LAWNICZAK	Gaussian Stochastic Processes with sample paths in Orlicz spaces
J.F. LE GALL	Quelques résultats de fluctuation pour la saucisse de Wiener

G. LETAC	Réciprocité des familles exponentielles sur $\mathbb{R}$
P. McGILL	Les sauts d'un processus de Lévy
M. MORA	Familles exponentielles naturelles sur $\mathbb{R}$ et leur fonction variance
D. NUALART	Equations d'onde stochastiques : Propagation des singularités
J. PICARD	Equation différentielle stochastique avec bruit coloré
R. RUSSO	Tribus séparantes et propriétés de Markov pour une classe de processus généralisés et ordinaires à n paramètres (gaussiens)
C. SAVONA	Approximation d'un problème de filtrage "linéaire par morceaux"
A.S. SZNITMAN	Propagation du chaos pour un système de splines browniennes se détruisant
P. VALLOIS	Intégration stochastique avec intégrant non adapté
S. WEINRYB	Application des résultats asymptotiques sur une ou plusieurs saucisses de Wiener localisées aux problèmes d'homogénéisation. Théorème central limite pour le potentiel associé.
L.M. WU	L'inégalité de Meyer sur l'espace de Poisson

EXPOSES 1987

L. ANDERSON	Prequantization of Infinite Dynamical Systems
J. ANGULO - R. GUTIERREZ	Sur des équations intégrales stochastiques de McShane dépendant d'un paramètre
M. BABILLOT	Fonctions harmoniques positives sur un espace riemannien symétrique
D. BOIVIN	Récurrence multiple et opérateurs linéaires
P. BOXLER	On a stochastic version of the center manifold theorem
W. CHOJNACKI	Non-trivial random cocycles
A. DERMOUNE	Décomposition en chaos et isométries définies par les intégrales stochastiques multiples
G. DEL GROSSO	Filtering for jump processes
G. DEWITT-MORETTE	Meiman's definition of "functional integrals as integrals on locally non compact groups with (rough) generalized measures"
W. DZIUBDZIELA	Limit laws for kth order statistics from conditionally mixing arrays of random variables
M. EMERY	Martingales dans les variétés
T. KOLSRUD	Dirichlet forms associated with boundary Dirichlet forms
T. KOLSRUD	Multiplicative random fields with applications in mathematical physics

P. KREE	Chaotic calculus, multiple stochastic integrals and Girsanov's formula
J. LAFFERTY	The index theorem and the action of a diffusion
J. LAFFERTY	Infinite dimensional geometry and stochastic mechanics
M. LAPIDUS	The Feynman-Kac formula with a Lebesgue-Stieltjes measure and Feynman's Operational Calculus
A.T. LAWNICZAK	RKHS for Gaussian measures on metric vector spaces
F. LE GALL	Mouvement brownien, cônes et processus stables
U. MANSMANN	About a random walk driven by a random media
M. PONTIER	Approximation d'un filtre avec observation sur une variété compacte riemannienne
S. ROSENBERG	An overview of the Atiyah-Singer Index Theorem
M. SANZ	Une application du calcul de variations stochastiques au problème du retournement du temps pour des diffusions
W. SZCZOTKA	Hereditary properties of queueing processes
A.S. USTUNEL	Quelques (petites) remarques sur les Diffusions Conditionnelles
J. VAN CASTEREN	Integral kernels and Schrödinger type equations : Pointwise inequalities for Schrödinger semigroups
W.D. WICK	Some recent results in the "hydrodynamic" theory of interacting particle systems
L.M. WU	Représentation de fonctionnelles et intégrale de Skorokhod sur l'espace de Poisson